T0327368

Bow-Tie Industrial Risk Management Across Sectors

Bow-Tie Industrial Risk Management Across Sectors

A Barrier-Based Approach

Professor Luca Fiorentini

This edition first published 2022

Registered Office
John Wiley & Sons, Inc., 111 River Street, Hoboken, NJ 07030, USA
John Wiley & Sons Ltd, The Atrium, Southern Gate, Chichester, West Sussex, PO19 8SQ, UK

Editorial Office
111 River Street, Hoboken, NJ 07030, USA

For details of our global editorial offices, customer services, and more information about Wiley products visit us at www.wiley.com.

Wiley also publishes its books in a variety of electronic formats and by print-on-demand. Some content that appears in standard print versions of this book may not be available in other formats.

Library of Congress Cataloging-in-Publication Data

Name: Fiorentini, Luca, 1976– author.
Title: Bow-tie industrial risk management across sectors : a barrier based approach / Professor Luca Fiorentini.
Description: First edition. | Hoboken, NJ : Wiley, 2022. | Includes bibliographical references and index. | Summary: "As stated by ISO 31000 "organizations of all types and sizes face internal and external factors and influences that make it uncertain whether and when they will achieve their objectives. The effect this uncertainty has on organization's objectives is a risk. All activities of an organization involve risk". ISO, together with the International Trade Centre and the United Nations Industrial Development Organization published a specific guide about the importance of the implementation of sound risk management practices in small and mid enterprises. Risk management is an integral part of all organizational processes and of decision making. It should be systematic, structured and timely. It also should be based on the best available information and tailored. It should consider human and cultural factors ("soft" factors) together with technical and organizational factors ("hard" factors)."– Provided by publisher.
Identifiers: LCCN 2021003284 (print) | LCCN 2021003285 (ebook) | ISBN 9781119523833 (hardback) | ISBN 9781119523826 (adobe pdf) | ISBN 9781119523673 (epub) | ISBN 9781119523857 (obook)
Subjects: LCSH: Risk management.
Classification: LCC HD61 .F49 2022 (print) | LCC HD61 (ebook) | DDC 658.15/5–dc23
LC record available at https://lccn.loc.gov/2021003284
LC ebook record available at https://lccn.loc.gov/2021003285

Cover Design: Wiley
Cover Image: © Nikolay_Popov/iStockphoto

C9781119523833_040521

To my wife Sonia, with whom, day by day and together, I always reach new important goals. Thank you for your support, patience and constant love, witnessed by the wonderful family we have.

Luca Fiorentini

Contents

List of Figures *ix*
List of Tables *xvii*
List of Acronyms *xix*
Preface 1 *xxi*
Riccardo Ghini
Preface 2 *xxiii*
Bernardino Chiaia
Preface 3 *xxiv*
Luca Marmo
Preface 4 *xxvi*
Giuseppe Conti
Preface 5 *xxvii*
Claudio De Angelis
Preface 6 *xxx*
Damiano Tranquilli
Preface 7 *xxxiii*
Enzo Matticoli
Preface 8 *xxxiv*
Salvatore Bagnato
Author Preface *xxxvii*
Acknowledgements *xxxix*

1 Introduction to Risk and Risk Management *1*
1.1 Risk Is Everywhere, and Risk Management Became a Critical Issue in Several Sectors *6*
1.2 ISO 31000 Standard *10*
1.3 ISO 31000 Risk Management Workflow *23*
1.4 Uncertainty and the Human Factor *53*
1.5 Enterprise Complexity and (Advanced) Risk Management (ERM) *58*
1.6 Proactive and Reactive Culture of Organizations Dealing with Risk Management *62*
1.7 A Systems Approach to Risk Management *76*

2 Bow-Tie Model *101*
2.1 Hazards and Risks *101*
2.2 Methods of Risk Management *102*
2.3 The Bow-Tie Method *134*
2.4 The Bow-Tie Method and the Risk Management Workflow from ISO 31000 *138*
2.5 Application of Bow-Ties *140*
2.6 Level of Abstraction *147*
2.7 Building a Bow-Tie *150*
2.8 Hazards *152*
2.9 Top Events *153*
2.10 Threats *154*
2.11 Consequences *156*
2.12 Barriers *156*
2.13 Escalation Factors and Associated Barriers *163*
2.14 Layer of Protection Analysis (LOPA): A Quantified Bow-Tie to Measure Risks *165*
2.15 Bow-Tie as a Quantitative Method to Measure Risks and Develop a Dynamic Quantified Risk Register *178*
2.16 Advanced Bow-Ties: Chaining and Combination *183*

3 Barrier Failure Analysis *185*
3.1 Accidents, Near-Misses, and Non-Conformities in Risk Management *185*
3.2 The Importance of Operational Experience *186*
3.3 Principles of Accident Investigation *189*
3.4 The Barrier Failure Analysis (BFA) *194*
3.5 From Root Cause Analysis (RCA) to BFA *208*
3.6 BFA from Bow-Ties *213*

4 Workflows and Case Studies *217*
4.1 Bow-Tie Construction Workflow with a Step-by-Step Guide *217*
4.2 LOPA Construction Workflow with a Step-by-Step Guide *243*
4.3 BFA Construction Workflow with a Step-by-Step Guide *250*
4.4 Worked Examples *265*

Conclusions *365*
Appendix 1 Bow-Tie Easy Guide *371*
Appendix 2 BFA Easy Guide *373*
Appendix 3 Human Error and Reliability Assessment (HRA) *379*
References and Further Reading *409*
Index *417*

List of Figures

Figure 1 Descent from Col du Chardonnet. Is it safe? *Source:* Luca Marmo archive photo. *xxiv*
Figure 2 Bas-relief depicting the god Kairos. *xxvii*
Figure 3 The epistemological meaning of security. *xxviii*
Figure 4 Swiss Cheese Model. *Source:* Reason, J., 1990. *5*
Figure 5 Top five global risks in terms of likelihood (2007–2020). *Source:* World Economic Forum, 2020. *8*
Figure 6 Top five global risks in terms of impact (2007–2020). *Source:* World Economic Forum, 2020. *9*
Figure 7 Different perspectives on risk. *11*
Figure 8 Definition of the scope of risk management. *12*
Figure 9 Relationship between principles, framework, and risk management process. *13*
Figure 10 The principles of RM according to ISO 31000. *15*
Figure 11 The RM framework. *19*
Figure 12 Components of a risk management framework. *20*
Figure 13 Risk management framework. *23*
Figure 14 Leadership and commitment. *24*
Figure 15 Internal and external context. *25*
Figure 16 Identify the requirements related to risk management. *26*
Figure 17 Implementing the risk management framework. *27*
Figure 18 Scheme of the risk management process according to ISO 31000. *28*
Figure 19 Relationship between the RM principles, framework, and process. *29*
Figure 20 Improving the risk management framework. *30*
Figure 21 The risk assessment phase in the context of the RM process. *31*
Figure 22 Level of risk. *34*
Figure 23 Frequency analysis and probability estimation. *37*
Figure 24 Risk acceptability and tolerability thresholds. *39*
Figure 25 Example of a risk matrix with level of acceptability regions. *40*
Figure 26 Prioritization of risk given impact and liklihood. *41*
Figure 27 Risk prioritization and the risk matrix. *42*
Figure 28 Matrix example for qualitative ALARP analysis. *43*

Figure 29 Achieving balance in risk reduction. *44*
Figure 30 Risk treatment activities. *46*
Figure 31 Residual risk. *47*
Figure 32 Risk management process continuous improvement. *49*
Figure 33 Documenting the risk management process. *50*
Figure 34 Skills and knowledge for a risk manager. *50*
Figure 35 Resources to be allocated for an effective RM. *51*
Figure 36 Understand the mission, objectives, values, and strategies. *51*
Figure 37 Risk control hierarchy and in practice. *52*
Figure 38 Thinking-Behavior-Result model. *Source:* Adapted from Fiorentini and Marmo (2018). *54*
Figure 39 Stimulus-Response model. *Source:* Adapted from Fiorentini and Marmo (2018). *55*
Figure 40 Two-pointed model. *Source:* Adapted from Fiorentini and Marmo (2018). *55*
Figure 41 Inverted two-pointed model. *Source:* Adapted from Fiorentini and Marmo (2018). *56*
Figure 42 Human factors in process plant operation. *Source:* Adapted from Strobhar (2013). *57*
Figure 43 The principles of RM according to ISO 31000. *59*
Figure 44 Main types of business risks. *62*
Figure 45 Most common enterprise risks. *63*
Figure 46 Culture maturity level in an organization. *64*
Figure 47 Safety culture levels. *66*
Figure 48 Quality of risk management approach. *68*
Figure 49 The pathological condition. *69*
Figure 50 The reactive condition. *71*
Figure 51 The bureaucratic condition. *72*
Figure 52 The proactive condition. *74*
Figure 53 The generative condition. *75*
Figure 54 The Deming Cycle PDCA. *77*
Figure 55 Swiss Cheese Model applied to a major industrial event. *88*
Figure 56 Maturity model. *Source:* Courtesy of EXIDA L.C.C. (USA). *90*
Figure 57 Feed line propane-butane separation column. *Source:* Adapted from Assael and Kakosimos (2010). *108*
Figure 58 Basic structure of a fault tree (horizontal). *117*
Figure 59 Basic structure of a fault tree (vertical). *118*
Figure 60 Basic Events. *118*
Figure 61 Example of the fault tree, taking inspiration from the Åsta railway incident. *Source:* Sklet, S., 2002. *119*
Figure 62 Gates. *120*
Figure 63 Fire triangle using FTA. *120*
Figure 64 Flammable liquid storage system. *Source:* Modified from Assael, M. and Kakosimos, K., 2010. *121*
Figure 65 Example of FTA for a flammable liquid storage system. *122*

Figure 66 Fault tree example. *123*
Figure 67 The structure of a typical ETA diagram. *124*
Figure 68 Event tree analysis for the Åsta railway accident. *125*
Figure 69 Pipe connected to a vessel. *126*
Figure 70 Example of event tree for the pipe rupture. *126*
Figure 71 Bow-Tie diagram structure. *127*
Figure 72 F-N Curve. *129*
Figure 73 Example of a risk matrix with acceptability regions. *130*
Figure 74 Calibrated risk graph. *131*
Figure 75 A typical Bow-Tie. *135*
Figure 76 Bow-Tie as the combination of an FTA and an ETA. *136*
Figure 77 The Swiss Cheese Model by James Reason. *137*
Figure 78 Bow-Tie project risk assessment. *140*
Figure 79 Bow-Tie diagram – transfer of a data center. *141*
Figure 80 Bow-Tie diagram on virtual classroom training. *146*
Figure 81 Level of abstraction. *147*
Figure 82 Zoom level and point in time. *148*
Figure 83 Example of point in time. *149*
Figure 84 Basic elements of a Bow-Tie diagram. *151*
Figure 85 Determining the threshold level to cause the top event. *155*
Figure 86 Barrier functions. *157*
Figure 87 Location of elimination and prevention barriers. *158*
Figure 88 Location of control and mitigation barriers. *158*
Figure 89 Barrier systems. *159*
Figure 90 Using the same barrier on either side of the Bow-Tie diagram. *160*
Figure 91 Classification of safety barriers. *Source:* Sklet, S., 2006. *162*
Figure 92 Barrier classification promoted by the AIChE CCPS Guidelines. *163*
Figure 93 The energy model. *Source:* Haddon, W., 1980. *164*
Figure 94 Generic safety functions related to a process model. *Sources:* Hollnagel, E., 2004. Barrier And Accident Prevention. Hampshire, IK: Ashgate; Duijm et al., 2004. *164*
Figure 95 Layers of defence against a possible industrial accident. *166*
Figure 96 A comparison between ETA and LOPA's methodology. *167*
Figure 97 Actions of a barrier. *172*
Figure 98 Misuse of escalation factors, with nested structure. *174*
Figure 99 Defining "activities" for a barrier. *175*
Figure 100 Quantifying a simplified Bow-Tie. *181*
Figure 101 Scale of the effectiveness of a barrier and the relationship between effectiveness and PFD (correct). *183*
Figure 102 Relationship between effectiveness and PFD (correct). *183*
Figure 103 Bow-Tie concatenation example. *184*
Figure 104 Difference between accident, near-accident and unintended circumstance. *186*
Figure 105 Principles of incident analysis. *187*
Figure 106 The importance of accident investigations. *188*

Figure 107 Steps in the analysis of the operational experience of organizations. *188*
Figure 108 Steps in accident investigations. *190*
Figure 109 The pyramid of conclusions. *192*
Figure 110 Example a Tripod Beta diagram. *193*
Figure 111 Possible Tripod Beta appearances. *194*
Figure 112 Example of a BFA diagram 1. *197*
Figure 113 Example of a BFA diagram 2. *198*
Figure 114 BFA core elements. *199*
Figure 115 General structure of a BFA diagram. *200*
Figure 116 Event chaining in BFA. *200*
Figure 117 Defeated barriers are not BFA events. *200*
Figure 118 Barrier identification in BFA. *201*
Figure 119 Correct and incorrect barrier identification in BFA. *201*
Figure 120 BFA analysis. *201*
Figure 121 Events types in a BFA diagram. *202*
Figure 122 Example of timeline developed for the *Norman Atlantic* investigation. *204*
Figure 123 Timeline example. *206*
Figure 124 The onion-like structure between immediate causes and root causes. *207*
Figure 125 Benefit of RCA. *208*
Figure 126 RCA Process. *209*
Figure 127 Levels of analysis. *211*
Figure 128 The Bow-Tie diagram. *213*
Figure 129 Bow-Tie risk assessment and incident analysis. *215*
Figure 130 Bow-Tie preparation workflow. *218*
Figure 131 From organization to critical tasks. *233*
Figure 132 Example of Barrier Criticality Assessment. *235*
Figure 133 Steps to identify critical barriers. *236*
Figure 134 Example of a barrier audit. *238*
Figure 135 Traditional audit: one element of the management system is analyzed at a time. *240*
Figure 136 Audit barrier-based: all elements of the management system identified as relevant to a specific barrier are analyzed. *240*
Figure 137 General workflow of LOPA. *244*
Figure 138 The general workflow of a survey. *250*
Figure 139 Incident barrier states and relation between barrier state and barrier lifecycle. *258*
Figure 140 Recommendations development and review. *264*
Figure 141 On the left: pier with a damaged downpipe; the concrete is wet and deteriorated. On the right: a similar pier with a safe downpipe; the concrete is in good condition. *267*
Figure 142 Effects of ageing and humidity on the concrete. The reinforcement bars are corroded and there are signs of rust on the beams. *268*
Figure 143 Concrete spalling on a Gerber support with a consequent capacity reduction. The cause of the damage has to be searched for on a damaged downpipe on the road joint (recently substituted). *268*

Figure 144 The spalling of concrete caused the corrosion to progress. The reinforcement bars broken due to the limited cross-section are causing a reduction of the capacity of the girder. *269*
Figure 145 Bow-Tie diagram for "Local reduction of the resisting capacity of a bridge due to ageing". *270*
Figure 146 Employee infected with COVID-19 virus. *271*
Figure 147 Fire in flight. *272*
Figure 148 BFA on food contamination (near miss). *273*
Figure 149 Web-based software development – Bow-Tie. *274*
Figure 150 IT systems protection Bow-Tie. *280*
Figure 151 Satellite view of Matera. *283*
Figure 152 Matera – Piazza Vittorio Veneto. On the right: steps. *Source:* Google LLC. *284*
Figure 153 Developed Bow-Tie to assess crowding-related risks – zooming the threats and preventive barriers. *285*
Figure 154 Developed Bow-Tie to assess crowding-related risks – zooming the consequences and mitigative barriers. *286*
Figure 155 Map to develop simulated scenarios. *287*
Figure 156 Different levels of service. *287*
Figure 157 Piazza Vittorio Veneto and the bottleneck in Via San Biagio, Matera. *288*
Figure 158 Impact of the soft obstacles on the pedestrian flow. *289*
Figure 159 Bow-Tie Risk assessment (whole picture). *290*
Figure 160 Helicopter loss of control Bow-Tie risk assessment. *291*
Figure 161 Treatment of critically ill patients. *292*
Figure 162 Treatment of patient with pain. *293*
Figure 163 Preparing parenterals (excluding cytostatic drugs). *294*
Figure 164 Administration of parenterals (excluding cytostatic drugs). *295*
Figure 165 Medication verification in handoff during hospital admission. *296*
Figure 166 Medication verification in handoff during hospital discharge (1 of 2). *296*
Figure 167 Medication verification in handoff during hospital discharge (2 of 2). *297*
Figure 168 Administration of medicines. *298*
Figure 169 Treatment of patients with acute coronary syndrome. *299*
Figure 170 Administering intravascular iodinated contrast media (excluding intensive care patients). *300*
Figure 171 Applying a central venous catheter (CVC). *301*
Figure 172 Operating on a patient. *302*
Figure 173 Hospitalization of vulnerable elders (> 70 years) (1 of 4). *303*
Figure 174 Hospitalization of vulnerable elders (> 70 years) (2 of 4). *304*
Figure 175 Hospitalization of vulnerable elders (> 70 years) (3 of 4). *305*
Figure 176 Hospitalization of vulnerable elders (> 70 years) (4 of 4). *306*
Figure 177 Performing surgical procedures. *307*
Figure 178 Elaboration of the threat "external corrosion" and main escalating factors and controls. *308*

Figure 179 Link between controls and the company HSE management system procedures. *309*
Figure 180 BFA of Flixborough (UK) incident. *310*
Figure 181 BFA of Seveso (Italy) incident. *311*
Figure 182 BFA of Bhopal (India) incident. *312*
Figure 183 BFA of *Piper Alpha* (UK – offshore) incident. *313*
Figure 184 BFA of Pembroke Refinery (Milford Haven) (UK) incident. *314*
Figure 185 BFA of Texas City (US) incident. *315*
Figure 186 BFA of Macondo (*Deepwater Horizon*) (US – Offshore) incident. *316*
Figure 187 BFA of Fukishima (Daiichi) (Japan) incident. *317*
Figure 188 Drug administration Bow-Tie. *318*
Figure 189 Area involved in the accident. Right, unwinding section of the line, left, the front wall impinged by flames. *Source:* Taken from Marmo, Piccinini and Fiorentini, 2013. *320*
Figure 190 The flattener and the area involved in the accident. Details of the area struck by the jet fire, view from the front wall. *Source:* Taken from Marmo, Piccinini and Fiorentini, 2013. *320*
Figure 191 Details of the hydraulic pipe that provoked the flash fire. *Source:* Taken from Marmo, Piccinini and Fiorentini, 2013. *321*
Figure 192 Map of the area struck by the jet fire and by the consequent fire. The dots represent the presumed position of the workers at the moment the jet was released. *Source:* Marmo, Piccinini and Fiorentini, 2013. *322*
Figure 193 Footprint of the jet fire on the front wall. *Source:* Marmo, Piccinini and Fiorentini, 2013. *322*
Figure 194 Timescale of the accident. F1 is the time interval in which the ignition occurred. F2 is the time interval in which it is probable that the workers noticed the fire. The group 5 and group 6 events are defined as in Table 28. *Source:* Marmo, Piccinini and Fiorentini, 2013. *323*
Figure 195 The domain used in the FDS fire simulations. *Source:* Marmo, Piccinini and Fiorentini, 2013. *326*
Figure 196 Simulated area, elevation. *Source:* Marmo, Piccinini and Fiorentini, 2013. *326*
Figure 197 Jet fire simulation results: flames at 1 s from pipe collapse. *Source:* Marmo, Piccinini and Fiorentini, 2013. *328*
Figure 198 Jet fire simulation results: flames at 2 s from pipe collapse. *Source:* Marmo, Piccinini and Fiorentini, 2013. *329*
Figure 199 Jet fire simulation results: flames at 3 s from pipe collapse. *Source:* Marmo, Piccinini and Fiorentini, 2013. *330*
Figure 200 Jet fire simulation results: temperature at 1 s from pipe collapse. *Source:* Marmo, Piccinini and Fiorentini, 2013. *331*
Figure 201 Jet fire simulation results: temperature at 2 s from pipe collapse. *Source:* Marmo, Piccinini and Fiorentini, 2013. *332*
Figure 202 Jet fire simulation results: temperature at 3 s from pipe collapse. *Source:* Marmo, Piccinini and Fiorentini, 2013. *333*

Figure 203 Scheme of the hydraulic circuits with two-position (a) and three-position (b) solenoid valves. *Source*: Marmo, Piccinini and Fiorentini, 2013. *334*
Figure 204 Event tree of the accident. The grey boxes indicate a lack of safety devices. *Source:* Marmo, Piccinini and Fiorentini, 2013. *336*
Figure 205 Damages on the forklift. *340*
Figure 206 Frames from the 3D video, reconstructing the incident dynamics. *342*
Figure 207 Bow-Tie diagram of the ThyssenKrupp fire. *343*
Figure 208 Twente stadium roof collapse Tripod Beta analysis. *345*
Figure 209 Water treatment Bow-Tie analysis. *346*
Figure 210 Timeline of the sample (developed with CGE-NL IncidentXP). *348*
Figure 211 Possible RCA of the sample (developed with CGE-NL IncidentXP). *348*
Figure 212 Possible Tripod Beta of the sample (developed with CGE-NL IncidentXP). *349*
Figure 213 Possible BFA of the event (developed with CGE-NL IncidentXP). *350*
Figure 214 Bow-Ties developed to assess fire risk in multiple railway stations. *352*
Figure 215 Fire load. *354*
Figure 216 Bow-Tie worksheet developed by TECSA S.r.l. and Royal Haskoning DHV to quantify a Bow-Tie scheme with a LOPA approach. Not real scores and data presented in the image. *356*
Figure 217 Barriers/protection layer scores. *357*
Figure 218 Weakest barriers and the public. *359*
Figure 219 Bow-Tie model for fire risk assessment in PV plants. *361*
Figure 220 Map of ceraunic density in Italy. *362*
Figure 221 Annual average temperature in Italy. *363*
Figure 222 Deming Cycle from a barrier-based perspective. *366*
Figure 223 Bow-Tie core elements and general structure. *371*
Figure 224 Bow-Tie guiding principles. *372*
Figure 225 BFA core elements. *374*
Figure 226 Incident barrier state. *375*
Figure 227 Incident barrier state decision support tree. *376*
Figure 228 BFA guiding principles. *377*
Figure 229 Classification of human failure. *380*
Figure 230 Fault tree analysis, current configuration (ANTE). *399*
Figure 231 Fault tree analysis, better configuration (configuration A). *400*
Figure 232 Fault tree analysis, the best configuration (POST configuration). *401*
Figure 233 Frequency estimation of the scenario "Oxygen sent to blow down, during start up of reactor of GAS1". *402*
Figure 234 The Swiss Cheese Model by James Reason. *Source:* Reason, 1990. *404*
Figure 235 Level 1: Unsafe acts. *405*
Figure 236 Level 2: Preconditions. *406*
Figure 237 Level 3: Supervision issues. *406*
Figure 238 Level 4: Organizational issues. *407*

List of Tables

Table 1 Applicability of tools for risk assessment. *32*
Table 2 Example of "what-if" analysis. *Source:* Adapted from Assael, M. and Kakosimos, K., 2010. *109*
Table 3 Guidewords for HAZOP analysis. *109*
Table 4 Extract of an example of HAZOP analysis. Adapted from Assael and Kakosimos (2010). *111*
Table 5 Subdivision of the analyzed system into areas. *112*
Table 6 Hazards and assumed event in HAZID. *113*
Table 7 List of typical consequences. *113*
Table 8 HAZID worksheet. *114*
Table 9 Different classification of barriers as physical or non-physical. *161*
Table 10 Comparison of defined hazards with insufficient detail and optimal degree for evaluation. *170*
Table 11 Comparison of defined top events with insufficient detail and with an optimal degree for evaluation. *171*
Table 12 Comparison of defined causes with insufficient detail and with an optimal degree for evaluation. *171*
Table 13 Comparison of defined consequences with insufficient detail and with an optimal degree of evaluation. *172*
Table 14 Barrier Types. *173*
Table 15 Quality scores and judgments on the effectiveness of barriers. *182*
Table 16 Standard Performance Scores (PS). *182*
Table 17 Definition of BRFs in Tripod Beta. *195*
Table 18 Example of spreadsheet event timeline. *203*
Table 19 Example of Gantt chart investigation timeline. *203*
Table 20 Barrier function score (FS). *234*
Table 21 Barrier consequence of failure score (CS). *234*
Table 22 Barrier redundancy score (RS). *234*
Table 23 Barrier criticality ranking. *235*
Table 24 Barrier criticality assessment example. *235*
Table 25 Interpretation of the barrier-based audit response histograms. *239*
Table 26 Survey team members should and should not. *252*

Table 27 General information about the case study. *319*
Table 28 Record of the supervisor systems (adapted from Italian). *Source:* Marmo, Piccinini and Fiorentini, 2013. *323*
Table 29 Threshold values according to Italian regulations. *Source*: Marmo, Piccinini and Fiorentini, 2013. *325*
Table 30 Summary of the investigation. *341*
Table 31 Example of calculating HEP with the SPAR-H Method. *392*
Table 32 PIF (current configuration) *397*
Table 33 PIF (Configuration A) *398*
Table 34 PIF (POST configuration) *398*
Table 35 Frequency of incidental assumptions considered. *403*

List of Acronyms

AHJ	authority having jurisdiction
AIChE	American Institute of Chemical Engineers
ALARP	as low as reasonably practicable
BCM	business continuity management
BCMS	Business Continuity Management System
BFA	barrier failure analysis
BIA	business impact analysis
BPCS	basic process control system
BRF	basic risk factor
BSCAT	barrier-based systematic cause analysis technique
BT	Bow-Tie
CCD	cause-consequence diagram
CCPS	Centre for Chemical Process Safety
COSO	Committee of Sponsoring Organizations of the Treadway Commission
ERM	enterprise risk management
ETA	event tree analysis
FARSI	functionality, availability, reliability, survivability and interactions
FMEA	failure modes and effects analysis
FMECA	failure modes, effects, and criticality analysis
FMEDA	failure modes, effects, and diagnostic analysis
FSMS	fire safety management system
FTA	fault tree analysis
GAMAB	globally at least as good
GIGO	garbage in, garbage out
HAZID	hazard identification
HAZOP	hazard and operability analysis
HEART	human error assessment and reduction technique
HEMP	hazard and effects management process
HEP	human error probability
HFACS	human factors analysis and classification scheme
HLS	high-level system
HSE	health, safety, and environment
HSEQ	health, safety, environment, and quality
ICT	information and communications technology

IE	initial event
IEC	International Electrotechnical Commission
IEF	initial event frequency
IPL	individual protection layer
IRM	The Institute of Risk Management
IRPA	individual risk per annum
IRT	independent protection layer response time
ISO	International Organization for Standardization
IT	information technology
KPI	key performance indicator
LFE	learning from experience
LOPA	layer of protection analysis
LOPC	loss of primary containment
MEM	minimum endogenous mortality
MGS	at least the same level of safety
MOC	management of change
NFPA	National Fire Protection Association
NMAU	not more than unavoidable
PDCA	Plan-Do-Check-Act
P&ID	piping and instrumentation diagram
PFD	probability of failure on demand
PHA	preliminary hazard analysis
PIF	performance-influencing factor
PPE	personal protective equipment
PSM	process safety management
QIQO	quality in, quality out
QRA	quantitative risk assessment
RA	risk assessment
RAGAGEP	recognized and generally accepted good engineering practice
RBD	reliability block diagram
RCA	root cause analysis
RM	risk management
ROI	return on investment
RPN	risk priority number
RRF	risk-reducing factor
SCE	safety critical equipment
SHIPP	system hazard identification, prediction and prevention
SIF	safety instrumented function
SIL	safety integrity level
SIS	safety instrumented system
SLC	safety life cycle
SLIM	Success Likelihood Index Method
SMS	safety management system
SPAR-H	Standardized Plant Analysis Risk-Human Reliability Analysis
THERP	technique for human error-rate prediction
TR	technical report

Preface 1

Riccardo Ghini

Quality Head Italy & Malta and South Europe Cluster, Sanofi

Risk assessment is a basic concept that has always accompanied me throughout my work and professional experience, so being able to contribute, albeit marginally, to the drafting of this monumental work fills me with pride and happiness.

Since the time of Legislative Decree 626/94, the ability to evaluate the probability of occurrence and the possible consequences of accidents and injuries at work has been a fundamental skill for me to develop, through the study of ever-more-refined methods and techniques of investigation. Finding all these useful analysis tools grouped in this way, brilliantly described and accompanied by real application examples, represents for me, and for all professionals, a unique opportunity for enrichment and deepening.

In fact, as my career continued, I soon realized how the concepts underlying this book can be effectively applied, not only in the field of work safety, but also in all areas of business activity, where words like "risk," "scenario," "analysis of the causes," and "continuous improvement" have become commonly used, as they are based on the very structure of the management systems developed in accordance with the various reference standards, now completely standardized.

Furthermore, we mustn't fail to mention the importance assumed by the methods of analysis, assessment, and operational management of the risks associated with the predicate offenses of Legislative Decree 231/2001 (administrative liability of companies and entities), which constitute the essential element in the preparation of a Corporate Organization, Management, and Control Model that effectively prevents the occurrence of the types of offense and, at the same time, constitutes a valid exemption in the context of a possible criminal trial.

The real cultural transition, however, takes place when the concept of risk assessment is adopted and is also applied outside the professional sphere, elevating it to a rational criterion to guide our daily choices: "do I overtake or not overtake the car that's in front of me?, "do I subscribe to this insurance policy or not?," "do I vaccinate my children or not?" These are all questions and situations we face every day, and for which it is very useful to identify the possible "top event," the "consequences" that can be generated, and the "causes" that can originate it, as well as to know what "barriers" we can implement in our defence.

This book is therefore much more than a scientific text for a few super-technicians and experts; it is a concrete and useful reference to all, to bring order and reasoning into our decisions, whatever they may be, in a world increasingly dominated by superficiality and disinformation.

I would also like to underline another aspect, often not adequately communicated: the concept of risk not only with a negative meaning, as a threat or weighting of an unfavourable event, but also, from the perspective of ISO 31000, as a positive deviation from the result expected, therefore, as an opportunity, to be evaluated and seized for the development of the organization. A better understanding of this dimension of risk would certainly facilitate a wider and more extensive use of the methodologies illustrated in the book.

At this point, before diving into reading and studying, I just have to applaud the authors, who represent all-Italian excellence, similar to Ferrari and Parmigiano Reggiano, in this scientific field traditionally the prerogative of Anglo-Saxon and American schools, and of which we must all be proud.

Preface 2

Bernardino Chiaia

Head of SISCON (Safety of Infrastructures and Constructions), Politecnico di Torino

The number and the magnitude of accidents worldwide in the industrial sector and in the realm of civil and transportation infrastructures has risen since the 1970s and continues to grow both in frequency and socioeconomic impact. Several major accidents in the industrial sector (see, e.g., the Seveso chemical plant disaster in 1976, the Bhopal gas tragedy in 1984, the Chernobyl nuclear accident in 1986, the *Deepwater Horizon* oil spill in 2010, the explosion in Warehouse 12 at the Port of Beirut in 2020) have been under the lens of the United Nations Office for Disaster Risk Reduction (UNISDR), which puts great effort in developing safety guidelines within the Sendai Framework for Disaster Risk Reduction 2015–2030.

At the same time, the number of infrastructure failures in developed countries rose dramatically since the beginning of the new millennium. This is partly due to ageing and poor maintenance of bridges, viaducts, tunnels, and dams, which were constructed mainly in the first 35 years after World War II. Moreover, traffic loads and required performances have increased 20 times the original design conditions. On the other hand, in underdeveloped countries there is clear evidence that industrial regulations are less strict and that a general lack of a culture of safety generally results in looser applications of the rules, thus producing a physiological higher percentage of accidents.

In this evolving context, the barrier-based approach named *Bow-Tie* represents a successful methodology to approach risk analysis in a consistent and robust manner. The method allows a synthetic and powerful control of multiple hazard scenarios, clearly differentiating between proactive and reactive risk management.

In this book Dr Fiorentini clearly shows the applicability and the advantages of the methodology to various situations. He shows that, once all the hazard scenarios have been correctly identified and well defined, the definition of the most appropriate barriers represents the core of the methodology to ensure risk reduction. In the non-standard case of civil engineering, for example, the Bow-Tie method shows how inspections and maintenance operations represent *preventive* control barriers against the risk of structural collapse, whereas retrofitting, traffic limitations, and active monitoring represent *mitigating* or *recovery* barriers.

The wide experience of Dr Fiorentini, along with his clarity and scientific rigour, make the book a unique and comprehensive essay on the Bow-Tie methodology of risk assessment.

Preface 3

Luca Marmo

Professor of Safety of Industrial Processes, Politecnico di Torino Department of Applied Science and Technology

In over 30 years of mountaineering and ski touring (see an example in Figure 1), I have done thousands of risk analyses, probably more than I have ever done in my professional career. Each preparation for a climb includes risk analysis. Imagine, or remember, if you have the same passion as me, a classic of European ski mountaineering, the high street Chamonix Zermatt. Climbing it takes three days if you are a pro climber, four if you are super-trained—better five or six if you are merely human—between glaciers, crevasses, overhanging rocks, and descents hanging from a rope with skis on your shoulders. 6,300 m of positive altitude difference, all between 1,600 and 3,800 m of altitude. Risky? Yes. Accidents, even fatal ones, in these environments are not so rare. However, those who do not practice mountaineering tend to overestimate the risks because they do not have the cognitive tools to evaluate them.

Figure 1 Descent from Col du Chardonnet. Is it safe? *Source:* Luca Marmo archive photo.

What will be the risk of causing an avalanche, or in any case of being hit by one? And the degree of coverage of the crevasses along the route? Will the snow be sufficient to guarantee the solidity of the snow bridges or will a chasm open under my skis when I least expect it? Will the weather be favourable or will I be surprised by a blizzard at 150 km/h on the glacier? And if so, will I have at my disposal a protective barrier, sufficient clothing, satellite device, material to take care of myself?

In mountaineering, risk assessments are based on often uncertain data. Weather forecasts are really reliable only within 72 hours, the state of a slope can be inferred from the historical weather data of the previous weeks, and the evaluation methods are often deductive and unstructured. Forecasting is fundamentally based on experience and knowledge of the environment. This is why it is wise to maintain substantial safety margins. In my career, undoubtedly many more times I gave up from a climb, evaluating the risks to be more excessive than they weren, than I really got into trouble.

We are luckier in our professional life. We have more reliable data, we are confronted with less uncertain situations. Therefore, we can apply more rigorous and schematic methods. We can clearly identify the functional relationships between the elements of a machine and outline a specific picture of the process knowing in detail the characteristics of the substances used.

The sharpness of the picture we can paint is exceptional when compared to the drawing based on which we decide whether to reach a peak. And so much clarity deserves a schematic and systematic approach. Bow-Tie and barrier failure analyses are excellent tools for describing the cause-and-consequence relationships of both simple and complex systems because they allow the precise identification of the relationships between the initiating causes and unwanted events. Unfortunately, I fear they do not apply to mountaineering, but I am convinced that if you have the patience to reach the end of this book, you will find that they are of great help in your professional activity.

Preface 4

Giuseppe Conti

Head of Legal and Corporate Affairs Italy, ENEL S.p.A.

Risk management and its related methodology are used to represent the main and fundamental tool for the conscious and measured prevention of a series of safety issues that every industrial operator may have to face.

The availability of scientific methodologies continuously updated and developed to facilitate this difficult task for companies is a resource of great value.

This relevance is clearly appreciable to the extent that the set of assessments carried out in advance allows the prevention of the risk of problematic events, or at least the reduction the probability that they will occur. Having implemented evaluation methods such as those covered by this treatise may also help to facilitate the reconstruction of the dynamics and root causes of the event itself.

This last aspect, through the perspective of a legal practitioner, be it a lawyer or a judge, represents an essential technical and scientific support.

The availability of a methodological and scientific approach since the preliminary phase of an event is a fundamental resource for the determination of the causes of a given event, of possible causes, and for the management of the related responsibilities.

The legal issues surrounding safety issues are inseparably linked to technical issues; only a correct scientific reconstruction of the events, causes, and any possible relevant element in the dynamics of the event can allow legal practitioners to manage consequential aspects such as the traceability of the event to the responsibility of one person or another with the consequent distribution of the related burdens, including economic ones, that result from it.

The collaboration between legal experts and experts of analysis, operational risk assessment is essential for the correct reconstruction of the events and for the proper conduct of investigations, checks aimed at the exclusion of responsibility, or the correct attribution of the same.

Preface 5

Claudio De Angelis

General Manager, National Fire Corp, Ministry of Interior, Italy

> *Hypocrites! You know how to evaluate the appearance of the earth and the sky; why this weather (ton kairon) can't you evaluate it?*
>
> *Luke 12, 56*

Krónos and Kairós were gods who impersonated the meanings that the Greeks attributed to time, one quantitative and the other qualitative.

Kronos is the abstract time that flows; it is made of seconds, hours, seasons, years; it is what marks the flow of life towards the end and the functioning of things until the end of their usefulness, the place where we are continually placed before our limit.

Kairos (Figure 2), on the other hand, is qualitative; it is the right time to live (or work); the right time to be or do.

Figure 2 Bas-relief depicting the god Kairos.

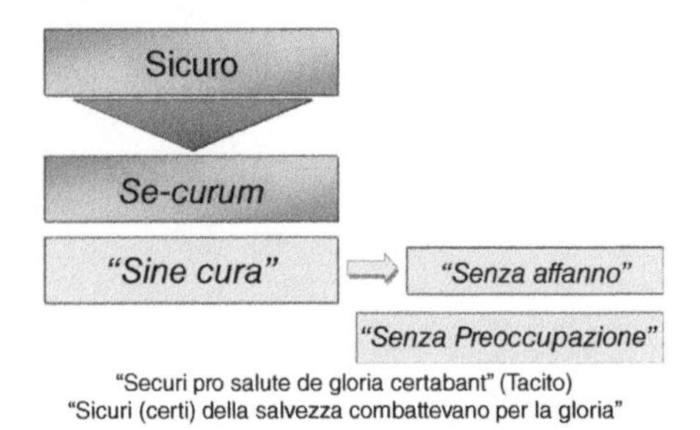

Figure 3 The epistemological meaning of security.

As can be seen in Figure 3, to be "safe," from the Latin *securum* [se (sine) + cure], means to be in a state without breathlessness, without worry. This is what must be guaranteed to the users of a building or activity in daily use.

This means that the safety of a building or activity lives in Kairos and not in Kronos, i.e. in the time for which it is designed, the "appropriate" time for which it is made, during which people's attention must be focused on what they have to do, certain that their safety is guaranteed if they respect simple ordinary ways of use.

The complexity of fire safety in construction means that the majority of designers see it as a specialization.

The emergence in the design of unusual technical-scientific problems, the complex problems related to the construction phase of the Opera, with the need to collect documentation and certifications during the project, the need to guarantee the expected performance of the protection measures over time (with reference to a service life of at least 50 years), and the economic value of these problems require analysis and study of unusual topics and problems.

For constructions of the past, for which safety in everyday use was a modest concern (it was considered sufficient to have a roof over one's head) and design methods were simple and inaccurate, even the incorrect and approximate definitions of safety were sufficient.

For today's sophisticated constructions rich in systems, where the theory of statics is replaced by theories of dynamics, where the description of fire as temperature is replaced by fluid dynamics models, and where the behaviour of materials is described by entropic theories, it is necessary to "chase" and better define the meaning of *safety*.

The correct measurement of safety through risk analysis is one of these issues.

Safety science is the discipline that studies risk in its various forms, direct and indirect, with the aim of reducing it to the minimum possible and controlling its consequences.

We speak of "reduction" of risk because, obviously, its complete elimination is mathematically impossible both because the risk variables are infinite and imponderable, and because zero risk is a theoretical limit that cannot be reached, the same as absolute zero for temperature.

As you know, the safety cycle is based on three parameters:

1) Risk analysis
2) The choice of prevention and protection measures
3) Security management

A correct and complete analysis is therefore the basis of the design.

In accordance with the RAMS (Reliability, Availability, Maintainability, and Safety) definitions proposed by the IEC 61508 standard, today we therefore define safety as the condition of a system characterized by a tolerable accident risk.

This volume is about the Bow-Tie method which, although very old, is still one of the most widespread methods for risk analysis and is well suited to approach risk analysis and management in a structured way and with a strong communicative impact because of its immediacy and because it is simpler than the combination of fault tree and event tree analyses, especially for non-experts and stakeholders in general.

The Bow-Tie is a risk analysis tool with a characteristic "bow tie" shape. The diagrams consist of a fault tree connected to an event tree. The junction point, the centre of the bow tie, represents the critical event under examination.

This method can also be used quantitatively and therefore validly used both in the world of industrial risk and for a more in-depth conventional fire risk assessment than possible with qualitative methods, such as those of the Fire Prevention Code or Legislative Decree 81.

This volume shows how you can easily set the basis for a more structured reasoning, regardless of the field of application. In every field of technology, risk has causes, consequences, and control measures (preventive and mitigative), which are generally referred to as "barriers" (the Swiss cheese slices from Reason).

The operation or non-operation of control measures affects the level of risk over time and the non-operation (or failure of a barrier) is in turn a function of a number of aspects: maintenance, information, training, calibration, and so forth. This allows the use of the paradigm to also support the management system.

By analyzing data relating to negative episodes (accidents, if we are talking about safety) it is possible to understand what the failed barriers are, how they link together, and, therefore, the causes of failure, facilitating the analysis of the root causes.

This approach can also be extended to ordinary fire prevention.

The application of this technique makes it possible to analyze a pool of fire scenarios considered representative and sufficiently exhaustive of fire safety risks. Each scenario can be characterized both in probabilistic terms and in terms of dynamics and potential direct and secondary effects (domino effect) on safety.

Starting from the description of the building under assessment, the procedure identifies and characterises the most representative fire scenarios both in terms of causes and effects (the two wings of the bow tie) and then guides the definition of the fire prevention strategy, as it highlights the most critical technical systems identified in the design phase.

In fact, the analysis with the Bow-Tie method (supplemented by a quantitative assessment for the definition of independent protection levels and for the estimation of the probability of failure) can integrate the risk assessment according to the flow referred to in Section G of the Ministerial Decree of 3 August 2015.

Preface 6

Damiano Tranquilli

HSE Manager, Grandi Stazioni Rail S.p.A., Ferrovie dello Stato

In our everyday experience, each of us faces choices and decisions and implements actions that are carried out in direct interpersonal relationships or more widely in the context of our social relations.

They may be instinctive or almost-instinctive choices, due to experience or volition or because, perhaps based on the limited scope of the effects, we believe that we can achieve them regardless of the consequences.

They can be the object of longer meditation because they are considered important or directing with respect to our personal sphere or in the economy of our relationships and because according to our way of being, perhaps also because of the conditioning of experience and training, we can be more or less inclined to reflect on our actions.

Even if with different intensity and purpose, in retrospect, we often find ourselves reconsidering choices, decisions, and actions that we are going to rethink for mere attitudinal inclination or for having to manage effects and consequences not initially foreseen.

In the same way we collect our experiences and, even if in the perspective of the different individualities of each one, we consolidate a meaning that informs our further choices, in the orientation of the decisions or in the ways of their implementation, even if only to "not do it again"!

In other words, we can say that risk management goes naturally alongside the complexity of our actions, taking shape and consistency according to our individuality.

We may be more inclined to analyze and weigh choices in a preventive manner or be more instinctive and strong-willed in acting, and perhaps more skilful in correcting our actions in progress.

One may want to try a certain action, regardless of any consequence, and accept the risk of having to react to the effects to the point of fatalistically trying one's luck.

In any case, in the organic unity of our cognitive process, risk management is an intrinsic component; the way we manage it characterises each of us, partly as a consequence of our way of being, partly as a result of our experiences.

The above considerations should, however, lead us to further reflection.

If, in fact, the risk management of our actions appears almost immediately associated with our actions in individual unity, the analysis of the process is much more complex than the organizational, decision-making, and operational processes of a positive organization, which, regardless of the relative degree of complexity, constitutes the sum, and not necessarily the synthesis, of multiple singular processes.

The risks that affect organizations relate to the different profiles in which the organization's actions are structured.

Let us consider, for example, the organizational and decision-making profile that inevitably influences the body's actions.

From this point of view, think, for example, of the possibility that technical and commercial skills and responsibilities may be allocated to different parts of the organization.

In fact, this is a possibility frequently found in social and economic realities which corresponds to the need, which can be absolutely shared, for specialisation in decision-making and operational processes.

Nevertheless, however, it cannot but be noted that this articulation necessarily corresponds to the risk that the maximization of revenues, on the one hand, and the comprehensible minimization of technical criticalities, on the other, may determine a functional imbalance with the risk of polarizing processes and preventing organizational balance and decisional synthesis if it is not mitigated through specific processes or internal coordination functions.

Still at an organizational level, a similar risk condition can be seen in organizations whose complexity is accentuated by relationships and constraints between bodies.

Consider, for example, a company subject to the management and control of another company/entity, perhaps within a more widely articulated group.

In this case, the management policies and objectives dictated by the parent company significantly influence the choices, decisions, and actions of the subsidiary, which nevertheless remains autonomously responsible in its legal subjectivity.

It is evident how, in the case in question, the condition described would end up accentuating the internal imbalance in the process of synthesis of the subsidiary's business decisions with the consequent repercussions on the operating processes and the management of related risks, making it necessary to strengthen the internal control systems from a preventive and corrective point of view.

Moreover, the rise of responsibilities on the parent company in the regulations on organizations subject to the management and coordination of other entities, determines for the parent company an additional risk that can be mitigated only by harmonising the decision-making processes between entities in a bidirectional manner.

In other words, if human subjectivity, at least in power, is naturally able to synthesise experiential and formative stimuli by widely adapting the different profiles of its action, the same process is not as natural in organizations.

In the first instance, it is necessary to frame risk management in an integrated manner within company processes and to understand its instrumentality with respect to the achievement of objectives and therefore the creation of value for the organization.

To this end, risk management must be taken away from all subjective and ethical connotations in order to be traced back to the strictly technical process.

Only in this way will it be possible to develop an autonomous sensibility of the organization, not of the single actors, but in terms of synthesis, superior and widespread culture, starting its virtuous growth through the recovery and transversal diffusion of operational experiences, the exploitation of synergies in interfering processes, and in the prevention and reaction to accidental events.

The first step, which together denotes the acquisition of this awareness in the organization and promotes its development, is the adoption of a policy and systemic risk management.

It is a process of a progressive nature for which it would not be lawful, nor would it be useful, to identify and define serial stages, but rather to search for representative behaviour that is widespread and oriented in the awareness of the individual contribution to the organizational synthesis.

But in order for this condition to be achieved, attention must be focused on risk management methodology.

If, in fact, the concept of risk itself appears to be an extremely vast and changeable, if not elusive, category, only through the definition and adoption of a management methodology is it possible to trace the process back to the concreteness necessary to establish an objective foundation in the company's dialect, to guarantee its effective and widespread perception by individuals, but above all to create a condition of intelligibility of the risk management process in the organization and therefore able to favour its effective participation.

But the importance of defining and adopting a methodology must be seen above all in the flexibility it offers to the supported process.

That experiential recovery and its re-elaboration in the company processes mentioned above, understood as an awareness of the organization's synthesis, is made possible only through a methodology defined and shared in the organization that allows the reading of experiences and the prospecting of new activities on the basis of uniform parameters that offer homogeneous measurement and comparison tools.

In this perspective, the importance of the methodology in the risk management process and in its projection with respect to the creation of value for the organization must be understood.

Preface 7

Enzo Matticoli

Renewable Energies HSE Director

When I met Luca Fiorentini the first time, I was attending his Bow-Tie method presentation at 2017 Safety Expo Bergamo. I saw a Bow-Tie diagram and I was impressed by the power of a tool (new to me) that would have supported both fire risk management and identifying the barriers to mitigate it. It was clear to me that a risk-based approach is essential.

In 2012, I moved from the Oil & Gas sector to Renewable energy. Moving from an international working environment (Europe, Middle East, etc.) to an Italian photovoltaic company, I experienced a very poor HSE culture together with a very bad HSE climate. In order to improve the HSE key performance indicators, I had to start from the basics. Nevertheless, I had the opportunity to build and certify HSE-integrated management systems based on the international standards. Having had continuous improvement as a driver, I was looking for new tools to reach the highest HSE standards.

So, in 2019, together with Luca Fiorentini and Rosario Sicari, we developed a photovoltaic fire risk assessment with the barrier-based Bow-Tie method. Some barriers were already in place; some others needed reinforcement or a different implementation. Stakeholder involvement, fire break zones, video surveillance systems, fire brigade training and close cooperation with O&M contractors, lightning protection system (LPS), insurance policies, automatic fire detection systems, inverter protection and safety systems, and so on are some of the proactive and reactive barriers identified to mitigate photovoltaic plants' fire risks, focusing on the top event.

As HSE professionals, the most difficult challenges are those regarding top management involvement. I will never forget that during the HSE senior management review, when I presented the Bow-Tie analysis, as soon as the Bow-Tie diagram was on the screen, I immediately got the directors' attention. They had an overview of multiple plausible scenarios in a single and organic picture. No doubt the Bow-Tie method helped directors address measures to avoid unpleasant consequences such as loss of life, business interruptions, reputational damage, and so forth. Without such a clear and powerful tool, it would have been much more difficult. A personal thanks goes to Luca and Rosario.

Preface 8

Salvatore Bagnato

Operations Manager – Ed.Ina Donje Svetice

In both the daily aspects of social life and in the management of issues relating to the life of companies, there is often confusion between the terms risk and hazard.

The meaning of the word *hazard* can be confusing. Often dictionaries do not give specific definitions or combine it with the term *risk*, which helps explain why many people use the terms interchangeably and, therefore, often incorrectly.

There are many definitions of danger and risk. Still, among them, we will propose the one relating to the new ISO 45000, since management systems today represent an essential aspect of the organization of any company that wishes to achieve its business objectives. Today, it is impossible to separate the economic issues from those concerning the health and safety of workers as well as respect for and protection of the environment.

ISO 45001 defines a hazard as a "source or situation with a potential to cause injury and ill health," while the risk is defined as the "combination of the likelihood of occurrence of a work-related hazardous event or exposure(s) and the severity of injury and ill health that can be caused by the event or exposures."

So, the hazard is the feature of the process that can harm an individual, and the risk is the likelihood that it will happen, along with how severe the consequences will be.

The risk assessment process is also, in this case, an intrinsic component of each person's life and allows, consciously or unconsciously, to manage risks of any nature (related to health, economics, etc.) by operating with preventive actions (to reduce the probability that an event is assessed to happen) or mitigative (to minimize the effects that this event can cause).

The process of risk assessment is somewhat informal at the individual social level but can become a sophisticated process at the strategic corporate level. However, in both cases, the ability to anticipate future events and create effective strategies for mitigating them when deemed unacceptable is vital.

Risk assessment is the process that:

- Identifies hazards and risk factors that have the potential to cause harm (hazard identification).
- Analyzes and evaluates the risk associated with that hazard (risk analysis and risk evaluation).
- Determines appropriate ways to eliminate the hazard or control the risk when the hazard cannot be removed (risk control).

In many companies now, the risk is fully integrated into the strategic planning process and in the context of the organization's performance. Risk management is permeated within the company, determining those changes that can transform the negative potential that could lead to crises or failures into growth opportunities.

In the scientific literature and the various standards of management systems, today, there are different definitions that all have in common the presence of three elements:

1) Hazards
2) Consequence seriousness
3) Frequency of a specific scenario

There is a risk if a given event is considered (with its probability of occurrence) and, depending on the type of impact, an opportunity, a loss, or the presence of uncertainty is determined. Events that have only negative consequences are indicated as pure risks. In general, a tolerability threshold is set for them and managed in such a way as to fall within this threshold.

Generally, there are three types of risk assessment, namely qualitative (Q), semi-quantitative (SQ), and quantitative risk assessment (QRA).

In QRA, numerical values are independently assigned to the various risk assessment components and the level of potential losses. When all the elements (threat frequency, safeguard effectiveness, safeguard costs, uncertainty, and probability, etc.) are quantified, the process is considered entirely quantitative.

In the SQ, frequency and severity are approximately quantified within ranges.

Finally, Q does not assign numerical values to the risk assessment components. It is based on the scenario. Several threat vulnerability scenarios are determined by trying to answer "what-if" questions. In general, qualitative risk assessment tends to be more subjective in nature

The lower levels of assessment (Q and SQ) are considered most appropriate for screening for hazards and events that need to be analyzed in greater detail. One approach to deciding the proper level of detail could be to start with a qualitative approach and to add for more detail whenever it becomes apparent that the current level is unable to offer an understanding of the risks, discrimination between the risks of different events, and so on.

It is possible to refer to the existing literature (widely used by many states in the promulgation of laws and regulatory provisions) to analyze the numerous methodologies implemented over time, in particular, to support sectors of activity characterized by elements of risk that could determine if the event were to occur, and serious consequences for people and the environment, as well as for the company itself (O&G, nuclear, etc,).

The first step for the risk assessment is hazard identification. Hazard identification (HAZID), hazard and operability (HAZOP), safety integrity level (SIL), failure modes and effects analysis (FMEA), what-if (WI), and safety checklist (SCL) are all examples of methodologies used worldwide.

The next steps are:

- Risk estimation and ranking of risk
- Risk evaluation
- Implementation of risk reduction

The output of the risk estimation should be a list of risks in ranked order for consideration. The process of risk evaluation starts with the highest risk. It proceeds down the list of identified potential risk reduction measures until it is evident that no further risk reduction measures can be justified. It should be demonstrated that risks are controlled to ensure compliance with the relevant provisions, and not intolerable. Risk estimation needs assessing both the severity (consequence) and frequency (likelihood) of hazardous events. For the Q or SQ approaches, a risk matrix is a convenient method of ranking and presenting the results.

A risk reduction measures study should be carried out by a multi-disciplinary brainstorming team with adequate experience, knowledge, and qualifications. The team will take each risk in turn and identify potential risk reduction measures, including any identified during the risk assessment.

Although the elimination of danger is the ultimate goal, it cannot be easy and is not always possible. A hierarchical approach to risk reduction involves: hazard elimination (the most effective hazard control); hazard substitution (replacing something that produces a hazard (similar to removal) with something that does not create a hazard); engineering controls (these do not eliminate hazards, but rather isolate people from hazards); administrative controls (changes to the way people work.); and personal protective equipment.

When discussing risk reduction and the hierarchy of risk control, it is essential to introduce the concept of the barrier. Barriers are functions and measures designed to break a specified undesirable chain of events. In other words, their function is to prevent a hazard from manifesting itself or mitigating its consequences.

Both control and recovery barriers are elements of the bow tie methodology that will be discussed in depth. The Bow-Tie method is a risk assessment method that can be used to analyze risk scenarios. It's named after its shape and contains eight elements: hazard, top event, threats, consequences, preventive barriers, recovery barriers, escalation factors, and escalation factor barriers.

Anyway, all the topics presented up to now will be discussed in detail later, including the concept of ALARP (which represents a critical element of risk management) and risk reduction in a "region of acceptability."

Author Preface

Luca Fiorentini

Director, TECSA S.r.l.

Risk, as per the ISO 31000 international standard, is defined as "effect of uncertainty on objectives," where the "effect is a deviation from the expected": risk is "usually expressed in terms of risk sources, potential events, their consequences and their likelihood."

When I first met the Bow-Tie method many years ago I classified it as a simple, immediate, funny notation to describe simple situations. At that time I could refer to my experience in the industrial risk and process safety domains with HSE cases built with the use of multiple and combined methodologies (fault tree, event tree, HAZID/HAZOP/FMEA, etc.) up to full quantitative risk assessments based on calculations. So I started using Bow-Ties to summarize the results coming from other methods, nothing more. Immediate (and coloured) notation of my Bow-Ties started enriching my executive summaries, my papers, my conference slides, and so on with a great reward in terms of appreciation from readers, students, colleagues, and customers.

Later I realized that I had discovered one of the main capabilities of the method: the clarity power of notation. It happened during the preparation of my Italian book on fire risk assessment, in which I described a number of methods (also risk matrices, structured brainstorming, LOPA, etc.) and I decided to describe with a Bow-Tie diagram a couple of real incidents: the Buncefield tank fire in the UK (from a description given by the UK Health and Safety Executive in an official report) and the Thyssenkrupp fire in Italy that became very famous for the number of fatalities (seven) and in which I was part of the technical consultant group working on behalf of the Public Prosecutor's Office since the beginning of the investigations. Both of the two incidents, described with a Bow-Tie, raised the usual interest and curiosity in the readers, but I did understand that Bow-Tie is the best way to deal with the essence of the various elements that define risk according to the ISO 31000 standard: hazard, deviation, threats, and consequences.

Deviations are raised from threats and could lead to potential impacts in a very simple and straightforward path. This flow can be interrupted by barriers (or controls) that can modify the likelihood and or the severity of the consequences. Simple enough!

The diagram can be modified to include more details (failure mechanisms of barriers, common causes of failure, roles and responsibilities associated to each control in place, . . .), a quantification (escalation factors, conditional probabilities, conditional modifiers, vulnerabilities, . . .), a link among different diagrams, the real-time status of the barriers in place, and also a simulated alternative that considers nice-to-have additional or modified controls – without losing the beautiful and clear overall picture that, at the end, could also be seen as a combination of a fault tree with an event tree, with the additional ability to conduct a LOPA on each threat-consequence pair, conduct a cost-benefit analysis on risk mitigation alternatives, associate probabilities derived from other calculations, investigate the failures of human-centred barriers, and so on, with the ultimate ability to validate the barriers and calculate the risk over time and to improve risk assessment using the results of learning from experience.

These limitless possibilities of enrichment are offered even if we are guaranteed that the risk picture is visible and always available also to non-technical parties, among them the stakeholders involved in risk-based decisions. Focus should be given to barriers over time and this process can be referred to "risk management as all the coordinated activities to direct and control an organization with regard to risk," whatever risk we should face to make decisions.

This book gives the reader my idea of the Bow-Tie method and barrier-based risk management as one of the best tools to deal with current complexity with a structured and consistent approach that allows for value protection and creation inside organizations of all types and sizes facing internal and external factors, including human behaviour and cultural factors. The principles described with a Bow-Tie diagram are those at the foundation of an effective and efficient risk management framework enabling an organization to deal with the effects of uncertainty on its objectives over time with a customized, integrated, yet dynamic approach based on the best available information. Complexity is increasing, day by day; let's keep our Bow-Ties simple and "smart."

Acknowledgements

First of all, I would like to express my thanks to Rosario Sicari, director of Chaos Consulting (www.chaosconsulting.it), with whom I share an important personal friendship as well as many professional interests, including accident investigation, forensic engineering and, especially, risk management with the barrier-based methods shown in this volume. Rosario, with his impressive analytical skills combined with a broad knowledge of risk analysis methods and risk management principles, guarantees an important and constructive daily confrontation that allows, together with a communion of intent and in complete operational harmony, to operate successfully in support of clients for a consulting activity oriented to complexity management. With him I shared the adventure of this book on methodologies that are much appreciated by both of us, discovering a skilled technician, a patient detail curator and reviewer, and a determined critic. Of course, as I have in the past, I will enjoy sharing with him some other engaging professional experiences and business opportunities.

I also would like to thank Damiano Tranquilli. He gave important support in the description of the proactive and reactive risk management culture that complex organizations should adhere to (ref. par. 1.6). He is a very good friend of mine and this friendship was born at the margins of a number of Bow-Tie brainstorming sessions related to the use of the method to define priorities for risk reduction measures to be implemented in some very complex railway stations following a long period of fire risk audits to verify fire strategies, fire regulation compliance, fire prevention and protection measures, and assets' future developments. Our project Bow-Ties, with me as a consultant and Damiano representing the customer, made me discover some of his unique qualities: he is very pragmatic and able to deal with complexity coming from the multiple, intercorrelated processes of very large companies, in a fascinating manner and without any apparent efforts. He is the "management systems man" since he knows how to design, implement, review new organizational processes, and assess the associated risks, with the precise "risk-based thinking" underlined by this book. I have to admit that I took advantage of his particular propension to a systemic and risk-based approach in many cases as a friendly personal consultant and I appreciated, for this book, his preface and his help in the chapter dedicated to complex organizations, which he masters very well.

A heartfelt thank you to CGE Risk for the valuable support offered in the writing of this book, for the many examples and for their beautiful images that make us appreciate even more the beauty of the Bow-Tie, as well as its usefulness for use in various business sectors.

For this reason, as I am unable to thank the entire CGE Risk team by name, my thanks go in particular to Jeroen van Dommelen, Iris Curfs, and Geert van Loopik, who I ask to extend my recognition to their valuable collaborators for everything we have done together so far and for what we will do in the future. Barrier-based methodologies discussed in this volume are challenging to apply to real-world cases where complexity should be described; CGE Risk is the leader in barrier-based software tools supporting these methods, giving the users a number of possibilities to be successful in engaging risk assessment in order to focus on the elements, on the relationships and, of course, on the outcomes of the risk assessment rather than on the drawing part that, in some cases, with traditional non-risk-oriented tools, could be a cumbersome experience, with the ability to also obtain a pretty advanced user interaction and clear colourful diagrams. I thank CGE Risk very much for having granted me the chance to use their tools to develop the pictures and examples of this book, giving the reader a uniform and consistent visual experience from the principles to the explanations to the examples and case studies. Their tools deserve a try!

Finally, my thanks go to those who have contributed to providing valuable examples of the methodologies described in this book. These are professionals from different business contexts, demonstrating the universal applicability of the methods discussed here. Therefore, I would like to thank: Bernardino Chiaia and Valerio De Biagi from Politecnico di Torino; Ian Travers from Ian Travers Limited; Jasper Smit from Slice; Tor Inge Saetre from AdeptSolutions; Salvatore Tafaro from the Italian National Fire Brigade; Emily Harbottle from Harbottle Hughes Risk Management; Ed Janseen; Arthur Groot, Process Safety Expert – Royal Haskoning DHV; David Hatch, process safety expert; Orazio Cassiani, risk manager in the healthcare sector; Prof. Luca Marmo from Politecnico di Torino; Paul Heimplaetzer; Annalisa Contos from Atom Consulting; Domenico Castaldo from TIM S.p.A.; Luca Toncelli and Francesco Fanelli from Human Factor Italia. I thank them very much for their efforts in giving some examples of barrier-based methods from their own sectors, giving valuable impressive content of real applications that enrich the volume in a way the reader can find immediate applications of the risk terms and definitions, principles from different perspectives, allowing also to take some ideas out the pages to start creating their first Bow-Ties or benchmark those they already have with some of the greatest barrier-based risk management experts from different countries I had the chance to meet during my professional experiences.

1

Introduction to Risk and Risk Management

Risk management is undoubtedly one of the most frequently used phrases in the current social and economic scenario, the importance of which has progressively expanded in line with the regulatory evolution which, in a consolidated manner, has taken on risk and its management as a criterion for the responsibility of the individual and the organization.

Apart from the subject of the interaction between the evolution of regulations and the development of risk policies, it can certainly be said that risk management is now an integral part of company processes, not simply as a legal necessity but increasingly as a factor and opportunity for consolidation and development of the organization and production processes.

From this assumption emerges the need to focus attention on risk management methods, not so much to find a definition – which, although effective, would not offer great applicative utility – but to substantiate its implementation with methodologies able to offer logical coherence to all phases of the process.

Completeness and effectiveness of risk management are in fact directly correlated to the capacity of the process to develop in phases and levels directly related to the articulation of the organization and the processes analyzed.

Considering the complexity and operational breadth of the company organizations, and therefore the network of processes, both internal and external, that condition the pursuit of results, it is quite clear that the risk management process is not immune from the risk of losing coherence of the parameters adopted in the analysis, altering the outcome of the assessment and jeopardising the possibility of consistent revision in the updating and comparison phases.

Of course, managing risk is not a question of "harnessing" it through the search for aprioristic rules that limit its potential in terms of flexibility and adherence to the specific realities analyzed, which is – a bit simplistically but effectively – the added value of the process.

It is only a matter of identifying methodologies that assist the development of risk management in a flexible but coherent manner in all phases and levels, guaranteeing its analytical and evaluative rigour, and its applicative and comparative replicability, ensuring its effectiveness with respect to the complexity of modern organizations and the network of relevant processes, in an organic vision of organizational, technical and management factors.

Bow-Tie Industrial Risk Management Across Sectors: A Barrier-Based Approach,
First Edition. Luca Fiorentini.

Risk management can have variously articulated perimeters and purposes of analysis.

Consider, for example, the risk linked to the safety management of a production unit located in a single location, on the outskirts of an average provincial town, characterized by a single operational process and in a context in which no interference between production processes belonging to different employers takes place.

Imagine instead the same production unit that is part of a wider production organization, belonging to a holding company that splits the operating processes between several subsidiaries, relocating them to several plants where different employers operate with interference between their respective suppliers.

Again, consider the hypothesis that the corporate development indicated above leads to the location of the production unit in the context of complex infrastructures in which several owners/managers/employers subject to different reference regulations operate, and not always coordinated from a technical and managerial point of view, such as, for example, in airports and railway stations where, moreover, the presence of third parties (the public) is predominant; to be practical, think of the employer of a catering activity within an airport.

The perimeter of the risk analysis would seem to be limited in the first case, that of the single production unit; moreover, the identification of processes relevant to risk management would be reduced to the only production process located in the production facility.

The perimeter of the analysis is more articulated in the second case and enormously more complex in the third.

But what would happen if, in the single production unit of the first case, radiographic materials or highly harmful chemical products were treated?

The perimeter of the analysis would be unchanged from a strictly technical point of view, although considerably more complex; the management profile would instead be enormously more complex and open to external factors where it is generally considered, in force of the law, that the employer must take appropriate measures to prevent the technical measures adopted from causing risks to the population or deteriorating the external environment by periodically checking the continued absence of risk.

But let us return to the catering production unit where the parent company, as in the second case, has introduced the administration of different product lines belonging to different subsidiaries and outsourced, for the common benefit, the cleaning and maintenance services of the common food court and the internal spaces in the legal availability of the individual subsidiaries.

The development has preserved the technical risk factors, implementing the management ones from the point of view of interference in operational processes and communication and coordination costs.

Finally, it is clear that the complexity of the risk management process is accentuated when our catering production unit moves into the station (equally, albeit on a smaller perimeter, where it moves into a multiplex cinema or shopping centre).

The technical and management process, and the corresponding risk factors, is significantly extended in correspondence with the management of the station premises, where other commercial, service and railway production units operate alongside the catering unit, without prejudice to the presence of the public.

The catering unit, like the others, is the bearer of the management of its own risks, which must be coordinated with those of the other production units.

But the same production unit is the client of the initial set-up of the space in use and the service and maintenance activities of the same, in which, however, it interferes with the supply of condominium services provided through other contractors on behalf of the station manager.

Likewise, again with regard to safety and fire prevention, the fitting out and operation of the unit must remain consistent and coordinated with the fire prevention design of the station, which in turn is developed in primary and secondary activities included in the building complex.

It is also necessary to consider the hypothesis in which the controlling company of our catering production unit arranges for the sale of products and the size of the space in use to make the point of sale subject to fire prevention controls with the need to coordinate, also in terms of time, the relevant documentation of the single unit with that of the station.

All of this is articulated in an extensive activity of cooperation and collaboration that sees the relationship between the station manager and the manager of the production unit installed, with charges of promotion of coordination from time to time with the client or promoter of the modification, management and coordination of the risks introduced, and relative diffusion to the other parties for the corresponding evaluation from the point of view of their respective interests.

Furthermore, development of the same process occurs over time in relation to the changes made from time to time by each of the parties in question (set-up, entrusting of services and works, etc.) and the repetition and updating of the cooperation and collaboration process described above.

To further accentuate the complexity of the process, consider the need for the station manager to guarantee adequate safety conditions with respect to third parties present in the station for whom there is no margin for process management other than those limited to signage in the areas.

Irrespective of the "colour" with which the three conditions represented above and the deliberately simplistic nature of the relative characterization have been represented, it undoubtedly emerges that risk management, already with respect to the technical and managerial safety profile alone, is a multi-stage process (work safety, fire prevention, administrative requirements, etc.) and is carried out on several levels of interaction between different but mutually relevant and interfering subjectivities.

In this sense, the need for a methodology that organically covers the different phases and connects the different levels with logical consistency and applicative replicability, also for updating and comparing the results in relation to the changes that have occurred, appears evident and of pressing necessity in order to guarantee the complete mapping of technical and management activities in the reciprocal interferences.

In the three cases represented above, with the development of our catering unit, no reference was made by chance to the decisions of the controlling holding company.

Paradoxically, and in purely theoretical terms, the scope of risk management analysis could be extended to infinity or almost infinity; to confirm this, we reiterate the need for a methodology that assumes consistent parameters and replicates them in all phases of the process.

Some correlations are, however, directly formulated by the legislator, where it provides (e.g. some specific regulations on industrial risk management) that the risk of the technical and management process goes back to the body in terms of responsibility, or when the body itself is presumed to be exempt from responsibility where it has adopted and effectively implemented an organization and management model ensuring a corporate management system with respect to specific obligations.

In addition, perhaps even more explicitly and with reference to security processes, the importance of organizational factors is laid down in the text of other Italian laws, which states that "the guarantee positions [...] also apply to a person who, though not having a regular office, actually exercises the legal powers relating to each of the persons defined therein."

In other words, and with reference to corporate risk management, organizational factors cannot not be included in the risk management process, even if it is from time to time aimed at more specific profiles such as safety or environmental risk assessment or in terms of sustainability.

Just think, for example, of the case in which the company organization divides management and commercial competencies and top management entrusts the safety function delegations to the manager in charge of management.

Any misalignment of processes shall not correspond to the same responsibilities with respect to any non-compliance.

In other words, with the risk management processes, the possibility of instrumental use of function delegations is in some way undermined and the assessment of risk as a criterion for making the organization responsible refers to the more detailed analysis of the consistency between organization and responsibility.

The above considerations, albeit in a concise and in many ways summary manner, are intended solely to highlight the assumption that risk management, due to the importance assumed in the current regulatory framework and the complexity of the reference scenarios, is required to be based on methodologies capable of replicating logical and evaluative consistency in all phases and levels of the process, regardless of the scope of the analysis.

With these premises, this book is intended to be a valid reference, for professionals and technicians in the risk management sector, to the rigorous approach of barrier-based risk management. The basic idea behind this approach is to integrate the traditional and well-defined risk management processes using a barrier-based perspective; basically, the one suggested by James Reason with his well-known Swiss Cheese Model (Reason, 1990) (Figure 4).

This integration leads to value-added risk management that, if pursued in every direction, becomes the solid foundation on which to build enterprise risk management (ERM). Several books have already been published on similar topics. This one wants to implement these key concepts in the daily business of every organization by transferring this information with the help of two barrier-based methods: Bow-Tie (BT) and barrier failure analysis (BFA).

Their advantages and disadvantages will be discussed in-depth, as well as how to create, step-by-step, a typical Bow-Tie and BFA diagram, regardless of the business sector of application.

The intentions of this book are:

- To present the contents of the technical standards about risk management (ISO 31000, *Risk Management – Guidelines*; ISO, 2018);
- To give a solid introduction to enterprise risk management (ERM) across different industrial sectors;

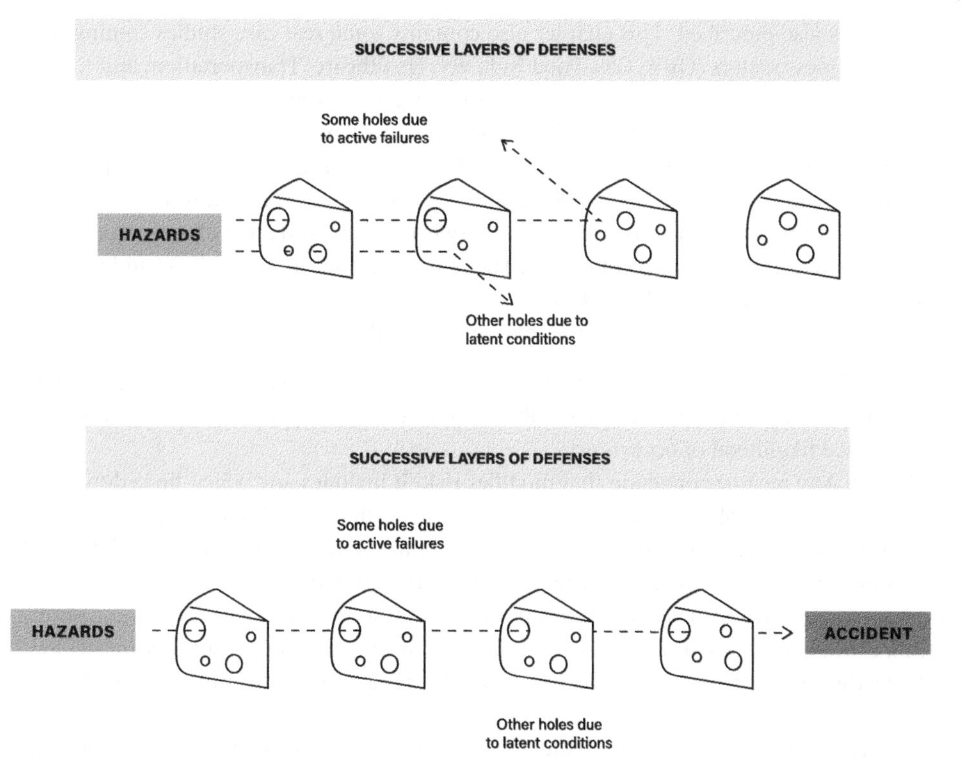

Figure 4 Swiss Cheese Model. *Source:* Reason, J., 1990.

- To suggest a risk-based perspective for some ISO standards;
- To introduce the barrier-based approach to risk management;
- To present the Bow-Tie method, its history, its elements, and how to get value from its implementation;
- To present barrier failure analysis, its elements, and how to get value from its implementation;
- To show how it is possible to link proactive and reactive phases of risk management (RM) using these methods;
- To offer some worked examples of both Bow-Tie and BFA methods;
- To help the reader with a step-by-step guide on how to implement the two methods.

The Bow-Tie method is widely adopted in the process industry, where it was born, and it has been developed to deal with industrial risks. This explains why some guidelines mentioned in this book, as properly referenced, come from the American Institute of Chemical Engineers – Center for Chemical Process Safety (AIChE-CCPS, 2018).

The first chapter is dedicated to a general introduction to risk and risk management: the ISO 31000 standard is presented, with its framework and process. The second chapter is where the Bow-Tie method is deeply discussed in theory, also showing some advanced usages; similarly, Chapter 3 is for BFA. Chapter 4 describes the Bow-Tie and BFA construction workflows with step-by-step guides; the layer of protection analysis (LOPA) construction workflow is also described. This chapter also contains some real case studies coming from several business sectors (Oil & Gas, Food Industry, Healthcare, Transportation, and Energy, just to mention a few). After the Conclusions, three appendixes end the book: the first two are simple, supportive material that the risk analyst or the incident analyst can use as a few-pages guide where only the key concepts are expressed, just as a fast refresh tool. A short presentation of the Human Error and Human Reliability Assessment is in Appendix 3.

Throughout this book, a consistent vocabulary is used to avoid any misunderstanding in using extensive and complex terminology. The main reference for this is ISO Guide 73, *Risk Management – Vocabulary* (ISO, 2009).

The following list is a glossary of the main terms used.

- *Risk:* Effect of uncertainty on objectives (either positive or negative deviation from what is expected). Often expressed as a combination of the consequences of an event and the associated likelihood of occurrence.
- *Control:* Any measure or action that modifies risk. It includes any policy, procedure, practice, process, technology, technique, method, or device that modifies or managed risk. Risk treatments become Controls or modify existing Controls once they have been implemented.
- *Risk Source:* Where a risk comes from. It has the potential to generate a risk.
- *Hazard:* A source of potential harm; present condition, event, object, or circumstance that could lead to or contribute to an unplanned or undesired event such as an accident.
- *Issue:* Risk with a probability of occurring of 100%; that is, it has eventuated into an existing issue.
- *Risk Assessment:* Process that is made up of risk identification, analysis, and evaluation.
- *Risk Identification:* Process of finding, recognizing, and describing risks involving the identification of risk sources, events, causes, and potential consequences.
- *Risk Analysis:* Process to comprehend the nature of risk and to determine the level of risk.
- *Risk Evaluation:* Process used to compare risk analysis results with risk criteria to determine whether a specified level of risk is acceptable.
- *Risk Treatment:* Process to modify risk that can involve avoidance, taking or increasing risk, removing the risk source, changing the likelihood, changing the consequences, sharing the risk, retaining the risk by informed decision.
- *Residual Risk:* Risk remaining after risk treatment.
- *Consequence:* Outcome of an event and affects objectives.

For the purposes of this book, it is assumed that the reader knows that risk is different from a hazard (the risk is the future impact of an uncontrolled hazard or, better, the future uncertainty created by the hazard). This is a fundamental difference since all methods are intended to assess the risks associated by inherent methods.

1.1 Risk Is Everywhere, and Risk Management Became a Critical Issue in Several Sectors

No human activity is risk-free. Its definition, so intimately connected to statistical and probabilistic assessments, imposes that the "risk zero" does not exist. Keep in mind that any business activity carries a risk of an entrepreneurial nature, that is, the inherent challenge of succeeding or not in that business. But, of course, there are also other types of risks that an entrepreneur must pay attention to, including operational, health and safety, environmental, and reputational risks, to name a few.

However, considering the presence of risk to be limited to entrepreneurial activities would be a mistake. In everyday life, we take opportunities by taking risks. There would be no opportunity to cross a road without running the risk of being overwhelmed by a running car, of driving a car without running the risk of going off the road, of undergoing medical treatments without running the risk of their failure and being ineffective, even of getting married without running the risk of having to face a divorce. Risk is everywhere, just looking at reality from a different perspective, as long as this change of point of view does not generate anxieties or fears, but calmly allows the acceptance of a true and incontrovertible fact: risk zero does not exist.

As shown in Figures 5 and 6, there is clearly no limit to the applicability of the concept of risk, which can also be scaled to a global dimension. For example, the Annual Global Risk Report by the World Economic Forum (World Economic Forum, 2020) lists the main global economic, environmental, geopolitical, social and technological risks perceived as priorities by the sample interviewed. Interestingly, in recent years there has been a greater awareness of the global environmental risks associated with extreme weather events, loss of biodiversity and the failure of actions to address what is known as the climate crisis, for which there is a high perception of risk both in terms of probability and magnitude, although the use of weapons of mass destruction and water crises continue to be of obvious concern from the point of view of impacts.

Developing a good perception of risk is, therefore, a fundamental piece for a more complete and comprehensive understanding of the world around us and its complexity. Indeed, equipping yourself with this awareness is often the key to the success in many organizations, whose managers, playing the role of true leaders, offer their expertise in guiding the organization through risks and opportunities.

Faced with this permeability of any human activity to risk, we understand the importance of risk management, as an operational and directional tool to ensure the sustainability of business processes and, more generally, human actions. Maintaining an "acceptable" (or at least "tolerable") state over time is, in fact, the ultimate goal of risk management, the complexity of which should force the establishment of well-structured processes and models capable of protecting the organization from dangers and finding added value where there is a risk of any kind.

The concept of risk is associated with several considerations. It is generally seen as the likelihood of an event being harmful, the possibility – more or less likely – that a threat exists and that it could harm an organization's objectives. But events can also have impacts that are positive, negative, or sometimes both. Some events bring with them a neutral conception of risk, such as weather forecasts. If these announce a certain chance of rain for next week, then for a farmer, it is an opportunity while for a tourist it is a threat.

From this brief premise, it is understood that it is necessary to detach on the basis of the mere negative conception of risk as a harmful event to marry, whenever possible, a positive vision (an opportunity to be seized) or at least neutral (probability of event). The definition suggested by ISO 31000, therefore, offers a 360-degree view of the concept of risk: "the effect of uncertainty on objectives."

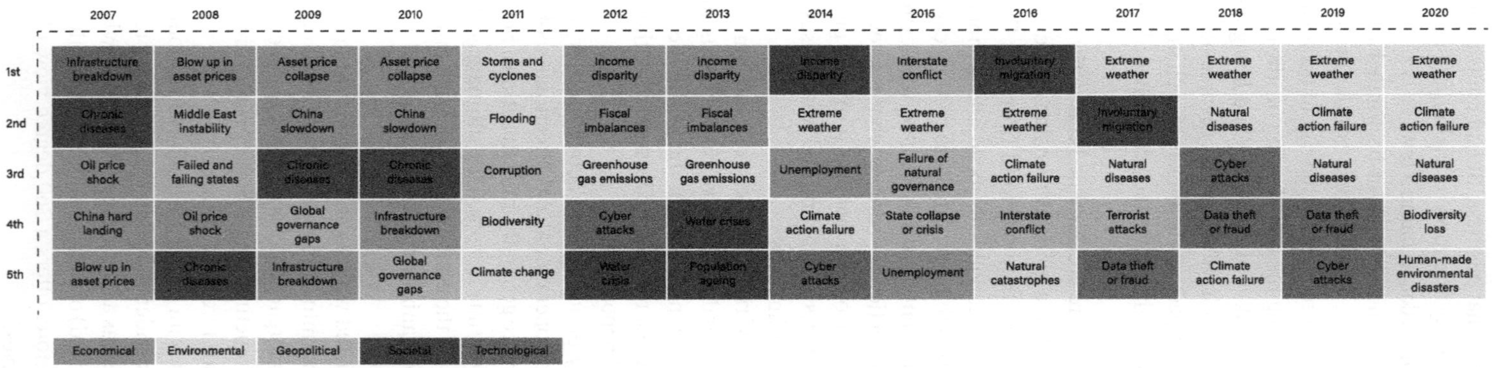

	2007	2008	2009	2010	2011	2012	2013	2014	2015	2016	2017	2018	2019	2020
1st	Infrastructure breakdown	Blow up in asset prices	Asset price collapse	Asset price collapse	Storms and cyclones	Income disparity	Income disparity	Income disparity	Interstate conflict	Involuntary migration	Extreme weather	Extreme weather	Extreme weather	Extreme weather
2nd	Chronic diseases	Middle East instability	China slowdown	China slowdown	Flooding	Fiscal imbalances	Fiscal imbalances	Extreme weather	Extreme weather	Extreme weather	Involuntary migration	Natural diseases	Climate action failure	Climate action failure
3rd	Oil price shock	Failed and failing states	Chronic diseases	Chronic diseases	Corruption	Greenhouse gas emissions	Greenhouse gas emissions	Unemployment	Failure of natural governance	Climate action failure	Natural diseases	Cyber attacks	Natural diseases	Natural diseases
4th	China hard landing	Oil price shock	Global governance gaps	Infrastructure breakdown	Biodiversity	Cyber attacks	Water crises	Climate action failure	State collapse or crisis	Interstate conflict	Terrorist attacks	Data theft or fraud	Data theft or fraud	Biodiversity loss
5th	Blow up in asset prices	Chronic diseases	Infrastructure breakdown	Global governance gaps	Climate change	Water crisis	Population ageing	Cyber attacks	Unemployment	Natural catastrophes	Data theft or fraud	Climate action failure	Cyber attacks	Human-made environmental disasters

Economical | Environmental | Geopolitical | Societal | Technological

Figure 5 Top five global risks in terms of likelihood (2007–2020). *Source:* World Economic Forum, 2020.

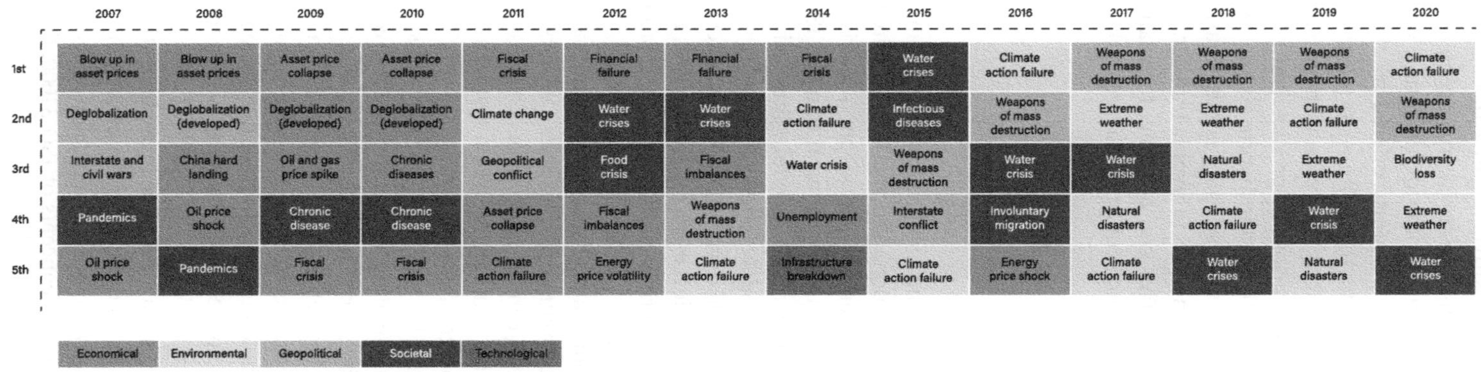

	2007	2008	2009	2010	2011	2012	2013	2014	2015	2016	2017	2018	2019	2020
1st	Blow up in asset prices	Blow up in asset prices	Asset price collapse	Asset price collapse	Fiscal crisis	Financial failure	Financial failure	Fiscal crisis	Water crises	Climate action failure	Weapons of mass destruction	Weapons of mass destruction	Weapons of mass destruction	Climate action failure
2nd	Deglobalization	Deglobalization (developed)	Deglobalization (developed)	Deglobalization (developed)	Climate change	Water crises	Water crises	Climate action failure	Infectious diseases	Weapons of mass destruction	Extreme weather	Extreme weather	Climate action failure	Weapons of mass destruction
3rd	Interstate and civil wars	China hard landing	Oil and gas price spike	Chronic diseases	Geopolitical conflict	Food crisis	Fiscal imbalances	Water crisis	Weapons of mass destruction	Water crisis	Water crisis	Natural disasters	Extreme weather	Biodiversity loss
4th	Pandemics	Oil price shock	Chronic disease	Chronic disease	Asset price collapse	Fiscal imbalances	Weapons of mass destruction	Unemployment	Interstate conflict	Involuntary migration	Natural disasters	Climate action failure	Water crisis	Extreme weather
5th	Oil price shock	Pandemics	Fiscal crisis	Fiscal crisis	Climate action failure	Energy price volatility	Climate action failure	Infrastructure breakdown	Climate action failure	Energy price shock	Climate action failure	Water crises	Natural disasters	Water crises

Economical | Environmental | Geopolitical | Societal | Technological

Figure 6 Top five global risks in terms of impact (2007–2020). *Source:* World Economic Forum, 2020.

In relation to the point of view (positive, negative or neutral) with which you look at risk, the organization will set a consequent risk management strategy. From a positive point of view (risk as an opportunity), the organization will tend to maximise the ability to take advantage of the opportunity offered by risk. For example, a lender decides to finance a project and will try to maximize its return on investment. According to the neutral conception, the resulting risk management strategy is to calculate the probabilities of the various risk scenarios and predict their performance. The organization that marries this approach will constantly process reports to monitor and review its risks. If, on the other hand, the view adopted is the negative one (risk as a threat), then the resulting strategy can only be to avoid, transfer, reduce or otherwise not increase the current set of risks. This can be done, for example, by avoiding unsafe technologies, implementing additional control measures, taking out insurance policies and so on.

1.2 ISO 31000 Standard

Regardless of the type, entity, size and complexity of an organization, many regulations and laws (both national and international) increasingly require the adoption of management systems that cover risk management. The adoption of a risk management model that over time complies with the international technical standard ISO 31000 can increase the effectiveness of this action: the organization's efforts are made consistent with a general model that is already consolidated, widely tested and used.

In fact, every organization has to deal with those factors, internal and external to its corporate structure, that make the achievement of its objectives uncertain. In other words, each organization must face its own risks, managing them appropriately in order to ensure the achievement of its objectives.

The ISO 31000 technical standard aims to provide principles and guidelines for risk management. The standard, adopted voluntarily, preserves a universal conception such as making it applicable to any company context, regardless of the nature of the risks associated with the organization's activities, adapting itself in a systematic, transparent and credible way, with the possibility of a progressive approach.

The ISO 31000 technical standard provides guidance to ensure adequate risk management in organizations. The content of the standard applies universally to any organization, regardless of entity, type, business sector and size; for this reason, the indications of the technical standard must then find appropriate customization depending on the specific context of application. Moreover, the document is valid throughout the entire life of the organization, depicting the whole life cycle for risk management. The standard contains proper references that can be applied to any activity within the same organization and embraces decision-making at all levels.

Technical Report ISO/TR 31004 is a valid guide to implementing ISO 31000 effectively. In particular, it provides a well-defined approach to ensure the proper transition, inside organizations, from their risk management arrangement to one that is consistent with ISO 31000. By explaining the underlying concepts of ISO 31000, the Technical Reports offers guidance to the principles, the framework, and the process described in the standard. These concepts are discussed in depth in the following pages; however, the interested reader is invited to consult both the standard and the technical report to go even deeper and learn extra details that are not treated here.

A third document that every good risk management practitioner has to know is the IEC 31010. This international standard contains a valid guide on the main and well-recognized techniques to identify, analyze, and assess risks. It also offers an approach to the selection of these techniques, depending on some input parameters. However, the document itself contains extra references to other sources that the reader can use to have a detailed description of the methods.

Before presenting the risk management framework, it is necessary to introduce the definition of risk. According to the vocabulary used in the sector worldwide, as stated in the international standards, risk is the "effect of uncertainty on objectives." This definition recalls the necessity to explain what an objective is. An objective, from an organizational point of view, can be defined as the business goals, thus including not only the need to maximize profit, but also the safety goals, reputational goals, environmental goals and so on. Having clarified what risk is, as stated in the previous paragraph, it is clear that risk can be seen from three different points of view (Figure 7):

- Positive view: the risk is seen as a potential gain;
- Neutral view: the risk is seen as the likelihood of events;
- Negative view: the risk is seen as some form of loss.

These three different perspectives reflect different perceptions of the risk. All of them share the occurrence of an event, whose effect can be positive, negative, or both. Indeed, an effect can be seen as a deviation from the expected. This deviation might result in opportunities or threats. In the end, the risk is usually described in terms of hazardous sources, potential events (scenarios), and their consequences and likelihood.

So, having defined what the risk is, the next question becomes: "What is Risk Management?" According to clause 3.2 of ISO 31000, it can be described as the "coordinated activities to direct and control an organization with regard to risk." The expression, widely used both in ISO 31000 and in ISO/TR 31004, refers to the principles, framework, and process that should be set up by every organization in order to manage risk effectively.

Described in this way, managing risks may appear extremely easy; but the reader should take into consideration that the concepts presented here are valid for every kind of organization, regardless of its type and dimension, so inevitably, the description is high-level and needs to be tailored for the peculiarities of each company.

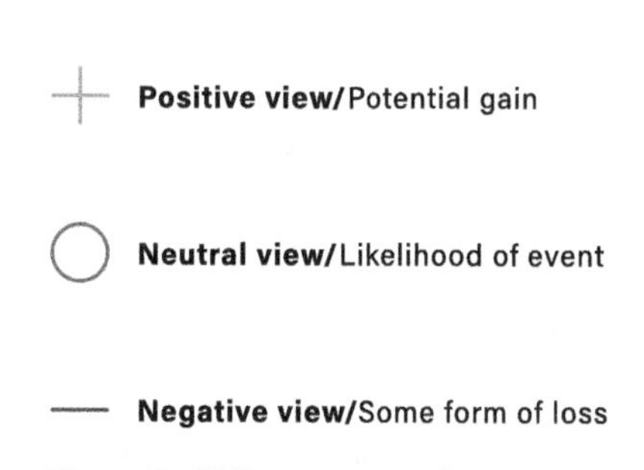

Figure 7 Different perspectives on risk.

Managing an organization means conducting a series of coordinated activities to direct and control the organization itself, with the unique goal of trying to reach its objectives. Therefore, risk management is part of management, because it involves the required activities to dominate, or at least control, the effect of uncertainty on those objectives. This is why it is important that risk management is fully integrated into the general management system of the organization; otherwise, its effectiveness is deeply compromised. In this way, it is possible to pursue the risk management purpose, i.e. the creation and protection of value (as stated in ISO 31000, clause 4), improving performance, encouraging innovation, and supporting the achievement of objectives.

Risk management is carried out in the identification, analysis, and evaluation phases, with the aim of identifying any corrective actions that may be necessary to meet the risk acceptability criterion that the organization has set for itself, as well as in relation to any legal requirements. Through this process, the organization communicates with stakeholders and monitors and reviews the risks and control measures put in place to ensure that no further action is needed to reduce the risk levels achieved. This logical process is described in detail by ISO 31000. The standard not only defines the risk management process to be adopted, but also outlines the principles, framework, and workflow that the organization must integrate with its values, policy, and strategy in order to ensure effective risk management and maintain it over time.

The risk management system may apply to all or part of an organization, depending on the objectives and expectations deriving from the application of this management model (Figure 8). For this reason, establishing the context of application of the risk management system represents the "zero" phase of this process: this means not only identifying the organization's objectives, but also the environment in which they are pursued, the boundary conditions, identifying the parties involved, and defining the risk acceptability criterion (or criteria, if diversifying the assessment in relation to the nature and complexity of multiple risks for the organization is intended). This fundamental action is also known as "context analysis."

The relationship between the principles underlying risk management, the framework in which it takes shape, and the operational management process is shown in Figure 9.

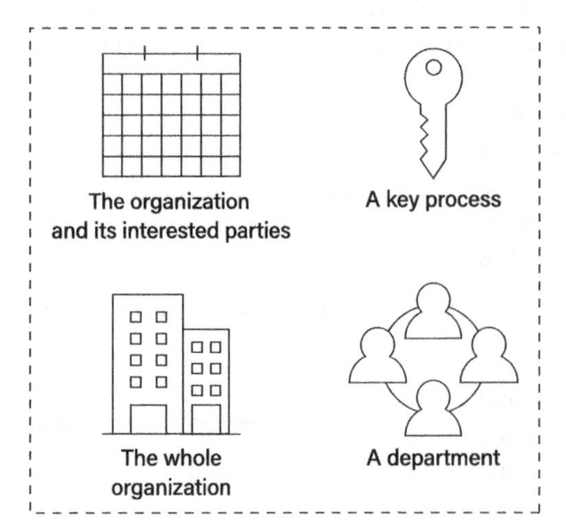

Figure 8 Definition of the scope of risk management.

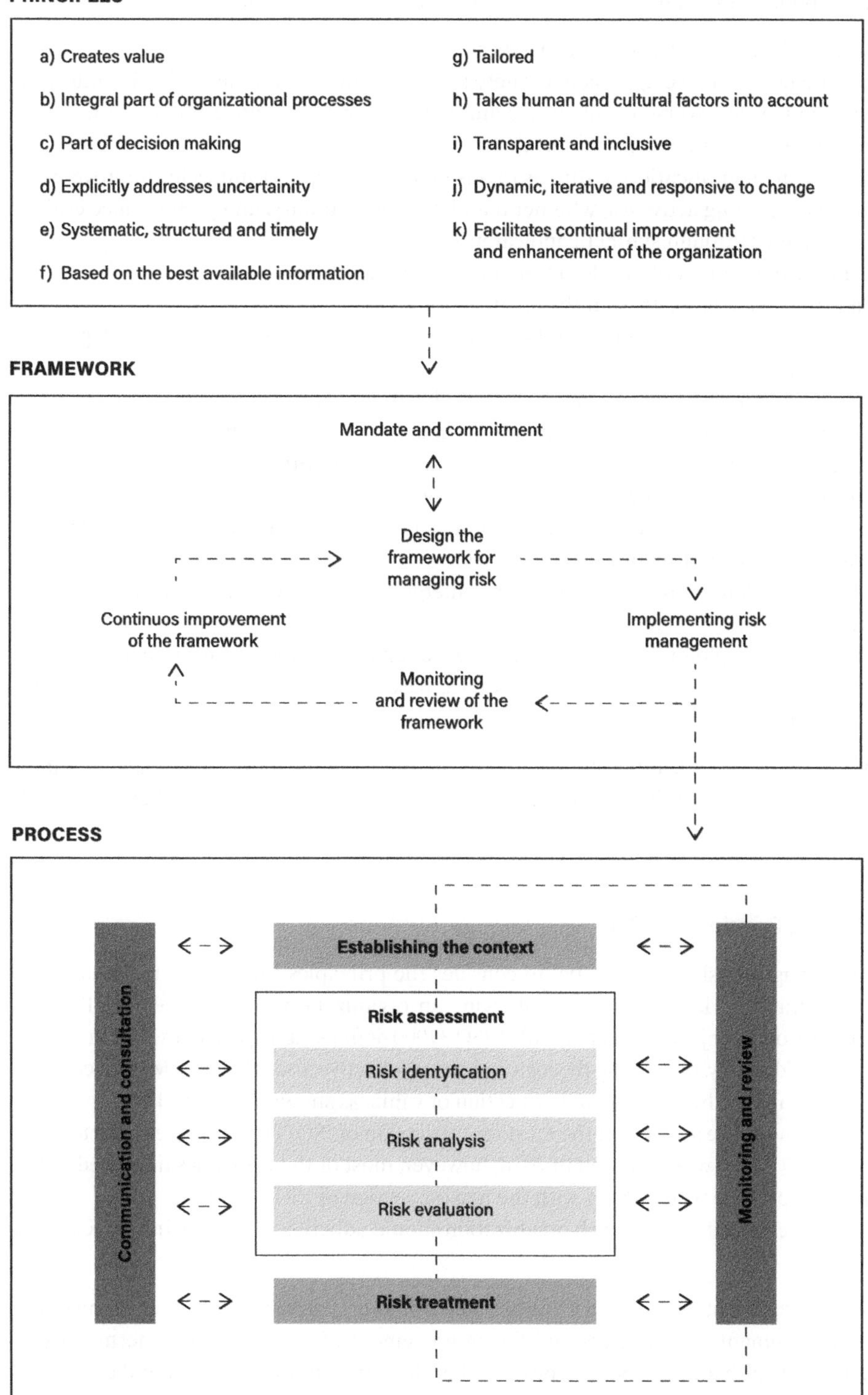
PRINCIPLES
a) Creates value
b) Integral part of organizational processes
c) Part of decision making
d) Explicitly addresses uncertainity
e) Systematic, structured and timely
f) Based on the best available information
g) Tailored
h) Takes human and cultural factors into account
i) Transparent and inclusive
j) Dynamic, iterative and responsive to change
k) Facilitates continual improvement and enhancement of the organization
FRAMEWORK
Mandate and commitment
Design the framework for managing risk
Continuos improvement of the framework
Implementing risk management
Monitoring and review of the framework
PROCESS
Communication and consultation
Establishing the context
Risk assessment
Risk identyfication
Risk analysis
Risk evaluation
Risk treatment
Monitoring and review

Figure 9 Relationship between principles, framework, and risk management process.

The adoption of the ISO 31000 standard allows each organization to:

- Increase the probability of achieving its goals;
- Encourage a proactive approach to safety from risk (and original hazards), i.e. intervening even before adverse events occur and not only in the reactive phase following an accident, near-miss, or non-conformities;
- Improve the identification of threats to this process and opportunities for improvement;
- Improve reporting activities, whether mandatory or voluntary, and performance evaluation activities through synthetic indicators (KPIs);
- Facilitate the fulfilment of relevant mandatory standards;
- Improve stakeholders' trust in the organization;
- Build a solid management foundation on which to base its decisions and plan its development;
- Improve the internal control system and be able to provide evidence of it;
- Allocate resources (mainly human and financial) efficiently to pursue risk mitigation and ensure over time that the risk reduction factor is identified as necessary;
- Improve operational effectiveness and efficiency;
- Minimize the effects of a negative event, such as an accident, near-miss, or non-conformities (including process anomalies);
- Improve incident management, from the emergency response phase to the investigation and analysis of the incident;
- Improve the exploitation of operational experience, developing recommendations or more general observations, effective as well as based on the "root causes" of and incident;
- Increase resilience.

In few words, organizations embracing a structured approach to risk management have the chance to improve their resilience to complexity and have a higher chance to reach their strategic objectives.

1.2.1 The Principles of RM

When managing risk, it is essential to consider the principles that enable the creation of solid RM framework and processes, allowing an organization to dominate the effects of uncertainty on its objectives. In particular, ISO 31000 defines eight principles that an organization should satisfy, describing the logic behind an effective and efficient RM. Indeed, their satisfaction allows the creation and protection of value, as shown in Figure 10.

Readers should be aware that the most recent version of ISO/TR 31004 goes back to 2013, whereas ISO 31000 was reviewed in 2018; however, most of the principles described in the technical report are well aligned with the newest version of ISO 31000.

To manage risk effectively, each organization should adhere to the following principles at all levels:

- Risk management creates and protects value. In other words, it contributes to the tangible achievement of objectives and the improvement of performance in terms of (for example) health and safety, environmental protection, quality, efficiency and continuity of operations and reputation.

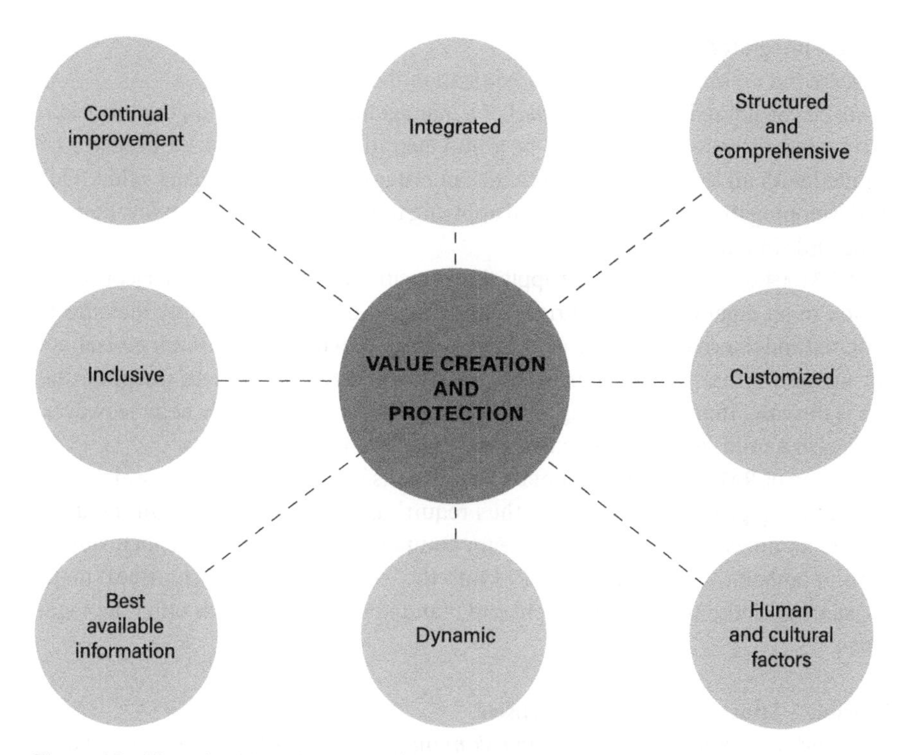

Figure 10 The principles of RM according to ISO 31000.

- Risk management is an integral part of all business processes. It is therefore not a stand-alone activity, but also integrates with project management, management of change and strategic planning.
- Risk management is part of the decision-making process. In fact, it helps decision-makers to make reasoned choices by prioritising the actions to be taken.
- Risk management explicitly focuses on what is uncertain. It is therefore good to be aware from the outset of the role of uncertainty in this area.
- Risk management is systematic, structured and timely. It contributes to operational efficiency and the achievement of consistent, comparable and credible results.
- Risk management is based on the best information available. Stakeholders should be informed of the limitations arising from the information and models used.
- Risk management is tailor-made, taking into account the internal and external context of the organization, including its articulation and complexity.
- Risk management takes into account human and cultural factors.
- Risk management is transparent and inclusive.
- Risk management is dynamic, iterative and sensitive to changes (these in particular must be subject to rigorous management of the change process).
- Risk management helps to pursue the continuous improvement of the organization.

The principles, which are described hereafter, should be tailored to the specific part of the RM framework under consideration and, even if described in general terms, need to be deeply understood and applied continuously to produce the positive effects of an effective RM system.

First Principle – Integrated

Risk management is an integral part of all organizational activities.

RM cannot be considered as a separate part of a management system for an organization, nor an extra administrative requirement or a bureaucratic task, but instead needs to be fully integrated with all the managerial activities in order to create and protect value. This is done by developing the RM framework and applying the RM process to the relevant decision-making and similar activities.

The ISO/TR 31004 suggests how to best apply this principle. In particular it underscores that every decision in an organization bring risks with it, because of the uncertainty that affects both the internal and external context, whose changes cannot be controlled by the organization.

It is not necessary that a management system is formalized in order to integrate the RM. If this is the case, than establishing a RM framework can significantly help. However, ISO 31000 is also a solid reference for existing management processes.

The integration of RM is also fundamental to avoid risks being understood only after that the decision-making process is concluded, thus requiring costly modification in the decision taken and an unacceptable waste of time for competitive companies. In conclusion, it is important to embed the components of RM into the existing (formal or informal) management system, with the same commitment and mandate that are spent for other managerial activities.

Second Principle – Structured and Comprehensive

A structured and comprehensive approach to risk management contributes to consistent and comparable results.

Reliable and successful results can be obtained with a consistent approach to RM. Indeed, satisfying the organization's objectives requires consistent risk criteria that cannot be decided depending on the mood of the day or on the specific needs that may require extra flexibility to reach the goal.

Therefore, RM must consider the time dependencies and be applied timely, at the right moment in the decision-making process; otherwise (if done too early or too late) it could be costly to change direction and the best opportunities could be lost.

The structured approach can be satisfied by the application of the RM process as established in ISO 31000, which suggests the proper activities to be implemented and their sequence.

Third Principle – Customized

The risk management framework and process are customized and proportionate to the organization's external and internal context related to its objectives.

Each organization has its own peculiarities and needs. There is not one specific design and implementation of the RM framework and processes, as they need to be customized for each organization, thus requiring flexibility. Size, business sector, culture, and management approach are elements that need to be considered when defining a way to manage risks.

Indeed, the technical standards that are used as a reference offer a generic approach to RM, being applicable to every type of organizations and risks. The necessity of tailoring the

RM may also emerge within the same organization, when dealing with different risks (e.g. operational, financial, reputational). Systems, methods, and criteria could be different, but the general approach should be consistent and compliant with ISO 31000. Moreover, the designed RM framework should also embed any legal requirements or other external obligations.

Tailoring does not simply mean to change one or more elements of the RK framework or RM processes, as they are described in ISO 31000. It also implies a specific tailoring effort during the design and improvement of the RM framework, not only its implementation. For example, an organization may need to take into account its internal issues like staff turnover or a massive hiring of inexperienced employees who need to be informed and trained on what is required by the RM.

Finally, it must be highlighted that a customized RM framework can be more easily integrated within the general management system of the organization, even if the reverse is always possible, i.e. to modify the decision-making processes to fit the structure of the RM framework.

Fourth Principle – Inclusive

Appropriate and timely involvement of stakeholders enables their knowledge, views, and perceptions to be considered. This results in improved awareness and informed risk management.

The involvement of stakeholders is a crucial point for effective RM. It is therefore important to build trust reciprocally in every step of the RM processes, otherwise some key elements for the design and implementation of the RM could be missed. When implementing this principles, it is important to take into account issues of privacy, security, and confidentiality, so that risk-related information is secured. To do so, it is generally recommended to separate the relevant information in risk registers.

Fifth Principle – Dynamic

Risk can emerge, change or disappear as an organization's external and internal context changes. Risk management anticipates, detects, acknowledges and responds to those changes and events in an appropriate and timely manner.

The environment in which an organization operates is subjected to continuous changes. Those changes in the internal and external context lead to some changes in risks too. Similarly, any modification in the objectives of the organization will change its risks. Therefore, it is important that the RM processes are able to reflect this dynamicity.

The ISO 31000 technical standard explicitly refers to monitoring and reviewing phases for both the RM framework and the RM process. They are two different activities. Monitoring means to observe continuously some key parameters in order to determine whether everything works as intended. Reviewing, on the other hand, is a structured process to check if the hypothesis at the base of the design and implementation of the RM remains unchanged or, if not, a review is required on the resulting decisions.

In conclusion, risks evolve continuously and organizations are far from being static: the RM framework needs to be monitored and reviews periodically, in order to pursue the continual improvement, real paradigm for every management systems.

Sixth Principle – Best Available Information

The inputs to risk management are based on historical and current information, as well as on future expectations. Risk management explicitly takes into account any limitations and uncertainties associated with such information and expectations. Information should be timely, clear and available to relevant stakeholders.

A correct understanding of a risk comes from the availability of the best available information. Indeed, information and data are affected by a certain degree of uncertainties: to know their sensitivity is crucial to develop clear and precise risk criteria. Every decision-making processes should rely on evidence-based information, but this is not always possible because of restraints on time or resources available. In this case, expert opinions should support the limited information available, always avoiding group bias and other human-related errors for this type of judgement.

Risk-related data help to develop supportive statistical predictions, but past evidence has been shown to not necessarily predict the future accurately. Where there is a lack of information, but there is the evidence of a potential harm, prompt action is required to avoid a dangerous situation escalating in a real risk scenario.

Of course, the design, implementation and improvement of the RM framework is based on the best available information. To obtain the best results, the quality of the data, and thus their reliability and accuracy, should be regularly checked and, if required, a review could be triggered.

Seventh Principle – Human and Cultural Factors

Human behaviour and culture significantly influence all aspects of risk management at each level and stage.

The design of the RM framework should consider the human and cultural factors, i.e. the cultural characteristics and level of knowledge of the involved stakeholders, by obtaining their views and analyzing their features. In particular, among the factors to be taken into account there are: social influences, politics, cultural background, and concepts of time. RM may fail to detect early warnings or recognize complexity, or remain indifferent to others' views. To avoid this, the RM framework designer should ask if the organizational structure is appropriate to the needs of the organization, if the formal accountabilities are clearly identified, if the job descriptions are clear and responsibilities are correctly linked, if the communication channels are effective, if the morale in the organization is periodically monitored, if the recruitment and remuneration policies are clear, if periodical internal and external audits are performed to look for unsafe human behaviours, if the procedures are aligned to the policies, and so on.

Eighth Principle – Continous Improvement

Risk management is continually improved through learning and experience.

The operative experience of the organization offers many inputs to correct the RM and improve the capacity to convert risks into opportunities. Indeed, the continual improvement for an organization is the real driving force of any management system that is based on the well-known Deming cycle, i.e. the continuous loop of the Plan-Do-Check-Act (PDCA) phases, as the RM is.

The continual improvement allows better-informed risk-based decisions to be made, helping to reduce uncertainty in achieving objectives, but overall, this principle allows an organization to remain alert to new opportunities that may arise both internally or externally, with the aim of improving RM efficiency as well.

Continuous improvement may include improving the quality of risk assessment, improving the RM framework, improving the decision-making process, and extending the range of action of RM, including new activities. What is important is that its goal is clearly identified in the RM policy and communicated both formally and informally.

In order to understand what needs to be improved, an organization should monitor both qualitative and quantitative indicators. Of course, some improvements may require long time to be achieved, such as when a specific budget needs to be allocated. Therefore, it becomes crucial to plan the improvement activities taking into account their priorities and benefits.

The application of the principles of RM, as described in ISO 31000, helps organizations in defining their strategy, achieving their objectives and making informed, risk-based decisions, contributing, in the meanwhile, to the improvement of the management systems.

Each organization should therefore adopt a structured approach to risk management: this is what ISO 31000 calls the framework (Figure 11). It ensures that the information

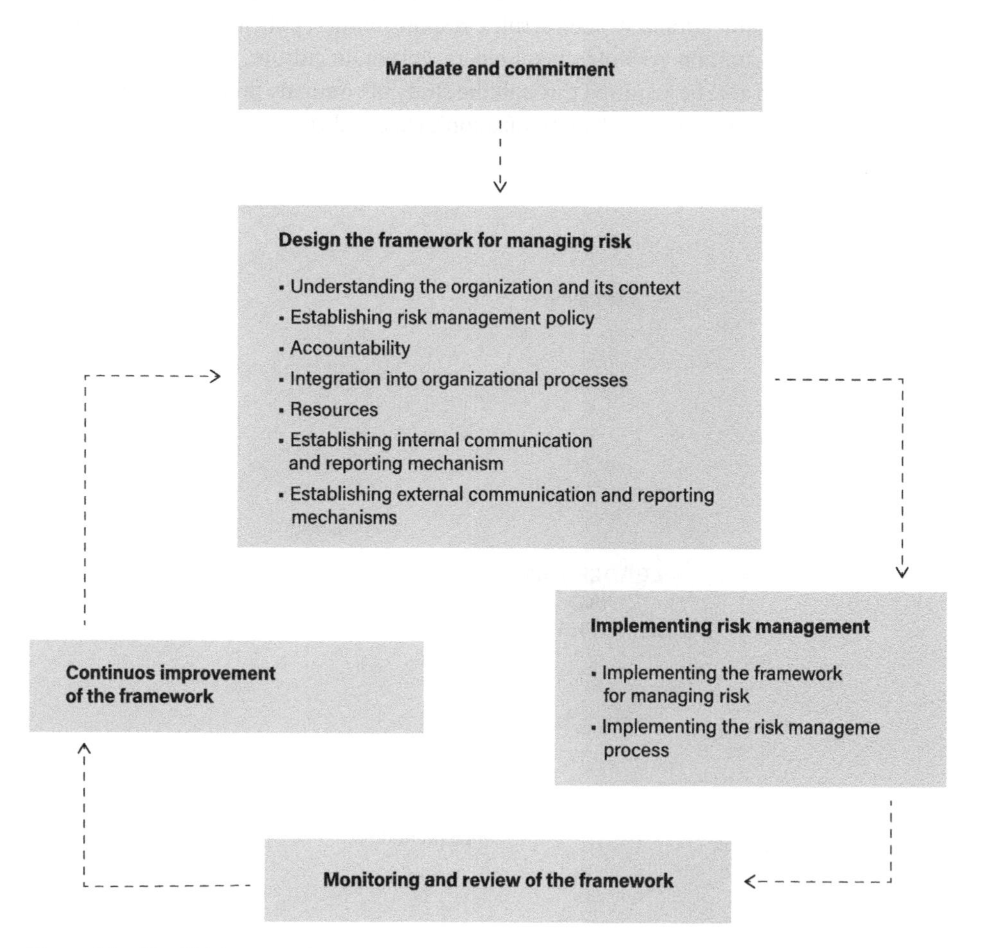

Figure 11 The RM framework.

derived from the risk management process is properly used as a basis for decision-making and the definition of responsibilities at all levels of the organization.

The structured approach to risk management must be based on solid foundations: these are the mandate and commitment, intended as dedication to the cause, of the organization. In other words, this means defining and supporting a risk management policy, ensuring that the corporate culture is always aligned with it. The commitment also implies the definition of performance indicators, strategies and risk management objectives, which must be aligned with those of the organization, always ensuring compliance with mandatory regulations. The corporate mandate requires that responsibilities be assigned to the different levels of the organization, which must also guarantee adequate resources (human and economic) for risk management; communicate the expected benefits of risk management to all stakeholders; and ensure that the structured approach to risk management remains appropriate over time.

Having established the above, it is necessary to design this framework, whose components are shown in Figure 12. To do so, a priority is to know the organization and the context in which it operates, both the external one (social, cultural, political, legal, financial, technological, economic, natural, environmental, international, national, local, etc.) and the internal one (organizational structure, roles, responsibilities, policies, objectives, strategies, resources, information systems, stakeholders, corporate culture, etc.). Knowledge of the organization can also be acquired through the study of company processes and any risk assessments already carried out. The organization's expected risk management objectives

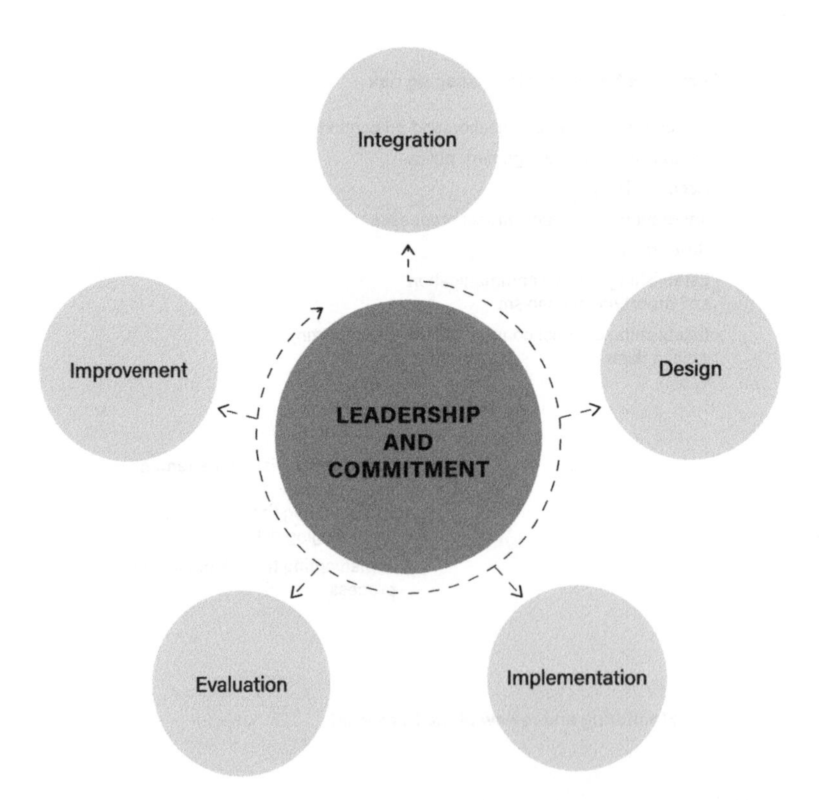

Figure 12 Components of a risk management framework.

must be clearly expressed in a risk management policy document. It must describe how the organization intends to manage risk, clarifying the links between the organization's different objectives and related policies, and also highlight the persons responsible within the organization for such management. A good risk management policy should also indicate how any conflicts of interest are addressed, establish measured performance standards, allocate the resources needed to assist risk managers in their task, and show a commitment to review and improve the policy and framework on a periodic basis or in response to an event or change in the environment. The risk management policy should be communicated in an appropriate manner to all stakeholders.

As part of the structured approach to risk management, each organization must ensure that responsibilities are distributed to the various parties involved in the risk management system, having previously verified that there is the authority to do so, together with an adequate level of expertise in risk management, thus ensuring the implementation and maintenance of the risk management process, and ensuring the adequacy, effectiveness and efficiency of any control measures put in place. This can be facilitated by clearly identifying the individual persons with authority to identify risk management responsibilities, the persons responsible for the development, implementation and maintenance of the risk management structure, and the other responsibilities of persons at all levels of the organization in the risk management process, as well as establishing performance standard metrics and a timely reporting system (also to be understood as documented evidence).

It is essential that this structured approach is integrated into company processes in an effective and efficient manner. The risk management process should be an integral part of these processes, not a separate one. In particular, risk management should be incorporated into business development policies, strategic planning and change management, ensuring that this integration reaches all levels of the organization and is incorporated into business practices, procedures and processes.

The implementation of such a risk management framework requires that the organization allocate adequate resources. In this sense, it would be necessary to question human resources, their skills, experience and competence; the financial resources required by each individual phase of the risk management process; the methods and tools to be used; documented procedures and processes; knowledge and information management systems; and training and updating programmes.

It is also essential to establish communication and reporting mechanisms inside and outside the company context. With reference to the internal context, these mechanisms aim to:

- Support and encourage awareness and understanding of business risks and the assumption of related responsibilities, in order to ensure that the key components of the risk management structure, and any subsequent changes to them, are adequately communicated;
- Support adequate internal reporting on the adopted framework, its effectiveness and results;
- Ensure that relevant information from the application of risk management is available at all levels, at appropriate times according to defined visibility in relation to role and responsibilities;
- Put in place processes of consultation with internal stakeholders in the business environment.

With reference to the external context, each organization should develop and implement a plan on communication with stakeholders outside the company if identified in the context analysis. This translates into a series of appropriately defined activities:

- Adequate involvement of external stakeholders, ensuring an effective exchange of information;
- Production of reports for the fulfilment of legal requirements and possible requirements of the organization;
- Use of communication (including periodic communication) to build trust in the organization and communicate in a structured and shared manner;
- Communication with stakeholders in the event of a crisis or emergency event, providing feedback and adequate reporting on the consultation and communication processes that have been put in place, disseminating information on corrective actions, including preventive ones, defined by a detailed analysis of the root causes and aimed at preventing the repetition of the negative event.

Once the rules with which the organization intends to manage its structured approach (the framework) to risk management have been defined, they must be implemented. In particular, the implementation of the organizational framework dedicated to risk management requires that the organization must, at least:

- Define the timing and implementation strategies.
- Apply the risk management policy and process to business processes.
- Comply with regulatory requirements and any obligations also arising from voluntary adherence to internal or external rules, standards, and best practices.
- Ensure that decision-making processes, including the choice of objectives, are aligned with the results of risk management processes.
- Conduct information and training sessions by implementing an appropriate information, training and education programme.
- Communicate with stakeholders and consult them in advance of key implementation stages to ensure that the risk management structure adopted remains appropriate.

The implementation phase of the framework is followed by the monitoring and review phases of the framework itself, with the aim of identifying opportunities for its continuous improvement.

Once the organizational structure supporting the risk management system has been implemented, it is possible to proceed with the definition of the specific risk management process to be adopted.

The process of communication and consultation with stakeholders is extremely important, as they express their assessments on the basis of their risk perception, experience and degree of involvement in the organization's processes. This perception may vary subjectively due to differences in the values, needs, assumptions, concepts and attention of stakeholders. Because of their potential impact, these perceptions should be identified, recorded and taken into account in the decision-making process. The communication and consultation process should thus facilitate the exchange of true, relevant, accurate, and understandable information, not forgetting aspects of personal integrity and confidentiality.

It is therefore essential to define a criterion to be adopted to assess the significance of the risk even before proceeding with the actual analysis. This criterion must reflect the organization's values, objectives and resources. Some criteria may be imposed, or derive from legal requirements to which the organization must be subject. The risk acceptability criterion must be consistent with the risk management policy that is defined at the beginning of each risk management process and is continuously reviewed. In defining the risk assessment criterion, the following factors must be taken into account:

- The nature and type of causes and consequences that may occur and how they may be measured;
- The nature and extent of the hazards faced by the organization for each of the processes, products and services to which risk management must be applied according to the defined implementation strategy;
- The way in which the probability of an event occurring is defined;
- The way in which the level of risk is determined (e.g. methodology);
- The stakeholders' point of view;
- The level at which the risk becomes acceptable or tolerable according to the internal and/or external criteria that are applicable for each type of risk;
- The possibility or not of taking into account combinations of multiple risks as well as secondary or "domino" effects, including "escalation factors" that may affect an incidental sequence.

All these concepts are discussed more in-depth in the following section, dedicated to the RM workflow.

1.3 ISO 31000 Risk Management Workflow

As defined in ISO Guide 73, the RM framework (Figure 13) is a "set of components that provide the foundations (i.e. the policy, objectives, mandate and commitment) and organizational arrangements (i.e. the plans, relationships, accountabilities, resources, processes and activities) for designing, implementing, monitoring, reviewing and continually improving risk management throughout the organization."

WHAT DOES IT REFER TO?	WHAT DOES IT INCLUDE?	CAPACITY (RESOURCE AND CAPABILITY)
The risk management framework refers to the arrangements (including practices, processes, systems, resources and culture) within the organization's system of management that enable risk to be managed.	The framework includes clear statements from top management on the organization's intent regarding risk management (described in ISO 31000 as mandate and commitment) and the necessary capacity (resources and capability) to achieve this intent.	Capacity does not exist as a single system or entity. Capacity comprises numerous elements integrated into the organization's overall management processes.

Figure 13 Risk management framework.

1.3.1 Leadership and Commitment

Leadership and commitment (Figure 14) ensure the integration of RM into all organizational activities. Indicators that demonstrate the leadership and commitment include:

- The customization of the RM framework components on the specific needs;
- The implementation of the RM framework components;
- The definition of an RM policy, establishing the related approaches, plans, and actions;
- A wide communication of the RM policy;
- The allocation of adequate resources to RM;
- The definition of responsibilities to RM at the appropriate levels within the organization.

Being an RM leader requires solid skills in predicting and accepting changes that may be involved in the behaviour, culture, and processes, and should be reflected within the policy, always monitoring the expected performance while managing risks, and in the RM framework. Of course, meeting the principles of ISO 31000 is among the best efforts for excellence in risk management.

1.3.2 Understanding the Organization and Its Context

In building the risk management process, it is also of paramount importance to prioritize the context (Figure 15).

Its definition allows the organization to articulate its objectives, define the internal and external parameters to be taken into account in operational risk management, and define the scope of the process and the criteria of risk acceptability. These activities are fundamental for the definition of an effective policy. Understanding the external environment is

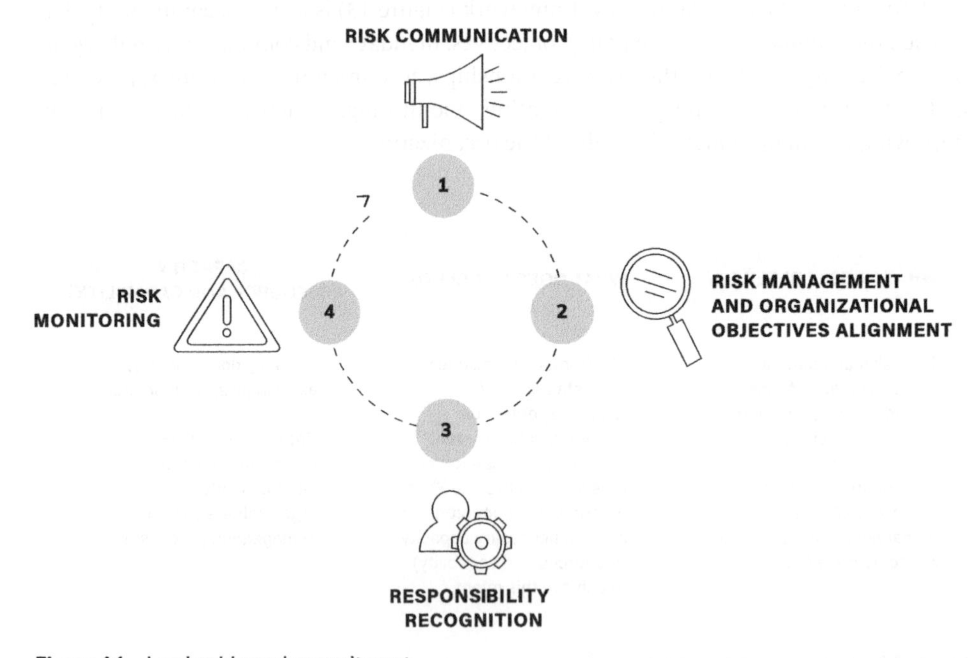

Figure 14 Leadership and commitment.

Figure 15 Internal and external context.

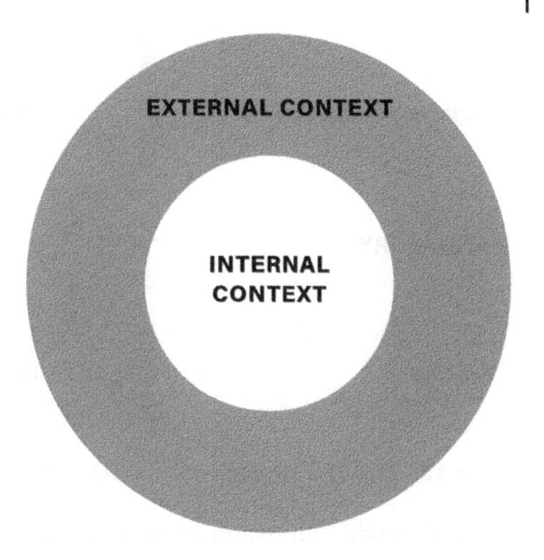

important to ensure that the objectives of external stakeholders are taken into account in the development of risk acceptance criteria. The external environment includes, for example, the social, cultural, political, legal, financial, technological, economic, natural, competitive, international, national, regional or local environment, but also key factors and trends that have an impact on the organization's objectives, relations with stakeholders outside the organization, their perceptions and values. It is essential that the legal requirements and regulations are taken into account when defining the criteria of acceptability and tolerability of risks. The internal context, on the other hand, represents the corporate perimeter within which the organization tries to achieve its objectives. The risk management process should be aligned with the strategy, structure, processes and corporate culture. The internal context includes everything within the organization that may influence the way the organization manages risk. It should be well identified, as:

- Risk management takes place in the context of achieving corporate objectives;
- The objectives and criteria related to the risk management of a particular project, process or activity should be considered in the light of the objectives of the organization as a whole and in compliance with the corporate policy defined by the top management;
- Some organizations are unable to recognize opportunities to achieve their strategic, project or business objectives and this affects the commitment, credibility, trust and value of the organization.

The internal context includes, for example, the organizational structure, roles and responsibilities, policies, objectives and strategies put in place to achieve them, available resources, such as time, capital, human and technological resources, relations with internal stakeholders, their perceptions, values, corporate culture, information systems, information flows and decision-making processes, whether formal or informal, guidelines and models also adopted by the organization through the signing of voluntary commitments. The requirements (both mandatory/voluntary or internal/external) related to RM are shown in Figure 16.

Obviously it is necessary to define the objectives, strategies, scope and parameters of the company activities where the risk management process is applied. This process should be

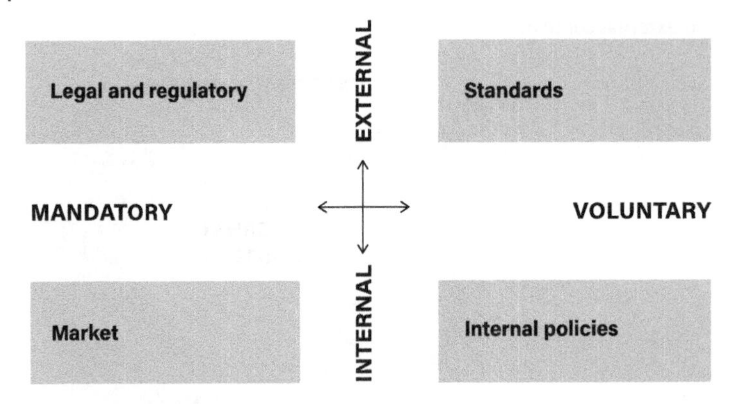

Figure 16 Identify the requirements related to risk management.

undertaken with full awareness of the need to justify the resources employed in the adoption of this organizational model. The context in which the risk management process is applied will vary according to the specific needs of an organization. Its identification may require:

- The definition of the objectives of risk management activities (including the risk aspects towards which the process is directed);
- The definition of responsibilities for the implementation of this process;
- The definition of the responsibilities distributed within that process, i.e. those of the parties involved in that process;
- The definition of activities, processes, functions, projects, products, services or assets in terms of time and space and all those resources that are intended to be used;
- The definition of the relationships between a particular project, process or activity and others within the organization;
- The definition of risk assessment methodologies (even more than one, commensurate with the types of risk and the degree of detail desired);
- The definition of the way in which the effectiveness and performance of the risk management process is assessed (e.g. indicators);
- The identification and specification of decisions to be taken and responsibilities for implementation.

Attention to these and other relevant factors helps to ensure that the approach adopted in risk management is appropriate to the circumstances, the organization, and the risks affecting the achievement of the objectives.

1.3.3 Implementation of the RM Framework

In order to implement the RM framework, an organization should:

- Develop an implementation plan that includes time and resources;
- Identify the features of any decision-making process within the organization (how, when, where, and by who decisions are made);
- Modify the plan if necessary;
- Be sure that the established arrangements (i.e. the RM framework) are clearly understood and put in practice.

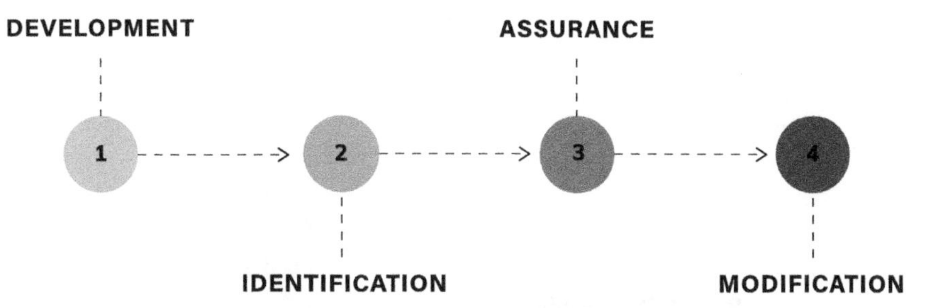

Figure 17 Implementing the risk management framework.

The implementation of the RM framework can be summarised in the four steps shown in Figure 17. It is important to highlight that a properly designed and implemented risk management framework will ensure that the changes in external and internal contexts will be adequately captured, allowing the RM, and by reflection the decision-making processes, to be changed accordingly.

1.3.4 The Risk Management Process

The risk management process should be:

- An integral part of management actions ("management");
- Incorporated into company practices and culture;
- Tailor-made, in relation to the specificity of company processes and the complexity of the organization itself.

This process includes the activities described in Figure 18:

- Communication and consultation;
- Context definition;
- Risk assessment, declined in the phases of risk identification, analysis of the reduction of the risk level starting from the identified hazards and comparison of the calculated risk level with the defined acceptability criteria;
- Treatment of the risk (measures to maintain or develop the level of risk identified, also in relation to the results of the acceptability assessment conducted);
- Monitoring and review.

The first phase of the risk management process is always the communication and consultation with all interested parties, whether internal or external to the company context. In fact, this activity should cover all stages of the risk management process. For this reason, its planning should take place at an early stage of the process. Effective communication and consultation with external and internal stakeholders ensures that those responsible for implementing the risk management process and stakeholders understand the basis on which decisions are made, and the reasons why particular actions may be required. This approach makes it possible to:

- Better establish the context by taking advantage of the contribution;
- Ensure that the interests of the parties involved are understood and duly taken into account;
- Ensure that risks are adequately identified in a shared and structured way;
- Bring together different skills and experience in risk analysis in order to ensure effective assessment and subsequent treatment;

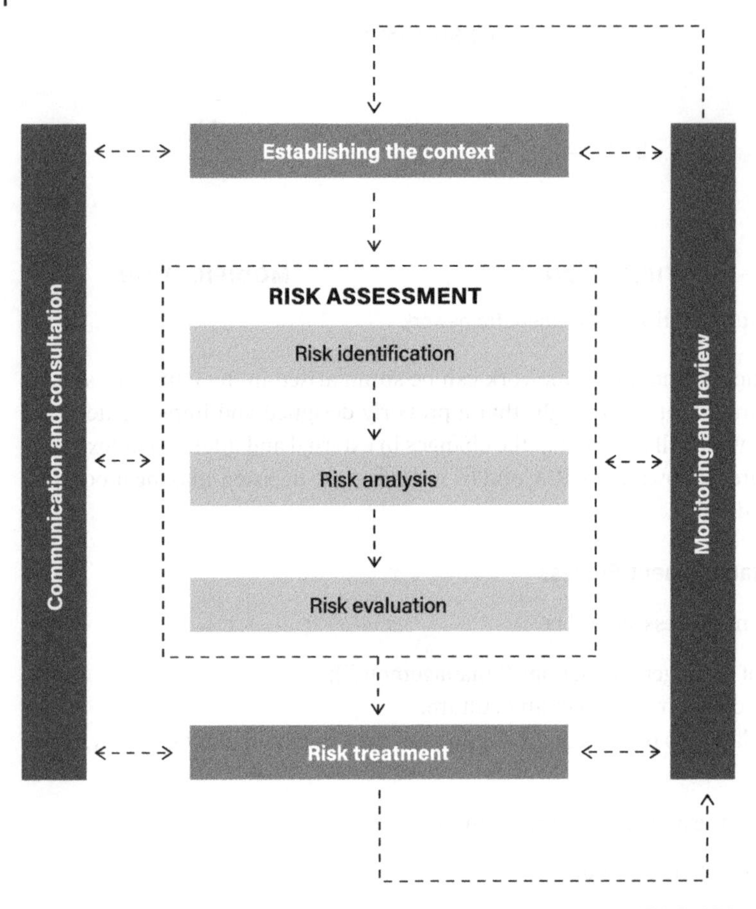

Figure 18 Scheme of the risk management process according to ISO 31000.

- Ensure that different views are duly taken into account when establishing risk acceptability criteria and risk evaluation;
- Develop an appropriate internal and external consultation and communication plan.

1.3.5 Relationship between the RM Principles, Framework, and Process

At the end, the ISO 31000 standard provides:

- The principles;
- A risk management framework;
- A risk management process that encompasses all the vital phases.

An organization wishing to succeed in implementing an effective RM should take into account the relations between them (Figure 19) and apply the principles at all levels; implement an effective RM framework, on the basis of those principles; and apply the RM process (divided in the steps identified by the framework) to the entire organization and modify it to reflect assets and contexts changes.

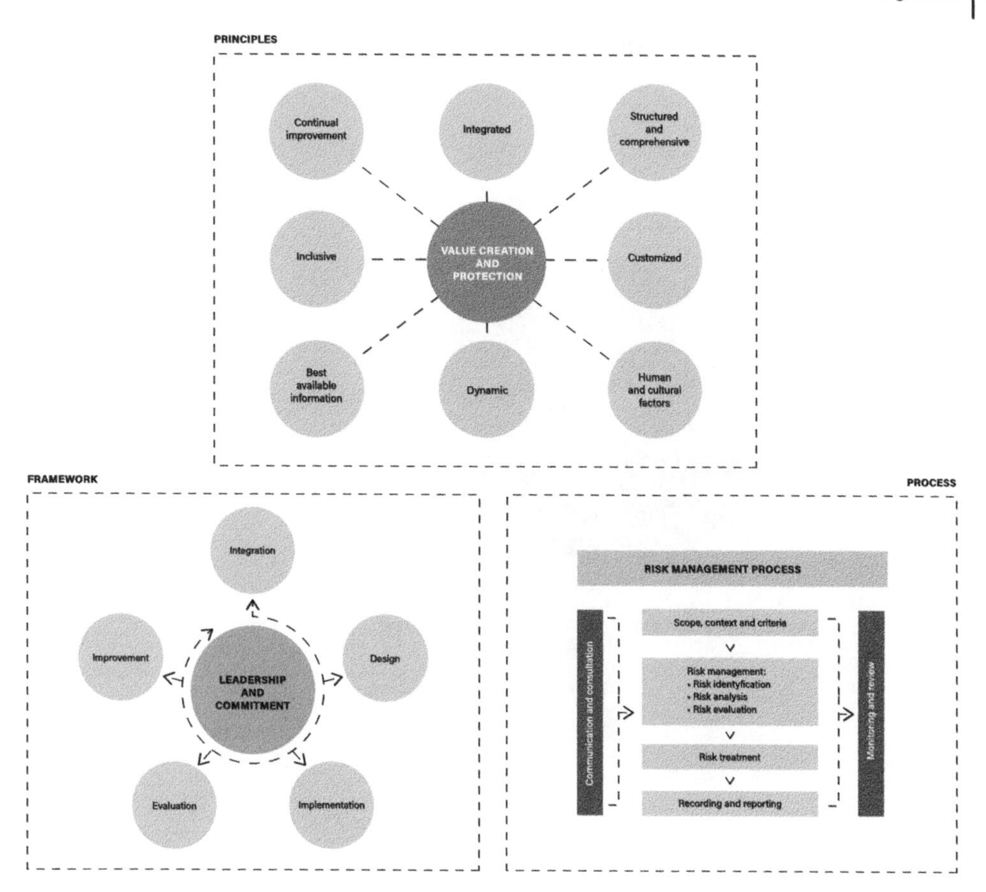

Figure 19 Relationship between the RM principles, framework, and process.

1.3.6 Evaluating and Improving the RM Framework

To evaluate the effectiveness of the implemented RM framework, and help the organization in better achiving its objectives, it becomes crucial to periodically measure its performance and compare it with what was expected. If deviations are monitored, then it is necessary to develop remedial actions to encourage the improvement of the system and facilitate the possibility to take opportunities where risk are present (Popov, Lyon and Hollcroft, 2016) in Figure 20.

1.3.7 The Risk Assessment Phase

Risk assessment is at the heart of the entire management process, as well represented in Figure 21. In particular, once the objectives, criteria, scope and degree of depth have been established, it constitutes the fundamental element for the construction, implementation and periodic assessment of the corporate risk management system. The risk assessment process consists of three phases: the identification of the risk, its analysis, and its evaluation. Some techniques to implement this phase are explained in ISO/IEC 31010 and discussed in the next pages of this book: they are listed in Table 1.

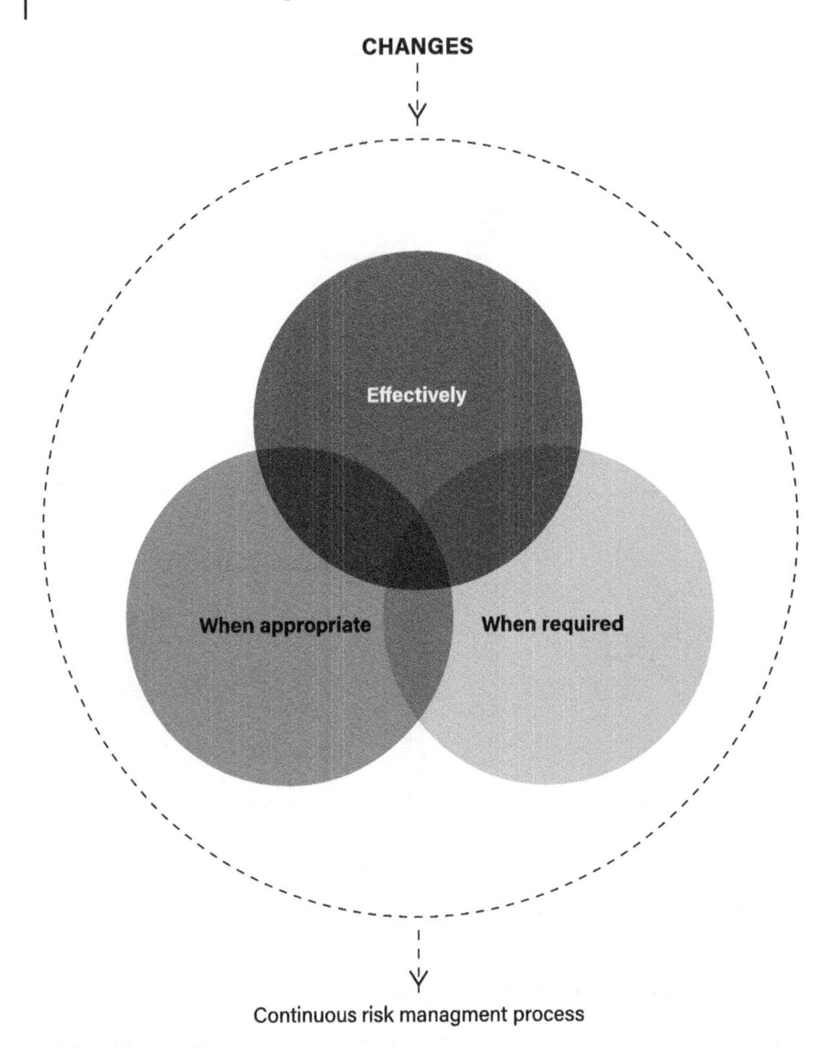

Figure 20 Improving the risk management framework.

This book will discuss some of the methods listed in the table from ISO/IEC 31010, convening a number of them among those having a wide applicability to various sectors and those that better could be implemented step-by-step by novice users. In particular, given a brief description of some commonly used risk identification methods (FTA, ETA, structured brainstorming, HAZOP/HAZID, matrices) that are among those strongly applicable to that initial step of the risk assessment, the book will focus on a barrier-based approach to risk analysis phase, discussing Bow-Tie, LOPA, RCA (via the BFA approach), HRA as single methods or in combination and applied to active and reactive phases of the RM cycle. Readers will notice that those have been judged to be among the best ways to be employed to fulfil the requirement of the risk analysis phase. At the end, some indications will be given for a cost-benefit assessment and consequence/probability matrix as methods to perform risk evaluation.

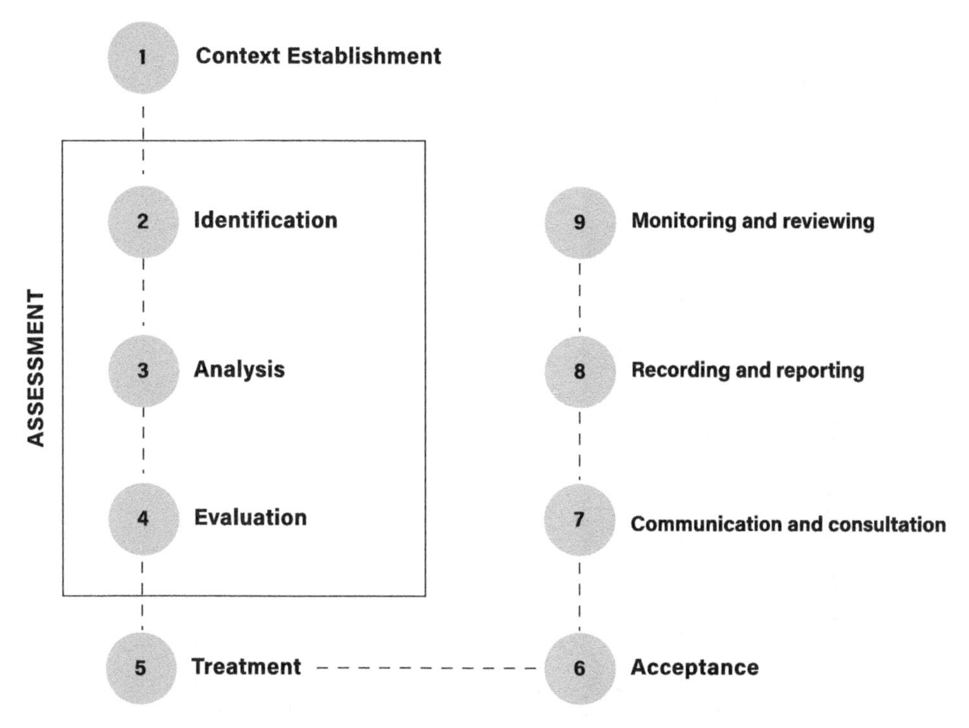

Figure 21 The risk assessment phase in the context of the RM process.

The risk assessment provides an understanding of the risks, their causes and consequences, and their probabilities (defined as the probability of the hazard as modified by present and planned measures). This provides important input to decide:

- If an activity must be performed;
- How to maximize opportunities for improvement;
- Which risks need to be treated, in order to reduce (possibly further) their level;
- How to prioritize risk treatment options;
- How to select the most appropriate risk treatment strategies capable of bringing high levels of risk to tolerable or acceptable levels, including in relation to the resources required (e.g. associated costs and time).

Risk assessment may often require a multidisciplinary approach, as the organization's risks can cover a wide range of causes and consequences and, as anticipated, combinations of risks and control measures often need to be considered.

This process should be properly documented and risks should be expressed in comprehensible terms so that the level of risk assessed is clearly communicated to stakeholders. Generally, the generated reporting should include at least the following information:

- Objectives and scope of work;
- Description of the relevant parts of the system and their functionality;
- Summary of the external and internal context of the organization and how it manages the system, the situation, or circumstance under assessment;

Table 1 Applicability of tools for risk assessment.

		Risk Assessment Process				
			Risk Analysis			
Tools and Techniques	Paragraph	Risk Identification	Consequence	Probability	Level of Risk	Risk Evaluation
Brainstorming	2.2	SA	NA	NA	NA	NA
Checklists	2.2	SA	NA	NA	NA	NA
HAZOP	2.2	SA	SA	A	A	A
RCA	3.5	NA	SA	SA	SA	SA
ETA	2.2	A	SA	A	A	NA
FTA	2.2	A	NA	SA	A	A
LOPA	2.14	A	SA	A	A	NA
HRA	A.3	SA	SA	SA	SA	A
Bow-Tie	2.3	NA	A	SA	SA	A
FN curves	2.2	A	SA	SA	A	SA
Risk indices	2.2	A	SA	SA	A	SA
Consequence/probability matrix	2.2	SA	SA	SA	SA	A
Cost/Benefit Analysis	2.2	A	SA	A	A	A

SA = Strongly Applicable
NA = Not Applicable
A = Applicable

Source: Adapted from IEC/ISO 31010, 2019.

- Applied risk criterion and its justification;
- Limitations, assumptions and justification of hypotheses and assumptions made;
- Methodology (or methodologies) chosen for the evaluation, reasons for the choice made, and degree of detail deemed necessary;
- Results of the hazard identification and consequent risks;
- Data used, their sources and validation;
- Results of the risk analysis and their evaluation;
- Sensitivity and uncertainty analysis;
- Discussion of the results;
- Conclusions and recommendations;
- Bibliographic references, with particular reference to the methodologies taken as reference and their applicability to the specific case.

1.3.8 Risk Identification

Each organization should identify the hazards associated with its processes, products and services, sources of risk, areas of impact, events (including context changes), their causes and potential consequences. The objective of this phase is to generate an exhaustive list of risks based on those events that could create, increase, prevent, accelerate, or delay the achievement of the objectives. It is also important to identify the risks associated with the failure to pursue opportunities for improvement, which will be discussed later in this book. A comprehensive identification of risks is essential, as a risk not identified at this stage will not be considered in the subsequent analysis.

The identification should include those risks whose source is or is not under the control of the organization, even when not immediately identifying their source or causes. Risk identification should include an examination of domino effects, identifying cascading and cumulative effects.

Each organization should apply the risk identification tools and techniques that best suit its objectives and capabilities, as well as the risks it faces. Relevant up-to-date information is important in identifying risks; therefore it is important to know the background information, where available.

Once the risk has been identified, the organization should identify any existing risk control measures put in place, whether related to design, behavioural aspects, or hardware or software processes and systems. Such control measures may be active or passive and in any case characterized by an intrinsic probability of failure in relation to the performance (degree of risk reduction) that they are supposed to operate.

Methods for risk identification include:

- Evidence-based methods, such as checklists and revisions of historical statistical data;
- Team approaches, where a team of experts follows a systematic process of risk identification by means of structured sets of questions or instructions;
- Inductive reasoning techniques, such as HAZOP.

Regardless of the technique used, at this stage it is also important to identify organizational and human factors. Therefore, deviations of human factors from what is expected must be included in the risk identification process in the same way as "hardware" or "software" events. For this reason, this volume devotes a specific appendix to the topic. While human action can be the cause of a negative event, it is also a mode of response and intervention. Consequently, specific investigations must be carried out in order to understand both mechanisms.

1.3.9 Risk Analysis

Risk analysis is the phase in which the understanding of risk is developed. It provides input to the following phase of risk evaluation and enables a decision to be made as to which risks need further treatment in relation to the established acceptance criteria, as well as identifying the most appropriate strategies and methods for doing so. This phase can also provide input to the decision-making process of choosing between several treatment options covering different types and levels of risk.

Risk analysis consists of determining, by combining them, the consequences and associated probabilities of risk events identified in the previous phase of hazard identification, taking into account the presence – or absence – and the effectiveness of existing control measures. The level of consequences and their likelihood of occurrence are then combined to determine the level of risk (Figure 22), whose definition provided by ISO Guide 73 is: "Magnitude of a risk or combination of risks expressed in terms of the combination of consequences and their likelihood."

A risk analysis is based on the causes and sources of risk, their consequences, and the likelihood of these consequences occurring. Therefore, the factors influencing the level of severity of the consequences and their probability must necessarily be identified. It should be borne in mind that a single event can have multiple consequences and have an impact on several targets.

This phase normally includes an estimate of the severity of the consequences (effects) that may arise from an event, situation or circumstance, and their associated probability of occurrence, in order to measure the level of risk. However, sometimes in simple cases where the expected consequences are not significant and the associated probability of occurrence is estimated to be extremely low, a single parameter may be sufficient to make an estimate.

The methods used for risk analysis may be qualitative, semi-quantitative or quantitative. The degree of detail required will depend on the particular application, the availability of

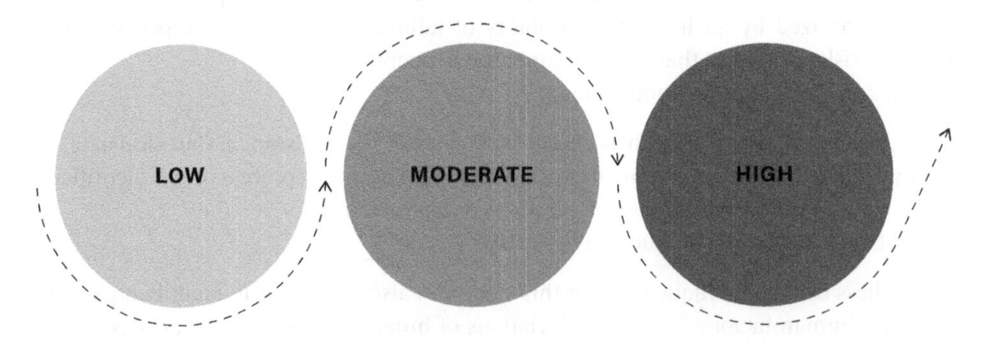

Figure 22 Level of risk.

reliable data, and the decision-making needs of the organization and of course the type of risk (and its potential effects). Some methods and the degree of detail of the analysis may be prescribed by a specific law with which the organization is obliged to comply.

Qualitative assessments define the consequences, the probability and the level of risk through levels of significance, such as high, medium, and low, thus assessing the resulting level of risk from the combination of consequences and probability through a qualitative criterion.

The semi-quantitative methods use numerical scales for consequences and probabilities and combine these levels to produce a level of risk by adopting a specific formula. The scales can be linear or logarithmic or have other types of relationships; the formulas used are the most varied.

The quantitative risk analysis (QRA) estimates, through numerical values often expressed according to codified parameters, including risk indices, the level of consequences and their probability, and produces values of the risk level in specific units defined in the previous phase of context identification. A quantitative risk analysis may not always be possible or desirable due to insufficient information on the system or activities analyzed, lack of data, complex influence of human factors, complexity and severity in terms of the extent of the complete survey, and so on. In fact, sometimes quantitative analysis is not applied because such an effort is not required and it is sufficient to adopt a semi-quantitative or qualitative analysis that is certainly synthetic but also immediately expendable and clearly effective.

In cases where qualitative analysis is used, it is good to provide a clear explanation of the terminology adopted and the reasoning behind the definition of the criteria used.

Even in cases where a quantitative risk analysis is adopted, it is good to keep in mind that calculated risk levels are always estimates. It must therefore be ensured that these risks are not attributed a level of accuracy and precision inconsistent with the accuracy of the data and methods used.

The results of this phase must be expressed in clear terms depending on the type of risk and in a form that provides easy input for the next phase of risk assessment.

1.3.10 Analysis of Control Barriers

Risk levels will depend on the adequacy and effectiveness of existing control measures. The analysis of control barriers requires that it highlight:

- The existing control measures for a specific risk;
- Whether these barriers are capable of dealing adequately with the risk so that it is managed at an acceptable level;
- Whether the control measures are operating as intended (as designed) and whether their effectiveness can be demonstrated (even over time).

The answers to these questions can only be provided with certainty if adequate documentation is available and if the way in which the processes under analysis are managed is well known, as well as the conditions under which the barriers can correctly reduce the risk assigned to them.

The level of effectiveness (or, using a synonym, of maturity) of a particular control measure can be expressed qualitatively, semi-quantitatively or quantitatively, according to the type of analysis adopted. Although it is generally difficult to express a measure of effectiveness in a highly accurate manner, it is nevertheless useful to express and record a measure of the effectiveness of the control barrier in such a way that it is possible to make an informed judgement as to whether it is convenient to improve the performance of the same barrier or to adopt a different risk control measure.

1.3.11 Consequences Analysis

The consequences analysis, to be combined for the risk estimation with the assessment of the probability of occurrence, allows the determination of the nature and type of impact that could occur assuming that a particular event could happen. The same event may have different impacts of different magnitude, affect the fulfilment of different objectives of the organization, and influence different stakeholders. The types of consequences to be analyzed and the targets that are affected must be established at the context definition stage.

Consequence analysis can range from a simple description of the findings to detailed quantitative modelling.

Consequences may have a low level of impact but a high probability of occurrence, or vice versa (a high level of magnitude and a low probability of occurrence) or intermediate values. In some cases, it may be appropriate to focus on risks with very severe consequences, regardless of their probability of occurrence. In other cases, it may be equally important to analyze risks with low-level consequences but whose impact is frequent or even chronic with cumulative or not negligible long-term effects.

The analysis of consequences requires that:

- Existing control barriers are taken into account, together with all relevant factors that contribute to having an effect on the magnitude of the consequences (including, of course, any escalation factors).
- The consequences of the risk are always related to the objectives of the risk management system originally defined.
- Consequences relevant to the scope of the risk assessment are considered, i.e. those in line with the defined scope and work perimeter.
- Secondary consequences (so-called domino effects) are also considered, such as those affecting systems, activities, equipment or organizations connected to the risk assessment scope, also taking into account the common causes of failure and the impact on critical systems and processes for the organization.

1.3.12 Frequency Analysis and Probability Estimation

Probability estimation is generally performed according to three approaches (Figure 23) that can be used individually or jointly:

1) Use of reliable and relevant historical data to identify those events that occurred in the past from which it is possible to extrapolate their probability of occurrence in the future. The data used must be relevant to the type of system, organization and activity considered. If the historical analysis reveals a very low frequency of occurrence, then any estimate of probability that can be derived from it will be affected by great uncertainty. This

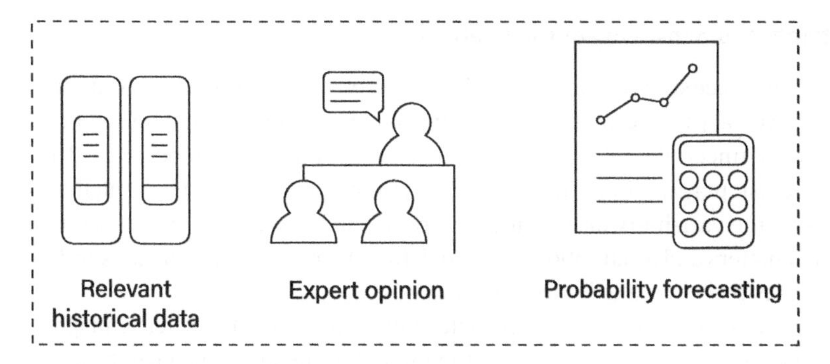

Figure 23 Frequency analysis and probability estimation.

is particularly true when there are no historical events of the type being investigated, so it cannot be concluded that such an event (or circumstance) will not occur in the future.

2) Use of predictive techniques such as the fault tree or event tree, which will be discussed more extensively in the next chapter. When historical data are not available or are not adequate, it is necessary to derive the probability from the analysis of the system, activity, equipment or organization, noting their states of success or failure. The numerical data are then combined to produce an estimation of the probability of occurrence of the adverse event under consideration. When using techniques of this type, it is important to ensure that the analysis is carried out with the right expertise, as common causes of failure on different parts or components of the system leading to the same cause may be found. Sometimes, some simulation techniques can be used to generate the probability that an equipment will fail due to ageing and degradation processes, taking due account of the effects of uncertainties.
3) Expert judgement used in a systematic and structured way for the estimation of probabilities. This approach has been formalized by a number of methodologies, some of which are also set out in ISO/IEC 31010.

1.3.13 Preliminary Analysis

It is important to identify the most relevant risks and exclude from the subsequent analysis phases those that are less significant in relation to the context, the complexity of the organization and the defined objectives. The objective is to ensure that resources are prioritized to the most significant risks for the organization. In doing so, due attention must be paid to low impact and high frequency risks that can have significant domino, cumulative, and long-term effects.

The selection of the "most significant" risks should be based on criteria defined in the previous context definition phase. The preliminary analysis determines the course of the following actions:

- Decide to treat the risks without further evaluation;
- Set aside those non-significant risks on which treatment would not be justified (also economically, according to an ALARP (as slow as reasonably practicable) approach);
- Proceed with a more detailed risk assessment in order to have more information for decision-making.

1.3.14 Uncertainty and Sensitivity of the Analysis

Risk analysis is often affected by a non-negligible degree of uncertainty. Understanding this uncertainty is necessary to interpret and communicate the results of risk analysis effectively. The analysis of uncertainties associated with data, methods and models used to identify and analyze risk plays an important role in their application. Uncertainty analysis involves determining the variability or inaccuracy of results as a outcome of the collective variation of the parameters and assumptions used to define the results. An analysis similar to that of uncertainties is the so-called sensitivity analysis.

Sensitivity analysis first of all must allow the determination of how the level of magnitude of the consequences of the risks changes as the input parameters vary. It can be used to identify those data that must necessarily be accurate and those that are less sensitive for which high accuracy is therefore not required.

The sources of uncertainty of the parameters that govern risk analysis with greater sensitivity must be adequately stated and communicated to all stakeholders, including, first and foremost, those involved in the assessment process.

1.3.15 Risk Evaluation

The objective of the risk evaluation is to assist the organization in the decision-making process, based on the results of the risk analysis, about which risks need to be treated, as well as identifying the priorities for the implementation of such treatment.

The risk evaluation is carried out by comparing the risk levels estimated in the previous phase of the analysis with thresholds of acceptability (or tolerability) defined by a pre-established criterion. Based on this comparison, the need for further action to reduce the level of risk is analyzed. These decisions should take into account the assumptions and models on the basis of which the previous phase of analysis was carried out, in order to properly consider the tolerances and sensitivities of the data obtained. Such decisions should also be made in accordance with legal requirements where defined.

Possible decisions include the following:

- Whether a risk must be subjected to the next stage of treatment;
- Set priorities for treatment;
- Which of the possible alternative solutions to achieve the desired result should be followed, as well as in relation to the resources that need to be employed.

Sometimes the risk evaluation may lead to the decision to undertake further in-depth risk analysis studies; other times it may lead to the decision to maintain existing control barriers, thus accepting the pre-existing level of risk. These decisions are influenced by the company's risk appetite and the criteria of acceptability and tolerability that the company has set when setting its objectives, in accordance with the risk management policy.

1.3.16 Acceptability and Tolerability Criteria of the Risk

The simplest approach to defining a risk acceptability and tolerability criterion is to divide risks into two categories: those requiring treatment and those not requiring treatment. This

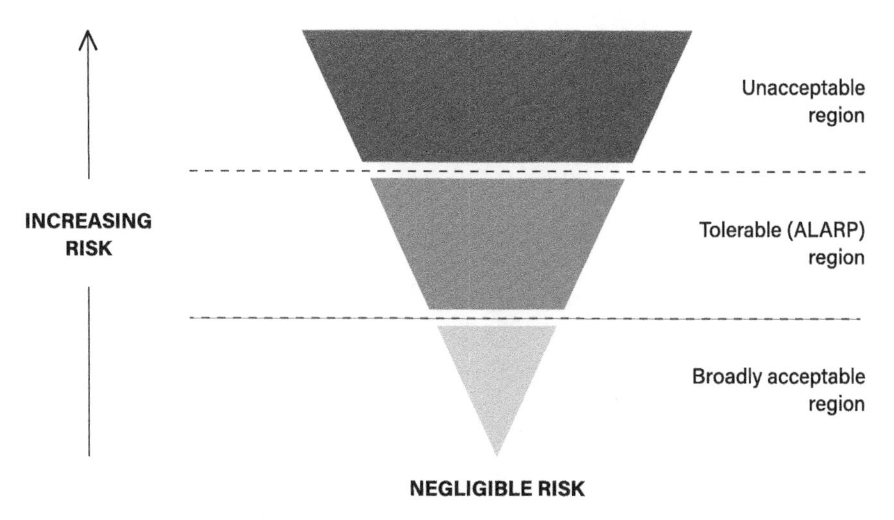

Figure 24 Risk acceptability and tolerability thresholds.

approach leads to seemingly simple results, but it does not reflect the uncertainties inherent in estimating risks and defining the boundary between risks that will need to be treated and those that will not. Once it is clarified that the risk acceptability and tolerability criterion must be defined by the organization on the basis of its own principles and risk sensitivity, the decision whether or not to treat a risk may depend on the costs and benefits of implementing any additional control measures. For this reason, a cross-sectoral approach across various industries is to divide risks into three bands (Figure 24):

- A higher band, where the level of risk is considered intolerable whatever the benefit of the risky activity. In this band, the treatment of the risk is essential, whatever the cost;
- An intermediate band, where the relationship between the costs and benefits expected from the implementation of additional measures is considered, comparing opportunities and potential consequences;
- A lower range, where the level of risk is considered acceptable, or such that no further treatment is required.

One of the most common methods for the representation of the risk acceptability and tolerability ranges is the risk matrix, within which reference ranges based on effect categories and/or probability classes can be identified.

1.3.17 The Risk Matrix

The risk matrix is a useful tool for the graphical visualization of risks and the combination of their magnitude and frequency levels (Figure 25). It is particularly used when a semi-quantitative analysis is carried out. This type of analysis, as already discussed, uses a numerical approach, typical of a quantitative analysis, together with simplifying and conservative assumptions about the evaluation of the level of severity of consequences, the evaluation of the frequency of occurrence of the initiating causes of an event, and the efficiency of control measures. Generally, the results of a semi-quantitative analysis are

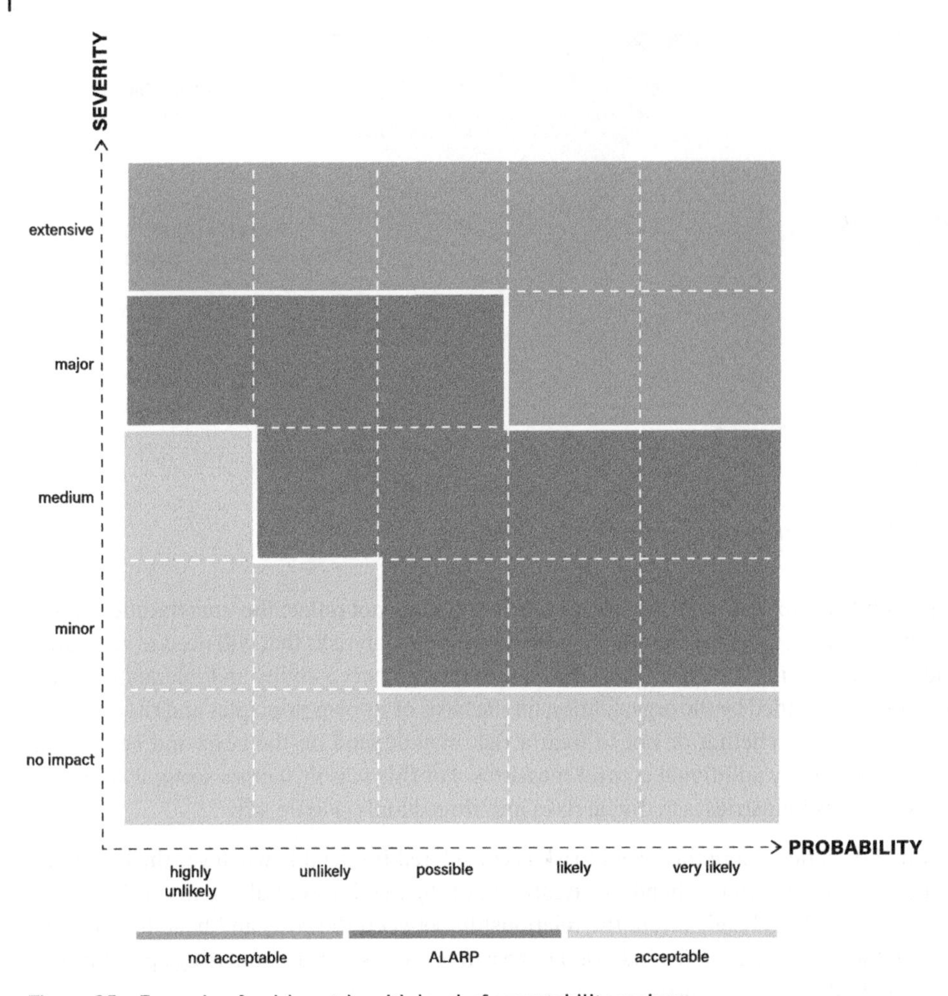

Figure 25 Example of a risk matrix with level of acceptability regions.

expressed in orders of magnitude. However, a risk matrix is also generally used for qualitative analysis, such as that in Figure 25. In it, both probability and severity are expressed in qualitative terms, which must be evaluated by an experienced team to assign the appropriate level of risk, given by the combination of a given severity class with a specific probability class. On the other hand, in a semi-quantitative analysis, the frequency of occurrence is generally expressed on occasions per year (occ/year) while the consequences are identified through a progressive level from 1 (the least severe) to 5 (the most severe), in relation to the severity of the expected consequence. In the example, the grey region defines the most severe risks. A risk falling in this region often requires the immediate stop of the organization's activity in this area, being absolutely unacceptable. The blue area of the matrix identifies a particular region of the matrix where the risk could be accepted (tolerable risk). The risks in this region require an ALARP study. Briefly, this is a cost-benefit analysis of the potential intervention required to further mitigate the risk in order to target the region of acceptability. Since mitigation may require an economic effort that is not justified by the

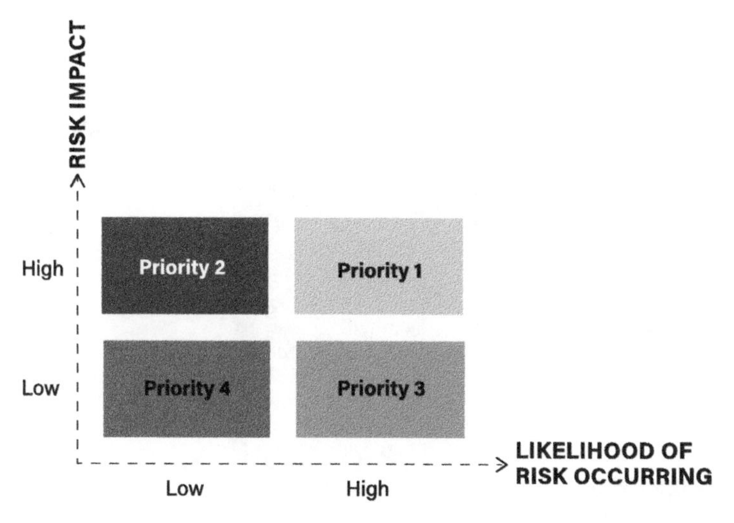

Figure 26 Prioritization of risk given impact and liklihood.

reduction of the level of risk, a risk falling within the ALARP region could be accepted as such: the managers (or, in general, whoever is responsible for the risk) will take responsibility for this choice based on a cost-benefit analysis.

Finally, the green region is about acceptable risks: no further ALARP mitigation or study is required for them.

ALARP is not the only risk acceptance criterion available. There are also other criteria, such as GAMAB (globally at least as good), MEM (minimum endogenous mortality), MGS (at least the same level of safety), NMAU (not more than unavoidable), but it is not the subject of this book to address them, so the ALARP study, one of the most widespread in the world of risk management, will be considered.

The graphic representation of the risk level through the matrix allows the introduction of the risk prioritization (Figure 26), i.e. the possibility of assigning a degree of priority to the various risk treatment options that will be offered, as better explained shortly.

The prioritization of risk allows the organization to make risk-based decisions, thus allowing itself to be guided in decision-making processes by the need to target an acceptable level of risk. In the case of the risk matrix shown in Figure 27, this means prioritising scenarios with a risk level close to the top right (very high impact). The proposed graphic representation guarantees a good communication of information to all stakeholders, while pursuing the principle of inclusiveness of risk management.

1.3.18 The ALARP Study

Each organization should support ALARP studies, helping to develop a comprehensive risk mitigation strategy for scenarios deemed credible, identifying and prioritising any viable risk mitigation action that can reduce the risk associated with a scenario up to an ALARP level. In addition, for those scenarios without the opportunity to further reduce the level of risk, each organization should provide a mechanism to document that the risk considered is already at a tolerable level (ALARP).

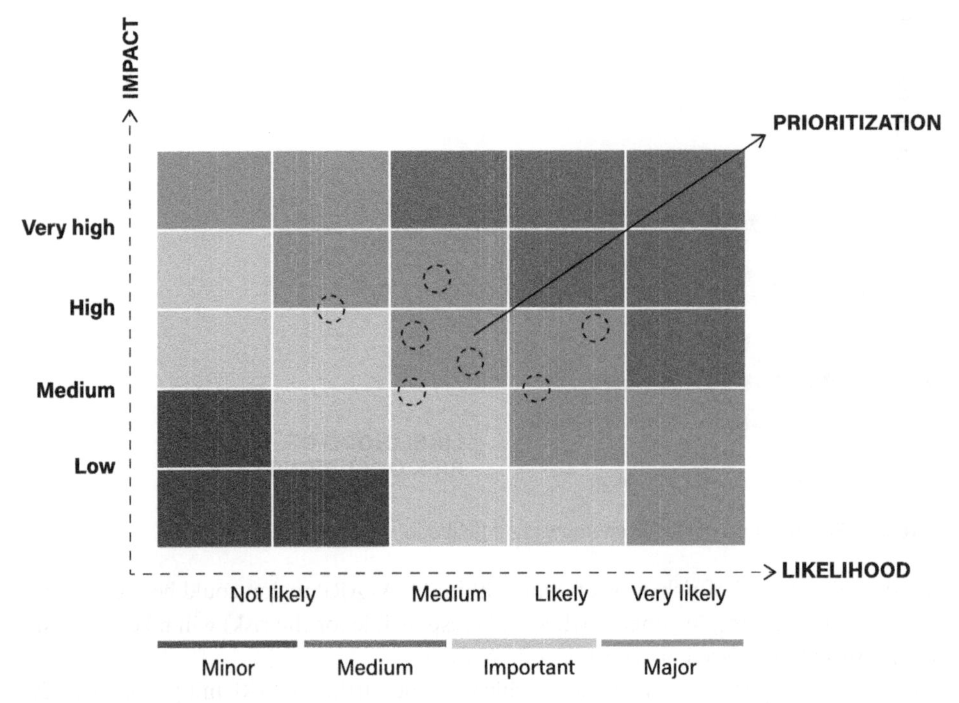

Figure 27 Risk prioritization and the risk matrix.

ALARP studies should be undertaken using a documented process and performed by personnel familiar with the details of the risk scenario being assessed. Typically, the personnel dedicated to ALARP studies are the same personnel who carry out the risk analyses.

Any potential risk mitigation opportunities identified within the ALARP process will be documented as follows:

- A recommendation;
- An option that is not recommended because of its impracticability in relation to the resources to be put in place (including those for maintaining the recommendation over time);
- An option that is not recommended because of other identified and deemed more favourable risk mitigation opportunities.

It should be noted that risk mitigation alternatives are often not mutually exclusive, so multiple recommendations could result from the ALARP study of a single risk scenario. Moreover, not all risk mitigation measures or recommendations in an ALARP study will generally be classified as control measures (barriers). Those recommended actions that cannot qualify as barriers, because they do not produce a reduction in the risk level by acting on the frequency of occurrence or severity of a scenario, will be collectively identified as "other measures." Examples of "other measures" could be – depending on the specific context – warning signs or labelling. "Other measures" are typically identified as actions to be applied to consequences whose risk is already acceptable.

An ALARP study aims to answer the following questions in a structured way, for each risk scenario assessed:

- What alternatives are available to eliminate, reduce or manage the risk?
- What factors determine the feasibility of any risk mitigation alternatives?
  - How much risk mitigation is substantially achieved by the measure?
  - What resources are needed to implement the measure?
  - What synergies would be achieved if the measure is implemented?
  - Should the measure also be implemented in similar facilities/workstations?
  - Is the measure congruent with the current practices adopted in the reference work sector?
  - What is the technical and operational feasibility of the measure?
  - How long would the implementation of the measure take?
  - What other risks would be impacted by the implementation of the same measure?
  - What is the availability of the measure?

In general, the risk reduction strategy imposes a hierarchy of possible options in the following order:

- Eliminate the hazard (e.g. by changing the activity, process, system under consideration and source of the risk considered).
- Reduce the hazard (e.g. by reducing the amount of flammable substances stored in the warehouse).
- Control the hazard through additional measures (barriers).

According to this hierarchy, for example, an action that eliminates the hazard will have a much greater benefit than the installation of an additional barrier to control it.

Before carrying out an ALARP study, an organization might ask itself whether the action required to reduce a specific risk has already been taken with positive feedback from other assets in the same organization or from other competitors. It is also necessary to ask first of all whether the suggested action is in fact a requirement arising from new mandatory regulations or standards or international best practice. In such cases, the organization may decide not to carry out any cost-benefit analysis, deciding a priori to implement the new measure.

Although there are different methodologies to perform a cost-benefit analysis, for the purposes of this book it seems sufficient to mention qualitative analysis.

It is based on the adoption of a matrix like the one in Figure 28. For each risk falling within the risk tolerability region (i.e. the ALARP region of the risk matrix in Figure 25), the analyst considers the costs (e.g. in financial and/or time terms) and benefits (in terms

		Expected benefits		
		High	Average	Low
Associated costs	High			
	Average			
	Low			

Figure 28 Matrix example for qualitative ALARP analysis.

of risk reduction) expected from the implementation of a particular option. The scale of benefits could for example be as follows:

- High: The risk leaves the ALARP region and becomes acceptable.
- Medium: The risk reduces in level, but remains in the ALARP region of risk tolerability.
- Low: The measure does not reduce the level of risk.

The definition of the cost level will typically depend on the sensitivity of the organization and other strategic policy criteria, both in economic-financial and temporal terms. According to the example matrix in Figure 28, high-cost and low-benefit options will not be implemented, while high-benefit measures will always be implemented, regardless of the costs associated with their implementation.

When calculating associated costs, it is necessary to consider both the resources required for the design of the further control and those connected with its implementation and maintenance over time. It is also useful to consider not only direct costs for implementation and operation but also indirect costs (inspection, testing, and periodic maintenance or information and training for use). In fact, like the risks they intend to reduce, the controls have their own life cycle, within which the resources for maintaining efficiency over time could play a crucial role and be a discriminating factor in the selection of the most appropriate treatment measure.

In conclusion, the objective of the ALARP study is therefore to offer a cost/benefit assessment, which can be summarized in Figure 29.

1.3.19 Risk Management over Time

Risk management does not end with a snapshot taken at a certain moment, but instead requires a dynamic approach. This is done first by following the risk assessment phase with the treatment phase. The risks analyzed will in any case be monitored and reviewed; as additionally, the entire structured approach to risk management (the framework) will be monitored and reviewed at predetermined intervals, thus reviewing the performance of the system.

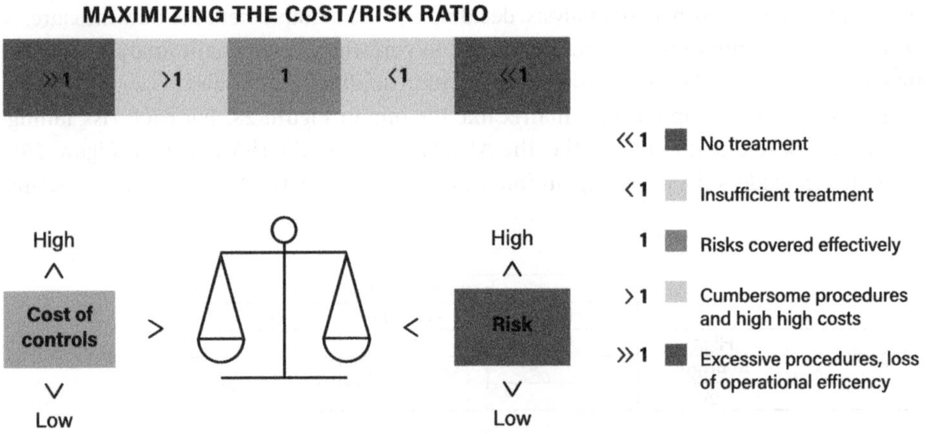

Figure 29 Achieving balance in risk reduction.

1.3.20 Risk Treatment

The risk treatment often requires the choice of one or more options to modify the risks assessed during the previous evaluation phase and then implementing these options. Once implemented, the risk treatment generally provides new control measures or changes to existing ones. This is a cyclical process consisting of the following steps:

- Assessment of risk treatment;
- Decision whether the residual risk, i.e. the level of risk obtained downstream of the implementation of the treatment, is tolerable;
- If the level of risk is not tolerable, generate a new risk treatment;
- Assess the effectiveness of this new treatment.

As discussed above, risk treatment options are not necessarily mutually exclusive or suitable for all circumstances. Generally, the possible options are as follows:

- Avoid risk by deciding not to start or continue the risky activity.
- Remove the source of the risk.
- Change the frequency of occurrence of the damaging event, with interventions aimed at reducing the frequency of its causes or reducing the probability of failure on demand of control measures.
- Change the level of expected consequences, for example by increasing the effectiveness of mitigating control measures.
- Share information on the risk level with other parties, including contractors and insurers.
- Accept the risk through an informed and aware decision.

Selecting the most appropriate risk treatment option often involves a cost-benefit analysis, already mentioned in the ALARP study previously introduced, always bearing in mind the legal requirements, the social responsibility, and the environmental protection. Each organization should also make decisions on risks that may require treatment that cannot be justified on economic basis, such as risks with very severe yet extremely rare consequences.

When choosing risk treatment options, the organization should consider the values and perceptions of stakeholders, who should be involved in the decision-making process, by adopting the most appropriate means of communication with them. Indeed, although equally effective, some actions may be more easily accepted by some stakeholders than others.

The corrective action plan should clearly identify the priorities in risk treatment. It should be noted that the risk treatment phase may itself introduce new risks, such as the failure or ineffectiveness of the treatment measures chosen. If new secondary risks are introduced during the risk treatment phase and need to be assessed, treated, monitored and reviewed, then these risks should be incorporated into the same treatment plan as the "original" risks, avoiding treating them with a different process.

The objective of risk treatment plans is to document the treatment options chosen, clarifying how they are implemented. At least the following information should be contained in a risk treatment plan:

- The reasons for selecting a particular treatment option, including the expected benefits of such implementation;
- Who is responsible for approving the plan and who is responsible for its implementation;

- The proposed actions;
- The required resources (both instrumental and organizational-managerial);
- The performance measures and any constraints;
- The monitoring and documentation requirements (including the recording of evidence);
- The scheduling of actions according to a pre-established plan and based on the established priority.

Treatment plans (and not just the individual options contained therein) should also be integrated with the organization's management processes and discussed with all stakeholders.

Decision-makers and other stakeholders should be aware of the nature of the residual risk after the treatment phase; therefore, the residual risk should be documented and subject to monitoring and review and, where appropriate, further treatment. The risk treatment activities are summarized in Figure 30.

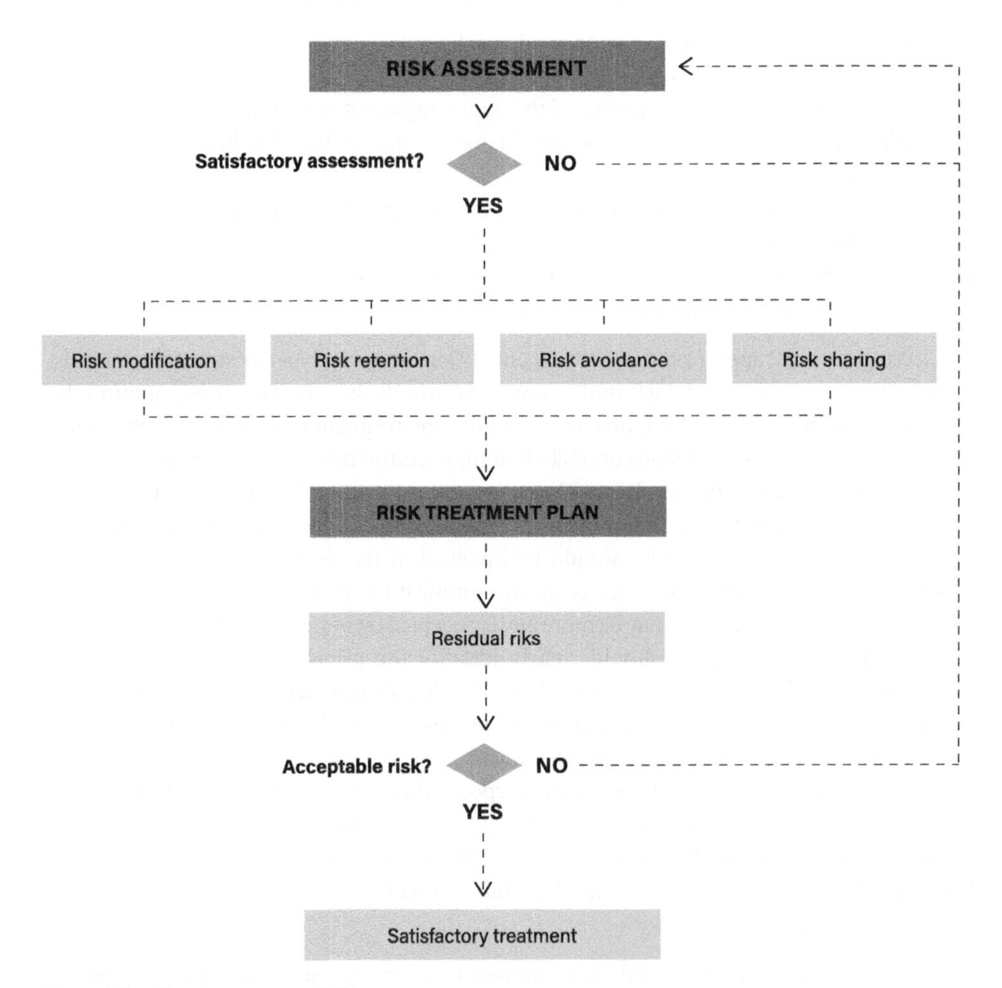

Figure 30 Risk treatment activities.

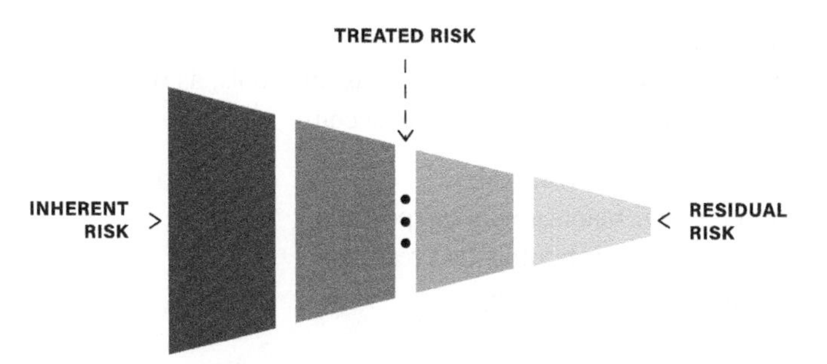

Figure 31 Residual risk.

At the end of the treatment phase, the inherent risk is reduced to a lower level, named "residual risk" (Figure 31). Residual risk can be defined as the risk that remains after the implementation of controls aiming to reduce the inherent risk.

1.3.21 Monitoring and Review

Both monitoring and review (also known as periodic review) are phases of the risk management process that involve, on a periodic or ad hoc predetermined basis, the verification and surveillance of the risks identified, analyzed, evaluated and treated and the related control measures. The persons responsible for the monitoring and review phases must be clearly identified within the organization. These phases should cover all aspects of the risk management process with the aim of:

- Ensuring that control measures are effective and efficient, both in design and operation;
- Obtaining further information to improve risk assessment;
- Analysing and learning lessons for improvement from events (including near-misses and anomalies), changes, trends, successes and failures;
- Identifying changes in the internal and external environment, including changes to the risk acceptability and tolerability criteria and the risks themselves that may require a review of corrective actions and related priorities;
- Identifying emerging risks, including in relation to the results of in-depth analysis of the root causes of adverse events.

The progress of the implementation of risk treatment plans provides a measure of performance to be monitored. The results can be incorporated into the management performance of the entire organization, internal and external documentation, and reporting measures and activities.

The results of the monitoring and review phases should be recorded and documented externally and internally when appropriate and should be used as input data for the review of the entire organizational risk management structure. This is in order to ensure that current risk management processes are commensurate with the size and complexity of the company as well as suitable for the defined objectives are maintained over time.

In general, all risk management activities should be traceable. In the risk management process, data and evidence are the basis for improving methods and tools, as well as the entire process. The decision to retain this data must take into account:

- The organization's needs for continuous learning;
- The benefits of reusing information for management purposes;
- The costs and benefits of creating and maintaining data records;
- Legal and operational requirements for data logging;
- How to access, retrieve and store multimedia data;
- The period of retention;
- The sensitivity of the information.

This includes periodic audit activities. The monitoring of systems through inspection (internal, second- or third-party inspection) guarantees the effectiveness and efficiency of the system.

1.3.22 Audit Activities

Regardless of whether third-party audits are applicable to the specific business context, the organization intending to adopt a risk management system must conduct, at planned intervals, internal audits in order to receive information useful to verify whether the risk management system is:

- Compliant with the organization's own requirements, including the risk management policy and the established objectives of the risk management system;
- Compliant with the requirements of ISO 31000 if selected risk management paradigm;
- Effectively implemented and maintained also in relation to changes that have occurred.

To verify this, the organization must:

- Plan, establish, implement and maintain one or more internal audit programmes, where frequencies, methods, responsibilities, consultation, planning and reporting requirements are clearly identified, taking into account the importance of the processes involved and the results of previous audits.
- Define the audit criteria and their scope.
- Select auditors and conduct audits ensuring objectivity and impartiality of the inspection process.
- Ensure that the results of audits are reported to relevant individuals such as managers, workers, and, if any, workers' representatives.
- Take action to address non-conformities arising from audits, thereby continuously improving the performance of the risk management system.
- Maintain documented information containing the results of audits and the implementation of the audit programme.

The interested reader can refer to the ISO 19011 technical standard which defines guidelines for management systems audits, where the competence required of auditors is also specified.

1.3.23 The System Performance Review

Not only must the specific risk management process be subject to periodic monitoring and review, but the entire corporate framework that supports these processes is also subject to review. The structured approach to risk management, in order to be effective and provide constant support in achieving company performance, must be monitored and reviewed. In this sense, the organization should:

- Measure the performance achieved in risk management through appropriate indicators, the appropriateness of which is periodically reviewed.
- Measure progress and deviations from expected targets on a regular basis.
- Periodically review whether the structure, policy and expected objectives are always appropriately defined, taking into account the external and internal context of the organization and related changes.
- Document risk, progress against expected objectives, and how effectively the risk management policy is implemented.

On the basis of the results of the monitoring and any revisions, the organisation is asked to express its opinion on any decisions to improve the framework, the policy, and the expected objectives in terms of risk management. Such decisions should lead to tangible improvements in terms of risk management culture and, overall, to the effective and efficient implementation of this organizational model, with small but certain steps, to fulfil the organization's policy and objectives (Figure 32).

The various technical regulations on management systems in any area identify this phase as a "management review." In this phase, senior management is required to periodically review the management system adopted. In doing so, it must also take into account:

- Status of actions resulting from previous reviews;
- Changes in internal or external context factors relevant to the management system, such as changed needs and expectations of stakeholders, change in legal requirements, or change in business risks;
- Level of implementation of the policy and objectives of the risk management system;

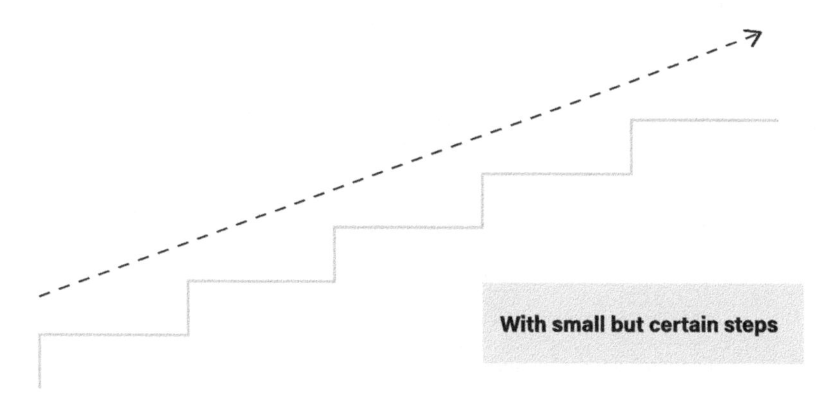

Figure 32 Risk management process continuous improvement.

- Information on the performance of the risk management system, including from the results of monitoring and measurement of performance standards;
- Adequacy of resources for the effective maintenance of the risk management system;
- Relevant communications with all stakeholders;
- Opportunities for the pursuit of the continuous improvement paradigm.

Appropriate documentation must be kept as evidence of the results of management reviews.

In order to make the overall review phase of the system more objective and structured, it is useful to define numerical indicators and monitor each process under review, from which trends and the achievement of the set objectives can be inferred.

At the end of the process, the risks need to be documented in a report (Figure 33), whose content should contain the information recommended by ISO 31000, whose format is consistent (date of emission, version number, author clearly identified, date of approval, and so on), and whose life cycle is properly managed.

The conclusion to this paragraph is dedicated to the person who is in charge with the risk management activities: the risk manager. Managing risk is a complex task and a risk manager should be appointed to ensure the required coordination. In this role, he/she should possess the following set of skills and knowledge, as identified in Figure 34:

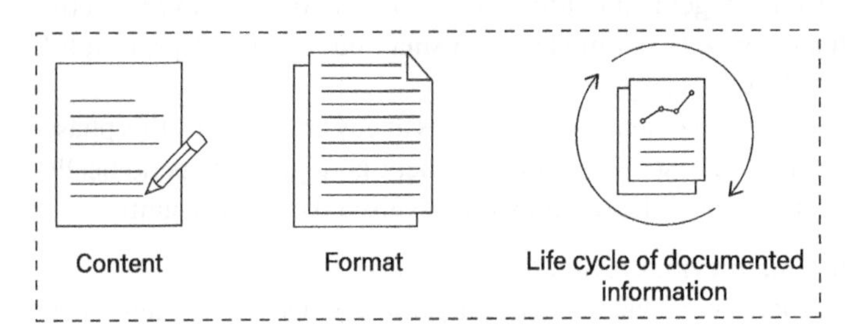

Figure 33 Documenting the risk management process.

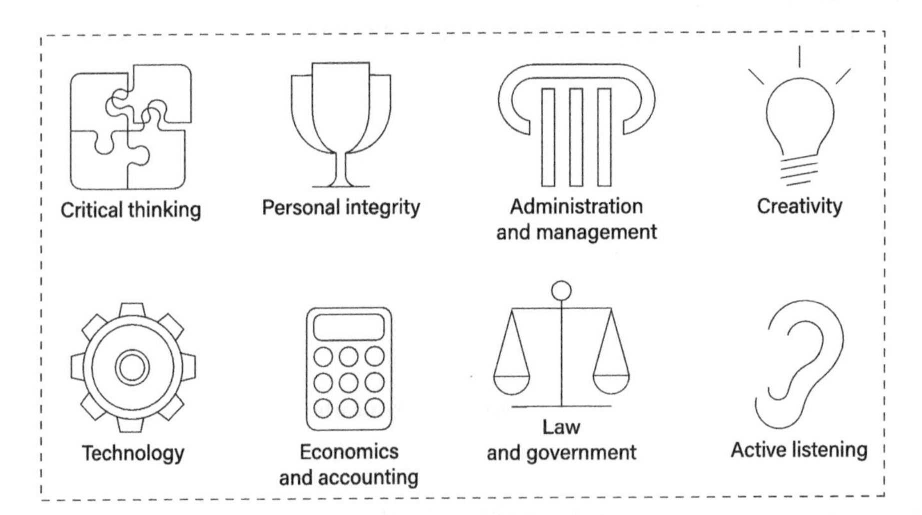

Figure 34 Skills and knowledge for a risk manager.

- Being a critical thinker, remaining logical and reasonable when identifying hazards, assessing risks, proposing solutions, or analyzing threats;
- Personal integrity, with a high level of morale;
- Being knowledgeable about administration and management;
- Being creative to finding smart solutions to challenging problems;
- Being knowledgeable and self-confident with technology;
- Being knowledgeable in finance and economics;
- Being knowledgeable in laws and politics;
- Being an active listener, ready to catch the opportunities coming from the comparison and the dialogue

In doing their job, regardless of the adopted work styles, risk managers should not make the mistake of feeling that they are solely responsible for managing the process. They are certainly the leader, but the proper implementation of the risk management system requires proper training of RM team members, a proper understanding of their responsibilities, and the creation of a sound infrastructure, which can be achieved only allocating the required resources (Figure 35).

A good risk manager ensures the strategic alignment between the RM objectives and the organization's goals (Figure 36). The consequent high level of consistency can be achived only through a deep understanding of the mission, objectives, values, and strategies of the organization.

At the end of this paragraph it is therefore worth highlighting how risk management is synonymous, in practical cases of risk treatment, with "risk control." Such risk control must follow the hierarchy and workflow that are summarized in Figure 37.

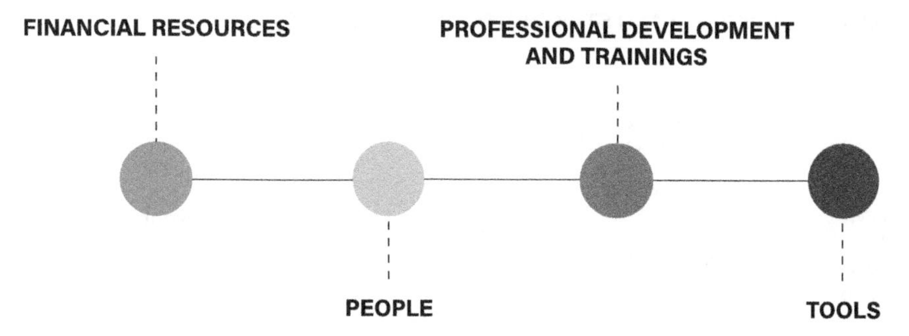

Figure 35 Resources to be allocated for an effective RM.

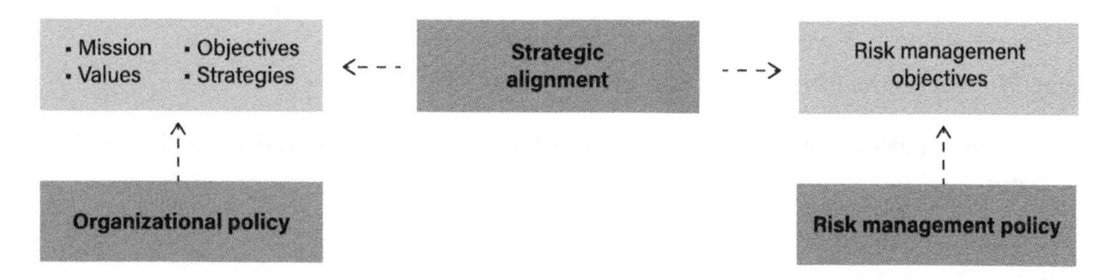

Figure 36 Understand the mission, objectives, values, and strategies.

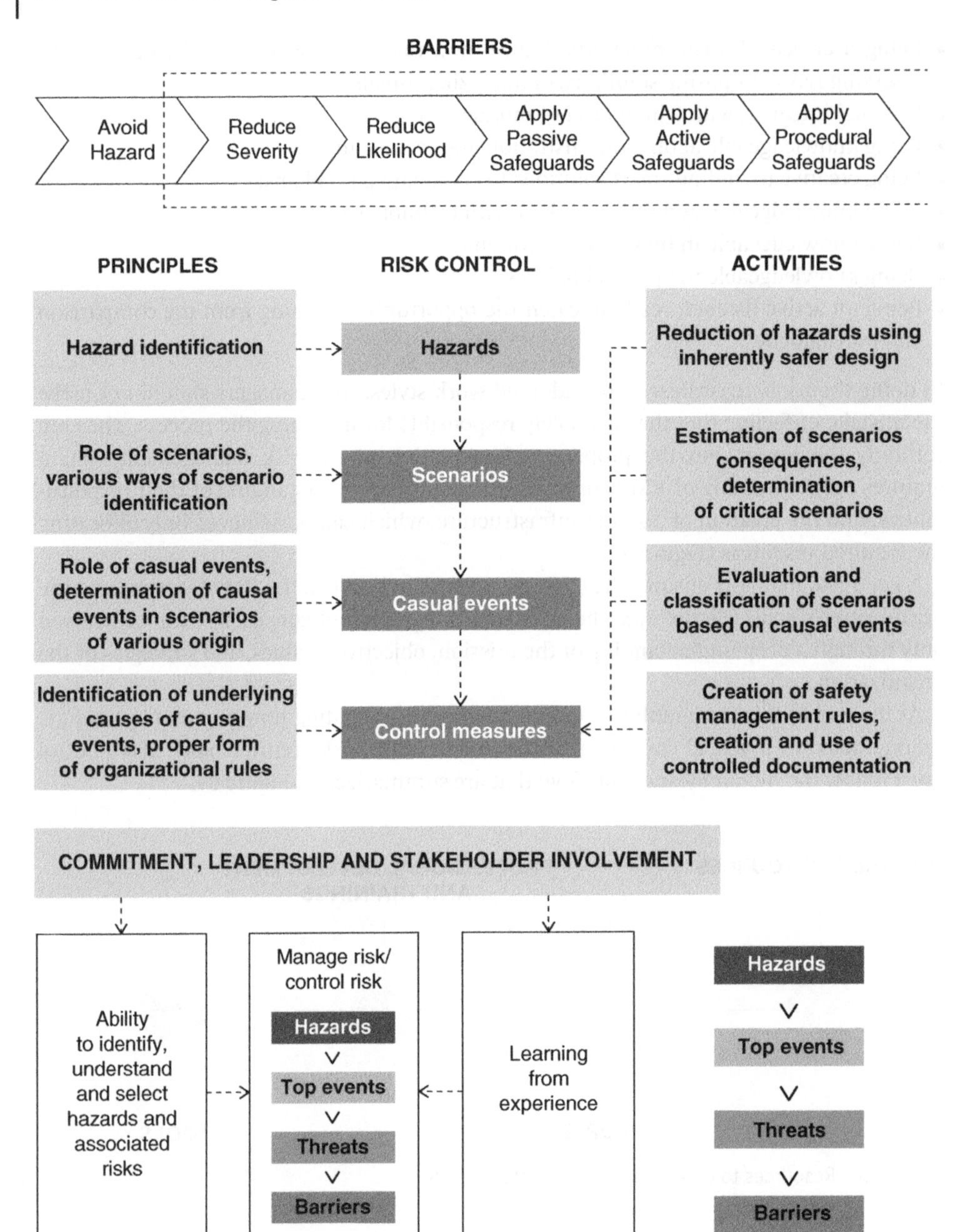

Figure 37 Risk control hierarchy and in practice.

From this perspective we observe Haddon's 10 strategies to control risks (Haddon, 1980): Pre-event

- Prevent the existence of the agent.
- Prevent the release of the agent.
- Separate the agent from the host.
- Provide protection for the host.

Event

- Minimize the amount of agent present.
- Control the pattern of release of the agent to minimize damage.
- Control the interaction between the agent and host to minimize damage.
- Increase the resilience of the host.

Post-event

- Provide a rapid treatment response for the host.
- Provide treatment and rehabilitation for the host.

1.4 Uncertainty and the Human Factor

The analysis of the human factor in risk assessment and incident analysis activities is fundamental to fully understand the impacts it can have on the organization, regardless of its type, scale, size and complexity or the role it played in real events or near-misses.

This chapter does not intend to provide a definitive description of the human factor (including human error), a subject for which numerous technical-scientific and popular publications are available at the international level, given the extent of all the topics related to this aspect and the ever-growing research in this field by institutes, researchers, experts, and consultants, as well as organizations, in the intimate discovery of their business processes. However, in the formulation of a risk assessment and a systemic approach to risk management, the study (qualitative or quantitative) of the human factor plays a fundamental role. This emerges powerfully by adopting a barrier-based approach for which, using intuitive graphical notations such as those proposed by methods like the Bow-Tie and barrier failure Analysis (BFA), the human factor is immediately recognized in some causes of risk, in the vulnerability associated with impacts and also in control measures (barriers constituted by the human response, barriers that do not guarantee their performance for aspects related to information, skills, or training, or fail totally due to human errors in the conduct of technical activities such as periodic maintenance on critical technical systems).

Often, human errors are recognized as the root cause of an accident, near-accident or non-conformities (including process anomalies), losing the opportunity for the organization to initiate an introspective analysis to find out how fertile ground could be provided for people, whose activity is part of a wider organizational structure, to make a mistake. Even tracing a cause back to the generic "human error" without further evaluation or investigation is absolutely reductive, if not totally useless. The failure to take into account human factors with an awareness of the point within the risk management and incident analysis activities, both provided for by ISO 31000, does not therefore allow the organization to follow up an effective plan of actions aimed at preventing the recurrence of similar events, thus denying itself the opportunity to pursue the rationale of continuous improvement. In order to do this, it is necessary to identify and evaluate certain performance factors, including ergonomics, workload, training, degree of competence and training, work organization and others.

It is therefore important to recognize, through examples collected from experience, the precursors of human error, identifying possible measures for their prevention that should be included in the risk analysis, and taking advantage of the barrier-based perspective increasingly established internationally.

The integration of human factors within risk management also requires that models and methods for the quantification of the probability of error are available, represented every day by methods such as THERP, HEART, SPAR-H and many others.

In the process industry sector, many authors have collected experiences and examples of human error within entire volumes. This is the case of Strobhar (2013), Woods et al. (2010) and Taylor (2016).

This chapter therefore aims to provide the basic knowledge to address the issue of human factors within organizations that intend to incorporate these aspects within the broader risk management according to the ISO 31000 standard (including other voluntary standards which, sharing the same structure, provide for the conduct of risk assessments, such as ISO 9001, ISO 14001, and ISO 45001) which, in context analysis, also explicitly requires the human factor to be considered.

The handling of such a complex topic is based on the knowledge of the main models of human information processing, together with the knowledge of behavioural patterns.

According to one of the most widespread human behavioural models (Forck and Noakes Fry, 2016), values and everything in which an individual believes (i.e. his opinions and faith) influence his or her way of reasoning and therefore his or her thoughts. These, in turn, influence the individual's behaviour, i.e. the way he or she acts. Behaviors can therefore lead to results, including an accident, near-accident or non-conformities. In order to modify the behaviour of individuals it is therefore necessary to change their mental model, their beliefs, and their values, as shown in Figure 38. It is hence necessary to act on the invisible in order to have results on the visible level.

It is then clear that the behaviour of an individual is peculiar to the individual him/herself with respect to a common stimulus (although, if we go deeper we can also say that the

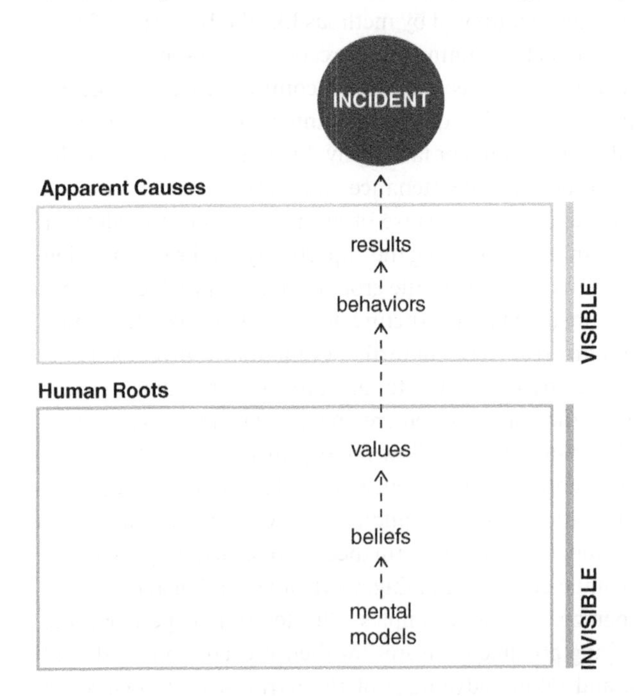

Figure 38 Thinking-Behavior-Result model. *Source:* Adapted from Fiorentini and Marmo (2018).

individual behaviour can be modified by the observation of the overall behaviour of a multitude of individuals, e.g. in the response of a mass of people in case of evacuation).

According to another model (Forck and Noakes Fry, 2016), the mental process is activated by a stimulus, which produces a response, i.e. it shapes human behaviour. The result is the consequence. In the incident analysis, the model is reversed as shown in Figure 39. Therefore, first the result (the accident) is analyzed by identifying what happened. Then, the analysis of the response clarifies how it happened. Finally, by reconstructing the mental process, it is possible to evaluate the stimulus that activated the sequence, thus establishing why the accident happened.

According to a third model (Forck and Noakes Fry, 2016), the success of human performance is achieved when certain internal and external factors affecting human abilities are met, as shown in Figure 40.

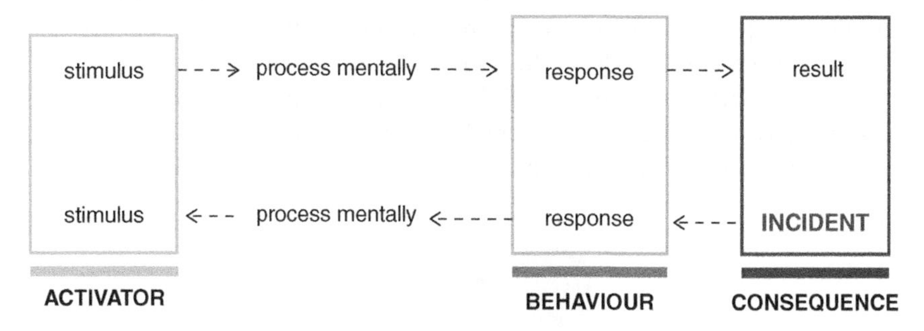

Figure 39 Stimulus-Response model. *Source:* Adapted from Fiorentini and Marmo (2018).

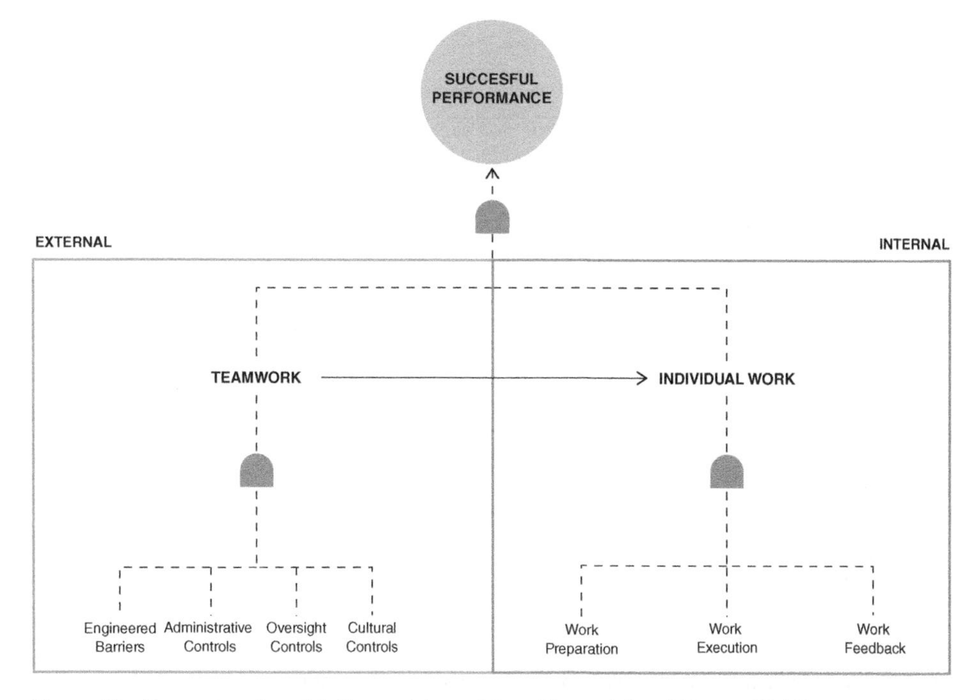

Figure 40 Two-pointed model. *Source:* Adapted from Fiorentini and Marmo (2018).

Similar to the previous model, this also, if used in reverse, becomes a tool for accident investigation as shown in Figure 41. Applying the bases of logic, AND logic doors become OR doors when they are "crossed" in reverse. Thus, starting from an incident, it is possible to have a sort of predefined logic trees (in fact very similar to fault trees) specifically addressed to the analysis of human factors.

An interesting model on human information processing is the one cited, among others in Strobhar (2013) and reported in Figure 42.

According to this model, a process that undergoes a generic disturbance *w(t)* produces a *y(t)* output signal. When the output signal diverges significantly from its reference value *r(t)*, then the operator detects the change *e(t)*, identifies the cause and compensates the deviation from the reference value, thus introducing a change *u(t)* to the process himself. What matters about this model of human information processing is the "black-box" represented by the operator, where the three fundamental actions of detection, identification and compensation take place. The most attentive reader will have noticed a certain overlap with the "Detect - Decide - Act" model at the basis of the barrier-based approach, so that the two models are in fact the expression of the same perspective that can be applied because the barrier-based approach also contemplates behavioural barriers (i.e. those that rely 100% on human intervention). Any inefficiency in one of these three phases, certainly

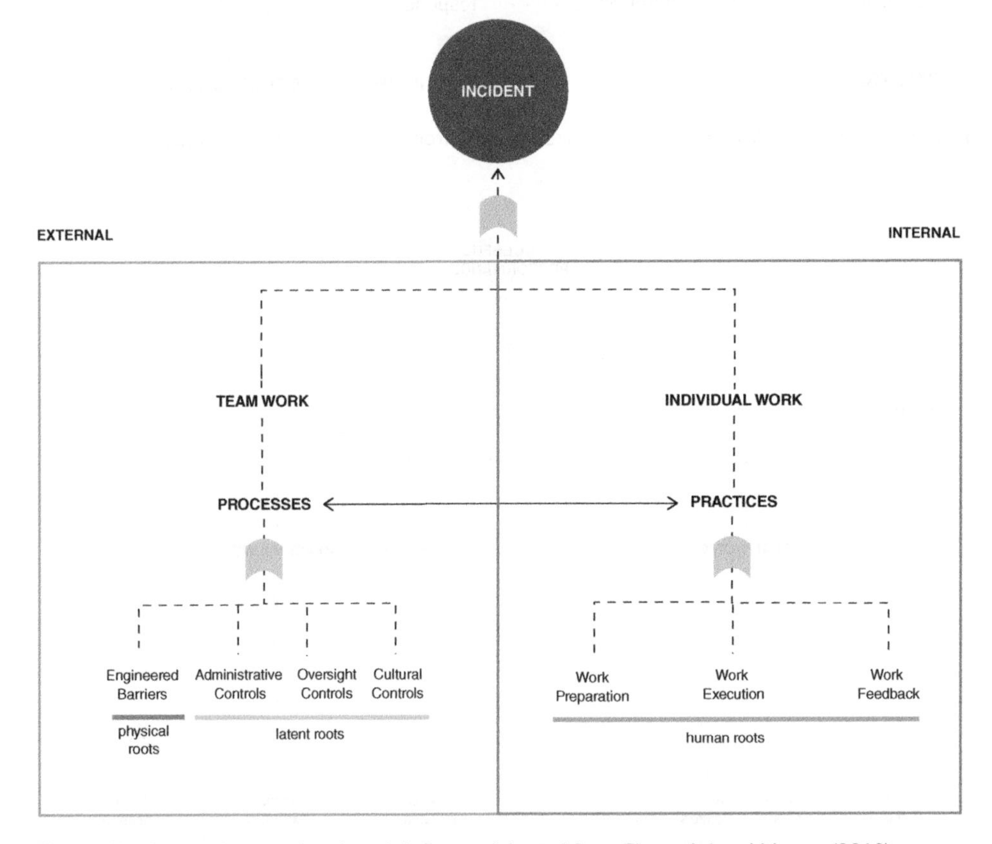

Figure 41 Inverted two-pointed model. *Source:* Adapted from Fiorentini and Marmo (2018).

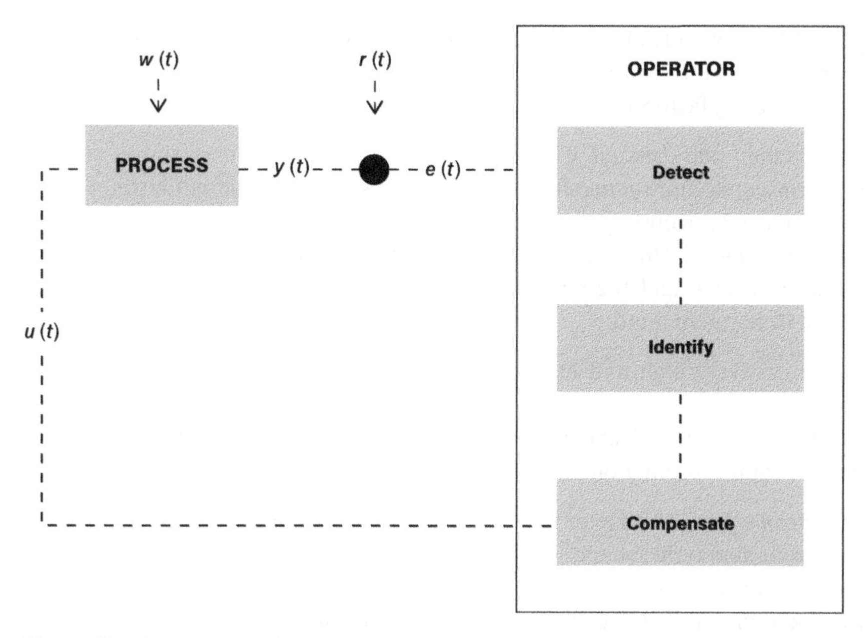

Figure 42 Human factors in process plant operation. *Source:* Adapted from Strobhar (2013).

conceivable since human intervention is far from being perfect, implies, like any barrier, including technological barriers, the assignment of a Probability of Failure on Demand value different from 0.

1.4.1 Performance-Shaping Factors

The multidimensional nature of human performance represents both an advantage and a disadvantage in the definition of human factors. How well a person performs a task cannot be attributed to a single factor, since the ultimate result of human performance is the product of several variables, often interacting with each other. One of the advantages of such a multidimensional approach to human factor analysis is the possibility of using a variety of tools to improve human performance, where required, to act on one or more specific variables. In many cases, deficient aspects in one variable can in fact be compensated for by modifying another dimension.

However, this approach brings with it at least two disadvantages. First, very often weaknesses in human performance are not directly addressed as they should be. Providing for additional training, adding alerts, or drafting longer procedures is a quick solution to solve an operator's performance problem, but only a thorough investigation can reveal the true root causes of performance degradation and thus offer cues for a more effective solution. Secondly, weaknesses in human performance rarely have simple and absolute answers because of the interaction between the various dimensions. For example, think of the interaction between training and ergonomics (i.e. the human-machine interface): the higher the level of information possessed by a user, the less information has to be replicated on a control system display.

Although there are a variety of ways to classify the variables that describe the dimension of human performance, it is common to refer to the following six dimensions (also known as performance-shaping factors):

1) *System automation and demand.* This factor takes into account, for example, control systems, alarms, and, more generally, the philosophies of automation of business processes by automated systems.
2) *Personnel and workload.* This factor includes the number of individuals involved in carrying out a given task and their relative workload, both mental and physical. Two aspects must be borne in mind:
   - Both an excessively high and excessively low workload can degrade the level of performance.
   - Increasing the number of staff reduces the individual workload, but increases the demand for team coordination.
3) *Man-machine interface.* This factor is intended to group together aspects related to the organization, content, and "shape" of the information needed by an operator to interface with a machine. This category includes aspects of ergonomics, rationalization of alarms, and design of command and control panels. The ultimate aim is to always provide the right information, in the right format, at the right time and with the right timing.
4) *Personnel selection and training.* This factor aims to examine aspects related to personnel selection procedures and methods, training programmes, skills and knowledge requirements for carrying out a task, as well as identifying training material and tools.
5) *Work organization.* Even the smallest task is part of a complex organizational process that may involve several functions within the company. It is in fact the organization that establishes expectations and tasks to be performed, therefore the way work is organized impacts on its performance. This factor includes the organization of personnel (e.g. in shifts), the creation of work teams, the identification of unitary operations and so on.
6) *Procedures and operational instructions.* This last performance factor plays a predominant role in what Rasmussen identified as the rule-based performance level (Rasmussen, 1983). In general, complex organizations today make great use of operational procedures to guide human intervention in their tasks. Therefore, since the procedures are themselves written by men (and therefore imperfect) it is essential to take this variable into account as well.

1.5 Enterprise Complexity and (Advanced) Risk Management (ERM)

Keeping in mind the principles of RM, shown in Figure 43, advanced RM can be introduced.

Enterprise risk management (ERM) is the set of organizational practices, corporate culture, and skills that an organization sculpts within its strategy and applies daily with the aim of managing risks by creating and preserving value. It is extremely important in strategic planning, because risks influence all departments and functions, aligning strategy and performance across the organization.

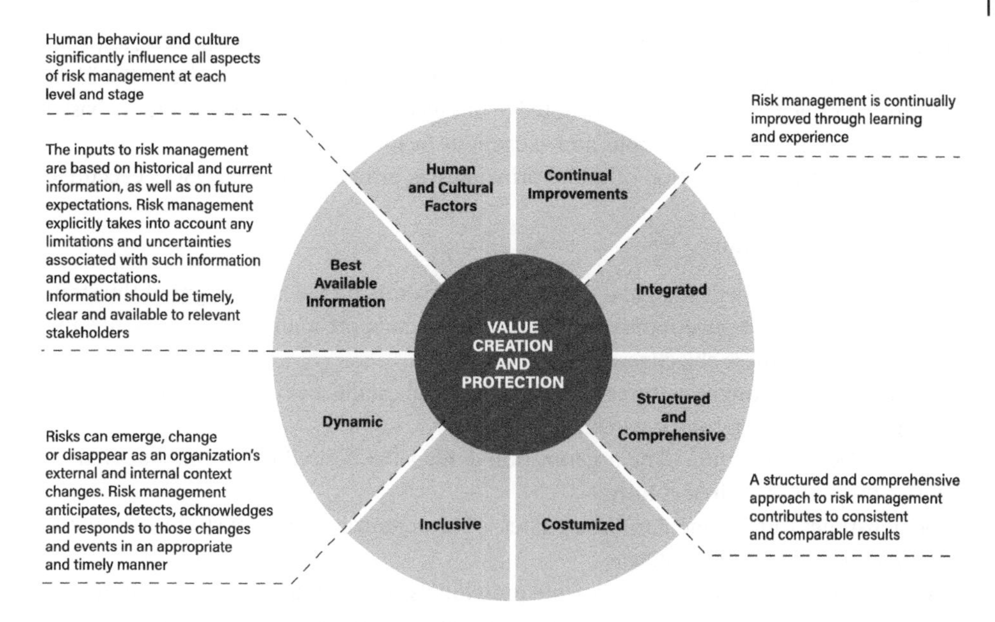
Human behaviour and culture significantly influence all aspects of risk management at each level and stage
The inputs to risk management are based on historical and current information, as well as on future expectations. Risk management explicitly takes into account any limitations and uncertainties associated with such information and expectations. Information should be timely, clear and available to relevant stakeholders
Risks can emerge, change or disappear as an organization's external and internal context changes. Risk management anticipates, detects, acknowledges and responds to those changes and events in an appropriate and timely manner
Risk management is continually improved through learning and experience
A structured and comprehensive approach to risk management contributes to consistent and comparable results
Human and Cultural Factors
Continual Improvements
Integrated
Structured and Comprehensive
Costumized
Inclusive
Dynamic
Best Available Information
VALUE CREATION AND PROTECTION

Figure 43 The principles of RM according to ISO 31000.

The main guidelines on ERM are offered by the Committee of Sponsoring Organizations of the Treadway Commission (COSO, 2017), to which the discussion of this paragraph adheres. ERM can be defined as a process that involves all people at each level of an organization and is applied on a strategic basis at each level or unit, with an inclusive character, in order to identify potential events that, if manifested, could damage the organization. The ultimate goal is therefore risk management according to risk appetite and tolerance, providing reasonable (but not absolute) assurance to top management, and achieving the organization's objectives in one or more categories.

To do this, the structure of the ERM is based on five solid components:

1) *Governance and culture.* The former includes the organization's effort in establishing the tone, the importance, and the responsibilities of ERM. The latter refers to the ethical values and the desired behaviours related to ERM.
2) *Strategy and objective-setting.* The ERM is aligned with the strategy and objectives set in the strategic planning process. This component is intimately connected with the risk appetite of the organization (i.e. the level of risk that an organization is willing to accept), which is established and aligned with strategy, while the objectives are the practical reflection of the basis for responding to risk.
3) *Performance.* The identified and assessed risks are prioritised according to the acceptability/tolerability criteria that have been established in the context of the risk appetite. The amount of risk that the organization has assumed is then reported to stakeholders.
4) *Review and revision.* The ERM performance needs to be periodically reviewed and revised when the ERM components are substantially changed over time.
5) *Information, communication, and reporting.* ERM is also based on solid and shared information which flows across the organization. This component is therefore as essential as the others.

These five components of the ERM structure are in turn based on 20 principles, applicable to any organization regardless of size, type or business sector. Thanks to them, the ERM structure as defined by COSO can be adapted to the specific needs of an organization, based on its context and other endogenous and exogenous factors.

The following is a description of all 20 principles as provided by the COSO's executive summary (COSO, 2017).

Governance and Culture

1) Exercises Board Risk Oversight—The board of directors provides oversight of the strategy and carries out governance responsibilities to support management in achieving strategy and business objectives.
2) Establishes Operating Structures—The organization establishes operating structures in the pursuit of strategy and business objectives.
3) Defines Desired Culture—The organization defines the desired behaviors that characterize the entity's desired culture.
4) Demonstrates Commitment to Core Values—The organization demonstrates a commitment to the entity's core values.
5) Attracts, Develops, and Retains Capable Individuals—The organization is committed to building human capital in alignment with the strategy and business objectives.

Strategy and Objective-Setting

6) Analyzes Business Context—The organization considers potential effects of business context on risk profile.
7) Defines Risk Appetite—The organization defines risk appetite in the context of creating, preserving, and realizing value.
8) Evaluates Alternative Strategies—The organization evaluates alternative strategies and potential impact on risk profile.
9) Formulates Business Objectives—The organization considers risk while establishing the business objectives at various levels that align and support strategy.

Performance

10) Identifies Risk—The organization identifies risk that impacts the performance of strategy and business objectives.
11) Assesses Severity of Risk—The organization assesses the severity of risk.
12) Prioritizes Risks—The organization prioritizes risks as a basis for selecting responses to risks.
13) Implements Risk Responses—The organization identifies and selects risk responses.
14) Develops Portfolio View—The organization develops and evaluates a portfolio view of risk.

Review and Revision

15) Assesses Substantial Change—The organization identifies and assesses changes that may substantially affect strategy and business objectives.
16) Reviews Risk and Performance—The organization reviews entity performance and considers risk.

17) Pursues Improvement in Enterprise Risk Management—The organization pursues improvement of enterprise risk management.

Information, Communication, and Reporting

18) Leverages Information Systems—The organization leverages the entity's information and technology systems to support enterprise risk management.
19) Communicates Risk Information—The organization uses communication channels to support enterprise risk management.
20) Reports on Risk, Culture, and Performance—The organization reports on risk, culture, and performance at multiple levels and across the entity.

The principles of ERM, as defined by COSO, can be observed from an ISO 31000 perspective. This provides an important opportunity for organizations to better integrate risk management and strengthen the activities related to the rest of business processes. Also, this integrated approach would allow ISO 31000 principles to provide an approach for the definition of the necessary risk-based actions to ensure growth and greater economic assurance for the organization.

From the ERM perspective, successful risk ranagement requires (Louisot and Ketcham, 2014):

- A full understanding of the emerging risks;
- The definition and understanding of the risk appetite;
- Accounting for extreme events, such as unexpected large deviations from the ordinary, like a fat tail (the probability distribution that displays a large skewness or kurtosis in comparison to a normal or exponential distribution) or a black swan (an event which can have high impacts, but whose probability of occurrence is low);
- Assessing and aggregating all risks, implementing a "portfolio approach";
- Considering qualitative tools and sound judgement;
- Broadcasting the risk culture across the organization.

There are four main types of business risks to take into account in the ERM (Figure 44). They are:

1) Strategic risk;
2) Financial risk;
3) Compliance risk;
4) Operational risk.

Strategic risks affect the capability of an organization to achieve its objectives. Examples include reputational risks, risks to market presence, lack of acceptantc in the marketplace of a business plan, change in customer demands, changes in technology, or competition with other businesses. Basically, strategic risks are associated with the correct (or incorrect) application of business strategies and the ability to change them when any changes in the external or internal context happen. Facing positively the strategic risks is the key of vey successful multinational companies.

Financial risks concern the organization's ability to fulfil its financial obligations. They are mainly divided into three families: the exposure to changes in market prices, the actions (and transactions) with other organizations, and the internal organizational failures (Bansal, Kauffmann, Mark and Peters, 1991). Investment, market risk, interest rate risk, exchange rate risk, liquidity risk, and inflation risk are types of financial risk that should be considered.

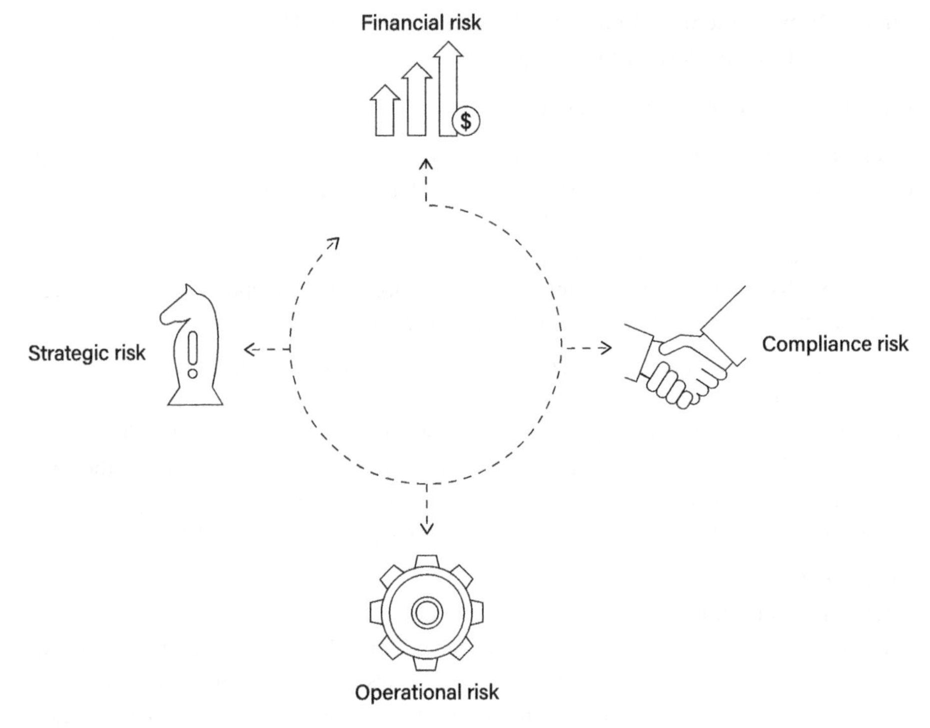

Figure 44 Main types of business risks.

Compliance risks are about the failure to act in accordance with the laws, regulations or internal policies. They are the legal penalties that an organization might be exposed to when compliance obligations are not met. This category includes (from a compliance perspective) the environmental risk, the occupational health and safety risk, the corrupt practices, and the social responsibility and quality issues.

Finally, the term operational risks refers to the failure to achieve the intended outcomes from the operations of the organization. They might come from inadequate procedures, policies or systems. Poorly faced operational risks may lead to reputational or other financial risks. Since operations are the daily basis of any organization, an effective operational risk management is essential to guarantee the prevention and mitigation of the most common type of risks, supporting the organization in achivieng its operational objectives.

A more exhaustive set of enterprise risks is shown in Figure 45.

1.6 Proactive and Reactive Culture of Organizations Dealing with Risk Management

1.6.1 Risk Management between Fulfilment and Opportunity

In order to understand the importance of the risk management culture (Figure 46), it is necessary to identify in advance its relationship with the processes to which it is (or should be) applied.

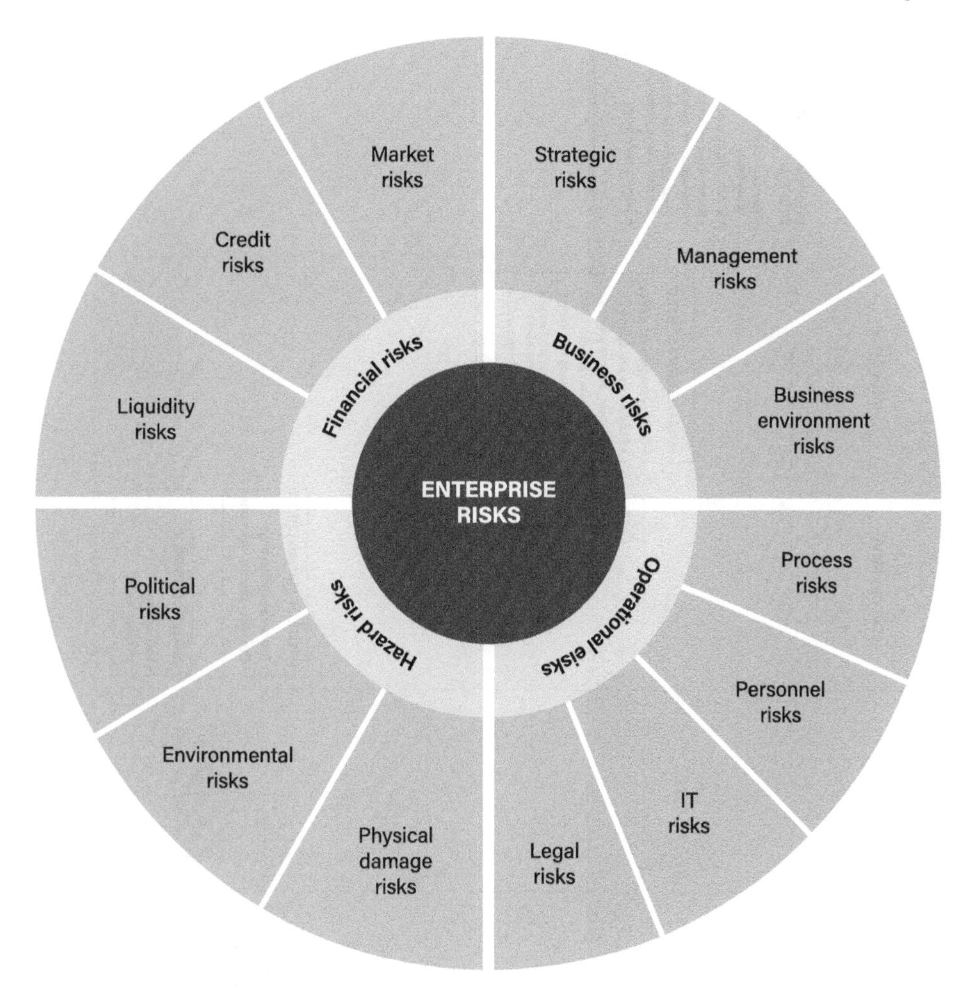

Figure 45 Most common enterprise risks.

If, in fact, the purpose of an organization is to pursue the maintenance and growth of its assets, whether material or moral, the quality level of risk management, and therefore the degree of commitment to its implementation, will be justified to the extent that it contributes value to the organization, in terms of ensuring conservation or growth.

In summary, and however cynical the statement may seem, the cost of risk management – as the use of means and resources – and therefore its level of quality, must find remuneration in the value it is able to guarantee to the organization.

It is probably easier and more immediate to associate the concept in question to an entrepreneurial organization where, through financial parameters, it can be easy to measure the cost of risk treatment and the expected and actual benefit on the managed processes.

The concept is indeed equally coherent for a moral organization; consider, for example, the reputational value and the negative effects that the conduct of a representative could generate if it is not in line with the institutional policies of the organization itself, even more so if the organization is not able to remedy it in a preventive or corrective manner.

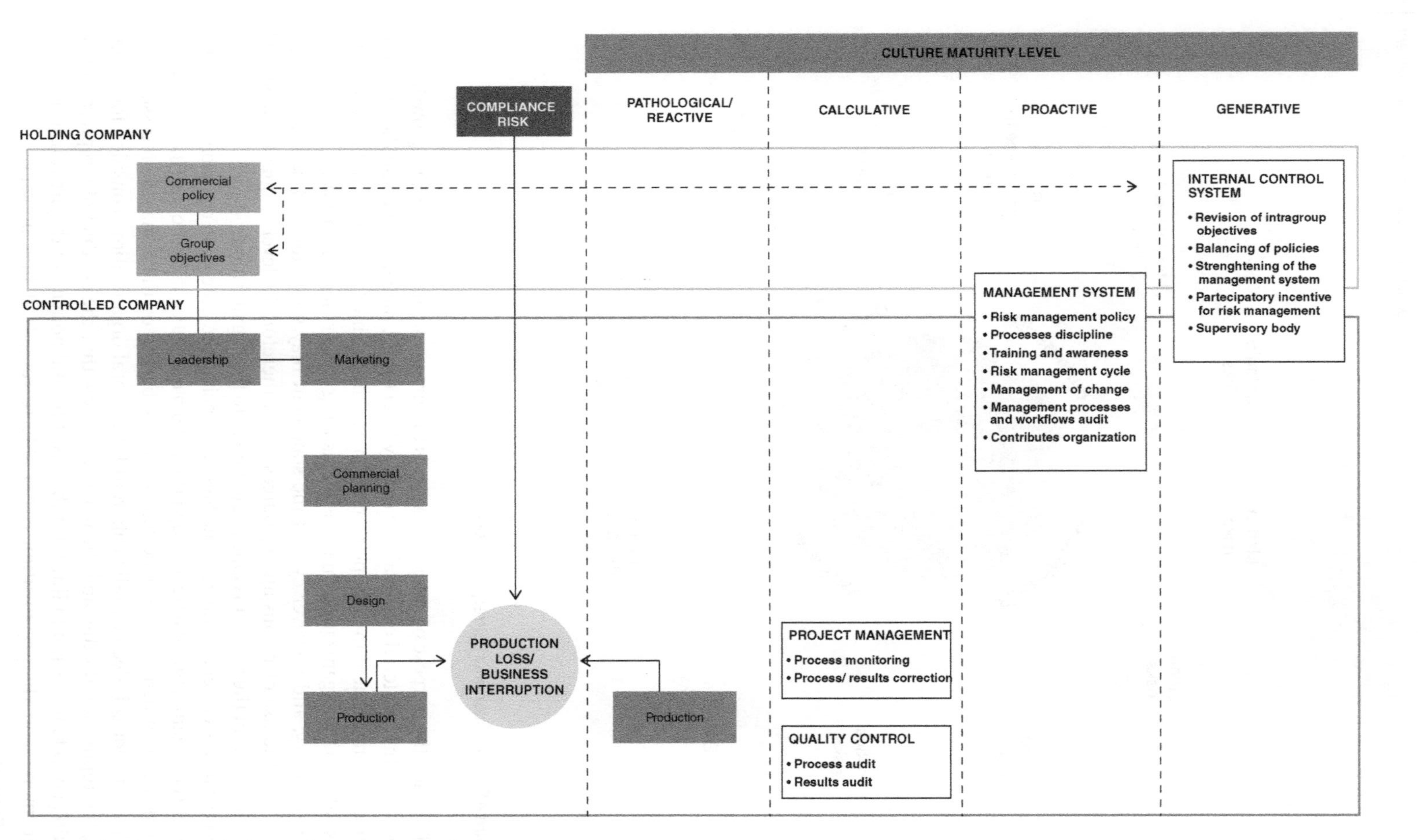

Figure 46 Culture maturity level in an organization.

From these considerations it is possible to derive, as a first and immediate consequence, that risk management can, or even better should, be framed as a normal component of all the technical and managerial management processes of the organization, rejecting any "ethical" approach to risk and ignoring that its treatment is limited to dedicated sectors of the organization but often "separate" from its ordinary operating cycle.

Talking about risk management as a separate process with respect to the functioning of the organization is in fact a risk in itself:

- On the one hand, it ends up depriving risk analysis of the technical contribution and vision in which a given objective is framed, focusing attention on the result and relegating the risk profile to an external factor to have to consider, perhaps on the basis of regulatory constraints.
- On the other hand, in the dialectic of the organization, it polarizes the role subjectivities, creating useless, if not potentially harmful, functional antinomies with respect to the achievement of the result.

Managing the processes of the organization, in an integral and inclusive way of risk profiles, both in the organizational and operational phases, allows the enrichment of the perspectives considered in the management process and, above all, the treatment of the risk to be balanced with all the profiles of the organization, triggering a virtuous mechanism of development of the quality of management.

Let's try for example to consider the process of recovery and commercial development of a building or the design and management of an industrial production plant.

From the first point of view, that inherent to the implementation of the process, both activities, although technically different, appear as extremely complex processes involving the treatment and convergence of multiple competences and organizational roles in the sub-processes respectively entrusted.

The progress of the activity requires the identification of profiles that give order to the complexity and promote choices and priorities in the comparison of alternatives.

It is also evident how the treatment of the risk profiles analyzed in relation to the alternatives considered, in the current implementation and management perspective, allow the enrichment of the process and the optimization of the choices.

Consider, for example, the management repercussions of the design choices in the fire prevention field with respect to the development of the projects or the operating costs associated with these choices; it is quite clear that during the design phase, the interest of the project team could favour the respect of delivery times or ease of implementation, leaving management with a heavy inheritance in terms of operational or financial risk.

On closer inspection, in addition to the risk of process fragmentation, there is also the risk that the number of factors considered (including risk factors) may slow down the process; however, in application of the considerations in question, the treatment of this risk at the organizational level can appropriately assess and mitigate the profile, foreseeing intermediate phases, i.e. the organizational figures, among those involved, responsible for promoting the overcoming of possible decisional deadlocks.

In any case, the process is significantly improved as any solution adopted will have been assessed, at least with an awareness of the degree of risk taken and the commitments to be made to mitigate its impact.

From the second point of view, including the treatment of risk profiles in the process management, already in the organizational phase, allows, on the one hand, anchoring the analysis to the specificity of the initiative and, on the other hand, stimulating the contribution and responsibility of all the actors involved, making the treatment of risk an added value of the organization.

In fact, the specificity of risk analysis, integrated in the management process and in the awareness of the actors involved, mitigates, at least at an organizational level, the risk of missing or inadequate assessment of operational risks or those of overestimating them, negatively constraining the planning and management of the initiative undertaken.

Therefore, in these terms, the treatment of risk profiles is integrated within the reference process, both organizational and operational, as a transversal qualitative requirement in all phases of the process itself, losing the connotation of separateness and speciality that often distinguishes it and is, at least in power, a factor of maintenance or creation of value in the organization and results. Considering safety-related risks in an organization could identify different levels of risk management that are in strict relationship with the safety culture of the organization itself (Figure 47).

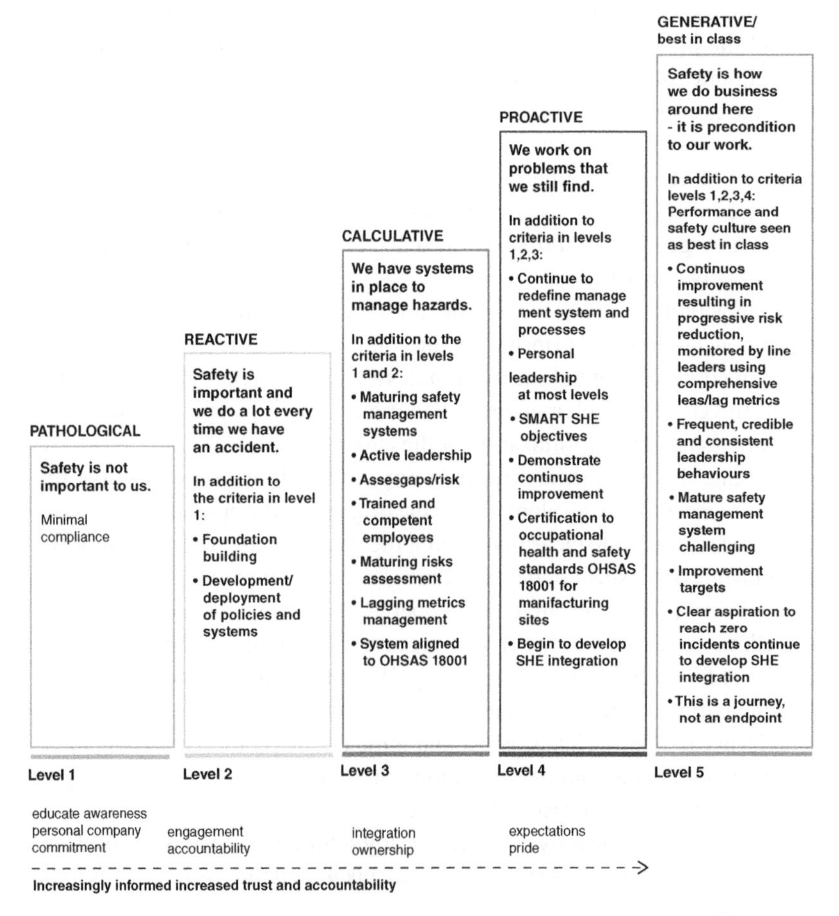

Figure 47 Safety culture levels.

1.6.2 Quality of Risk Management

With the considerations expressed, we wanted to affirm that the treatment of risk contributes to the conservation and increase of the value of the organization and in coherence with this contribution, risk management can be a qualitative component of the organizational and operational processes of the organization.

In this perspective, risk management on the one hand implements and enriches the contents of the organization's processes and, on the other, creates the conditions for a participatory approach to risk management that mitigates the risk of subjective polarization in the implementation of operational processes and, above all, creates the conditions to produce further value through the recovery of existing production.

Let us now try to make this last profile explicit and by applying the concept with respect to the path of growth of a human being, aware of all the approximation of the comparison but confident of the support that an empirical reference to a case of common knowledge can offer.

At a few years old a child starts walking; he or she stumbles or collides with objects, either stationary or moving, on his or her path because they are not properly considered.

Progressively, with a series of falls and some crying, the child increases his skills, and reacts by learning to recognize and evaluate objects on his path to avoid them.

Therefore, the child achieves a degree of awareness in which he or she observes the rules dictated on the basis of experience and walks with reasonable confidence in the home environment.

Furthermore, the child learns to apply the rules of experience autonomously and to interpret heterogeneous signals for walking in environments other than domestic and habitual ones.

In the end, the child pursues his or her will to walk by applying independently and automatically the rules of experience acquired and increases his or her skills by running and playing.

The example in question, whose summary and approximate character is reiterated, is however certainly useful in focusing on some concepts mentioned in the previous pages.

One learns not to fall, to walk indoors, then outdoors and then again to run and play, processing together the objective pursued and the associated risks.

In other words, an entrepreneurial organization pursues economic objectives set out in the articles of association and does not a priori manage risk; however, it is clear that the treatment of the risks associated with the objective is a component of the management process aimed at achieving the result.

Treating the risk is impossible only in association with the objective identified and the process necessary to pursue it. We can say that risk does not exist without the objective of which it remains as a sort of negative predicate.

In this perspective, the treatment of risk brings value to the organization.

In the child's developmental process, as simplistically described, objective and risk are treated in direct association, initially unconscious and mutually interfering, with the result of enabling the child to achieve the objective and setting the conditions for further development of his or her abilities.

In this path, objectives and risks continue to be treated jointly and to feed each other in a process of capacity and awareness growth that continuously recovers the bases of experience to rework them with respect to the new objectives and to identify new risk profiles to be evaluated with respect to the new objectives.

In this perspective, at a methodological level and even more so in an organization, the treatment of risk integrated in the management process brings the added value of fluidity and process efficiency through preventive and corrective remedies already integrated along the way.

Finally, we cannot but observe how, in the organic unity of our child, the rules of experience are acquired and made available to all the sensory and cognitive abilities of the person, evidently affected by the same developmental path.

We certainly do not dwell here on the interactions of this unity, but we cannot but note how the same perspective applied within an organization, or even more so within complex and mutually interfering organizations, can further contribute to the growth of the organization through the recovery and diffusion of the rules of experience, the application of the same rules on different management processes of the same organization, and the stimulus to the identification of new signals or new scenarios.

In other words, the treatment of risk, as a speculative predicate of the objectives pursued, as a process culture can offer further added value to the organization.

We therefore try to apply risk treatment in the developmental stages related to the child to an organization, classifying a series of "standard" conditions, as shown in Figure 48. In the following paragraphs, fire safety level has been used as an example of different culture levels of an organization to clarify the importance of the attitude of the organization towards risk management. This could be extended to any kind of risk that threatens the organization.

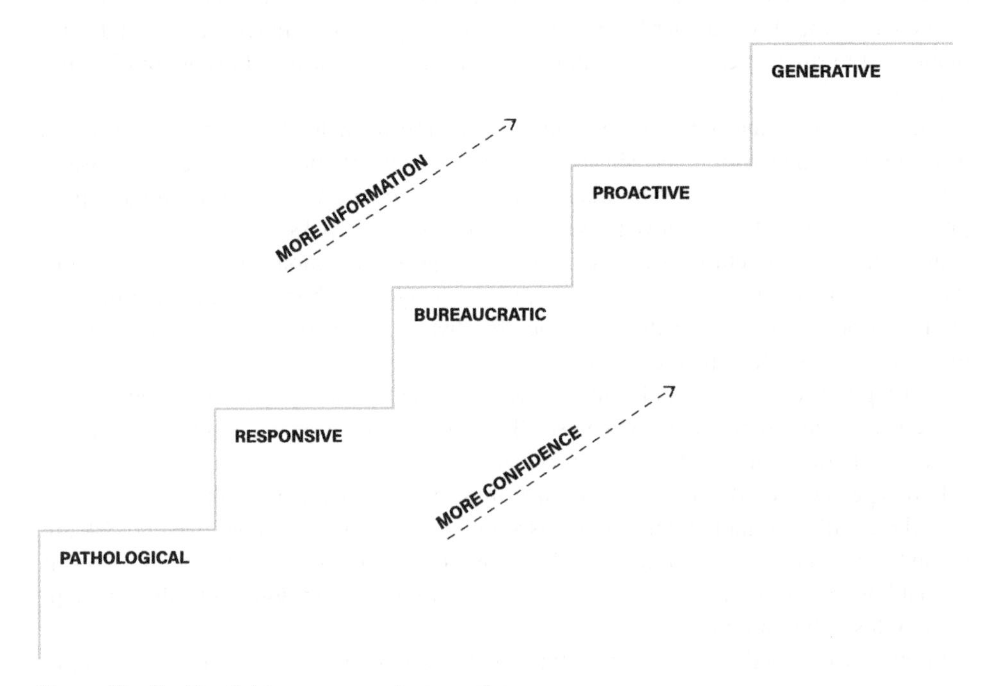

Figure 48 Quality of risk management approach.

1.6.3 The Pathological Condition

The "pathological" condition (Figure 49) corresponds to the absence of preventive risk treatment, a condition in which the organization promotes its objectives through processes aimed exclusively at obtaining results.

The organization identifies the objective and develops in a serial manner the activities deemed necessary to obtain it without consideration of the interfering factors – if not limited to those that can offer opportunities for direct maximization of the result – and therefore without the identification of preventive or subsequent corrective factors.

The management process is necessarily poor because the lack of risk treatment prevents the implementation of the analysis of the operational process, jeopardizes the possibility of developing synergies between company processes, and is characterized by the lack of information disseminated throughout the organization.

Organization objectives should be clearly identified through a specific process that also identifies the expected results; after all, the pathological condition pervades the same organizational structure, preventing a priori the preliminary analysis necessary to define the process and the expected results.

In other words, the organization wants to "walk" and starts to do so without considering the ability to sustain the necessary commitment over time, without identifying the obstacles on the path or alternative trajectories or verifying the opportunity to achieve the same objective in different and alternative ways.

The organization does not treat the risk until it manifests itself in a dangerous condition that results in damage that jeopardizes the achievement of the result.

The organization reacts within the limits of what is necessary with respect to the incidence of the damage with respect to the organization and the restoration of the operation

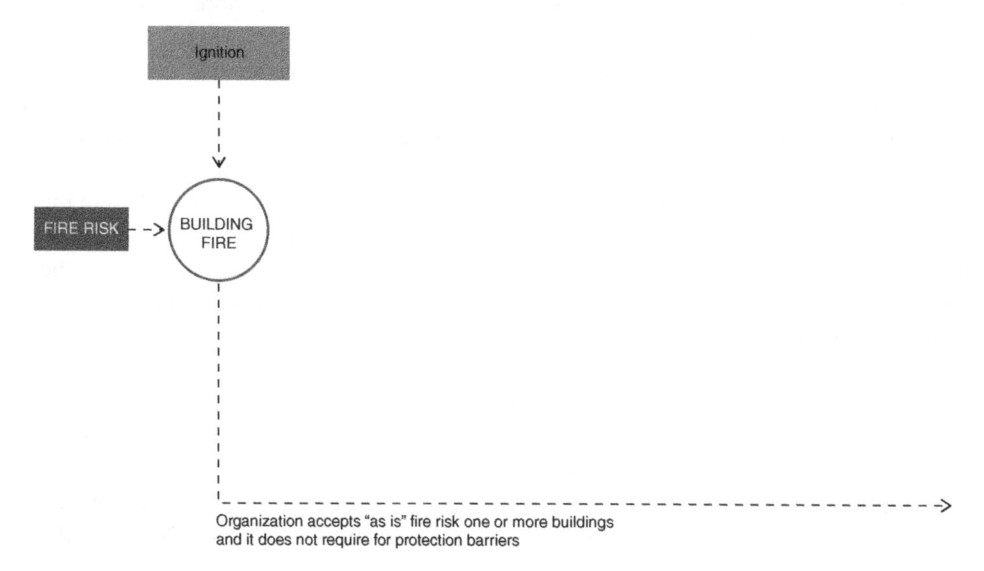

Figure 49 The pathological condition.

of the process without, however, re-evaluating its suitability and persecutability and, above all, without "capitalizing" the event by transforming it into a rule of experience to be evaluated with respect to the organization and the complexity of its processes.

One might conceptually wonder whether this is acceptance of the risk or whether the organization ignores it a priori.

The pathological condition is typical of organizations operating on processes and scenarios characterized by limited technical production constraints, market scenarios with a low competitive level, a low level of outsourcing of operational processes, and regulatory contexts that tend to be prescriptive, which would make the value contribution of risk management appear negligible.

On closer inspection, in reality, the risks of business continuity or the risk of inadequate calibration of production objectives, for example, are quite significant even in expected monopoly markets that could generate significant revenue losses or penalties with respect to the failure to meet demand.

In short, in order to answer the question about the organization's awareness, we can consider that the pathological condition – which, as indicated earlier, pervades the organizational structure and processes – is at the same time the result of a risk acceptance that is not fully aware by an organization that, even knowing it, does not accept the stimuli and does not use them to evolve, evidently endorsed by a technical, regulatory, and market context that is not particularly demanding.

It is certainly possible to affirm in this sense the correlation between the degree of quality of risk management in the organization and the reference context.

1.6.4 The Reactive Condition

If an obstacle is hit when trying to walk, the reaction will be all the more important the more significant the damage and pain suffered.

This in brief can be a description of the "reactive" condition in risk management (Figure 50).

This is a condition immediately superior – if possible identifying a line of qualitative development of the risk management culture – to the pathological condition.

The organization not only suffers damage for which it may even be sanctioned, as in the pathological condition, but in the face of the importance of the damage suffered it reacts immediately by making significant efforts to eliminate the damage suffered and align the process with the management of the risk that has occurred, integrating the process itself with corrective measures that had not been previously assessed.

This is a strong reaction, the result of the awareness acquired by the organization as a result of the damage suffered but limited to the event that occurred.

The reaction does not feed the evolution of risk management either with respect to organizational or operational processes.

As with the pathological condition, it can be assumed that, in the reactive condition, the organization has accepted the risk although with the greater awareness reflected in the intensity of the reaction.

After all, if in the organic unity of the person, evolution is dictated by the natural ability to elaborate stimuli, the satisfaction of needs, and the action of external educational and training factors, the ability of the organization to understand, acquire awareness, react to

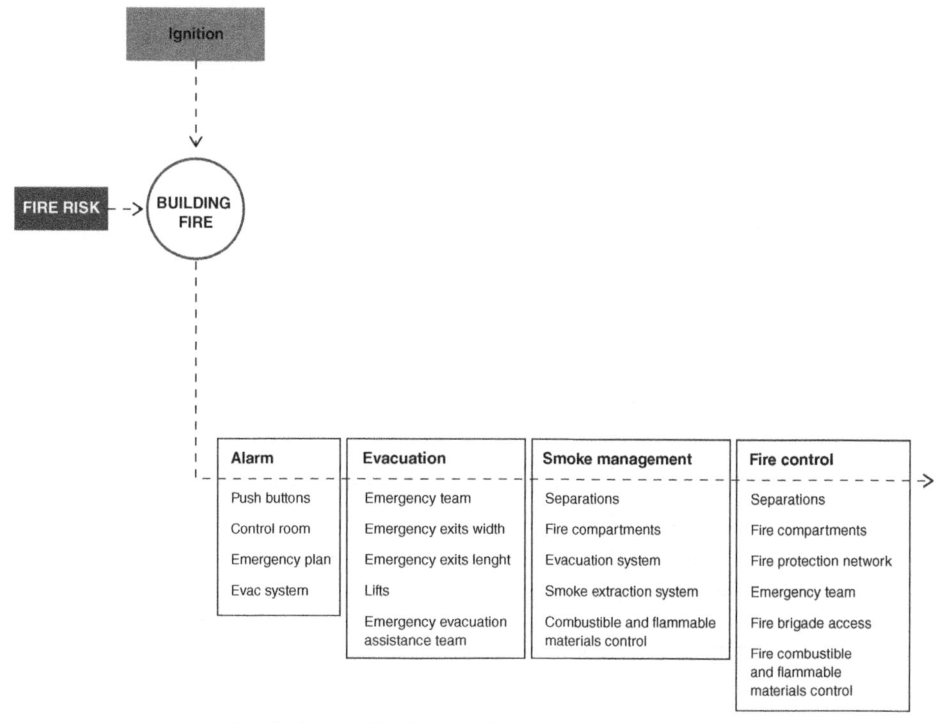

Figure 50 The reactive condition.

stimuli, and gradually evolve to "superior" conditions is the result of an artificial construction strictly dependent on organizational policies, the subjectivities responsible for management roles, and the technical, regulatory and scenario constraints in which the organization operates.

In this perspective it is possible to affirm that the reactive condition can be endorsed by prescriptive regulatory contexts in which the cases of possible non-compliance are typically outlined and identifiable and to which the strong but precise reactive approach of the organization in case of accident is adapted.

This context probably does not favour the development of risk management as a generative process of added value for the organization.

1.6.5 The Bureaucratic Condition

In the "bureaucratic" condition of risk management (Figure 51), we can clearly identify a first concrete step in the line of qualitative development of risk management in the organization.

In the bureaucratic condition the risks are consciously treated by the organization which, according to their identification, dictates the risks as preventive corrective factors.

Not only that, the assumption of the rules by the organization implies that they are disseminated to all members of the organization who must comply with them in order to prevent non-conformities and accidents.

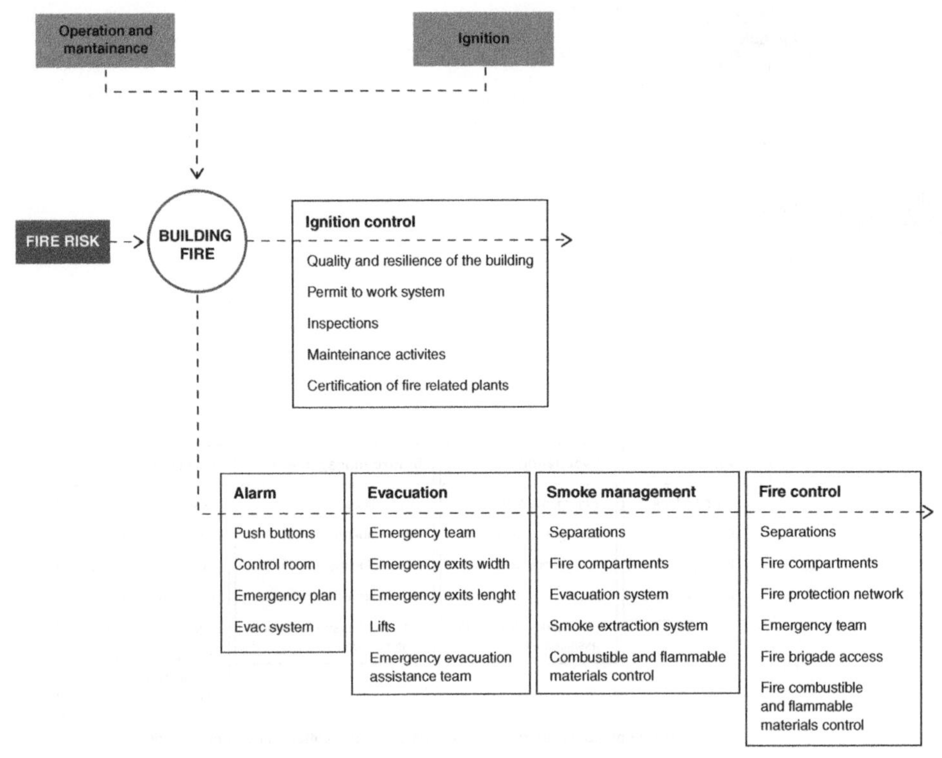

Figure 51 The bureaucratic condition.

The qualifying step of the bureaucratic condition lies in the affirmation of a risk management policy by the organization, which in turn informs and shapes the operational processes.

It is only through the affirmation of a policy that risk management takes on a systemic character in organizational and operational processes, attesting to the awareness of the organization and removing the treatment of risk from the individuals who make it up.

The bureaucratic condition definitely overcomes the grey area of pathological and reactive conditions in which risk acceptance is based on the limited awareness of the organization.

Of course, it is not that the individual members of organizations defined as pathological and reactive cannot be able to identify and assess the risks of the activities from a technical and managerial point of view; simply, that type of organization does not require it.

In the bureaucratic condition the organization manages the risk, identifying the profiles and dictating the rules that everyone must implement to avoid the dangers.

In other words, an organization in bureaucratic condition is like a child who has acquired the necessary rules to walk inside the house; it is an organization that moves with relative safety within a given perimeter and as long as all its actors respect the rules.

On closer inspection, however, it is an organization whose risk management is not flexible and does not adapt to changes in the scenario, does not produce added value, and at most guarantees the preservation of existing value as long as the context does not change.

Furthermore, it is an organization that does not seize the opportunities associated with the commitment required of its actors to comply with the rules and misses the opportunity to take and treat critically external stimuli to enrich processes or simply exploit the synergies between homogeneous processes in dealing with the corresponding risks.

Finally, precisely because of this rigidity, such an organization would not be safe in those regulatory contexts that base responsibility on risk management and that introduce the adequacy of the evaluation with respect to the accentuated complexity of the current technical and management processes, ending with the admission of an a posteriori judgement on the goodness of the management process and therefore on the conduct of the actors involved.

1.6.6 The Proactive Condition

In the proactive condition (Figure 52) there is widespread attention of the actors in the organization to the danger signals.

The organization has taken an important step forward compared to the conditions described previously. In the bureaucratic condition in fact:

- The policy has enabled awareness of the organization of risk and its treatment to be taken into account.
- Awareness generated the analysis and assessment of the risk, which in turn led to the definition of the rules.
- The rules are deliberately spread throughout the organization and everyone must follow them to avoid dangers.

In the proactive condition, on the other hand, risk awareness extends beyond the boundaries of the organizational process and informs operational processes: the actors not only follow the rules to avoid dangers and accidents, but are also attentive in advance to danger signals.

They identify the signals and react by bringing them back into the management process for their treatment and for implementing the risk management and knowledge assets.

If the precondition of the proactive state lies in the existence of a risk policy on the part of the organization (as already in the bureaucratic condition), the maturation towards the proactive condition is dictated by the organization's need to grow.

As the child, after learning to walk in the house, ventures outside into an unfamiliar environment, the proactive organization needs to identify danger signals to learn to move in new scenarios, markets, constraints and contexts, developing the process and handling of the associated risk.

One might wonder whether the detection of the danger signal is the implementation of a rule or the result of the organization's stimuli.

Probably, the delimitation between the sphere of having to do it and wanting to do it is not identifiable in a clear demarcation line.

More likely, observing an organization in concrete terms, one would end up observing a range of behaviours in which they both alternate in profiles but whose orientation, over time, towards one or the other of the spheres in question could certainly be an indication of the organization's maturity.

Of course, to speak of "will" is to be understood in any case within the framework of a system of rules dictated by the organization; it would be in other words a will progressively

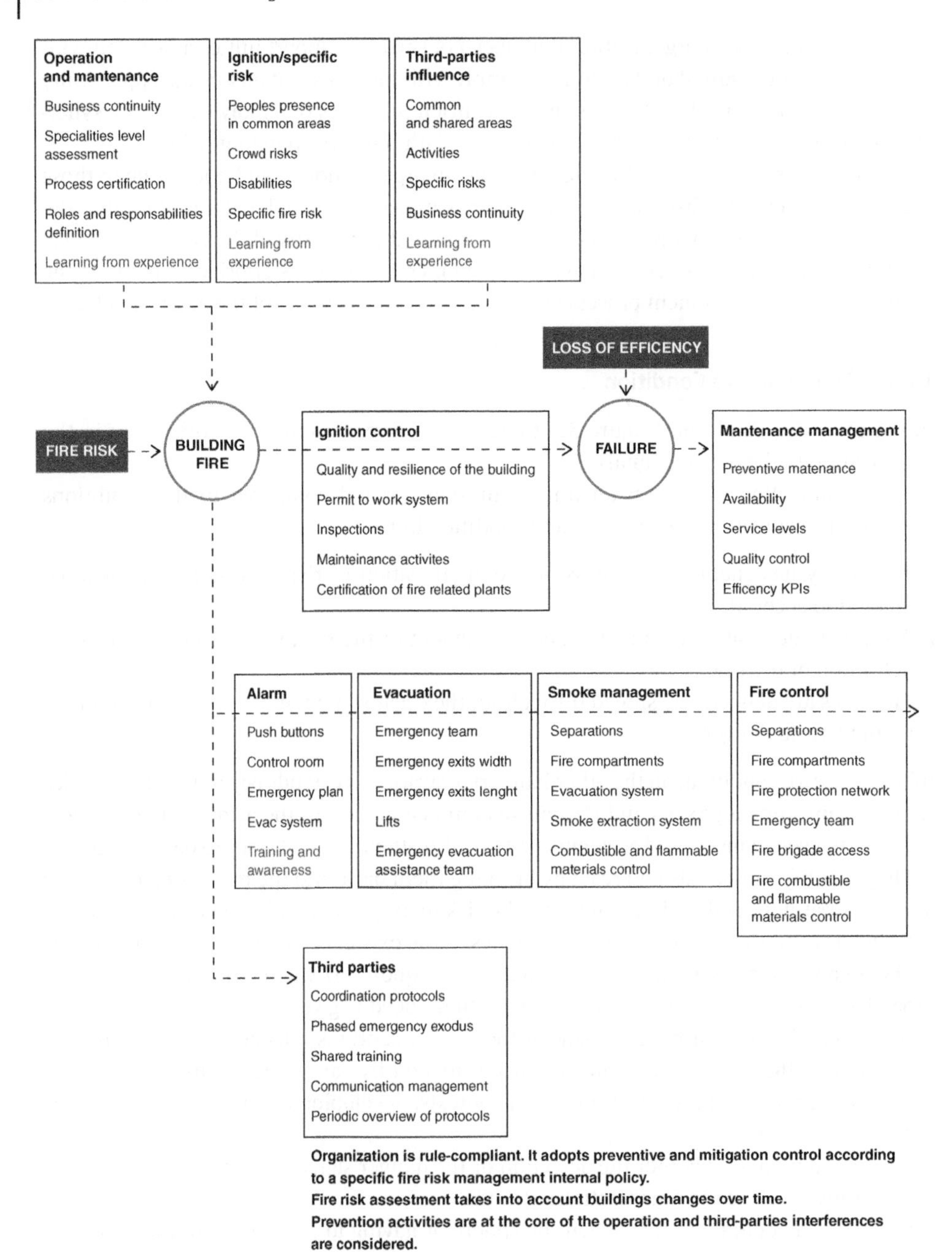

Figure 52 The proactive condition.

induced or easily obtainable by the organization as a result of the development and implementation of a policy oriented towards sharing and widespread participation in the organization.

In any case, it is clear that, as shown at the beginning, risk management in the proactive condition is able to generate value for the organization.

1.6.7 The Generative Condition

The generative condition (Figure 53) represents in some way the complete maturation of the organization, the one in which risk management is an integral part of company processes, and a factor competing with the production of value in the perspective indicated at the beginning of this discussion.

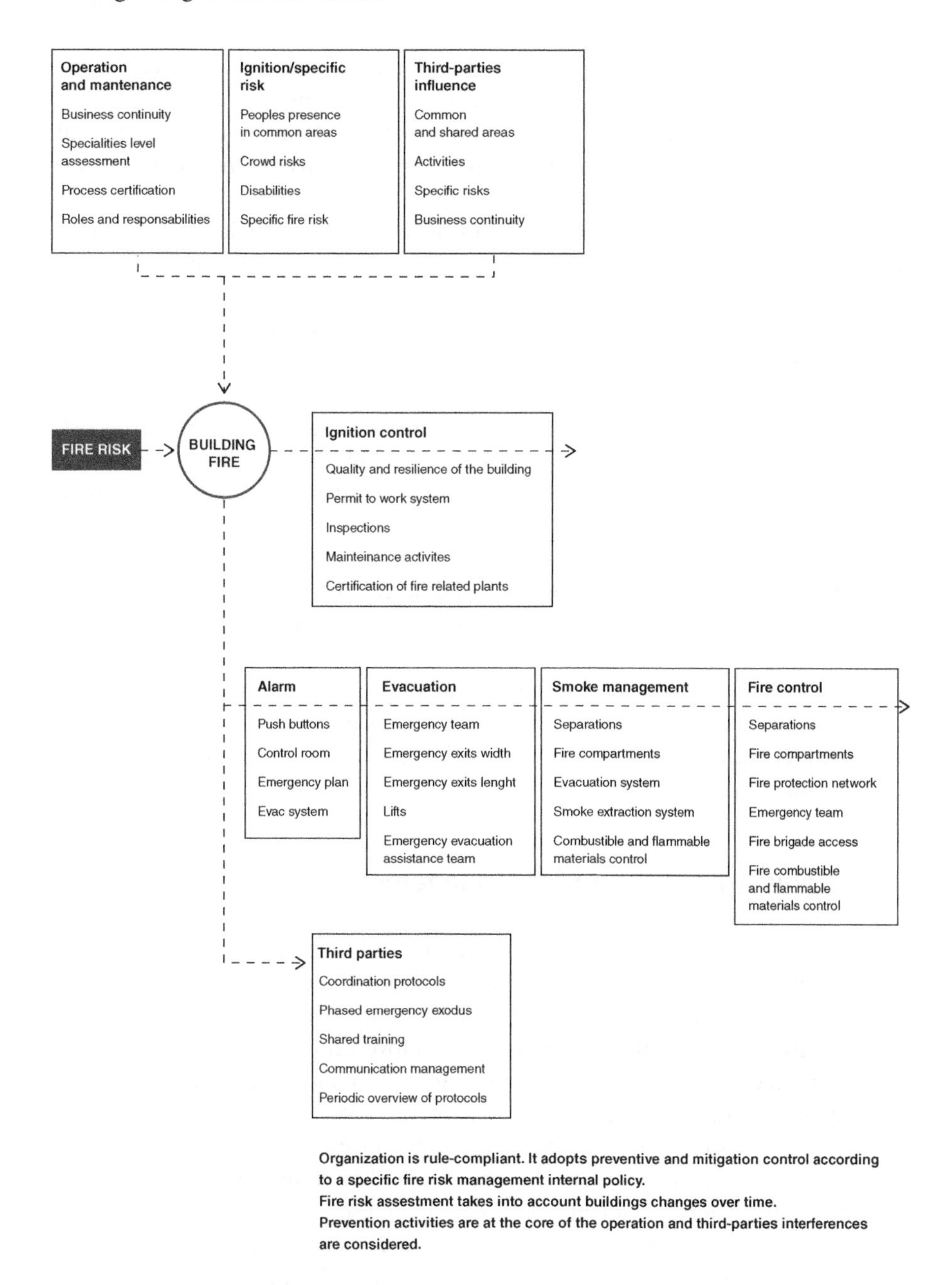

Figure 53 The generative condition.

For the condition in question – taking up again the comparison with regard to the duty of care and voluntary behaviour – the actors of the organization, whether they are involved in organizational or operational processes, naturally consider risk management to be an integral part of their activities, working, on the basis of the sensitivity induced by the organization and the training factors provided, to identify risk profiles, to implement containment actions and to contribute to updating the analysis and evaluation and, therefore, the processes.

The organization naturally develops a path of continuous regeneration based on the widespread contribution of the actors, which ends with self-feeding.

The generative condition allows the organization to adapt to changing scenarios, to deal with regulatory development with greater flexibility, and to overcome the technical constraints of operational processes.

But above all, the generative condition elevates the organization by depersonalizing it from the actors that make it up and characterizing it in terms of subjective synthesis superior to the policy pursued.

In other words, through policy and generative maturation, the organization protects itself from the risk of application deviations that can ultimately characterize the individual conduct of its actors.

In conclusion, the descriptive articulation of the degree of culture that characterizes risk management in organizations through the proposed classification is obviously a mere methodological expedient aimed at affirming the diversity and orientations that can be revealed in practice.

However, it is immediately evident that these are not imminent conditions but rather orientations of the organization in a more or less fluid condition from one to the other, depending on the degree of consolidation of the policy in its adequacy of application within the organization.

Nevertheless, the conditions described can be read as a term of reference or comparison that can offer support to the analysis of concrete situations from which to move to orient the development of a given organization towards the opportunities offered by the correct application of risk management in the perspective of value creation.

1.7 A Systems Approach to Risk Management

For many years every aspect of the organizations were treated with a deterministic approach with the precise aim of governing them with rules and detailed requirements. Organizations were seen as systems of single components. In this framework performances of organizations were directly linked with the continuity and improvement of single components of the entire system: each one with a different importance given its criticality and the effects on the outcomes. This initial approach survived up to the 1990s. Since then organizations have been recognized as entities governed by a number of different processes (primary, auxiliary and supporting processes). This new approach allowed the identification of the main flows of materials, information and data, energy, and so on across the organization, intended as a system with specific objectives and processes organized in areas and sectors (commercial process, production, customer support, etc.), each one having its own key performance indicators (KPIs) to be reviewed periodically. Organizations became systems composed of processes and elements. Processes could be governed by procedures rather than with specific prescribed requirements to be followed to achieve performance. Organizations were requested to learn from negative occurrences and deviations from intention and to measure the increase or decrease related to the objects in one or multiple domains. Increases have been identified with the improvement to be maintained over time in a circular scheme known as Plan-Do-Check-Act (PDCA) (see Figure 54), where the failures from planned intentions should be avoided with preventive actions and non-conformities management, both raised by a specific audit program whose results should be taken to the attention of top management and any eventual stakeholders.

This systemic approach to processes (internal and external) governed organizations up to 2015, when it was completely replaced by a risk-based approach. Processes are the atomic unit of an organization, but deviations (with the limitation in the extent of achieving the desired performance) are connected to a number of internal and external conditions; it is quite difficult to associate each single deviation to the failure of a single element of the process and the deviation from the organization objectives is likely to be linked to a number of perturbed situations rather than the failure of a single process (even if a primary and vital one). Single deviations could lead to disruptions in other correlated processes and small ones, cumulatively, could lead to severe problems with a significant loss of performance. The initial theory of component failures can describe the single element, and the subsequent process theory can describe deviations inside the primarly affected process and those few directly linked (in input and in output) to that. Real incidents, deviations, and anomalies are useful in describing the local problem. In any case real episodes with impact cannot be easily understood and organizations are not able to improve taking advantage of the learning from experience transversal knowledge process. Lack in efficiency has been discovered to reside in the impossibility to deal with complexity that characterize every domain of application. Complexity poses risks with a potential impact on components and processes, and both of them, and risks, take advantage of shared vulnerabilities and the human factor (including the error probability of human-centered processes, tasks,

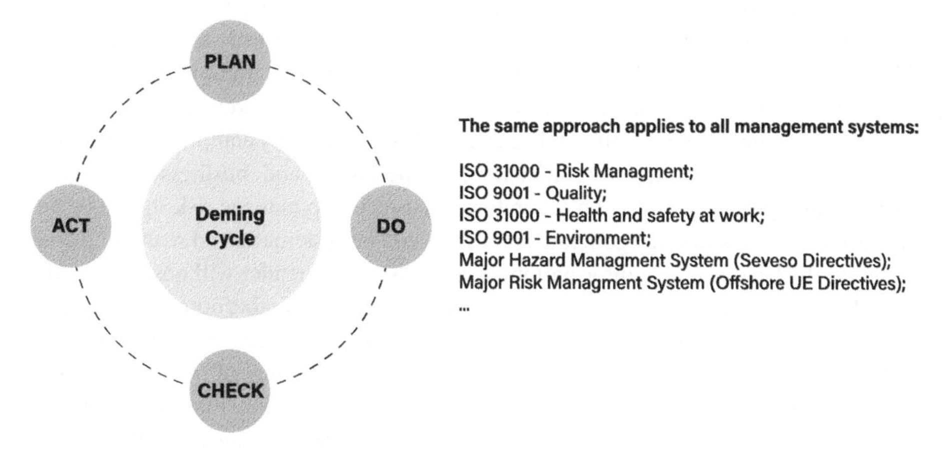
PLAN
DO
CHECK
ACT
Deming Cycle
The same approach applies to all management systems:
ISO 31000 - Risk Managment;
ISO 9001 - Quality;
ISO 31000 - Health and safety at work;
ISO 9001 - Environment;
Major Hazard Managment System (Seveso Directives);
Major Risk Managment System (Offshore UE Directives);
...

Figure 54 The Deming Cycle PDCA.

operations). With such complexity it is not possible (not only because it is too time-consuming) to identify all the possible outcomes (especially impact) of all the possible deviations. Impacts also could be directed at the same time towards different vulnerable receptors with a different degree of severity. This awareness raised the need to deal with complexity (in an always-evolving reality) using a risk management approach. The ISO 31000 standard, especially in its more recent edition, became the kernel of organization management across different domains. Organization management systems turned from a process-centered perspective to a risk-centered perspective and the latest objective of each single organization is risk management as the core activity to avoid failures, disruptions, and so forth. Successful organizations are those that can prove to be risk-resilient to top management and stakeholders. The risk management process embraces all the processes of the organization and is conceived with the right focus and zoom level to manage systematic issues and single-element performance, considering their relative criticality in the overall picture. Improvement comes from risk-based thinking and actions are identified, selected and put in place according to their estimated risk reduction factor. Since risk is an effect of uncertainty on objectives (deviation from expected), two pillars should govern modern organizations that are willing to manage risks to take better decisions: risk assessment (composed of risk identification, analysis and evaluation phases) for the risks to be treated and periodic monitoring and review, including reporting and communication activities. With this risk management process, it is possible to design a strategy for complexity and it is also possible to learn from real experience, as well as modify the risk management approach considering new, emerging, and underestimated risks. Complexity management means risk reduction and control for those risks arising from the known internal and external contexts, against specific risk criteria defined by rules (in certain domains) or by the company (according to the willingness to preserve assets and create more value). A systematic approach to organizations defined as systems has to be based on risk analysis. The purpose of the analysis is to comprehend the nature of risk and its characteristics including, where appropriate, the level of risk. It involves a detailed description of a number of factors: likelihood of events and consequences, the nature and magnitude of impacts, complexity and connectivity, time-related factors (including volatility), the effectiveness of existing controls, sensitivity and confidence level. Subsequent risk evaluation should support decisions: Are available controls sufficient? Should new controls be put in place? Should existing controls be maintained? Risk management creates and protects value. It contributes to the demonstrable achievement of objectives and improvement of performance in, for example, human health and safety, legal and regulatory compliance, environmental protection, product quality, efficiency in operations and business continuity, governance and reputation, and so on. Therefore, the process to manage risk should be an integral part of the activities that generate risks. It should be systematic and structured and also based on the best available information. If controls, as the reader will observe in this book, are the barriers to avoid or lower the probability of an undesirable outcome, then the risk management process should be dedicated to barrier management in a way that both physical and human barriers act on risks and technological and human threats are managed over time to avoid any undesired impact on vulnerable assets. Risk management avoids the failure to recognize complexity and allows the organization to improve even during real incidents, since for each of them, root cause analysis of the factors that affected one or more risk control failure are identified. Risk management (aka barrier management) is dynamic, iterative and responsive to change. It continually senses and responds to change. As internal (and external) events occur, internal (and external) context change, monitoring and reviewing risk take place, new risks emerge, some change and others disappear. Controls are reviewed to ensure their ongoing effectiveness in response to change. Monitoring and review activities are fundamental to keeping the risk management process in place and effective.

Specific and distinctive activities should be put in place to assure that assumptions, results and related decisions remain valid over time and, if not, that a management of change process is in place and considers risk management activities. These activities are known as "monitoring" and "review" of the risk management process.

- Monitoring involves the routine surveillance of actual performance and its comparison with expected or required performance. It involves continual checking or investigating, supervising, critically observing, or determining the status in order to identify change from the performance level required or expected, as well as changes in context.
- Review involves periodic or impromptu checking of the current situation for changes in the environment, industry practices, or organizational practices. It is an activity undertaken to determine the suitability, adequacy, and effectiveness of the framework and process to achieve established objectives. Reviews should consider the outputs from monitoring activities.
- An audit is a process of evidence-based, systematic review against pre-determined criteria. While every audit is a review, not every review is an audit.

As defined in Annex E of the ISO/TR 31004 guide, risk management is (and should be) part of an organization management system. In modern organizations the integration of a risk management process into the organization's system of management should ensure that risk assessment is used as the basis for risk-informed decision-making at all levels of the organization. Many internationally recognised standards deal with management systems in general or, as readers will see, with regard to a specific content. Formalised management systems consist of requirements that provide a consistent framework in which the organization can establish sound practices and procedures to direct and control its activities. Risk management adoption in modern management systems increases the focus of top management and stakeholders on the organization's objectives and enable all risks in an integrated management system under the principles of the ISO 31000 standard that are fully compliant with the PDCA approach of the majority of recognized management systems. Nowadays it is possible to affirm that risk assessment is a formal key element of modern management systems, as a means to deal with specific complexities the organization should face to achieve the intended performance.

In the following paragraphs a glimpse of the most adopted management systems is given.

As the descriptions of those management systems show, risk assessment is an activity to be fully integrated in the management process and it comprises the core elements of the risk management process; it contains the following elements:

- An established context;
- Risk assessment (identification, analysis and evaluation);

- Risk criteria;
- Risk treatment;
- Monitoring and review;
- Communication and consultation.

These elements are pillars of each management system and, as readers will observe in the next chapters, the Bow-Tie is a simple diagrammatic way of describing and analysing the pathways of a risk from threats to impacts and reviewing controls. It allows a quantitative consistent output. According to this root cause analysis (especially in the format of barrier failure analysis) is a simple way to understand contributory causes and how the system (and associated processes) can be improved to avoid such future losses. This analysis considers existing controls/barriers (in place at the time of the deviation) and how they can be improved. It also allows a quantitative evaluation in terms of probabilities of failure on demand (PFDs) and risk-reducing factors (RRFs). Both the tools could be coupled with a LOPA, also called barrier analysis, that allows controls and their effectiveness to be evaluated. Those three techniques, referenced in IEC 31010, combined can help in satisfying the requirements of the modern management standards that specifically require, in the domain of application, a significative risk-based thinking.

If decisions arise from the consideration of results of a risk assessment, risk is a key element that can be considered common to all the management systems in place and included in each specific domain.

The "risk" element becomes part of all the core business processes: this raises and underlines the need to create interaction between all the management system approaches, among them quality, environment, compliance, energy, management, and so on, supported by their domain-specific management systems. The individual management system, considering risk management as a common and shared fundamental element, should form a single and integrated management system where a single policy could face several aspects and objectives are those of the entire organization. Integration can take advantage of a common structure of the latest edition of the domain-specific standard on management systems: this approach could lead to lean implementations and to important savings with more effective and documented results. Furthermore it is possible to notice that some threats have an impact on multiple different domains (e.g. a safety incident determining a loss in terms of people, environment, assets, business continuity, a cyber threat with an impact on different organization processes such as customer relationship management, supply chain, etc.): having a single shared organization risk register would lead to a better understanding of the risk profile of the organization, towards the adoption of an enterprise risk management framework. A single risk register could be easily maintained and updated and it could become a single point of access to all the stakeholders, eventually through specific KPIs and dashboards. In any case a single risk register doesn't request that a single methodology be used for the assessment of the various specific threats, while it could be used the receptor of normalized results, whereas, on the contrary, barrier-based risk management methods and tools (such as Bow-Tie and barrier failure analysis) could be seen as an opportunity to deal with a number of risks with a standard and ISO 31000 compliant approach, both during the design phase and over time.

Compliance with legal requirements (including an organization's internal standard) is mandatory for all the management systems: controls (barriers) must be established to ensure that it is a vital part of the management system regardless the themes to be managed. Furthermore, quality aspects should be guaranteed even in the risk management framework. Given these two examples it is obvious that risk-based thinking is the key element of various management systems, but to be successful it should be as practical as possible: if old prescriptive codes were too specific and not able to face the change, they were simple and straightforward. The success of risk-based management systems is strictly related to the underlying risk management process and its maintenance over time: non-identified risks or non-evaluated risks could pose severe impacts to a number of organization processes. Risk-based thinking and risk-based decisions require a strong commitment from leadership and the full involvement of human resources, with all the associated issues, as the need for competence and training, their periodic assessment, and so forth.

It is important to notice that risk-based standards ease this integration of risk management practices adopting a common structure, composed by a number of common elements that work in the PDCA general framework. A management system is a set of policies, processes and procedures used by an organization to ensure that it can fulfill the tasks required to achieve its objectives. These objectives cover many aspects of the organization's operations (including financial success, safe operation, product quality, client relationships, legislative and regulatory conformance and worker management). Many parts of the management system are common to a range of objectives, but others may be more specific. A complete management system covers every aspect of management and focuses on supporting the performance management to achieve the objectives. The management system should be subject to continuous improvement as the organization learns (return of experience).

Elements may include:

- Leadership involvement and responsibility.
- Identification and compliance with legislation and industry standards.
- Employee selection, placement and competency assurance.
- Workforce involvement.
- Communication with stakeholders (others peripherally impacted by operations).
- Identification and assessment of potential failures and other hazards.
- Documentation, records and knowledge management.
- Documented procedures.
- Project monitoring, status and handover.
- Management of interfaces.
- Standards and practices.
- Management of change (included project management).
- Operational readiness and start-up.
- Emergency preparedness.
- Inspection and maintenance of facilities.
- Management of critical systems.
- Work control, permit to work, and task risk management.
- Contractor/Vendor selection and management.
- Incident reporting and investigation.
- Audit, assurance and management system review and intervention.

Management systems should be documented with a specific set of evidences. The number and content of management system documents is not predefined, but rather is defined on the basis of the needs identified (in this case using a risk-based approach with the assessment of the impact following the lack of a document). The hierarchy of the selected documents is quite common and from top to bottom it is built around policies, the system manual, procedures, standard operating procedures (SOPs) and work instructions, records and forms (templates and filled forms). This generally adopted hierarchy does facilitate the integration of various management systems.

1.7.1 ISO 9001 (Quality) / ISO 45001 (Occupational health and safety) / ISO 14001 (Environment)

Among the most recognized and adopted management systems, specific attention should be given to ISO 9001, ISO 45001, and ISO 14001 management systems, which can be certified by third parties. They face aspects that are quite common to the majority of the organizations, respectively, quality, occupational health and safety, and environment. These standards will be discussed, for the goals of this book, together since, in their latest edition, they share common structure, definitions, and language.

The main objective of any modern organization is the creation of value within the company and its protection over time. The path towards this goal requires facing internal and external factors that influence, negatively or positively, the organization's ability to achieve it on a daily basis and making decisions on the consequent strategic actions to be implemented.

The decision-making tool that the organization uses to achieve this objective is the management system; according to the definition of ISO 9001 a management system is a "set of related or interacting elements ... aimed at establishing policies, objectives and processes to achieve these objectives." According to the given definition, the management system supports and guides the organization, usually its leadership, in the identification of these internal and external factors that influence its daily action, in the identification of significance, in terms of influence on the ability of each one to achieve the objective and, ultimately, as a decisional tool, in the identification of actions to be implemented to maximize positive influences and minimize negative ones.

In general, the ability, and the possibility, to achieve any objective is affected by a margin of uncertainty (and risk has been defined as the effect of uncertainty on objectives).

According to the definition of the latest version of these common ISO (9001, 14001, 45001, 50001), standards aligned with the HLS – High Level Standard Systems, the effect of this uncertainty, intended as a positive or negative deviation from what is expected (note 1), is the "risk."

Incidentally, note 4 to the definition provides a less qualitative indication of how this risk is made: "... Risk is often expressed in terms of a combination of the consequences of an event ... and the associated likelihood ... of occurrence."

The strategic action of the organization, or of its leadership and top management, must necessarily include knowledge and awareness of the uncertainty and, therefore, of the associated risk. To this end, within the harmonized approach introduced with HLS, the concept of risk-based thinking is implemented through clause 6.1, "Actions to address risks and opportunities," which we find in all standards, regardless of the specific considered domain (quality, safety, environment, or energy).

Some considerations on how to comply with the requirement:

- In all standards, the requirement is made explicit in general terms "... the organization must determine the risks and opportunities that need to be addressed ..." and therefore the organization, i.e. the top management, is basically free to adopt and apply specific methodologies, or not.
- In all standards, the organization is required to "maintain" (e.g. ISO 14001) or "preserve" (e.g. ISO 45001) documented information on the risks and opportunities that need to be addressed.
- The Appendices to the standards themselves do not provide concrete indications of the methodologies to be adopted since they directly refer to the ISO 31000 standard on risk management and related documents (as ISO/IEC 31010, which contains specific indications about methodologies and tools able to support a risk management framework or part of it, as, for example, the risk assessment step). No methodologies are mandatory, since the method, the eventual support tool, the approach (qualitative, quantitative, or mixed), the documentation, and the detail of the risk assessment should be defined considering any eventual regulation, the internal and external context, the competencies, the nature and detail of available data, the results of real incidents and near-misses, and so forth.

According to the third bullet point:

- ISO 9001 in Section A.4, Risk-Based Thinking, indicates that "... there are no requirements that require formal risk management methods or a documented risk management process....";
- ISO 14001 in Section A.6.1.1, General, indicates that "... there are no formal risk management requirements or documented risk management process....";
- ISO 45001 under Section A.6.2.2.2 OSH risk assessment and other OSH management system risks indicates that "... an organization may use different methods to assess OSH risks as part of its overall strategy to address different hazards or activities....".

From this, it would seem that it is possible to deal with risks and opportunities with any approach or even without a structured analytical approach. As anticipated, the risk approach should be defined according to a number of factors and in any case it should satisfy the requirements given by ISO 31000; effective risk management should be integrated, structured and comprehensive, customized, inclusive, dynamic and based on the best available information, coherent with human and cultural factors, and continuously improved. Nonetheless, the rules themselves require the organization to investigate the subject of risk management, based on the complexity of its activities, the context in which it operates, and the level of maturity of the company management system and to manage it with instruments commensurate with the risks.

Also in this case, the only indication is that the instrument, managerial or analytical, must be commensurate with the risks, leaving the organization free to choose, and the best reference is ISO/IEC 31010, which suggests a number of methods (describing both strengths and limitations). The risk-based thinking approach, underlying all the management systems, requires the evaluation of the influence the absence of management tools has on the achievement of objectives; it is therefore natural to look for an analytical tool for risk management, in order to limit the negative influence of its absence and evaluate objectively whether it is commensurate with the risks themselves.

A careful reading of the standards makes it possible to identify more detailed indications as to how any instruments to be used for risk management could, or should, be structured:

- ISO 9001 in Section A.4, Risk-based Thinking, indicates that "... organizations can decide whether or not to develop a more extensive risk management methodology than required ..." by the standard itself, for example "... through the application of other guides or other standards ..." (such as ISO 31000, Risk Management).
- ISO 14001 in Section A.6.1.1, General, indicates that "... the risks and opportunities related to environmental aspects can be determined as part of the assessment of significance ..." and that although it is up to the organization to select the method it uses to determine its risks and opportunities, the method selected "... may involve a simple qualitative process or a complete quantitative evaluation according to the context in which it operates. ...".
- ISO 45001 in Section A.6.2.2, OSH Risk Assessment and Other OSH Management System Risks, indicates that "... the method and complexity of the assessment (OSH risks) do not depend on the size of the organization but on the hazards associated with the organization's activities. ...".

Additionally:

- ISO 9001 in Section 6.1.2 states that "... actions taken to address risks must be proportionate to the potential impact on the compliance of products and services....".
- ISO 14001 in Section 6.1.2, Environmental Aspects, indicates that "... the organization shall determine those aspects that have or may have a significant environmental impact ..." using "... criteria established. ...".
- ISO 45001 in Section 6.1.1.1, General, requires that risks should be determined in order to "... prevent or reduce undesirable effects. ...".

All these principles lead to meeting the requirements of the standards through an analytical approach to risk management that starts from the definition of risk acceptability criteria and the identification of objective analytical tools that allow concrete demonstration of the achievement of the organization's strategic objectives. Valid help for the organization can be found in the framework proposed by ISO 31000, *Risk management – Guidelines* with clause 5 and in application guide ISO 31010, *Risk management – Risk assessment techniques*, which provides a complete overview of the main and most widely used analytical techniques for risk assessment as well as concrete indications for the choice of the most suitable technique (clause 1.1, Scope: "This International Standard is a supporting standard for ISO 31000 and provides guidance on selection and application of systematic techniques for risk assessment").

Barrier-based risk management methods (such as Bow-Tie and LOPA) as well as investigation methods (such as barrier failure analysis), given an initial hazard identification

(HAZOP/HAZID/HAZAN or structured brainstorming or even preliminary hazard analysis), are a suitable approach to implement the requirements of management systems since they operate on basic elements defining risks: possible hazards, threats, top events, associated impacts, and preventive and protective controls (or barriers, or independent protection layers). Furthermore, periodic risk monitoring could be associated with the status of the controls, as the investigation of the root causes of a (real) deviation from intended conditions could be associated with the failure evidences of one or more barriers. This allows risk management, over time, the ability to take into account human factors (as barriers, threats, escalation factors and vulnerable targets given an exposure level, i.e. the condition of being unprotected from hazards, thus leading to risk), and the chance to conduct qualitative and/or quantitative risk assesssment and modify the risk register on the basis of the results of the management of change (MOC) process. Such implementations are able to raise the concept of maturity model of the system considering the risks and their modification in terms of RRFs associated to the control measures.

Irrespective of the actual method used it is important to recognize human and organizational factors, since often investigations reveal that these factors (including individual factors, working environment, organizational factors and physical factors) play a fundamental role in the origin, the development and also the outcome of events.

Also, a risk model can be built with the aforementioned methods regardless of the type: with a consistent single approach it is possible to assess strategic risks (they can have an impact on the organization's ability to achieve its overall objectives and goal), compliance risks (they refer to the organization's exposure to legal penalties and economic losses as a result of non-compliance with applicable laws, regulations and even internal policies) and operational risk (they refer to losses resulting from inadequate procedures, policies, systems, and processes). This inner connection among barrier-based methods and risk management elements allow the organization to have comparable and reproducible results to be discussed with operators, to be illustrated to the stakeholders and to be demonstrated to authorities having jurisdiction and/or third parties (as certification bodies). Comparable results is a key element for prioritization of risks: the organization, given limited resources, needs to prioritize the risk treatment in a way that it considers whether the impact of risks is low or high. This prioritization activity compares scenarios while using multiple criteria (including cost-benefit analysis, feasibility, regulation constraints, level of concern of stakeholders, etc.). Bow-Tie, supported by a quantified LOPA approach, could help in comparing different strategies (different pools and combinations of barriers) and ante, post-modification situations.

Since every management system operation requires the definition of specific roles and responsibilities along with knowledge management activities and competence and training, barrier-based method notation could be easily employed to increase risk awareness and behavioural skills of the stakeholders at various level. The appropriate involvement of people, whose competencies are being developed as part of the training process (educational process aiming to develop knowledge, skills and behaviour in the specific management system domain), may result in them feeling a greater sense of ownership of the process and they will assume more responsibilities. In some cases Bow-Tie diagrams and BFA schemes are used to train people in tabletop exercises: this also guarantees the participation of people with different training levels and different competencies to risk review activities, increasing communication (internal and external) and awareness, and even to comply with specific legal requirements (e.g. safety risks communication, environmental policies, food quality inspections).

Given this, it appears clear that simplicity of such methods could better comply with the documentation requirements of the management systems: if documentation is not mandatory in extension (there is no specific requirement on how to document processes and operational measures and control but documented information is highly valued) it should be available and suitable for use, where and when it is needed. If documents embrace the risk concepts they could be reduced to a minimum and the organization could easily address the following activities:

- Distribution, access, retrieval and use;
- Storage and preservation;
- Control of changes (according to the MOC process);
- Retention and disposition (according to the risk life cycle in the risk register).

In this way the document life cycle aligns with the risk life-cycle described in ISO 31000 with a strong emphasis on the risk-based approach of the latest edition of the ISO standards.

Risk assessment and its documention become the input element for a series of other documents/processes:

- An emergency and response plan should be prepared against the scenarios defined in the risk assessment phase and periodic drills should validate the PFD of the mitigative barriers that are resources for the emergency response.
- Procurement activities related to critical elements (preventive and mitigation barriers) should comply with performance standards that maintain the PFD declared in the risk assessment.
- Competency and training of contractors with a critical role in the risk diagram are fundamental.

All the processes should be defined and assessed in terms of risk, but also monitored and reviewed over time. These pillars of the management systems could be achieved linking KPIs to barriers and to pathways from threats to impacts, as well as linking root cause analysis results to barriers to identify weak points in the processes and activate preventive/corrective actions and general improvement of the management system in the spirit of the PDCA global framework, to be discussed in the periodic management review, in which, via a risk-based decision process, it is possible to define any eventual needs to mitigate the risk of deviation from defined processes goals and strategic objectives of the entire organization via a specific action plan, whose results could be simulated via a generation of specific diagrams to compare actual and future configurations of the controls in the overall picture, along with the associated RRFs, resources, responsible people and documents. Improvement of the management system does not only rely on documentation update but also in factors of change consideration (specifically those affecting the risk register and the risk assessment). Continuous improvement leads to:

- Better process performance, enhanced organizational reputation and increased stakeholder satisfaction;
- A greater ability to react to risks (even new and emerging);
- Ensuring that people acquire the required competency level;
- Recognition and acknowledgement of performance improvements.

1.7.2 Industrial Safety (Major Accidents)

Barrier-based safety management systems are now widely adopted because of the advantages of their barrier-based perspective. Among the different methods that are available to implement them, the Bow-Tie is one of the most recognized ones. Leveraging its persuasive visual communication, it allows all the stakeholders of the risk management process to have, in a single shot, the complex picture of the relations linking threats, preventive barriers, top events, mitigative barriers, consequences, and escalation factors. There is a strong relationship between the four pillars of process safety management systems (commitment to process safety, understanding hazards and risks, risk management, and learning and improvement), the standard steps of the Deming safety life cycle and the Bow-Ties, also making use of the first official guidelines by AIChE CCPS. The payoff is tremendous: risk assessments come to life. Instead of being forgotten and archived, risk assessments are used because they are relevant in day-to-day operation. Thanks to the "Work Bow-Tie," operators know exactly the risks they are going to face when performing ordinary or extraordinary maintenance, also helping significantly in the evaluation of interferential risks and work-permit management. Furthermore, the aggregation of various data sources allows a level of understanding and insight into risks, which is unprecedented in risk management until now. The "Do" phase is followed by the "Check" one. This is when the organization looks for occasions for improvement: internal audits and incident analysis are part of this step. Their results are enhanced and appreciated thanks to the Bow-Tie approach, because they are now performed from a barrier-based perspective, meaning that the barriers are monitored in their effectiveness (audits) and eventual failings (incident analysis). This results in positive feedback for the "Plan" step of risk assessment that is now adjusted with the data coming from the real operative experience of the organization. Finally, audit recommendations, safety observations and incident actions are then evaluated and implemented during the "Act" phase, and the risk assessment is once again updated accordingly. The EU Directive on Major Accidents Prevention (also known as the Seveso Directive), EU Directive 2012/18/EU dated 4 July 2012, request for operators to have a specific major accident prevention policy (MAPP) supported by an appropriate safety management system (SMS) for controlling major accident hazards. This SMS has specific pillars, very similar to those of the OSHA process safety model: organization and personnel, identification and evaluation of major hazards, operational control, management of change, emergency planning, performance monitoring, audit and review activities. In order to demonstrate that everything necessary has been done to prevent major accidents, and to prepare emergency plans and response measures, the operator should provide the competent authority with information in the form of a safety report (updated on a periodic basis, usually every five years). This safety report should contain details of the plant, dangerous chemicals and their quantity, possible major accident scenarios and risk assessment, prevention and intervention measures (controls or barriers), and the SMS in place in order to prevent and reduce the risk of major accidents. Minimum data to be considered include, in a risk management approach: identification and accidental risk analysis and prevention method, major accident scenarios and their likelihood and causes (operational, external and domino, Na-Tech, cyber) along with the consequences, review of past accidents and incidents (including near-misses), controls in place (preventive and protection measures), and emergency plan. Authorities having jurisdiction should verify that the risk-reducing factors (RRFs) declared in the safety report (by the plant owner) are valid and maintained during the life cycle of the plant, by the support of the safety management system put in place. Declared PFD of the barriers shouldn't exceed data used in the risk assessment. Risk management workflow is completely inspected with two different audits: on the safety report content (assumptions, results, risk mitigation plans) and on the SMS in place (is the major accident risk from the safety report kept under control over time?). Since risk management also fails when a modification is put in place, MOC should be considered an ineludible part of the management system. Authorities should verify that the MOC process is included and actuated for each plant modification (even temporary) and that the plant life cycle coincides with the control measures (barriers) life cycles. In their safety report plant operators should answer three specific questions:

1) Do we understand what can go wrong?
2) Do we know what our systems are to prevent this happening?
3) Do we have information to assure us they are working effectively?

These answers are strictly related to threats and barriers in place to control and possibly avoid an impact or loss to vulnerable targets (people, environment and assets in Seveso plants) and the combination of the safety report with SMS is a good example of applied barrier thinking and risk management over time. In fact, an effective way to obtain a low probability of a major accident occurring is to use a system composed of multiple levels of protection, also called "defense in depth" and well represented with the Swiss Cheese Model by Reason, applied, as exemplification, to a major industrial event in Figure 55.

The LOPA approach is a quite common technique used during scenario screening. LOPA allows the assessment of the IPLs for each "threat-impact" pair that define a scenario. IPLs considered are those that control the deviation from the normal behaviour of a process value that is governed by the process control layer: process alarms, trip level alarms, emergency shutdown systems and safety instrumented functions, mechanical devices (relief

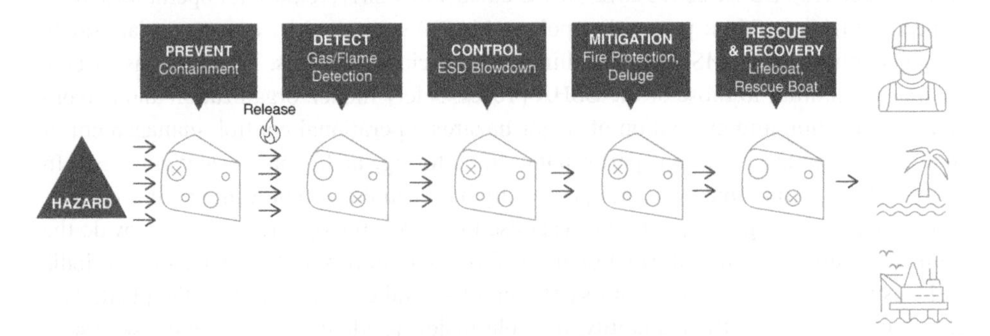

Figure 55 Swiss Cheese Model applied to a major industrial event.

valves, rupture disks), passive protection layers (dikes), emergency response layers (emergency plan, fire protection systems, etc.). A safety management system strategic goal is to keep barriers in place over time. Barriers could fail and the human factor plays a fundamental role, and each barrier could be assigned a specific PFD depending on a condition of the barrier. Human response to alarm, for example, is the first line of defence: it can be linked to a RRF of 10 (PFD = 0,1) in case specific conditions are verified for an effective control room operator response: written procedures are available, an operator is present 24 hours a day, and the operator has an indication of the problem, has enough time to act, andhas received proper information, education, and training on a periodic base.

The plant owner should demonstrate the availability of barriers and associated RRFs, as they have been identified and declared in the risk assessment of the safety report; the owner could use recognized and generally accepted good engineering practices (RAGAGEP) standards. Barriers, during inspections by the authorities, are verified in terms of condition and associated (declared) RRF/PFD, since these data come from a performance- and risk-based approach (analysis, design and operation), eventually integrated with the life cycle of RAGAGEP requirements in the framework of a specific organization culture, as recommended by the fundamental principles of ISO 31000.

1.7.3 Functional Safety and RAGAGEP Standards

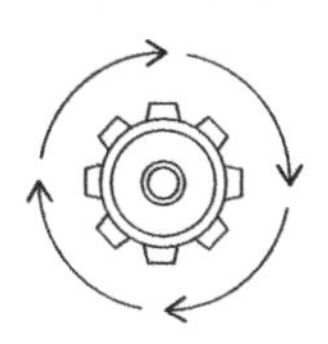

When it comes to process safety, most companies focus on the functional safety life cycle and compliance with the international standard IEC61511, "Functional safety: safety instrumented system for the process industry sector." The functional safety life cycle (SLC) provides the foundation for building a set of processes and procedures that support risk reduction and risk management, as defined within the standard itself. If followed correctly, the SLC provides a consistent methodology for achieving repeatable results, based upon a set of performance targets to manage and mitigate process risk within acceptable boundaries. The SLC can be applied to any hazardous process, whether it be chemical, petro-chemical, oil & gas, pharmaceutical, food and beverage and/or machine applications. A key objective of IEC61511 is the definition of the safety integrity level (SIL), 1/PFD, to be associated to critical safety instrumented functions (SIFs) employed to mitigate the industrial risks together with the other IPLs in place. Examples of SIFs are alarm systems, emergency shutdown systems, fire and gas detection systems, interlocks, andso on. Risk reduction factors needed to achieve an acceptable risk (with an ALARP approach) should be maintained over time considering architectural redundancy, inspection and test intervals, proof tests, maintenance operations, and so on. Among the methods suggested, implementing the SLC Bow-Ties and LOPA assessments plays a fundamental role as risk control focused methods. However, with the advent of the industrial internet of things (IIOT) and the growing use of wireless technologies, it is becoming more important to consider cybersecurity and the consequences of control and safety systems being compromised due to a cyber-related incident.

Similarly, if there is no defined alarm philosophy and rationalization of alarms, operators can miss vital alarms and lose valuable time in being able to respond to an incident due to the volume of alarms being generated. Accidents, such as the Buncefield and Texas City explosions, highlighted the problem, with operators being confused by too many alarms. Since the standards for functional safety (IEC 61511), alarm management (IEC 62682) and cybersecurity (IEC 62443) all follow a similar phased lifecycle (PDCA), it makes sense to consider all three together when it comes to process safety. As such, an integrated life cycle approach will help in reducing overall operational costs via increased efficiencies. The success of this implementation relates directly to the company's maturity and culture when it comes to adopting industry best practices. Synergistic benefits can be gained by considering all three together, when combined with the company's maturity level, that is strictly related to the quality of the culture and of the risk management process of the company. In the framework of an advanced maturity model in risk management practices, common elements of life cycles (usually built on a three-phase approach Analysis-Design-Operation similar to the more general PDCA), coupled with requirements of RAGAGEP guidelines could lead to an holistic composition, as described in Figure 56, where alarm management, cybersecurity and the functional safety life cycle are combined to support the plant operator in the management of the industrial risk associated with his plants and assets. In this sense:

1) Adopting RAGAGEP standards is the best way to assure a holistic composition towards an advanced maturity model.
2) Plant life cycle should be verified against the pillars of the SMS.

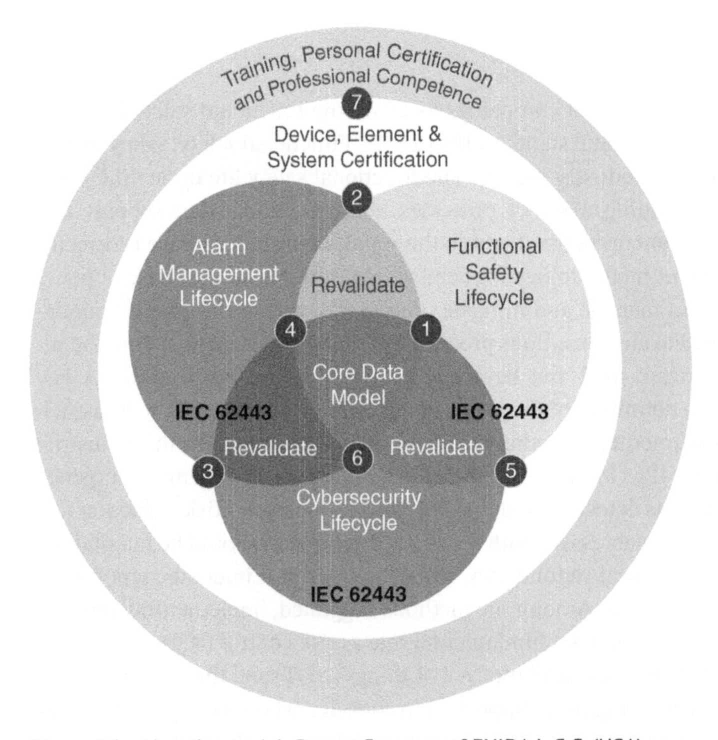

Figure 56 Maturity model. *Source:* Courtesy of EXIDA L.C.C. (USA).

3) Barrier life cycles should comply with the life cycles requested by RAGAGEP standards.
4) Plant owners could take advantage of a single integrated approach to deal with known risks and face emerging risks (e.g. cybersecurity).
5) RAGAGEP standards and supporting methods or tools could demonstrate the results to AHJ: credible, trackable, documented and justified performances in a risk-based framework coherent with ISO 31000 risk management principles.

1.7.4 ISO 55000 (Asset Management and Integrity)

ISO 55000 is an international standard covering management of assets of any kind. Before it, a Publicly Available Specification (PAS 55) was published by the British Standards Institute (BSI) in 2004 for physical assets. This standard focuses on helping the organization develop a proactive life-cycle asset management system. This supports optimization of assets and reduces the overall cost of ownership while helping the organization to meet the necessary performance and safety requirements. An asset management system provides a structured, best-practice approach to managing the life cycle of assets, with:

- Reduced risks associated with ownership of assets, including safety-related risks and business continuity risks;
- Improved quality assurance for customers/regulators, where assets play a key role in the provision and quality of products and services;
- New business acquisition – stakeholders recognize that top management put in place a strategy to ensure that assets meet the necessary safety and performance requirements (including environmental compliance).

This standard complies with the ISO 72 guidelines and justification for the development of management system standards. These guidelines outline the common elements of policies, planning, implementation, operation, performance evaluation, improvement and review by management. It also establishes that the management systems should be developed under the PDCA approach. ISO 55000 highlights that assessing the value of the organization's assets takes risk into consideration (as defined in the ISO 31000 referenced standard). It's important to underscore that ISO 55001 is not a standard on reliability and maintenance management (better described in a number of other documents, guidelines, books, etc.), but that doesn't mean that maintenance and reliability doesn't have an important role in it. Annex A (informative) for example explicitly mentions condition monitoring, life-cycle cost, non-destructive tests, and so on. Asset management, as per ISO 55000, covers the asset's life cycle, which considers all the different life steps of an asset: concept, design, procurement, construction/installation, commissioning (including pre-startup safety reviews), operation, maintenance, decomissioning and final disposition. In all the identified phases, a proper MOC process should be in place and documented.

ISO 55000 is applicable to any industry sector and considers different types of assets: material assets such as infrastructure, plant, equipment, buildings and other tangible objects, but also intangible assets such as good-quality business data, information systems, licenses, and other intangible assets such as brand, reputation, image, and customer loyalty of the organization.

However, physical assets are the lifeblood of all production processes. If a critical asset should fail, it could not only cause a security or environmental problem, but also disrupt business until it is repaired or replaced. By applying the standard, companies can gain a complete view of the integrity of the entire plant by removing the watertight compartmental approaches that exist in many facilities. They can also create a strategic plan for resource utilization and maintenance so that repairs and replacements are scheduled with minimal disruption to production.

It is quite obvious that integrity is a performance of the asset that should be maintained during the lifetime of an asset: in some cases a loss of integrity could result in severe impacts (e.g. loss of containment of vessels operating with a toxic chemical in a chemical facility, loss of stability of a bridge, etc.). Risk assessment plays a fundamental role in asset management in all its components: initial and periodic risk assessment and root cause analysis of real incidents to understand the causal relationship with the controls in place to avoid or mitigate the impact. Also, a single top event (loss of containment or integrity) can be related to threats involving different specific domains such as loss of containment due to overpressure, ageing, or human error. A top event could lead to an impact on several vulnerable receptors (people, environment, or assets due to domino effects). Given this example, it is clear that asset management could gain a serious advantage from a barrier-based approach that, for each control in place, defines failure mechanisms that determine the probability of failure on demand, to be related with the likelihood of the impacts. But, due to the complexity of asset-related threats and their evolution in impacts, the advantage could also be recognized in the ability to consider human factors, design (poor) conditions, time-related mechanisms to be monitored during time (e.g. corrosion, erosion, fatigue, creep, inspections effectiveness, etc.), escalation and contributing factors (as well as seasonal risk) or conditional modifiers (as those used in LOPA), and so forth. This approach, often represented via a fishbone diagram, has been effectively used to describe the portion of an asset management system, integrated with Seveso risk assessement, related to the ageing issues for integrity of vessels and equipment in chemical plants.

1.7.5 ISO 22301 (Business Continuity)

ISO 22301, *Societal security – Business continuity management systems – Requirements*, is an international standard related to business continuity management (BCM), which defines the requirements necessary to plan, establish, implement, and operate a documented management system, and to monitor, maintain, and continuously improve the management system to protect, reduce the possibility of occurrence, prepare, respond to, and restore destabilizing events for an organization when they occur.

The standard specifies requirements to implement, maintain, and improve a management system to protect against, reduce the likelihood of the occurrence of, prepare for, respond to, and recover from disruptions when they arise.

The requirements specified in the standard are generic and intended to be applicable to all organizations, or parts thereof, regardless of type, size, and nature of the organization.

The extent of application of these requirements depends on the organization's operating environment and complexity.

This document is applicable to all types and sizes of organizations that:

- Implement, maintain and improve a BCMS;
- Seek to ensure conformity with stated business continuity policy;
- Need to be able to continue to deliver products and services at an acceptable predefined capacity during a disruption;
- Seek to enhance their resilience through the effective application of the BCMS.

The standard helps organizations to:

- Identify and manage current and future threats to business processes or to the entire organization;
- Adopt a proactive approach to minimize the impact of accidents;
- Keep critical functions active during periods of crisis;
- Minimize downtime during accidents and improve recovery time;
- Demonstrate the company's resilience to customers, suppliers, and requests for quotation.

Therefore there is clearly a link between business continuity issues and risks and a strict correlation among the risk register and the business impact assessment (BIA).

The organization is requested to define, implement, and maintain a formal and documented process for business impact analysis "AND" risk assessment that:

- Establishes the context of the assessment, defines criteria, and evaluates the potential impact of a disruptive incident;
- Takes into account legal and other requirements to which the organization subscribes;
- Includes systematic analysis, prioritization of risk treatments, and their related costs;
- Defines the required output from the business impact analysis and risk assessment, and specifies the requirements for this information to be kept up-to-date and confidential.

The organization, with reference to the risk management system supporting the BCMs, is requested to:

- Identify risks of disruption to the organization's prioritized activities and the processes, systems, information, people, assets, outsource partners and other resources that support them;
- Systematically analyze risk;
- Evaluate which disruption-related risks require treatment;
- Identify treatments commensurate with business continuity objectives and in accordance with the organization's risk appetite.

BCM is a vital process to assure resilience over time: since disruptions could be determined by new and emerging risks it is fundamental to keep in place a dialogue with the risk management process and also learn from experience, even considering soft events, and performance monitoring during real events and drills. A barrier-based approach can induce the assessment of common cause failures that could affect more organization processes or determine the failure of several controls.

In this particular domain field, an easy notation, like the notation used by Bow-Tie/LOPA and by BFA, could be very effective in communicating relationships and describing business continuity procedures and in defining an incident response structure against identified disruptive events, with all the documents, roles and responsibilities associated (these elements guarantee the RRF associated to each control). Results of the assessment can be documented in business continuity plans and exercising and testing plans.

1.7.6 ISO IEC 27001 (Information Security)

ISO/IEC 27001, *Information Technology – Security Techniques – Information Security Management Systems – Requirements*, is an international standard that defines the requirements for setting up and managing an information security management system and includes aspects related to logical, physical and organizational security. Since information is an asset that adds value to the business, and since most information is now stored on computer media, every organization must be able to ensure the security of its data, in a context where the IT risks caused by breaches of security systems are constantly increasing. The objective of the most recent edition of the ISO 27001 standard is precisely to protect data and information from threats of any kind, in order to ensure its integrity, confidentiality and availability, and to provide that the requirements adopt an adequate information security management system aimed at a proper management of sensitive company data. The standard is applicable to companies operating in most commercial and industrial sectors, such as finance and insurance, telecommunications, services, transport, and government sectors. The approach of the ISO/IEC 27001 standard is consistent with the general framework ISO 31000 on risk management, based on the process approach, structured in safety policy, identification, risk analysis, risk assessment and treatment, risk review and re-evaluation, the PDCA model, use of procedures and tools such as internal audits, nonconformities, corrective and preventive actions, and surveillance, with a view to continuous improvement.

In the design phase of the information security stategy, ISO 27001 requires a risk assessment, which can be schematized in the steps well described in the ISO 31000 standard:

- Risk identification;
- Analysis and evaluation;
- Selection of control objectives and control activities for risk management;

- Assumption of residual risk by management (calculated via an ALARP approach);
- Definition of the Statement of Applicability.

The last point specifies the control objectives adopted and the controls implemented by the organization with respect to a list of control objectives provided by the standard itself. Fundamental to the standard is the informative Annex A, *Control objectives and controls*, which contains the 133 controls with which the organization that intends to apply the standard must comply. They range from security policy and organization to asset management and human resources security, from physical and environmental security to communications and operational management, from physical and logical access control to incident monitoring and handling (related to information security). Management of business continuity and regulatory compliance completes the list of control objectives. The organization must justify which of these controls are not applicable within its ISMS, for example, an organization that does not implement e-commerce within its ISMS may declare as not applicable the controls 1-2-3 of A.10.9 that relate to e-commerce activities.

1.7.7 ISO 19011 (Audit)

ISO 19011 provides guidance on the audit of management systems, including the principles of audit activity, management of audit programs, and the conduct of management systems audits, as well as a guide for assessing the skills of the people involved in the audit process. These activities include the person(s) managing the audit program, auditors, and audit teams. The standard is applicable to any organization that needs to plan and conduct internal or external management system audits or to manage an audit program. Since the publication of the latest edition of this document, a number of new management system standards, many of which have a common structure, identical basic requirements, common terms, and fundamental definitions. It follows the need to consider a more extensive approach for the audit of management systems, as well as to provide a more general guide. The biggest differences compared to the previous edition are:

- The addition of the risk-based approach in the principles audit;
- The extension of the program management guide;
- Audits, including the risk of the audit program;
- The extension of the audit conduct guide, in particular the section on audit planning;
- The extension of the general competence requirements of auditors;
- The arrangement of terminology;
- The extension of Annex A to offer guidance on the (new) audit concepts, such as context of the organization, leadership and commitment, virtual audits, legislative compliance, and the supply chain.

The organization should conduct internal audits at planned intervals to provide information on whether the management system(s) in place conforms to the organization's own requirements and the requirements of the reference standard and to provide information on whether the system is effectively implemented and maintained over time.

In this sense the organization should:

- Plan, establish, implement, and maintain an audit programme(s), including frequency, methods, responsibilities, planning requirements, and reporting. The audit programme(s) shall take into consideration the importance of the processes concerned and the results of previous audits.
- Define the audit criteria and scope for each audit.
- Select auditors and conduct audits to ensure objectivity and the impartiality of the audit process.
- Ensure that the results of the audits are reported to relevant management.
- Retain documented information as evidence of the implementation of the audit programme and the audit results.

The risk-based approach should influence the planning, conducting, and reporting of audits, in order to ensure that the latter are focused on issues that are significant for the audit client and to achieve the objectives of the audit program. One of the main goals of an audit is the maturity model of the organization in the risk management workflow, especially in the risk assessment steps and in the following activity to define actions, plans, and priorities. Since audit is a fundamental activity that guarantees the monitoring and review of the maturity model, it should be conducted in a way that critical elements of the management systems (and related recordings) are verified. In a risk-based approach to management systems for each scenario, the organization should identify control elements (preventive and mitigative barriers). During audit the inspection team should verify the controls identified by the risk assessment: presence and correct operation (as intended). They also should verify any eventual records related with near-misses and incidents in order to associate the root cause analysis results with the affected barriers and following decisions to avoid a recurrence. A risk-based audit is quite different from a traditional audit, the latter aimed to the exclusive and simple verification of the satisfaction of precise and predefined requirements. A risk-based audit is more complex since with document sampling and discussions the inspection team should evaluate the entire maturity model of the audited management system/systems and the underlying risk management common framework. This requirement calls for a different competency level of the auditors, who should also be proficient in risk management principles. While traditional audits were characterized by a single-focus approach on a specific aspect (e.g. training and competence, maintenance, work procedures, time programming, management review), risk-based audits are based on performance and are focused on controls or barriers identified as critical (and eventually recorded in a specific register) and on their failure mechanisms, since PFD is strictly connected with the sum of the causes (internal, external, common, etc.) that can determine partial or total failure, considering also human factors. Considering the results of a fire risk assessment, the internal emergency plan is a critical protection barrier since it helps in reducing the impact of a fire scenario to people, assets, and the organization's business continuity. This barrier could fail due to different causes, related to different aspects: the emergency plan is not available (work procedures), the emergency team doesn't know the plan (training), periodic drills are not organized (time programming), the plan is not updated (maintenance), the operator does not have the skills or time to put in place emergency procedures described in the plan (human factor), or the plan is available but its content is wrong (risk assessment). Given this example, the audit process is an opportunity to review the risk management process focusing on the elements that have been identified by risk assessment as critical elements for the organization.

1.7.8 ISO 39001 (Road Traffic Safety)

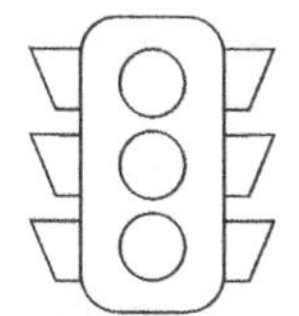

Road traffic safety is a global and well known problem. Road accidents and accidents at work are closely related: in Italy more than half of the accidents at work recorded by the Italian National Institute for Workplace Incidents Insurance (INAIL) statistics in 2011 (436 deaths, equal to 50.6% of the total) belong to the road traffic categories or ongoing travels, with or without means of transport. Alongside the growing attention to health and safety issues in the workplace, which has seen an increase in laws and decrees in recent years, with the increase in road traffic there has also been a growing awareness of road safety, both on the part of institutions and road users. ISO 39001 certification provides a global approach to road safety. ISO 39001 defines the necessary and useful elements for proper management of good practices aimed at road safety, with a specific focus on the actions taken and the expected results achieved with a view to improving prevention. The road traffic safety management system can be integrated or made compatible with all other management systems. The certification against ISO 39001 consists of assessing the dynamics of the organization's processes with respect to the transport system and road safety, verifying the acceptability of risk exposure and analyzing the actions taken in order to reduce both the probability of incurring a road accident and the severity of the same. Safety is strictly related to organization and its context; a special case are organizations where transportation is a usual activity of core business processes. The organization shall determine external and internal issues that are relevant to its purpose and that affect its ability to achieve the intended outcome(s) of its road traffic safety management system. In particular, each organization should identify its role in the road traffic system, identify the processes, associated activities and functions of the organization that can have an impact on road safety and also determine the sequence and interaction of these processes, activities and functions. The organization should prepare and update a risk assessment with the actions needed to prevent, or reduce, undesired impacts. A fundamental role in this initial risk assessment process is the identification of risk exposure factors, risk outcomes (consequences in terms of fatalities and injuries), and barriers available to prevent and mitigate the consequences such as:

- Road design and safe speed, especially considering separation (oncoming traffic and vulnerable road users), side areas and intersection design;
- Use of appropriate roads, depending on vehicle type, user, type of cargo and equipment;

- Use of personal safety equipment, especially considering seat belts, child restraints, bicycle and motorcycle helmets, and the means to see and be seen;
- Using safe driving speed, also considering vehicle type, traffic and weather conditions;
- Fitness of drivers, especially considering fatigue, distraction, alcohol and drugs;
- Safe journey planning, including consideration of the need to travel, the amount and mode of travel and choice of route, vehicle and driver;
- Safety of vehicles, especially considering occupant protection, protection of other road users (vulnerable as well as other vehicle occupants), road traffic crash avoidance and mitigation, roadworthiness, vehicle load capacity and securing of loads in and on the vehicle;
- Appropriate authorization to drive or ride the class of vehicles being driven or ridden;
- Removal of unfit vehicles and drivers or riders from the road network;
- Post-crash response and first aid, emergency preparedness, and post-crash recovery and rehabilitation.

Controls should be used in the risk assessment and this initial activity should be completed by a specific training and awareness program and by an incident investigation of negative impacts (road traffic crashes and incidents). Investigations should understand missing, poor, or non-effective controls as well as determine the underlying factors that the organization can control and/or influence that might be causing or contributing to the occurrence of those incidents.

1.7.9 ISO 19600 (Compliance Management Systems)

Organizations that aim to be successful in the long term need to maintain a culture of integrity and compliance, and to consider the needs and expectations of stakeholders. Integrity and compliance are therefore not only the basis, but also an opportunity, for a successful and sustainable organization. Compliance is an outcome of an organization meeting its obligations, and is made sustainable by embedding it in the culture of the organization and in the behaviour and attitude of the people working for it. While maintaining its independence, it is preferable for compliance management to be integrated with the organization's financial, risk, quality, environmental, and health and safety management processes and its operational requirements and procedures. An effective, organization-wide compliance management system enables an organization to demonstrate its commitment to compliance with relevant laws, including legislative requirements, industry codes, and organizational standards, as well as standards of good corporate governance, best practices, ethics, and community expectations. Embedding compliance in the behaviour of the people working for an organization depends above all on leadership at all levels and clear values of an organization, as well as an acknowledgement and implementation of measures to promote compliant behaviour. Non-compliance risks could pose a number of consequences to business continuity, to the organization leadership on safety (e.g. safety protection measures are not compliant with specific laws). Compliance with applicable rules, laws and regulations, but also with standards in general terms, assure risk reduction; therefore, compliance risk should be considered in the assessement of failure mechanisms of barriers, including the human factor (e.g. failure in period training of emergency team staff could result in a fine but also in a reduced performance of the emergency team during a fire). Compliance risk can be characterized by the likelihood of occurrence and the consequences of non-compliance with the organization's compliance obligations (including mandatory and volountary commitments).

2

Bow-Tie Model

2.1 Hazards and Risks

In everyday language the terms "hazard" and "risk" are often used synonymously, but they are not; they refer to completely different aspects. It is therefore fundamental to give their different definitions. A *hazard* is an action that has the potential to cause harm to human health or the environment or economic loss. A *risk*, instead, is a measure to express the probability that a certain event, in general terms an unwanted event, appears with a specific magnitude (that is to say, its level of severity). Being a combination of these two factors, risk can be expressed in a formula as follows:

$$\text{Risk} = \text{Event Likelihood} \times \text{Event Magnitude}$$

The hazardous nature, if we take the industrial context as an example, generally refers to the safety issues connected with used substances, including toxic and flammable ones. Taking inspiration from AIChE-CCP (2016), process hazards include chemical reactivity hazards, transportation of chemicals, static electricity, material properties concerning fires and explosions development, and many others. Several factors can lead to a hazard: equipment failures, human factors, or operational or managerial problems. Similar to the entropy law, the investigated systems tend to increase their "disturbance," causing an accident to occur. This happens until a sufficient amount of energy is added to the system under consideration: in our case, this "energy" is the object of the risk assessment and its management (Assael and Kakosimos, 2010). Performing a risk management means:

- Identifying the hazards involved with the specific chemical plant, including a prevision of the incident scenarios;
- Analysing the risks related to the identified scenarios, including evaluating the consequences of the scenarios;
- Developing the safety management system (SMS), in order to prevent the identified scenarios and/or mitigate their consequences.

Examples of some methods related to the probability of occurrence (like the fault tree analysis or the event tree analysis) are described in this Chapter, whereas the broad topic of the human factor effects is discussed in Chapter 1 and in Appendix 3.

Bow-Tie Industrial Risk Management Across Sectors: A Barrier-Based Approach,
First Edition. Luca Fiorentini.

Qualitative techniques are used by the experienced team to evaluate the hazards of an existing technology, with documented past experience. Differently from the other methods, including the Bow-Tie, they do not generally allow the conceptual link between risk assessment and incident investigation.

When talking of hazards and risks, the contraposition between the two ancient conceptions of the word must be mentioned. On the one hand, there is the deterministic approach, offspring of the scientific method where everything is rigorously obtained from logic processes in a complete and satisfactory way. It is the method of the time reversibility where the objects – intended as physical reality – are put in the centre of the reasoning. On the other hand, there is the probabilistic approach, offspring of the complex theory where certainties drop in favour of a likelihood of occurrence, of a limited knowledge because relations among objects are now preferred rather than their physical reality. As a consequence, the concept of *uncertainty* strongly imposes itself and leads to another important related concept: the one about risk. *Risk* can be defined as a measure of economic loss, human injury, or environmental damage or reputation regarding both the incident likelihood and magnitude of the loss, injury, or damage. To establish the likelihood of an event occurring, a frequency assessment is performed, while the definition of the magnitude requires a consequence assessment. The consequence is the ultimate result of an initiating event, a single deviation or multiple deviations, intended as a change in a state beyond specified limits, conditions or status (whose boundaries are monitored by the performance indicators).

It is interesting to note how the perception of risks may be different from how they actually are (Sanders, 2015). Most people fear trivial risks and underestimate the significant dangers of everyday life. Probably, this happens because risk perception is driven by emotions, being the human response guided by survival. To have an idea, according to the United States Department of Labor (Bureau of Labor Statistics), it is safer to work in a US chemical plant than at a grocery store. Indeed, the chemical industry established excellent safety records in the past decade. It is impressive how dangerous it is to be a timber cutter, a fisher, or a structural metal worker. The approach to industrial risks requires an open mind, free from prejudices. This is especially required of a forensic engineer, in order to carry out a correct investigation following a real event.

The risk acceptability is a criterion intimately connected to both the company policies and the compliance with the national laws and the technical standards worldwide recognized. Thus different companies may or may not accept the same risk, depending on their own managerial choices, even if a minimum level of risk acceptability comes from the company's compliance with standards.

2.2 Methods of Risk Management

This volume is intended to be a guide to risk management principles within organizations, especially industrial ones, and across different sectors. Risk assessment is a fundamental element of the entire organization divided into its process (main ones and auxiliary ones) and it needs special attention, including the aspects associated with the incoming elements and the findings to be made available for informed decisions.

As ISO 31000 outlines the principles and guidelines for risk management, the associated IEC/ISO 31010 technical standard reviews some techniques to be used for risk assessment. The focus, therefore, is on that phase of the risk management process that aims to answer the following questions:

- What can happen and why (risk identification phase);
- The consequences and probabilities of the hypothesized events and the factors that mitigate their severity or reduce their probability (risk analysis phase);
- What the final risk level, in relation to the tolerability and acceptability thresholds used (risk evaluation phase), is.

These steps are the fundamental activities in a risk assessment.

ISO 31010 is general in its nature, so it is a guide to be used for any type of organization, regardless of the type of risks being assessed. There may be specific standards for particular areas of industry, where some risk assessment methodologies are indicated that are preferable to adopt, for which a satisfactory level of consensus has been achieved among professionals working in that specific industrial sector. If these standards are consistent with the contents of ISO 31010, then it is generally sufficient to adopt the specific standard.

ISO 31010 fully supports ISO 31000, meaning that a risk assessment conducted in accordance with ISO 31010 contributes to the effective and efficient satisfaction of the other risk management process activities specified in ISO 31000. The ISO 31010 standard introduces a number of techniques, with specific references to international standards where concepts and the application of techniques are described in greater detail.

It should be made clear that ISO 31010 does not provide specific criteria for identifying the need to perform a risk analysis, nor does it specify the type of risk analysis method to be adopted for a particular application.

The standard also does not list all available techniques, and excluding a technique from ISO 31010 does not mean that it is invalid. Moreover, the fact that a particular technique is applicable to a given circumstance does not imply that that particular technique must necessarily be applied.

Like ISO 31000, ISO 31010 is not explicitly about security but, more generally, about any risk to the organization, although security references are sometimes more cited within standards for purely informational purposes and not to define the scope of the requirements and information reported there.

The following will show some of the risk assessment techniques in ISO 31010. Their first classification concerns the possibility of being applied to a particular stage of the risk assessment process. Techniques applicable to the preliminary identification phase of hazards, those at the risk assessment phase, and, finally, risk assessment techniques will be distinguished. The selection of risk assessment techniques to be used can be influenced by several factors, such as:

- The complexity of the problem and the methods required to analyze it;
- The nature and degree of risk assessment uncertainty, measured by taking into account the quantity and quality of the information available and the uncertainty requirements required to meet the objectives;

- The availability of required resources, in terms of time, level of experience, data, and costs;
- The need to have quantitative output.

It should be made clear from the outset that the classification proposed here is not to be understood as absolute, but is presented only for educational purposes. In fact, as also mentioned in Table 1 of ISO 31010, the applicability of the same risk assessment tool can extend to several stages of the process: it is not uncommon to use a technique that is applicable for both the risk identification and risk analysis phases, or for both risk analysis and risk evaluation, or even for all three phases. In relation to the specific objectives of the risk assessment process, it will then be possible to identify the application limits of those multi-purpose methodologies.

The purpose of this chapter is to offer readers some insights about the methodologies most used – a net of specific tools related to particular aspects or types of risks. Readers, on their own, can then construct personal insights in relation to their own peculiar abilities and needs.

Since risk assessment is a fundamental (and critical) process for the organization, where the methodology to be followed is not a prescriptive requirement, it must pay particular attention to selection in relation to the context, type of risks, skills, and objectives to be pursued.

ISO 31000 applies to organizations having different complexities, sizes, resources, and operating in different contexts. It is also the basis of modern high-level systems (HLS) for management. Therefore the methodologies and, eventually, associated tools supporting the risk assessment process and the risk management methods should be selected considering the characteristics of the organization but also some other key factors, such as the maturity level to implement those methods, the competence and proficiency of internal resources, and/or the availability of external consultants and advisors, the goals and the strategic objectives, the past experience of undesirable events and accidents (nature, number, main root causes already identified). In the following paragraphs some methods useful for the steps of a risk assessment are described, as an introduction to the Bow-Tie and barrier-based risk management framework.

2.2.1 Risk Identification

This section presents some methodologies generally used in the preliminary identification phase of hazards while retaining their applicability (more or less recommended) to the other phases of the risk assessment.

Brainstorming

Brainstorming stimulates and encourages free conversation between a group of people familiar with the subject for analysis in order to identify possible ways of failure and associated dangers, risks, criteria and options for the next stage of treatment. The term "brainstorming" is often misused to denote any kind of group discussion. In fact, true brainstorming uses special techniques to try to get a participant's thoughts to be stimulated by other members of the group in order to support each other in the cognitive processing process. For this reason, effective facilitation is very important in this technique. It is achieved by stimulating the start phase of the discussion (the so-called kick-off), guiding the group at regular intervals through several relevant discussion areas and capturing the emerging aspects (typically in real time).

Brainstorming can be used in combination with other risk assessment methods, or it can be used alone as a technique to encourage thought at every stage of the risk management process and at every stage of a system's life cycle. It can also be used for high-profile discussions when issues are clearly identified, or for detailed insights into particular issues.

Brainstorming places great emphasis on imagination. Therefore, it is especially useful when you want to identify the risks associated with new technology, for which there is no historical data available or solutions already identified in other contexts.

What is needed for brainstorming is just a team of people with an appropriate level of knowledge about the organization, system, process or application being evaluated.

The process can be formal or informal. Formal brainstorming is more structured, with participants preparing ahead of the session, whose purpose is defined and the ideas being questioned are evaluated. Informal brainstorming is less structured and often ad-hoc. In particular, in a formal process:

- The facilitator prepares suggestions in advance to stimulate and enrich the discussion.
- The objectives of the session are clearly defined, and the rules of the session are explained to all participants.
- The facilitator starts the discussion with a non-stop speech, and each participant explores their own ideas to identify as many aspects as possible. At this stage, there is not much discussion about whether a danger should or should not be on the list, in order to avoid inhibiting the uninterrupted discussion. All stimuli are welcome, and no one is criticized; the group's work advances rapidly to allow ideas to stimulate so-called lateral thinking;
- The facilitator can direct participants to a new topic of discussion, or change their perspective when the discussion deviates too much from the field they want to explore, or the thoughts are now devoid of added value. In any case, the idea is to gather as many ideas for the next phase of risk analysis.

The result of this process depends on the specific phase of the risk management process to which this method is applied. When applied to the preliminary hazard identification phase, which by far is the most frequent area of application, then the result will be a list of dangers or control measures put in place.

Among the advantages of this technique is that it is possible to include its ability to stimulate creative thinking, thus helping to identify new risks and new solutions. It typically involves stakeholders with key roles and therefore supports communication between them. In addition, it is a fast and easy technique to implement.

However, brainstorming has some obvious limitations. The main ones are related to the quality of the discussion and therefore, to the participants: they may not have the skills and knowledge required for their contribution to be effective. The methodology is also poorly structured, and it is, therefore, difficult to prove that all potential dangers have been taken into account with this process. Finally, particular group dynamics may arise where participants with valuable ideas remain silent while other participants dominate the discussion.

This limit could be exceeded by using a computer forum, where participants speak on condition of anonymity, to avoid being subjected to personal or political pressure that could prevent the free expression of their ideas. Similarly, the participants' ideas could be sent anonymously to a moderator and then discussed by the whole group.

Checklists

Checklists are a list of dangers (or risks or ways a control measure fails) generally developed from experience, both as a result of previous risk assessment and as operational experience in terms of past failures.

A checklist can be used to identify hazards, or risks, or to assess the effectiveness of control measures. This technique can be used at every stage of the life cycle of a product, process, or system. Checklists are sometimes used as part of another risk assessment technique but are used more to ensure that the entire scope of assessment has been covered downstream from the use of a more imaginative technique capable of identifying new problems.

The application of this technique requires a clear identification of the scope. Only then is the checklist that adequately covers the entire scope selected. For example, a barrier checklist cannot be used to identify new hazards or new risks. The process ends with the use of the checklist by a person (or a team of people) who delves into every single element of the system that is being analyzed and verifies that the items on the checklist are (or are not) present.

Checklists have the undoubted advantage of being able to be used even by non-experts. In fact, if well-designed, they manage to enclose in a simple tool a wide range of skills, making them available to even the least experienced. This ensures that the most common issues are not forgotten.

In contrast, checklists tend to inhibit the imagination during the risk identification phase, due to their highly schematic approach to the problem. In addition, they focus on the problems (in this case, the dangers) that are known, neglecting those knowingly and unknowingly unknown. By their nature, they tend to encourage a prescriptive attitude to risk management, passing on the idea that it is enough to tick a box to solve each problem. We must, therefore, be aware of these disadvantages.

What-if

The "what-if" technique is a systematic team study that sees a facilitator use a set of words or phrases during a session to stimulate participants to identify hazards. Typical phrases are used in combination with system/process/plant keywords being analyzed to investigate how they are affected by deviations from normal operating and behavioural conditions. Normally, this technique is adopted at the system level and not at the individual elements that compose it, with a lower level of detail than – for example – a HAZOP.

This methodology was originally designed to identify hazards in the chemical and petrochemical industry; it is now widely adopted to systems, procedures, plants and organizations in general. In particular, it is used to analyze the consequences of the changes in the risk assessment subject and the risks associated with them.

The method allows the identification of hazards by recursively asking such structured questions: "What would happen if . . ." (what-if) a certain system/process/equipment did not work as it should? The method also applies to procedures that govern processes.

Before starting the study, the system (or procedure, plant, etc.) that is being analyzed must be carefully defined, taking care to establish internal and external contexts through interviews with colleagues, documentation study, floor plans and drawings. It is also important that the level of expertise and experience in the working group is carefully selected in order to avoid poor results.

The methodology applies according to the following process:

- Before the study begins, the facilitator prepares a list of guiding words or phrases that can be based on predefined sets or be created ad-hoc to ensure extensive coverage of hazards.
- During the session, the internal and external context of the system (or process, equipment, etc.) that is being analyzed is discussed and shared.
- The facilitator asks participants to speak and discuss:
  - Known risks and dangers
  - Past experiences and incidents
  - Known and existing control measures
  - Any legal requirements and other constraints
- The discussion is facilitated by means of questions that use the structure "What if . . ." and a keyword/subject. Phrases like "What if" can also be "What would happen if . . .," "Could someone or something . . .," "Someone/something has ever. . .." The aim is to stimulate the study group to explore potential scenarios, their causes, and consequences.
- The risks are therefore summarized, and the group takes into account the control measures put in place (barriers).
- The group confirms the description of the risk, its causes, consequences, and expected barriers, and then records the data.
- The working group considers whether the control measures are adequate and effective and agrees on a measure to monitor their effectiveness. If it is not satisfactory, the group considers the possibility of further dealing with the risk, possibly defining potential additional barriers.
- During the discussion, further "What-if" questions are asked, in order to identify further possible risks.
- The facilitator consults the list that was previously created to monitor the status of the discussion and suggests any scenarios not considered by the team.

The output of this technique is generally a register of dangers (and, if desired, the risks associated with them). If the assessment has gone so far as to identify corrective actions, then they will form the basis for the risk treatment plan. In its objectives, the technique is similar to HAZOP and FMEA. The substantial difference is that it is less structured.

The advantages of this technique lie in its broad applicability to all forms of systems, installations, situations, circumstances, organizations, or activities. It allows the relatively quick identification of the dangers and risks associated with them. This technique is also oriented to the "system," allowing participants to observe the system's response to deviations rather than simply examining the consequences of a component's failure. The what-if study can also be used to identify opportunities for improvement, identifying those actions that lead to an increased probability of success of processes and systems. In addition, the involvement of colours that are required to ensure and monitor the effectiveness and efficiency of existing barriers reinforces their sense of responsibility. Although the technique lends itself to a qualitative or semi-quantitative approach, it is often also used to identify those hazards and risks that will later be analyzed through a quantitative study.

However, the "what-if" analysis requires that the facilitator be competent and experienced in the scope of the evaluation, and this could be a limitation. It also requires careful preliminary preparation so as to avoid an unnecessary waste of time during the course of the session. It is also necessary to be aware that the high-level application of this technique may not bring out complex causes, which can only be found with a detailed analysis. Finally, not only the facilitator but the entire working group must have a broad knowledge and experience base in order to avoid the failure to identify certain dangers.

In order to have a comparison between "what-if" and HAZOP, the following scheme will be analyzed with both of the two methodologies to underline pros and cons for each. The example takes inspiration from Assael and Kakosimos (2010). Let us consider the flow diagram in Figure 57. It is a feed line of the propane-butane separation column. Initially, the mixture enters in vessel D-1; then it is pumped (through P-1) towards the column T-1. An FRC valve controls the flow rate and the heat exchanger E-1 pre-heats the mixture before entering in T-1. In Figure 57, the following abbreviations are used:

- RV: relief valve;
- LI: level indicator;
- LLA: low-level alarm;
- FRC: flow recorder controller;
- TIC: temperature indicator controller.

In Table 2 there are two very easy examples of "what-if" scenarios.

HAZOP

The hazard and operability analysis (HAZOP) was invented in the UK during the 1970s. It is a very structured technique, allowing the identification of those hazards related to process deviations of parameters with respect to the normal range of activity. A HAZOP

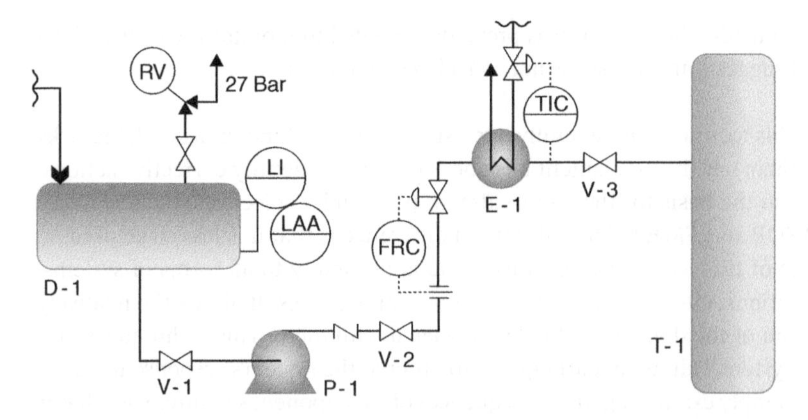

Figure 57 Feed line propane-butane separation column. *Source:* Adapted from Assael and Kakosimos (2010).

Table 2 Example of "what-if" analysis.

"What-if" Question	Consequence	Recommendation
. . . pump P-1 shuts down?. . . valve V-1 is accidentally closed?	• The liquid level rises in D-1 • Feeding T-1 is interrupted, causing operational upset	• RV will open if LI fails
. . . the FRC valve is leaking?	• Possible fire due to flammable mixture	• Schedule more frequent maintenance • Replace it with a double-seal system

Source: Adapted from Assael, M. and Kakosimos, K., 2010.

Table 3 Guidewords for HAZOP analysis.

Guideword	Meaning
NO	Complete negation, fully absent of
LESS	Quantitative decrease
MORE	Quantitative increase
REVERSE	Logical opposite
OTHER THAN	Complete substitution

analysis can be used for every kind of process. The structured path to identify the possible deviations consists in:

- Establishing a list of keywords, intended as a parameter's modifier. A typical set is shown in Table 3.
- Establishing a list of parameters, intended as those physical dimensions whose setting affects the process (like pressure, level, temperature, flow, composition, and so on).
- Combining each keyword with all the parameters, to identify all possible deviations. Obviously, some resulting deviations might have no sense or might be not applicable in the specific context being analyzed.
- Determining all the possible causes for each deviation.
- Determining all the possible consequences for each cause.
- Developing the necessary corrective actions to face (avoidance or mitigation) the hazardous scenarios being identified.

When performing a HAZOP analysis, balanced team composition is crucial to obtain good results. A HAZOP team requires:

- A team leader, to guide and help the team in reaching the objectives of the analysis. It is not necessary for the team leader to know technically the specific process being investigated since other members are required to bring that knowledge.

- A scribe (eventually), to write down the results emerging from the brainstorming activities of the team.
- Operator(s) experienced with the process being analyzed and its standard and emergency procedures.
- Technical specialists, like instrumental, electronic, mechanical or plant operator(s), depending on the specific process/plant (this category may include technologist engineers).

It should be well kept in mind that the primary goal of a HAZOP analysis is hazard identification: this means that engineered solutions must not be found during a HAZOP session, thus avoiding wasting time. Clearly, if the corrective action is obvious, then the HAZOP team may recommend it; instead, when the solution is not so immediate as to be reached, the task must be left to the engineering team.

Even if it is preferable for the HAZOP analysis to be carried out at the earlier stages of the design, so as to positively influence it, on the other hand, an already complete design is required to perform an exhaustive HAZOP. The compromise could be carrying out the HAZOP as a final check, once the detailed design is ready.

A HAZOP may also concern an existing facility, and it is generally used to identify hazards due to plant modifications or to propose modifications in order to reduce risks. Being a structured method, HAZOP is widely used in the process industry.

Taking inspiration from the already discussed example in Figure 57 about the feed line propane-butane separation column, Table 4 presents scenarios corresponding to the outcomes of the "what-if" analysis in the previous paragraph, to make the difference immediately visible. The example concerns the parameter "flow" and, one more time, is restricted to the subset of causes found in the "what-if" analysis. A full HAZOP is more extended to that presented in Table 4.

The primary goal of a HAZOP analysis is to identify hazards, which means that engineering solutions should not be found during a HAZOP session, thus avoiding wasting time. Clearly, if corrective action is obvious, then the HAZOP working group can recommend it. If, on the other hand, the solution is not immediately reached, this activity should be left to the engineering department after the conclusion of the HAZOP session.

Although it is preferable to conduct the HAZOP session in the early stages of the project, so as to positively influence it, it should be noted that a complete project is also necessary to perform a comprehensive HAZOP analysis. The trade-off is, therefore, to use a HAZOP study when a complete view of the process is available, but detailed design changes are still allowed.

A HAZOP can also cover an existing plant and is generally used to identify hazards associated with plant changes or to promote changes in order to reduce risks. Being so strongly structured, HAZOP is widely used in the process industry.

To perform a HAZOP study, you must have drawings, datasheets, logical and process control diagrams, know the layout, maintenance and operational procedures, emergency response procedures, and so on.

The objectives of the HAZOP analysis are met thanks to a systematic examination of how every single part of the system, process, or procedure will respond to changes in key parameters, using guidewords.

Table 4 Extract of an example of HAZOP analysis. Adapted from Assael and Kakosimos (2010).

Guideword	Deviation	Cause	Consequence	Recommendation
NO	No flow	Pump P-1 shuts down (failure or power loss)	The liquid level rises in D-1	There is already an RV. Place a high-level alarm (HLA) on D-1
			Feeding T-1 interrupted, causing operational upset	
		Valve V-1 is accidentally closed by an operator	Pump overheats: possible mechanical damage of the seal, leakage and fire	Point out the error in the operating procedures
			Feeding T-1 interrupted, causing operational upset	
LESS	Less flow	The FRC valve has a minor leak	Hydrocarbons in the air, possible fire	Schedule more frequent maintenance
				Install a double-seal system

This technique offers significant advantages, such as:

- A systematic and thorough examination of the system, process or procedure (with ample opportunity to conduct an expert assessment of the human factor by expert judgement);
- The involvement of a multidisciplinary group that also includes those people with direct operational experience on what is being discussed;
- An ability to generate solutions (risk treatment actions);
- Applicability to a broad spectrum of systems, processes, and procedures;
- The explicit consideration of the causes and consequences of human error;
- The allocation, as output, of a written register containing on time all the information analyzed and organized by criteria (e.g. by process or phase, by deviation, by consequent scenario, by the need of integration, etc.).

On the other side, HAZOP is an extremely time-consuming technique, and detailed analysis requires excellent documentation from which to extrapolate the information needed to start the risk identification process. Often, there is also a risk that the discussion will focus on the details of design, neglecting the context, and even the generality, of the problem analyzed. For this reason, it is essential to set the scope of analysis and objectives together with the working group before the session. The whole process is heavily based on the expertise of those present and designers in particular, for whom it may be difficult to remain objective enough to find any problems in their project. For this reason, it is recommended that the working group include people not directly involved in the project (system, process, procedure, etc.) under review.

HAZOP analysis is similar to FMEA, in that both techniques identify ways of failure of a process, system or procedure, with their causes and consequences. It differs, however, by taking into account the unseeded effects and deviations from the expected operating conditions, working backwards to find the possible causes and ways of failure, while the FMEA takes its cue precisely from the identification of ways of failure.

HAZID

A different tool that can be used in order to perform a preliminary hazard analysis is hazard identification (HAZID). The HAZOP is generally used late in the design phase; therefore, the identified safety and environmental issues can cause project delays or costly design changes. Instead, HAZID is a structured brainstorming (guideword-based) that is generally carried out during the early design phase, so that hazards can be easily avoided or reduced. The objective of HAZID is to identify all hazards associated with a particular concept, design, operation or activity (as stated in ISO 17776). Typically, the structured brainstorming technique involves designers, project management, commissioning, and operation personnel. Like a HAZOP, HAZID is based on inductive reasoning, so it is necessary for the analyst to have sufficient experience in "safety," and to think as widely as possible in order to ensure that predictable major accident hazards are not overlooked, including low-frequency events. To do so, the analyst focuses his/her attention on:

- The hazardous substances used as inputs, as intermediates, and as outputs;
- The used chemical processes;
- The equipment, components and materials being used;
- The plant layout;
- The environment surrounding the plant;
- The safety systems;
- The inspection, control, and maintenance activities.

To conduct a HAZID analysis, it is advisable to subdivide the scope of the analysis into homogeneous areas or functional groups, as from the process schemes (Table 5).

Then, the hazards associated with the performed activities are taken into account (fires, explosions, toxicity, and so on). It is not necessary to list all the possible causes for each incident; indeed, it is sufficient to identify a significant number of them to determine the probability of its occurrence. Pre-defined lists of hazards are generally used, like the ones provided in ISO 17776 related to the hazards that can be encountered in the petroleum and

Table 5 Subdivision of the analyzed system into areas.

Area	Designation	Detail	Flammable inventory	Toxic inventory	Comments	PDF
1	First-stage separator	First separator	Hydrocarbons	–	–	#
2	Crude booster pumps	Crude booster pumps, process area	Hydrocarbons	–	–	#
3	...	...	...	...	...	...

natural gas industries. The approach should be applied to each area and hazard guideword, asking the following questions:

- Is the guideword relevant?
- Is there something similar that should be identified?
- What are the causes that could lead to a major accident?
- What are the credible potential consequences?
- What are the preventive and mitigating barriers already specified (or expected)?
- Are there any additional barriers that could be proposed?
- Are human barriers (if any) reasonable?
- Is further (quantitative) analysis required to better understand the consequence of the hazard?
- What recommendations can be made?

An example of a list that can be used during the preliminary analysis is shown in Table 6. Finally, the analyst identifies the consequences of each assumed event, considering the most conservative scenario. The consequences can be pre-defined too, as shown in Table 7, as well as the preventive barriers and the mitigating measures.

The results of a HAZID analysis are arranged using HAZID worksheets, showing clear linkages between hazardous events, hazards, underlying causes, and control measures/safeguards (if any) as well as the corrective actions. An example of HAZID worksheet is shown in Table 8.

This methodology is particularly suitable to identify the dangers to be analyzed later by Bow-Tie. In fact, the results of a HAZID session can be superimposed on the input data

Table 6 Hazards and assumed event in HAZID.

Hazards	Assumed event
Hydrocarbons under pressure	Leakages
Toxic substances	Leakages
Lifting facilities	Falling parts
Transportation/traffic	Collision
Utility facilities	Loss of function

Table 7 List of typical consequences.

Consequences
Pool fire
Jet fire
Toxic gas cloud formation
BLEVE
VCE
Others

Table 8 HAZID worksheet.

Node: **P&ID #:**					**Date:** **Revision**
No.	**Deviation**	**Cause**	**Consequences**	**Safeguards**	**Recommendation**

required by a Bow-Tie analysis that, as best illustrated in the dedicated paragraph, include precisely the danger (associated with the guideword of HAZID), the top event (associated with the deviation analyzed), causes, consequences, and barriers. It is therefore convenient to set up a HAZID session prior to a Bow-Tie analysis. Note that this tandem coupling of these methodologies is also referenced in some regulatory contexts, such as the regulations related to major accidents prevention such as Seveso and the Offshore European Directives managing the industrial risk respectively associated with on-shore and off-shore industrial chemical plants.

It should be noted that, like HAZOP, the HAZID method, with the appropriate changes in the taxonomy used and in the way homogeneous areas are identified, can well be applied in areas other than that of originality coinciding with the oil & gas world and for the risks different from those characteristics of the installations of the chemical process industry.

FMEA/FMEDA/FMECA

Failure modes and effects analysis (FMEA) is a technique used to identify ways in which components, systems, or processes can fail to achieve their project intent.

The FMEA identifies:

- All potential ways of failure of the various parts of a system (a way of failure is what is observed to fail or perform incorrectly, including reduced efficiency compared to what is expected);
- The effects that these failures can have on the system that contains them or related systems;
- Failure mechanisms;
- How to avoid bankruptcies, and/or mitigate the effects (or possibility) of failures on the system.

Failure modes, effects, and criticality analysis (FMECA) adds to an FMEA the ability to assign a score to each identified failure, in relation to its own importance or criticality. This critical analysis is often qualitative or semi-quantitative, but can also be quantitative if real failure instalments are used.

Failure modes, effects, and diagnostic analysis (FMEDA) also essentially takes into account the operation of diagnostic systems compared to the simplest methodologies from which it derives, thus offering itself as a tool for the definition of system target fault instalments, the creation of the reliability block diagram (RBD) and, therefore, valid support for functional safety analyses related to the determination, allocation, and verification of safety integrity levels (SILs) defined under IC 61508 and IEC 61511. Like FMEA and FMECA, FMEDA is often used in reliability engineering and a detailed description is not required here but only an overall overview, as is the case with the other methodologies described in this chapter.

There are several FMEA applications: for components and products, for systems, for production and assembly processes, and for services or software.

FMEA, FMECA and FMEDA can be applied during the design, implementation or exercise of a physical system. Techniques can also be applied to processes or procedures, for example, to identify potential medical errors in the health system or ways of failure in maintenance procedures.

In general, methodologies can be used to:

- Provide support in choosing high-dependency project alternatives;
- Ensure that all systems and process failure modes and their effects on operational success are taken into account;
- Identify human errors and their effects;
- Provide a basis for planning maintenance and testing operations of physical systems;
- Improve the design of procedures and processes;
- Provide qualitative or quantitative information for risk analysis techniques, such as the fault tree.

For their application, you need enough data to understand how failure can occur. This information can be derived from drawings, analyzed process flow diagrams, environment details and other parameters that can affect the process, and historical-statistical information such as failure rate, where available.

The FMEA runs according to the following process:

- Definition of the scope of study and expected objectives;
- Training of the working group;
- Understanding of the system/process subject to FMEA analysis;
- System/process breakdown into its components/phases;
- Definition of the function satisfied by each component/phase;
- For each component or phase listed, identify:
  - How can any party reasonably fail?
  - What mechanisms could produce these ways of failure?
  - What effects could result from failure? With what consequences?
  - How is the fault diagnosed?
- Identify actions to change the design to compensate for detected failures.

For the FMECA, the working group continues further by classifying each failure identified in accordance with its own criticality. There are several ways to do this, including:

- The mode criticality index;
- The level of risk;
- The risk priority number (RPN).

The mode criticality index is a measure of the probability that the failure mode considered will actually cause a system-leading failure. It is defined as the product of the frequency of occurrence of the specific mode of failure and the probability that it will follow a system failure for the time the system is operational. Typically, this index is applied to equipment failures, for which each term of the product can be defined quantitatively, and the failure modes all have the same consequence.

The level of risk is achieved by combining the consequences of a failure mode with the likelihood of failure. Usually, it is used when the consequences of different fault modes are different from each other. The level of risk can be expressed qualitatively, semi-quantitatively or quantitatively.

RPN is a semi-quantitative measure of criticality obtained by multiplying numbers of scale graduates (typically from 1 to 10) chosen for the consequences of failure, the probability of failure and the ability to detect the problem (a mode of failure increases in priority if it is difficult to detect). This method is widely used in the quality studies of a system/product.

Once the failure modes are identified, corrective actions are identified, implementing them from the most critical fault modes.

The FMEDA, on the one hand, pushes the analysis towards the proper times of functional safety. The interested reader will be able to consult the IEC 61508 to learn the details.

The FMEA analysis is documented in a report that contains all relevant information such as:

- Details of the analyzed system;
- How the analysis is performed;
- Preliminary recruitment;
- Data sources;
- Results;
- Criticality (if FMECA) and methodology used to define it;
- Any recommendations.

The advantages of using these methodologies are many:

- They apply widely to the ways of failure of equipment, systems, and even humans, hardware, software, or procedures.
- They identify the ways in which components fail, their causes and effects on the system, and their presentation in the form of reports that are easily readable even by non-experts.
- They avoid the need for costly plant modifications, identifying problems in the early stages of the project.
- They identify individual modes of failure and the consequent need for redundant security systems.
- They provide input data for the development of monitoring programs, highlighting key features to monitor.

On the other hand, these methodologies can only be used to identify individual ways of failure and not their combinations. Finally, their application often takes a long time and can be difficult and tedious for particularly complex systems.

2.2.2 Risk Analysis

This section presents some methodologies generally used in the risk analysis phase while remaining applicable (more or less recommended) to the other phases of the risk assessment. Since risk analysis activities generally require a quantitative approach, a number of methods, cited by ISO 31010, have been selected, that allow for numerical insights, including, possibly a deepening of initial qualitative studies such as HAZID and HAZOP.

Fault Tree Analysis (FTA)

This method, created in the Bell Laboratories in the early 1960s, intends to reconstruct the exact sequence of primary and intermediate events leading to a top event failure. It is, therefore, useful to recognize those situations that may give rise to undesired consequences when combined with specifically identified events. The main structure of a fault tree is shown in Figure 58. An alternative way is to draw it vertically, as shown in Figure 59, whose basic events are in Figure 60. They are:

- Base events – further analysis not useful
- Undeveloped event – events not analyzed further at this time
- Top or intermediate event – events that are further analyzed

The FTA is an analytical deductive technique to analyze failures. It focuses on a particular undesired event and attempts to determine its causes. The undesired event is known as a "top event" in a fault tree diagram: it is generally a complete, catastrophic failure of the system under investigation (Assael and Kakosimos, 2010). The top event is the final effect, but it is also the starting point of a fault tree analysis. This explains why the formulation of the top event must be punctual and exhaustive: this will ensure the goodness of the outcomes provided by the FTA. The FTA diagram is a graphical representation of both parallel and sequential chains of failures that cause the predefined undesired event (i.e. the top event) to occur. Usually, each fault in an FTA is the combination of system failures (mechanical failures or human error), and the ineffective/missing/failed safeguards put on to stop the chain of failures, but that revealed incapable of doing so, for a determined reason.

It is important to underscore that a fault tree does not take into account all possible system failures or all possible causes potentially at the base of the event. Indeed, every fault

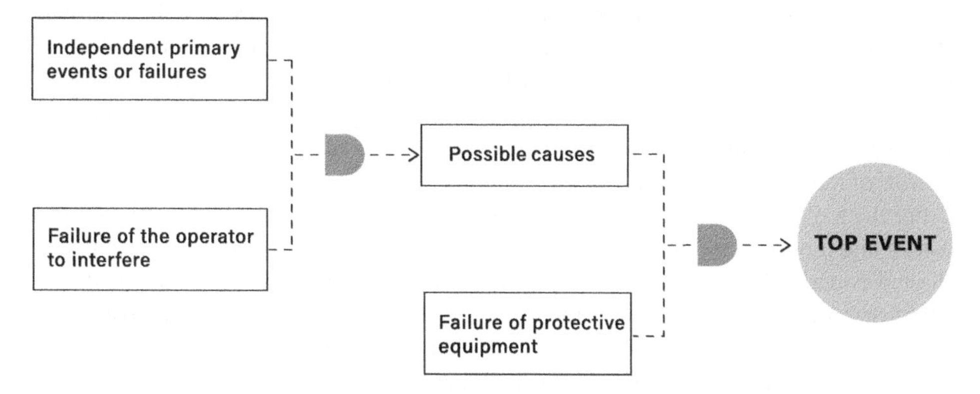

Figure 58 Basic structure of a fault tree (horizontal).

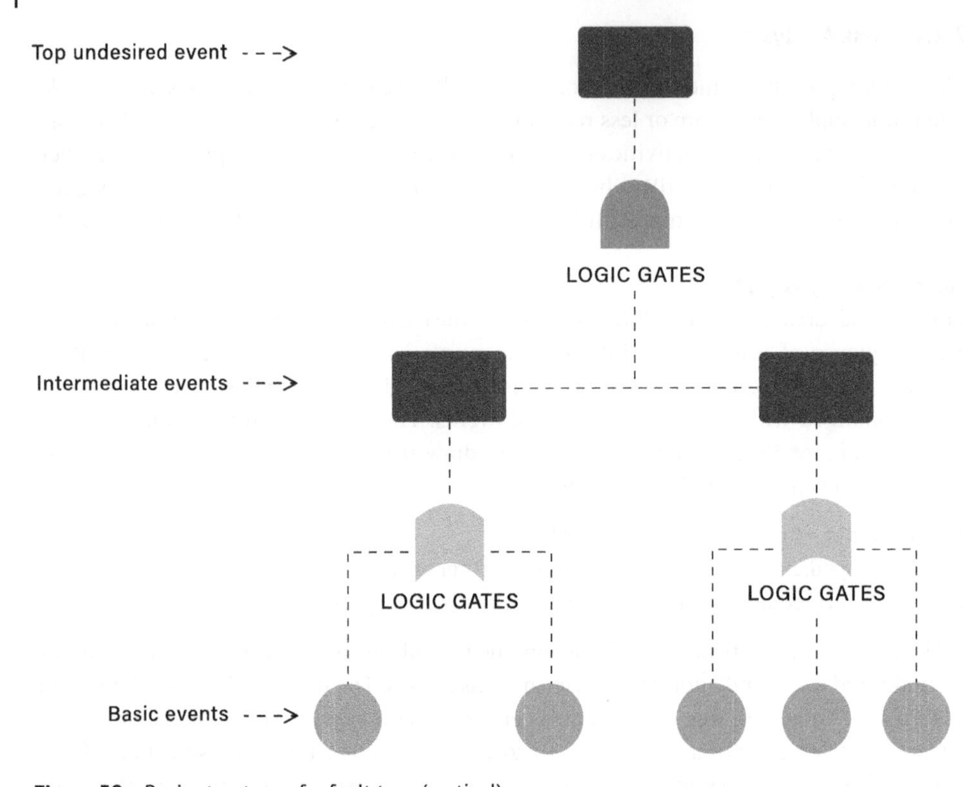

Figure 59 Basic structure of a fault tree (vertical).

Figure 60 Basic Events.

tree is designed for a particular top event, which represents the starting point that will define the development of the rest of the tree: only the contributing failure modes are indeed considered. Every single fault in an FTA is combined with the others through AND/OR logic gates. Other logic Boolean gates can be used, but generally, they are not required. The usage of logic connectors is useful when a single event could be caused by one or more factors that must act at the same time.

After the top event is identified, the analysis of the faults proceed level by level: first the possible and most general immediate causes are considered, always finding support in the collected evidence. Potential failures that are eliminated by matching with the evidence are then further investigated, thus finding the second level of causes. The iterative approach continues until the found causes are considered sufficiently detailed to stop the investigation.

It could also happen that more than one path between the same faults at the origin and top event is found. When this happens, and the tree is fully drawn in a flowchart, the more realistic path between the final failure and a specific set of causes is called the "minimum cut-set" and represents the shortest path between the two (Noon, 2009).

An example of a fault tree (in its upper part) is shown in Figure 61, taking inspiration from Sklet (2002) and the Åsta railway incident.

Fault trees, which may be considered as reversed FMEAs, are used to guide the investigative resources in the most probable causes. Up to now, the description of the FTA seems to define a qualitative analysis. Even if it is possible to leave the fault tree without any number, significant advantages are gained if it is used in a quantitative approach. Data about the failure probabilities are taken from historical databases (when available) or the guides provided by the manufacturers or independent publication. Obviously, there is software that performs a computer-based fault tree analysis. The numerical data about the probabilities of human errors, component failures, or environmental factors are combined using the mathematical rules for probability, depending on the logic gates on which multiple causes converge. Usually, the logic gates are those in Figure 62, where:

- OR gate – the output event occurs if ANY of the input events occur.
- AND gate – the output event occurs if ALL of the input events occur.
- TRANSFER gate – the output event transfers to or from another part of the fault tree.

For instance, an AND gate means that all the previous factors must be fulfilled to generate the subsequent event. From a probabilistic point of view, this is translated into a single probability of the subsequent event, obtained by multiplying all the probabilities of the single causes each other (according to the combined probability rule). Instead, if the connection is through an OR gate, the likelihood of the resulting event is equal to the sum of the probabilities of the single causes. Carrying on in this way, the probability of occurrence

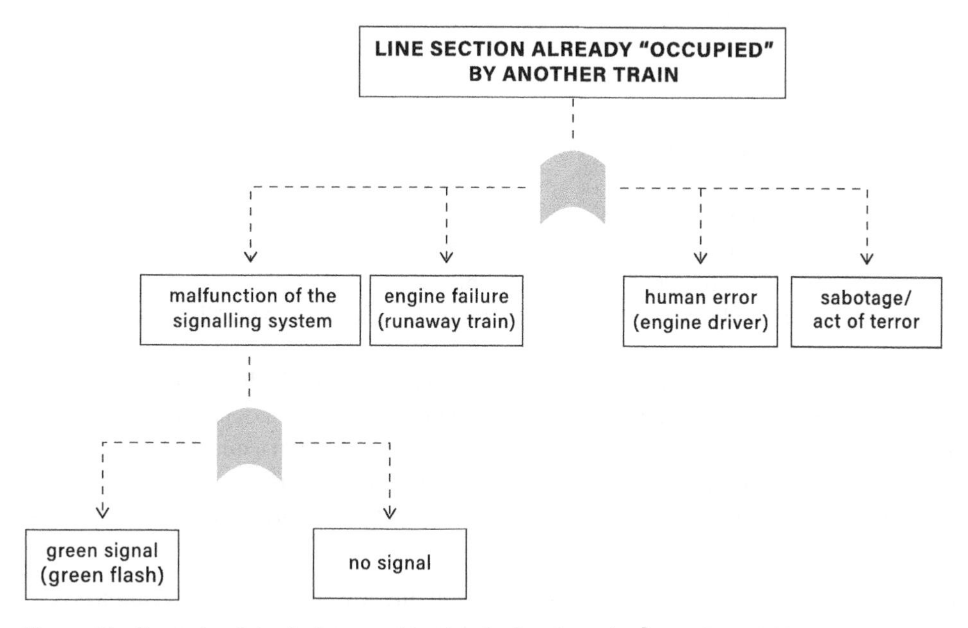

Figure 61 Example of the fault tree, taking inspiration from the Åsta railway incident. *Source:* Sklet, S., 2002.

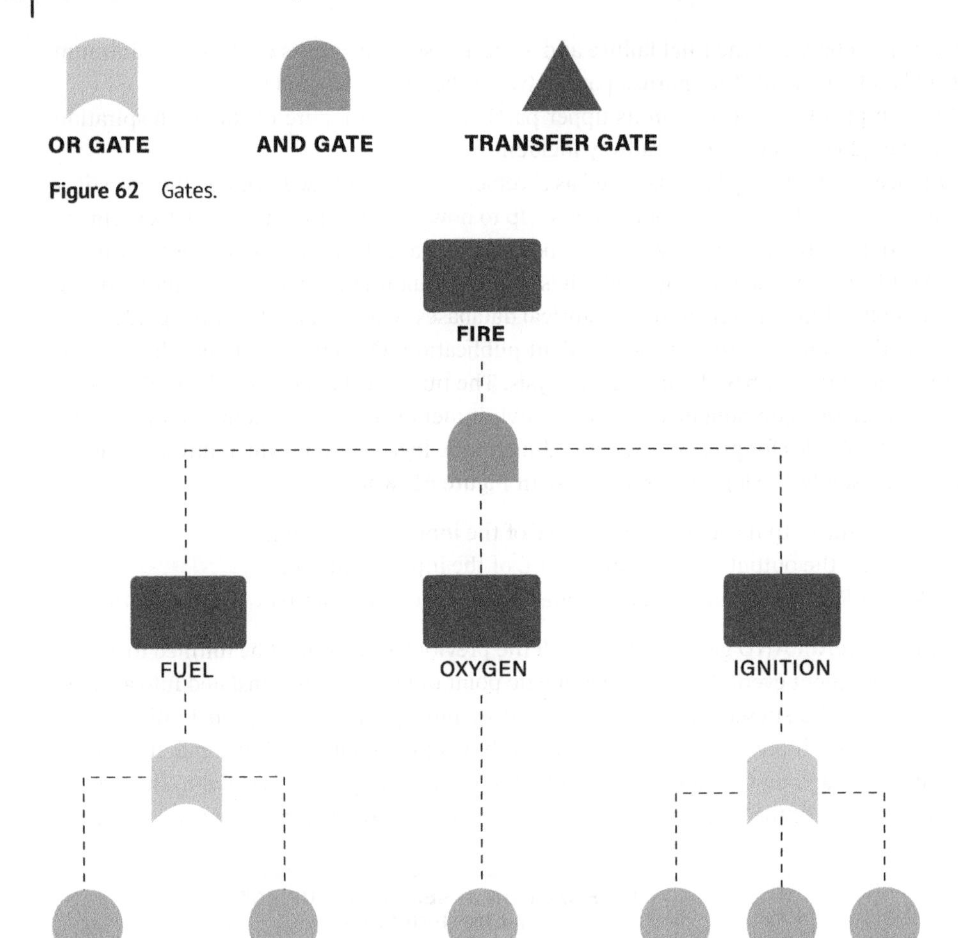

Figure 62 Gates.

Figure 63 Fire triangle using FTA.

of the top event is found. In risk assessment, this information is combined with the severity of the event (for instance, obtained through numerical simulation, in a quantitative approach) to assess the final risk of the top event.

For example, the well-known fire triangle can be transferred into an FTA using the diagram in Figure 63.

FTA is clearly an analytical tool for establishing relations; it does not provide any direct information about how to gather evidence (Katsakiori, Sakellaropoulos and Manatakis, 2009). The strength of the fault tree, even when used in a qualitative approach, is its ability to break down an accident into root causes (Sklet, 2004).

The following is an outline of how to conduct a fault tree analysis (Forck and Noakes Fry, 2016):

- Develop the problem statement (i.e. the top event, the reason for the investigation).
- Identify the first layer of inputs for the incident, considering basic components or procedures. Remember to consider all the possible inputs of failures for the considered equipment or procedures.

- Define the relationship between the top event and the first layer inputs through a logic gate.
- Evaluate each first-layer input by identifying its second-layer inputs.
- Define the relationship between a single first-layer input and its second-layer inputs through a logic gate.
- Continue with other layer inputs, until the required level of investigation is reached (typically, when the root causes are found, the iterative procedure stops).
- If required, gather additional information to complete, support, or eliminate some branches or single inputs of the tree.
- Document and report the result of the FTA, also highlighting the minimum cut-set path and the probabilities related to it.

A further example is now shown, taking inspiration from (Assael and Kakosimos, 2010). Consider the flammable liquid storage system in Figure 64: it is kept under pressure by nitrogen, and a pressure controller is used to maintain the pressure between certain limits; otherwise an alarm is sent to the control room. Relief valve RV-1 opens to the atmosphere in case of emergency. Consider the tank rupture due to overpressure as the top event: the corresponding fault tree is shown in Figure 65.

Another example of a fault tree is shown in Figure 66.

Event Tree Analysis

The event tree analysis (ETA) determines the potential consequences in terms of undesired incident outcomes, starting from an initiating event (i.e. equipment or process failure). The aim of the ETA is therefore complementary to an FTA goal. Indeed an FTA explains how

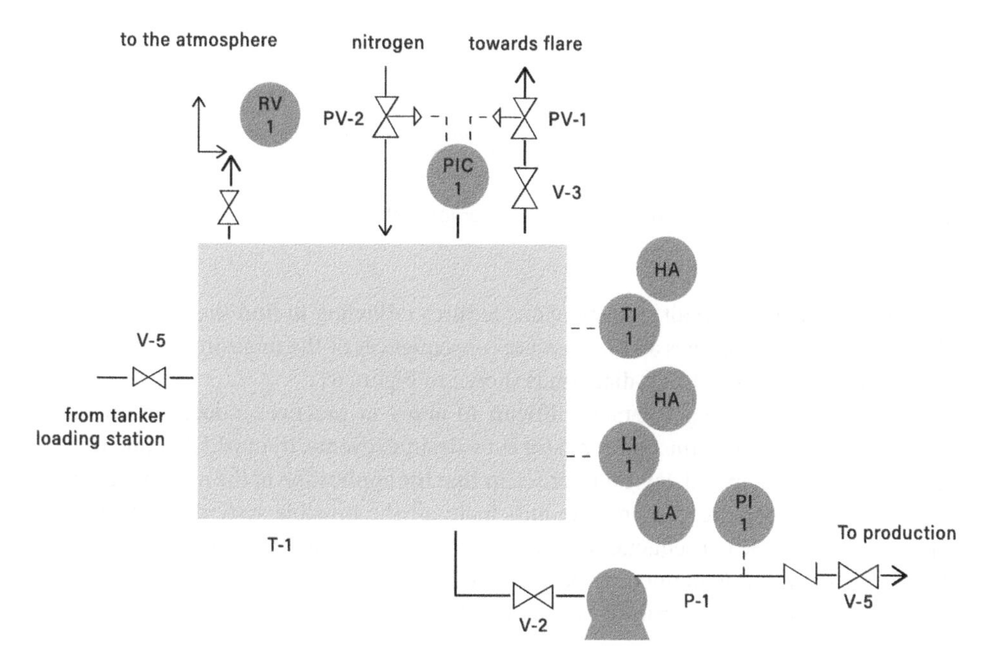

Figure 64 Flammable liquid storage system. *Source:* Modified from Assael, M. and Kakosimos, K., 2010.

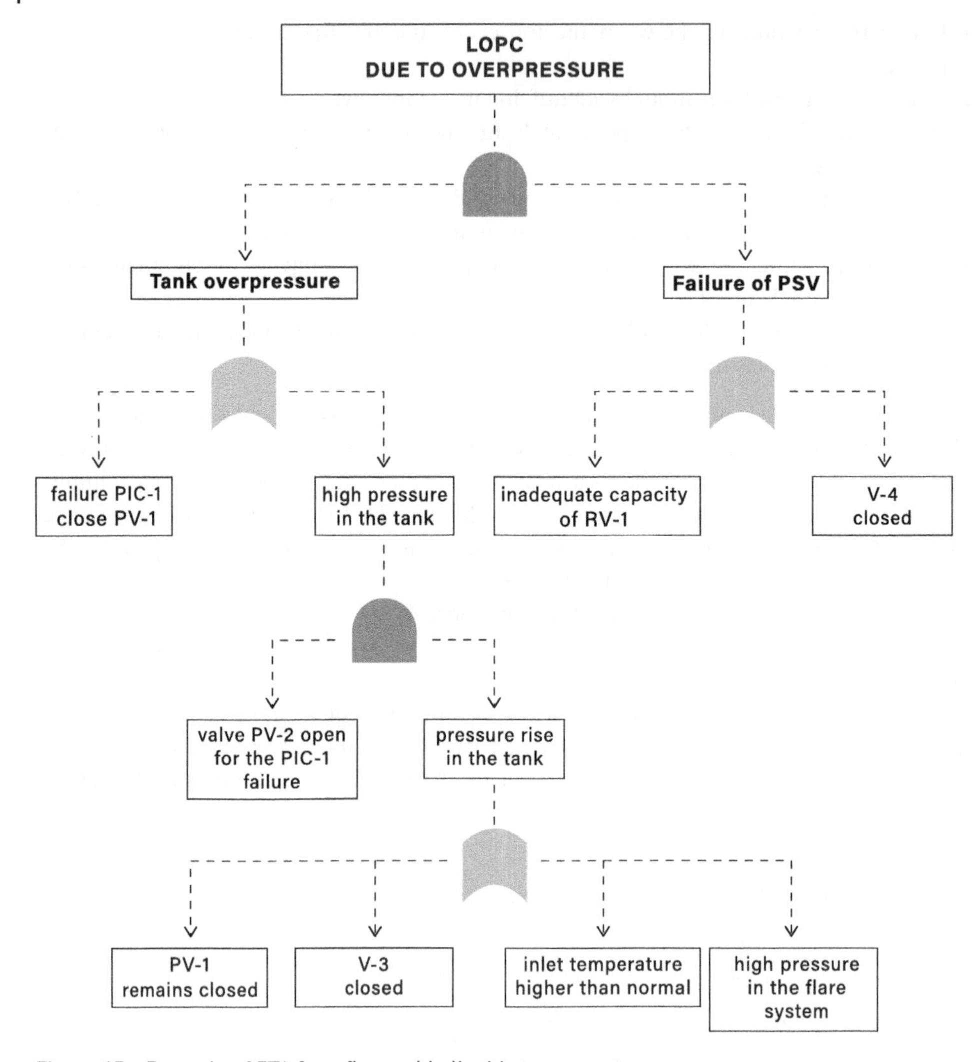

Figure 65 Example of FTA for a flammable liquid storage system.

an undesired event can result from previous failures (allowing to find the root causes as well), while an ETA examines all the possible consequences of the undesired event.

The structure of a typical ETA diagram is shown in Figure 67.

This technique is among the most difficult to apply in practice. Indeed, meaningful results are obtained only if the undesired (or even desired) events, from which branches are created, are fully anticipated. It is therefore clear that the application of the method requires strong practical experience, in order to anticipate all the possible system events and to explore all the possible consequences of those events (Assael and Kakosimos, 2010).

The event sequence is defined by barriers that could be either successful or not. An example of ETA for the Åsta railway accident is shown in Figure 68 (Sklet, 2002). The tree

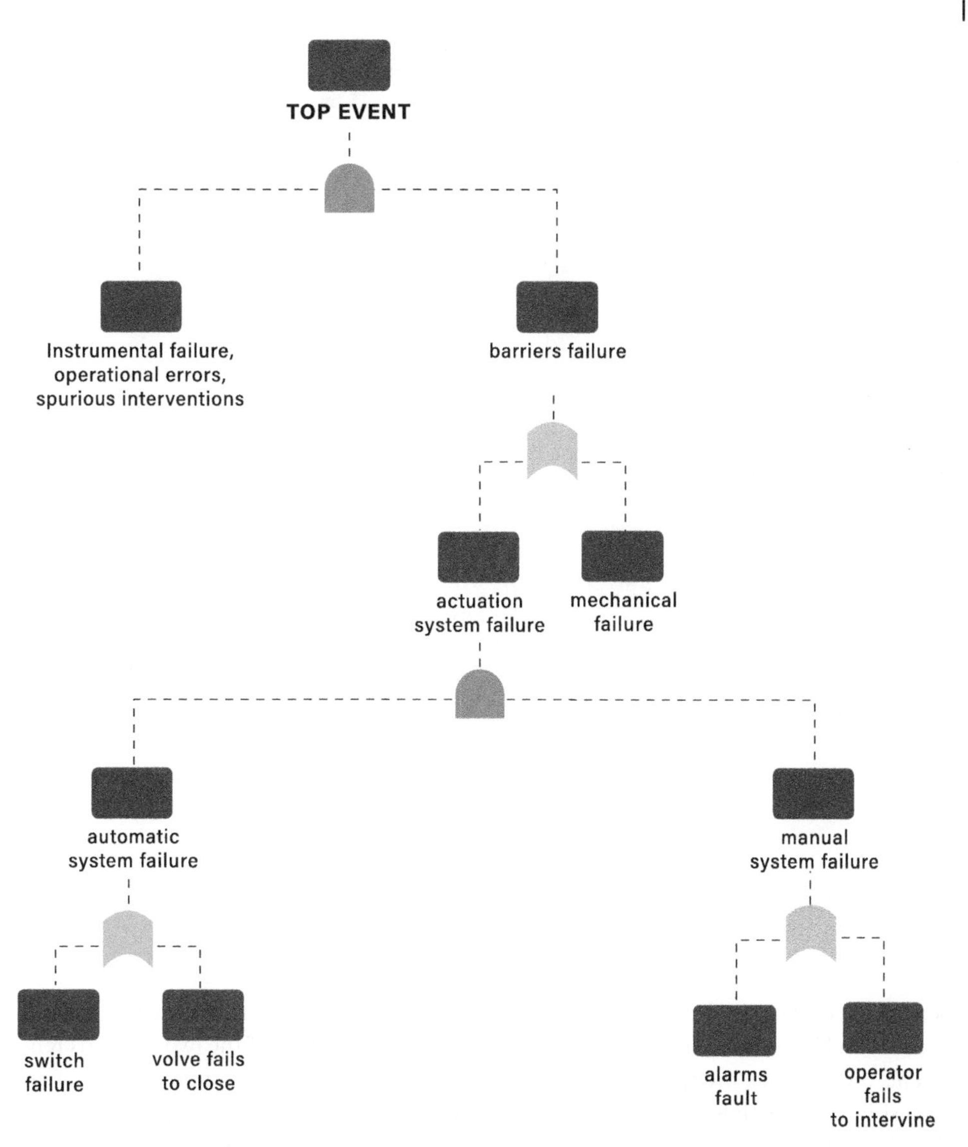

Figure 66 Fault tree example.

underlines how likely the occurred event was. The ETA is an excellent method for risk assessment, being used to identify possible event scenarios. In the incident investigation, the actual incident path may be underlined among all the possible ones. For instance, the real incident path of the Åsta railway collision is highlighted in Figure 69 with a thicker line.

From a quantitative point of view, the frequency of occurrence of each scenario is determined starting from the likelihood of the initial event and combining it with the probabilities of failure of the barriers put in position to create the nodes for diverging branches of the tree. As usual, combined probability rules are followed. Often, the probability of

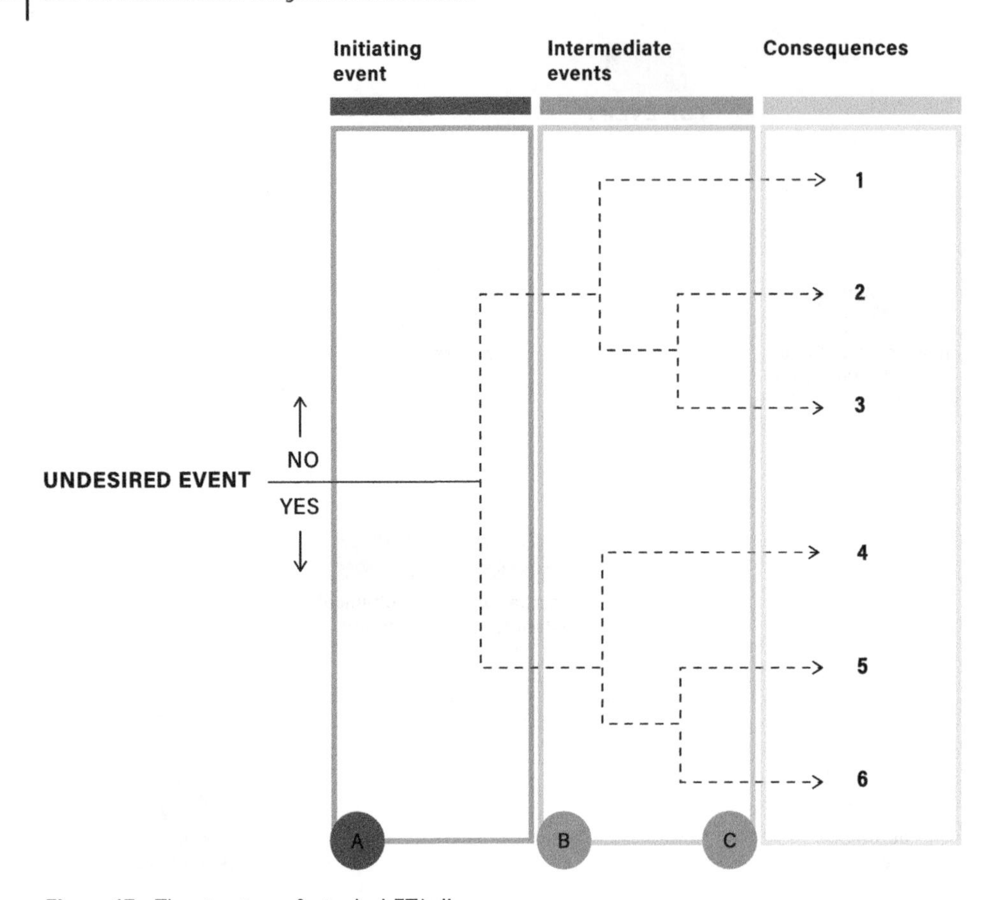

Figure 67 The structure of a typical ETA diagram.

occurrence of the initial event is obtained from a fault tree analysis: in other words, the top event of an FTA is the starting point for an ETA. The combination of the two methods represents the Bow-Tie, the further risk assessment that aims in the full comprehension of an undesired event both looking for its causes (FTA) and all its possible consequences (ETA). Bow-Ties are powerful tools to view, at a glance, both preventive and mitigating measures.

Once the probability of failure of a specific barrier is known (from the historical databases, experience, and so on), the likelihood of being successful is complementary to the unit (e.g., the probability that the barrier will fail or will not fail).

Let us consider a pipe connected to a vessel, as shown in Figure 70, taking inspiration from Assael and Kakosimos (2010). The possible consequences of a rupture of the pipe in point P needs to be found. The system is equipped with an excessive flow valve (EFV) and a remote controller isolation valve (RCV). The resulting event tree is shown in Figure 71. In the event tree, the pipe rupture is the undesired event (A), with a probability PA, while the EFV failure is an intermediate event (B) with the probability PB, like the RCV failure (C) that has a probability PC.

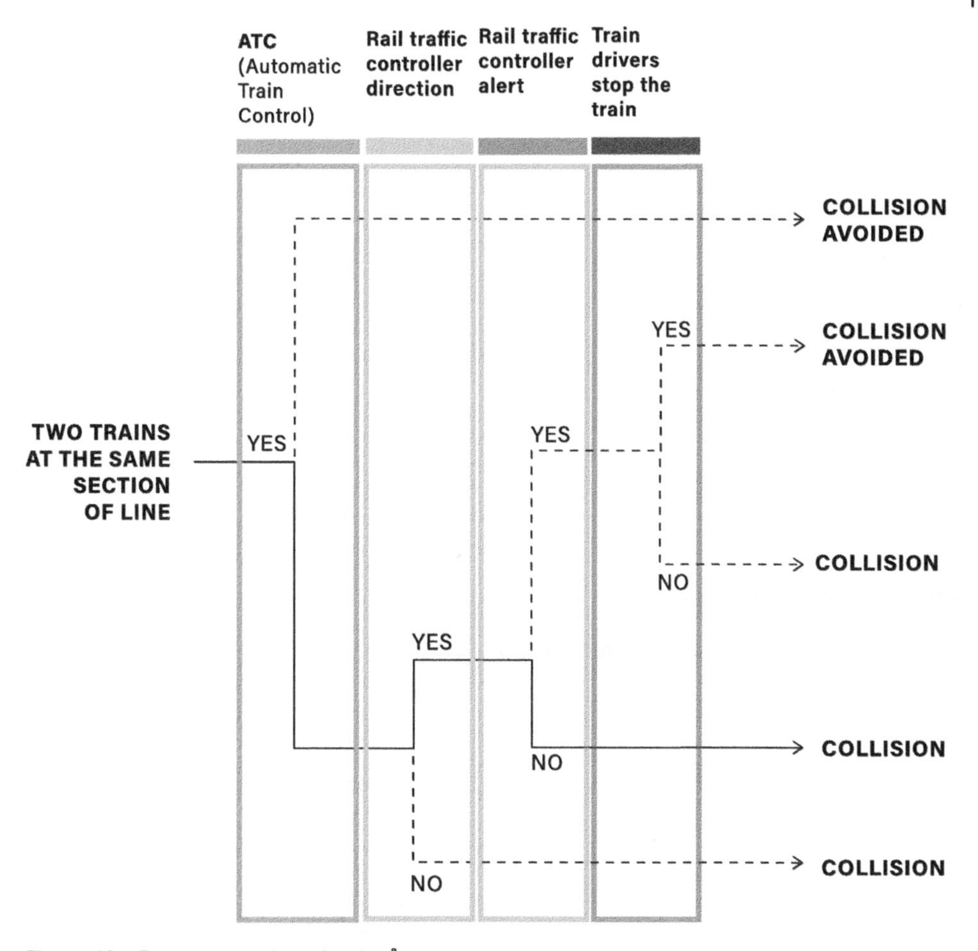

Figure 68 Event tree analysis for the Åsta railway accident.

According to the resulting event tree, the probability of a continuous leak is given by PA × PB × PC; the likelihood of leakage until RCV is closed is PA × PB × (1 – PC); finally, the likelihood of a minor leak is PA × (1 – PB).

2.2.3 Barrier-Based Methods

Risk management, once context analysis and risk assessment are completed, results almost entirely in the management of risk control measures, or barriers. Managing risk over time, ensuring that it remains within certain acceptability values, effectively means ensuring that the measures put in place by the organization to contain the risk remain intact, effective, and efficient over time. Starting from a simple reflection, two risk analysis methods based on the concept of barriers are presented here. They, therefore, offer a different perspective on risk management, positioning themselves as tools to support an entire risk management system that intends to adopt the barrier-based perspective that is emerging

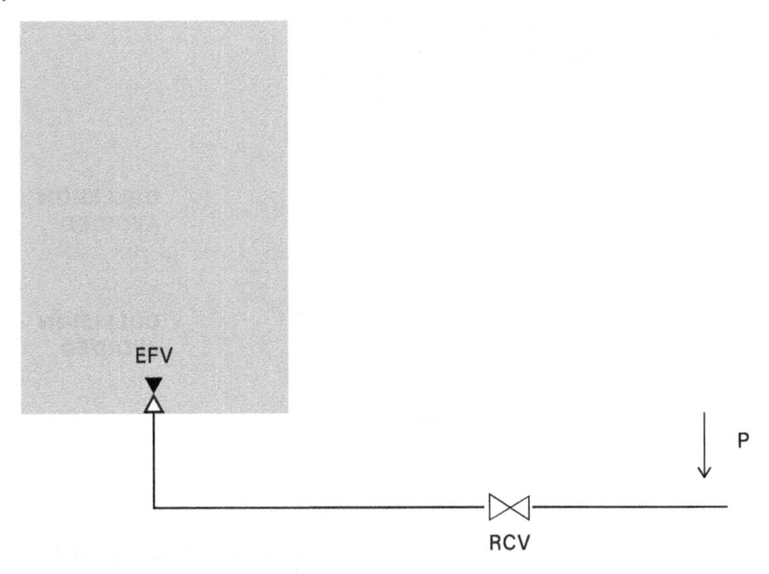

Figure 69 Pipe connected to a vessel.

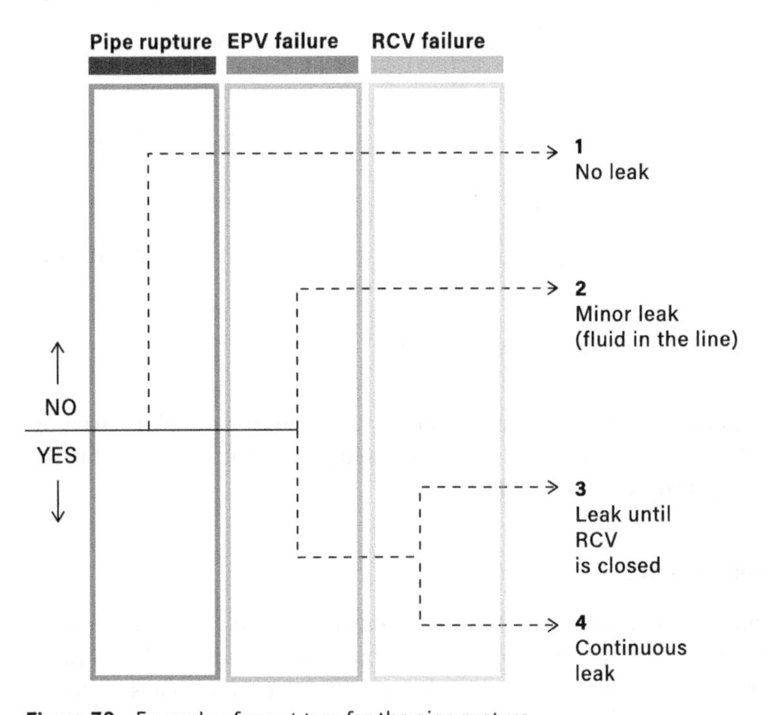

Figure 70 Example of event tree for the pipe rupture.

not only in the technical standards of voluntary adoption, such as those mentioned in Chapter 1. This approach is completely compliant with the framework of the ISO 31000 standard for risk management and can be adopted for the implementation of modern risk-based HLSs. The barrier is, therefore, a measure of control or grouping of control elements

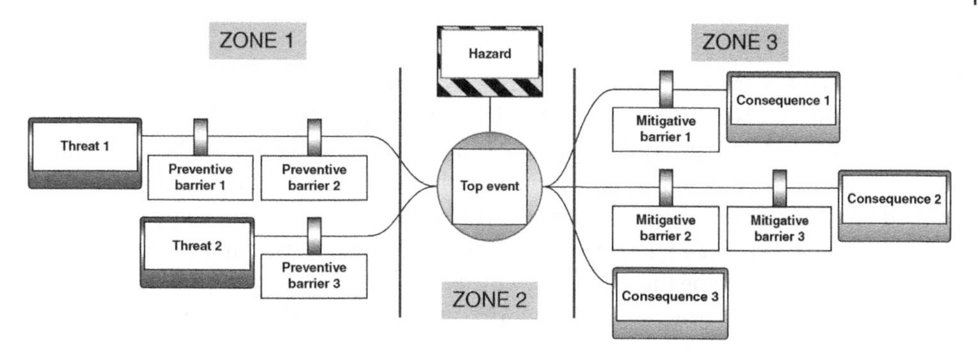

Figure 71 Bow-Tie diagram structure.

that, in itself, can prevent the development of a cause in a top event (preventive barrier) or can mitigate the consequences of the top event once it has manifested itself (mitigating barrier).

Bow-Tie

The Bow-Tie, thanks to its powerful ability to communicate information graphically, has established itself as the main technique of risk analysis based on the concept of barriers. The chapter is dedicated entirely to it, as a preferred methodology by the author of the book for the illustration of the principles of ISO 31000. In this paragraph, we provide a simple introduction to the method.

The Bow-Tie technique involves the development of logical flow diagrams developed in three distinct zones. An example of a Bow-Tie diagram is shown in Figure 71:

- Zone 1 (prevention) is represented on the left side of the diagram; it identifies all causes (blue rectangles) that can be associated with the unwanted event and, for each of them, highlights all the specific protection systems (both plant and operational control) that help prevent the unwanted event. Zone 1 can be considered equivalent to a simplified fault tree.
- Top event is represented in the centre of the diagram and uniquely identifies the danger considered (yellow and black striped rectangle) the primary incidental event called Top event; this event can, in turn, evolve, based on the dynamics of the incident in alternative incidental scenarios.
- Zone 3 (protection) identifies all potentially generated incidental scenarios (e.g. burning jet fuel, explosion, flash fire, etc.) and the combination of all the elements that allow its development, including all protection systems that can mitigate its effects. Zone 3 can, for all intents and purposes, be considered equivalent to a simplified event tree.

The Bow-Tie technique allows the identification and evaluation of the frequencies and consequences associated with scenarios and the quantification of the contribution of protective and mitigating systems (barriers) under normal production conditions. This aspect is also discussed later in this book and requires the introduction to the methodology of analysis that underlies the frequency quantification of Bow-Ties: layer of protection analysis (LOPA; see Section 2.14).

2.2.4 Risk Evaluation

Risk assessment is the last of the three phases of the more general risk assessment process. Thanks to risk evaluation, the level of risk obtained through the previous analysis phase is translated, through the methods described in this paragraph, into indices or values that can be compared with the thresholds of acceptability and tolerability defined at the preliminary stage together with the objectives of the analysis and the general objectives of the organization, in order to determine whether the risk can be accepted (and therefore be found in the context of continuous improvement), tolerated (and therefore falls into the so-called ALARP region) or whether it should be treated further.

FN Curves

F-N curves, often used in industrial safety linked to the significant risks associated with fires, explosions, and flashes, are a two-dimensional graphical representation of the cumulative frequency of the expected adverse consequences in relation to the extent of their magnitude (often expressed in terms of fatality numbers).

The construction of an F-N curve perils are known (note the different notation between the lowercase "f" and the uppercase "F"):

- N is the number of people estimated to be harmed (e.g. deceased) as a result of each event.
- F is the frequency attributed to each event.

Before building an F-N curve, it's good to make sure you have only one f value for each value of N since in the early stages of very complex analysis there could be multiple events with the same number of fatalities. So, if more than one event has the same number of fatalities, then you combine (sum) the frequencies to have a single value. When you are finally ready to produce an F-N curve, you identify the f-N pairs.

However, diagramming the f-N relationship does not produce the desired result, as the graph obtained is poorly informative. The result is not satisfactory even if you use a bar chart for groupings of N (e.g. No. 1–5, No. 6–10, etc.) because the result is highly dependent on arbitrary grouping performed.

It is therefore essential to work with the cumulative frequency F. The number F, for a given N, is the sum of the frequencies of all events whose fatality value is N or greater. In practice, the F-N curves report in order the expected frequency F that damage is given interests more than N people and in a sharp number of people, N. If N is the number of deaths, then the information provided by the F-N curve is related to the frequency of an incidental event capable of causing at least N deaths. Clearly, the estimate of the number of deaths is produced downstream of the studies on the analysis of the consequences, which are not the subject of this book. The analysis of the consequences should take into account the number of occupants, the extent of the damaged area, the time of the occupants' stay in the damaged area, and so on.

Given the orders of magnitude in play, F-N curves are often represented on logarithmic axes. Within the Cartesian space (Figure 72), areas of tolerability, acceptability, and unacceptable risk are identified. If an F-N curve intersects the region of unacceptability, then the risk is considered unacceptable.

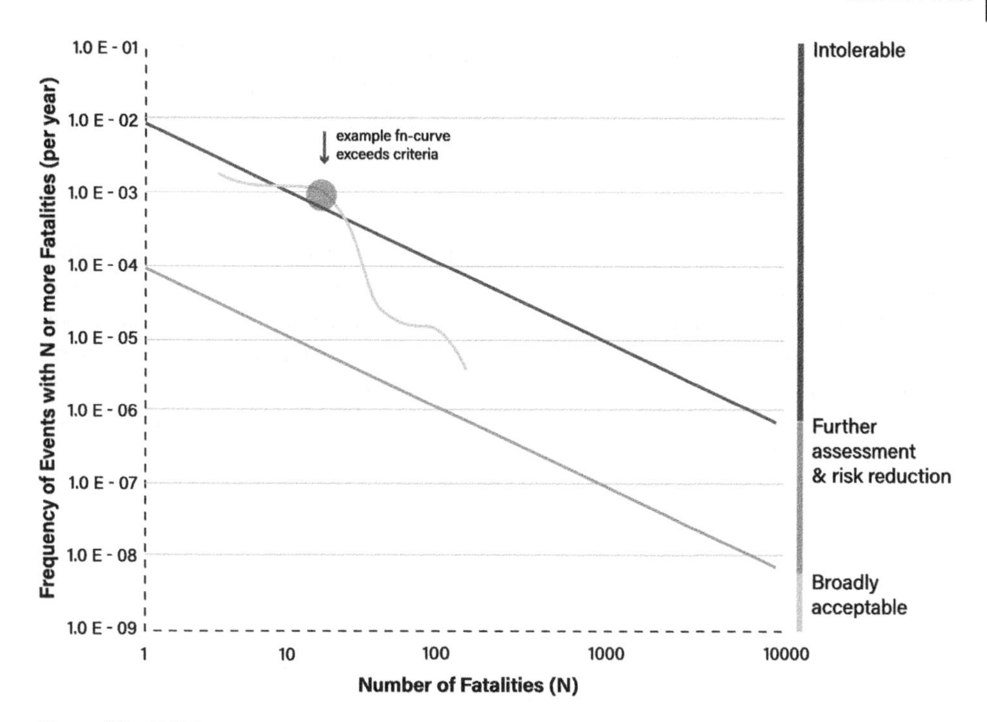

Figure 72 F-N Curve.

Risk Indices

The risk index is probably one of the simplest and most intuitive risk assessment methods. The risk index R is defined by the product of the frequency of occurrence f and the magnitude of the N consequences:

$$R = f \times N$$

The risk index can then be defined from both quantitative assessments of the f and N parameters, as well as qualitative assessments.

For example, the frequency of an event could be defined using the following qualitative scale:

- $f = 1 \rightarrow$ very unlikely event
- $f = 2 \rightarrow$ unlikely event
- $f = 3 \rightarrow$ very likely event
- $f = 4 \rightarrow$ extremely likely event

Similarly, a qualitative scale such as the following could also be arbitrarily defined for magnitude:

- $N = 1 \rightarrow$ no impact
- $N = 2 \rightarrow$ minor consequences
- $N = 3 \rightarrow$ major consequences

In this case, the R risk index is a value between 1 and 12. The higher the value, the higher the risk. The lower the value, the lower the risk.

Of course, the risk index, regardless of its definition in a qualitative, semi-quantitative (range by order of magnitude) or quantity, must then be compared with the acceptable risk index and tolerable risk index, values that will be defined from the beginning of the risk assessment.

Risk Matrices

Risk matrices are also a tool for risk assessment. For their treatment, please refer in full to Section 1.3 in Chapter 1. A risk matrix can be interpreted as the two-dimensional graphical representation of the risk index. Thresholds for acceptability and tolerability of risk are now translated into regions of the two-dimensional space, making the comparison between the level of risk reached and the aforementioned thresholds more immediate. An example of risk matrix is shown in Figure 25, copied here for convenience and shown as Figure 73.

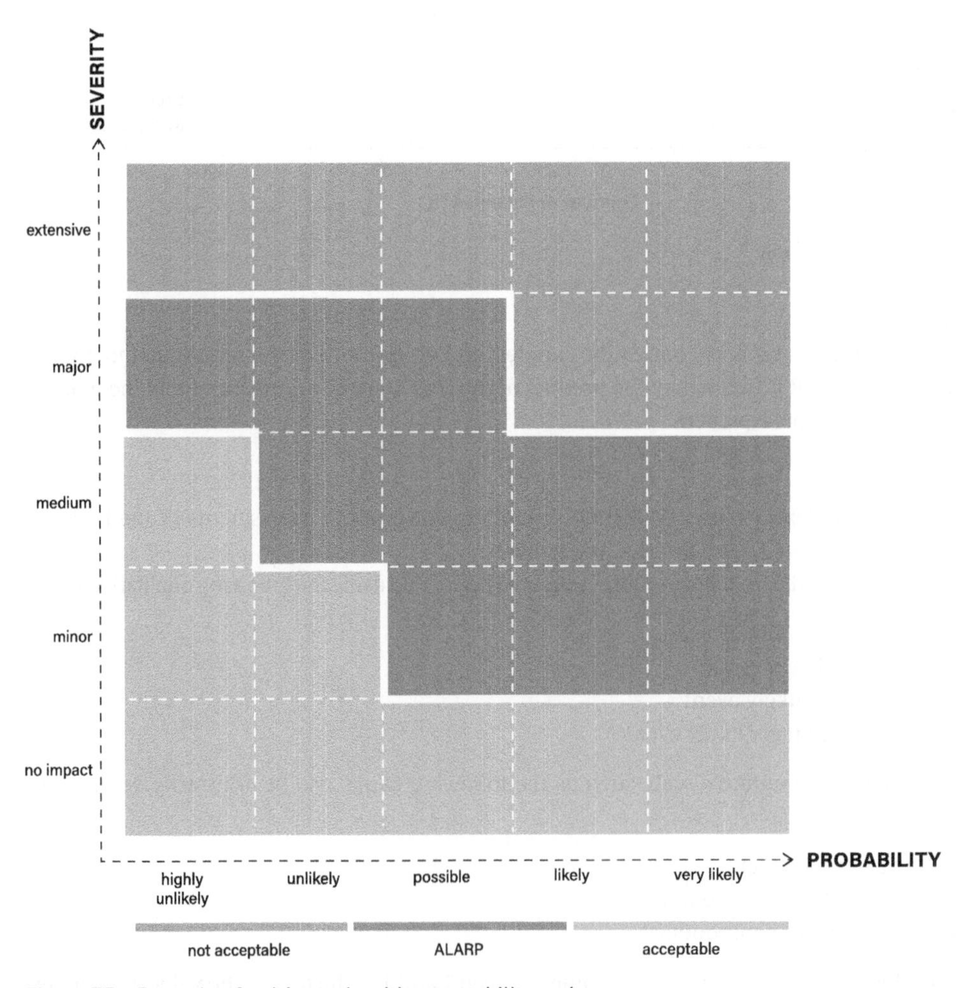

Figure 73 Example of a risk matrix with acceptability regions.

Calibrated Risk Graph

The calibrated risk graph method, given the general common structure presented in Figure 74, is generally applied in the SIL studies for process industries carried out under the technical standards IEC 61508 and IEC 61511 for aspects of functional integrity and safety with specific reference to the degree of risk reduction that must be associated with critical technical systems (IPLs) that act on demand and are generally characterized by a definable "pending" state.

In particular, this semi-quantitative method is used to determine the safety integrity level (SIL) to be assigned to safety instrumented functions (SIFs) put in place by safety instrumented systems (SISs). Through the application of the methodology, the required SIL is then classified, that is, the level of risk reduction that must be guaranteed by the presence and action in the event of a security function demand.

The starting point for SIL classification is the identification of the causes and consequences of process deviations that describe the incidental scenarios associated with a plant or section of it in a given plant configuration (including the expected process control systems).

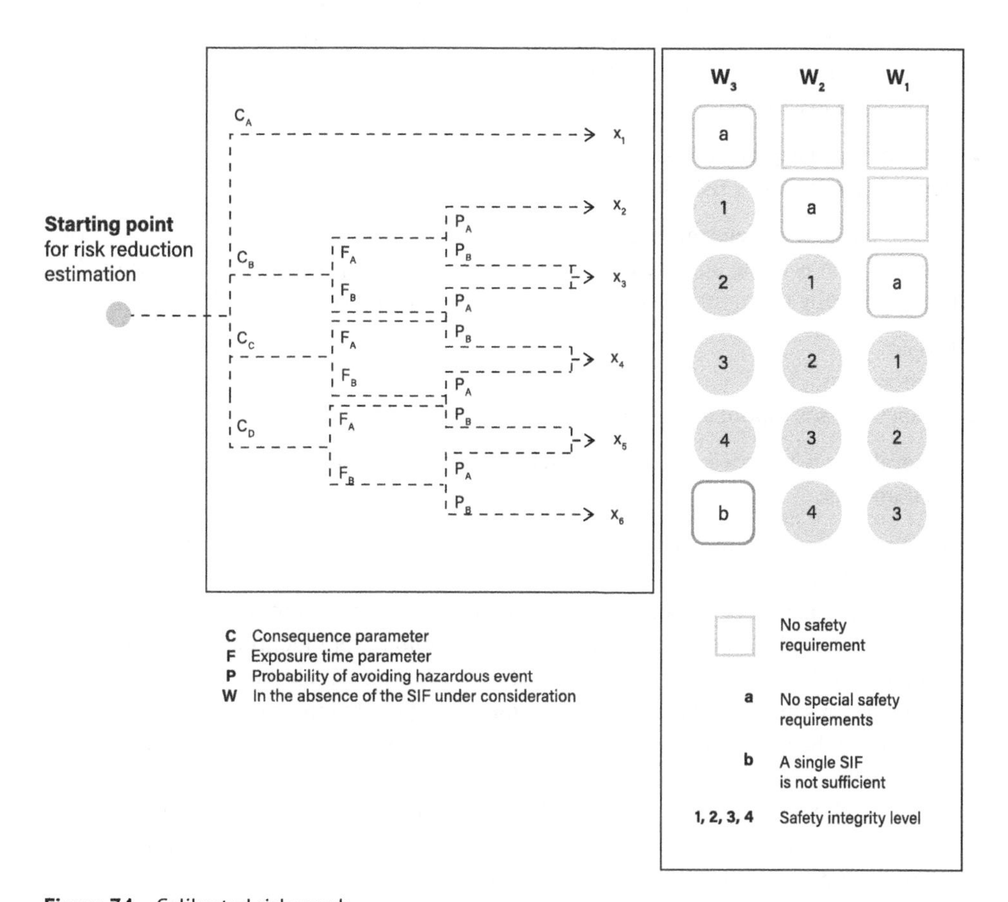

Figure 74 Calibrated risk graph.

In order to conduct a SIL classification, it is, therefore, necessary to identify incoming information to analyses aimed at determining the level of security integrity to be associated with each instrumental security function and in particular:

- Initiator event discovery: For example, basic control process system (BPCS) failure.
- Independent protection layer (IPL):
  - Operational interventions on alerting;
  - The intervention of safety integrity functions (SIFs);
  - Physical protections (PSV, rupture discs, etc.) and passive protection systems;
  - Active protection systems (fixed fire-fighting systems).
- Qualitative analysis of the expected consequences of the occurrence of a given event (meaning the occurrence of an initiator event and non-intervention of independent levels of protection in contrast)

The determination of the SIL, according to the selected methodology, is based on the use of parameters that, together, describe the nature of the dangerous situation that comes to be determined in the event that the instrumental security systems fail or are not available. Each parameter is chosen from a set of four possibilities, and the selected parameters are then combined to determine the required level of integrity. The calibrated risk graph approach is useful in identifying risk reduction needs.

The parameters allow:

- Determining a degree of risk;
- Representation of key risk factors.

For the purpose of determining the level of security integrity required for each security instrument that is being analyzed, the consequences of failing to intervene on demand of SIFs are generally taken into account, resulting in consequences for:

- People;
- Environment;
- Financial losses (optional).

The evaluation is carried out, as can be seen from the diagrams above, on the basis of four key parameters:

1) C (E): Parameter of the consequences (on people (C), environment (E)) associated with the occurrence of a detected incident.
2) F: Exposure parameter associated with the probability that an individual is present in the area affected by the consequences of the incidental event identified. This factor is a function of the fraction of time compared to the year in which the operational staff is normally present. It is also necessary to consider a possible increase in staff present to investigate the causes of an anomaly of the plant, which can give rise to the incidental event precisely during the search for the failure by the operators.
3) P: Probability of avoiding an incident by exposed persons, in case the security instrumentation does not work; this probability relies on the presence of warning systems in the field, independent of the instrumental security functions, which intervene promptly, and in any case before the occurrence of the incidental event and the escape routes present to ensure a rapid departure from the area potentially affected by the next event.

4) W: Frequency of occurrence of the causes leading to the estimated incidental event in the absence of the instrumental safety systems. This frequency is estimated considering the concatenation of all causes that may lead to the event itself, excluding independent instrumental protection systems.

In order to conduct the SIL classification activity of the instrumental security functions, i.e. the assignment of the degree of risk reduction that must be guaranteed by them and the following identification of their degree of availability in the face of the estimated frequency of initiator events, it is necessary to proceed with the calibration of the risk graph in view of the risk acceptability criterion (risk matrix). The calibration process is resolved by assigning numeric values to the parameters described above.

In the risk calibration phase, it is important to consider:

- The client's safety policies (e.g. risk acceptability matrix, security system redundancy, etc.);
- The minimum requirements imposed by the applicable regulation (e.g. Seveso and Offshore EU directives on industrial risk associated with chemical plants and offshore installations).

Cost-Benefit Analysis

The methods presented so far, once the tolerability and acceptability of risk criteria or the thresholds that identify these regions have been defined, make it possible to compare the level of risk actually present with these thresholds. These methods also allow, under appropriate working assumptions, to determine the extent of the risk reduction that may be necessary to reach an at least tolerable level of risk.

However, sometimes the decision to implement an additional control measure (i.e. a barrier) in order to cover a certain level of risk can be constrained by reasoning about the economic convenience of such an operation. In this case, a cost-benefit analysis is used to help the organization make a decision on that determination.

Assume that, in the absence of additional barriers, the occurrence of an injured event will result in an estimated damage of \$2,500,000 (total unmitigated hazard cost [TUHC]). The frequency of such a harmful event is known as the unmitigated event frequency (UEF), and, for example, suppose that an event can occur once every 15 years (i.e. UEF = 1/15). It is clear that the annual cost of the unmitigated hazard (AUHC) will be:

$$AUHC = TUHC \times UEF$$

In the AUHC sample, 2,500,000 to 1/15 is \$167,000 per year.

The risk reduction capacity by a barrier can be translated into quantitative terms using the risk reduction factor (RRF) parameter. The RRF to be met to ensure a minimum tolerable risk is expressed by the ratio of the frequency of the unmitigated event (UEF) to the maximum tolerable mitigated event frequency (TMEF):

$$RRF\_goal = UEF / TMEF$$

This implies that the risk tolerance criterion must, of course, be expressed in terms of TMEF.

With this premise, the RRF attributable to a specific barrier, its probability of failure on demand (PFD), results:

$$RRF = 1 / PFD$$

Known as the RRF of a specific barrier, the annual mitigated hazard cost (AMHC) is assessed, which is the reduced annual cost of danger attributable to the implementation of a control barrier.

$$AMHC = AUHC / RRF$$

In the example, assuming you want to implement a barrier with PFD plus 0.09, then you have AMHC plus 167,000/(1/0.09).

This means that the implementation of this barrier saves an annual amount equal to the difference between the unmitigated and mitigated annual cost.

$$\text{Hazard Cost Reduction} = AUHC - AMHC$$

In the example, this parameter is equal to 167,000 – 15,030 = $151,970 per year.

Now that it has been calculated how much money the barrier would save each year, you need to calculate how much the same barrier will cost each year. This cost will depend on the time and money it takes to design it, install it, test it, maintain it, any interest rates to finance it, and so on. The cost of a barrier is the barrier annual cost (BAC).

At this point, we calculate the benefit of the implementation of the barrier:

$$\text{Annual Barrier Benefit} = \text{Hazard Cost Reduction} - BAC$$

This process must, therefore, be repeated for each alternative barrier whose cost effectiveness of the installation is to be assessed. The barrier that offers the highest annual barrier benefit is the most cost-effective barrier to install.

Alternatively, you can also reason in terms of the relationship between hazard cost reduction and BAC. Reports above the unit indicate that the overall balance is positive and therefore the implementation of the pre-selected barrier is economically sustainable; lower-unit reports indicate that the cost of the barrier is greater than the expected benefits and that such assumptions should be discarded, at least from the point of view of the economic convenience analysis.

2.3 The Bow-Tie Method

The Bow-Tie risk assessment methodology owes its name to the typical shape that the Bow-Tie diagram generally takes: that of a "papillon" (Figure 75). Its strong ability to immediately transmit complex information through its powerful (though simple) notation and graphic design has made it one of the most widely used and appreciated risk analysis methods worldwide, regardless of type, size, and complexity of the application.

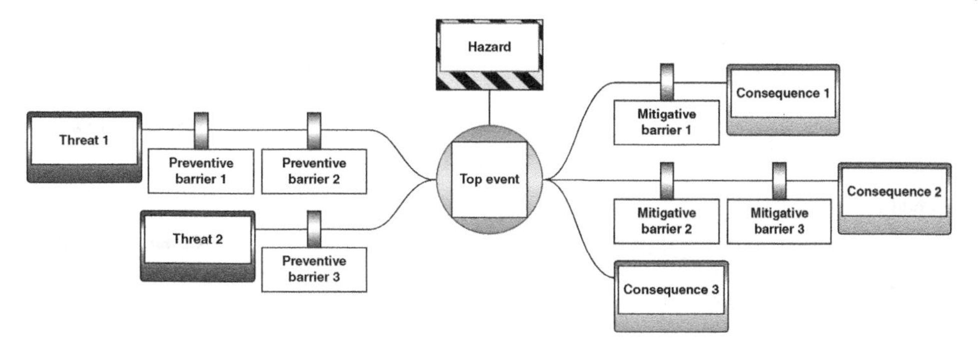

Figure 75 A typical Bow-Tie.

Its applicability is extremely wide and allows the analyst to deal with different types of risks with the same method (possibly combined with other methods), standardizing their treatment, documentation, communication, and discussion within the organization, which are peculiar aspects of the risk management process required by the ISO 31000 standard.

However, reducing its potential in the (albeit powerful) graphic notation alone is very reductive and risks diminishing its value within a larger overall framework that is risk management.

Although it is unclear when and how the method had its exact origin, it is widely recognized that Bow-Tie-based methods originated in the 1970s from the best-known cause-effects diagrams (CCDs) later adapted (in 1979, by David Gill, Imperial Chemical Industries) for use in post-incidental investigation.[1]

They became very widespread in the early 1990s when, following the accident onboard the *Piper Alpha* platform (North Sea, 6 July 1988, 167 deaths), the Royal Dutch/Shell Group developed a technique to improve the management of risk analysis activities in the years to come.

The application of the methodology quickly extended to other companies and other fields and first of all, in the Northern Europe world, to all those cases where the complexity of the reality to be assessed and the peculiarities of some installations (e.g. transport infrastructures), such as the presence of multitudes of people, special sub-services, and so on, were evident.

In order to offer a first definition, the Bow-Tie diagram can be understood as the fusion between a fault tree (FTA) and an event tree (ETA) (Figure 76). The junction point between the two trees, i.e. the centre of the Bow-Tie, is the so-called top event. In one fell swoop, the Bow-Tie therefore provides an exhaustive overview of everything that could cause an unintended event (the top event), through the branches to the left comparable to those of a cause tree, and the possible consequences of the top event, through the branches to the right superimposed on those typical of an event tree.

However, it should be noted that in the "simplified" definition of Bow-Tie as the mere union of a cause tree and an event tree, most of the inconsistent and incorrect uses of the Bow-Tie methodology fall. There are in fact several and also important differences between

1 The original BT diagrams were first introduced in 1979 on a course (Imperial Chemistry Industry) lecture about hazard analysis (HAZAN) held at the University of Queensland in Australia.

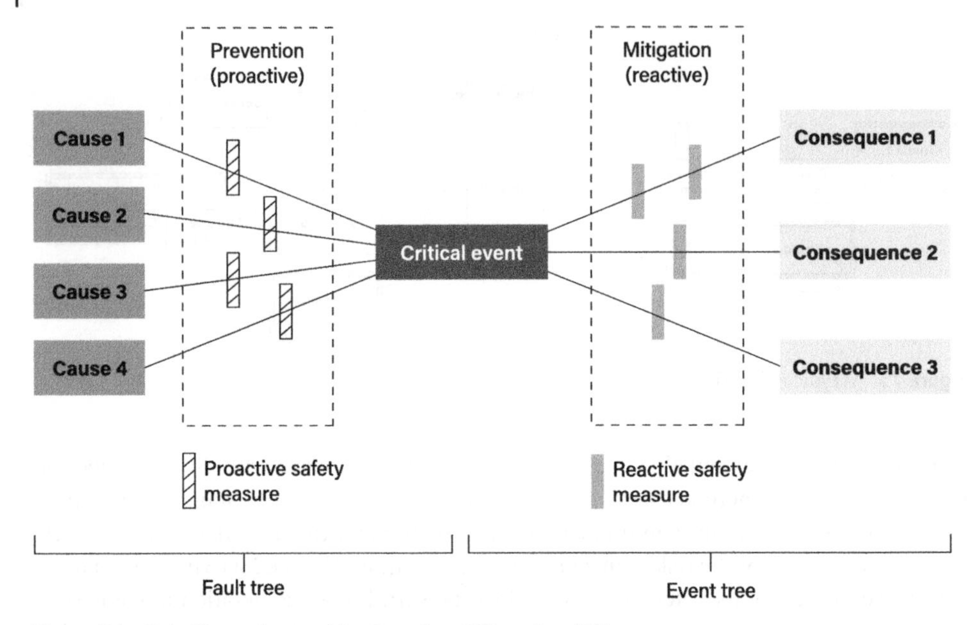

Figure 76 Bow-Tie as the combination of an FTA and an ETA.

the logic behind a cause tree and the left side of a Bow-Tie, first of all the independence between the causes and their direct ability to determine the top event. In other words, there is no possibility of considering the top event as the result of two or more causes that must occur at the same time; from a logical point of view it would seem that the causes in Bow-Tie are connected only by OR doors, unlike the cause trees where both OR and AND doors are used. As far as the right-hand side of the Bow-Tie is concerned, the representation of the typical intermediate events of an event tree does not translate so immediately into a BT diagram, which does not explicitly take intermediate events into account, although the experienced risk analyst knows that a reasoned and careful use of barriers could bridge these differences. What is discussed here represents only some of the reasons that led the author, with a strong sense of self-criticism, to present the parallelism between Bow-Tie and the union between FTA and ETA as a simplified definition, useful for novice analysts to understand the basic concepts of Bow-Tie. In general, this definition is not to be considered fully accurate and that is why the Center for Chemical Process Safety (CCPS) of the American Institute of Chemical Engineers (AIChe) decided to publish the first official guidelines for a correct use of the Bow-Tie method, in order to gain more important advantages and reduce cases of misuse. The content of these guidelines is discussed in detail in Section 2.14.2 where the main principles of the AIChE CCPS approach are discussed.

The Bow-Tie provides a comprehensive answer to the three basic questions that any risk manager knows and considers the basis of their work:

- Do you know what could go wrong?
- Do you know the systems (i.e. barriers) that are in place to prevent this from happening?
- Do you have enough information to say that these systems are working effectively?

These questions, which will be presented several times to the reader due to their importance, could be referred both to the initial and periodic risk assessment phases and the continuous management phase. They summarize the principal issues an organization should face if it wants to start dealing with undesired events that may arise during business operations.

The first question is therefore related to hazard identification and risk analysis, the second to barriers, and the third to their effectiveness. It is clear that from a barrier-based perspective (i.e. the IPLs of other methodologies), these are the questions that need to be answered in order to assess whether or not business risks are properly controlled.

Like other methods, the Bow-Tie diagram – with its explicit, complete, and intuitive graphical notation – is nothing more than a model used to interpret the complexity of reality. It is therefore not the tool that every organization must necessarily adopt, but rather the model that, better than many others, makes it possible to read and interpret the complex reality of risks and their management, in the opinion of the authors of this volume, who have used it in different situations, with different types of risks, and with the need to illustrate the results of consulting activities to various types of stakeholders, both internal and external to the organization. The method has also played a leading role in teaching activities on risk analysis and management, with particular reference to courses based on the principles of ISO 31000.

It, like other methods based on the concept of barriers, is based on the well-known Swiss Cheese Model by James Reason (Reason, 1990) (Figure 77). According to this model, which is widely used every day to explain the malfunctioning of complex technical and/or organizational-management systems, the measures to control a hazard (i.e. barriers) can be metaphorically compared to slices of Swiss cheese, which are placed between the hazard and the accident.

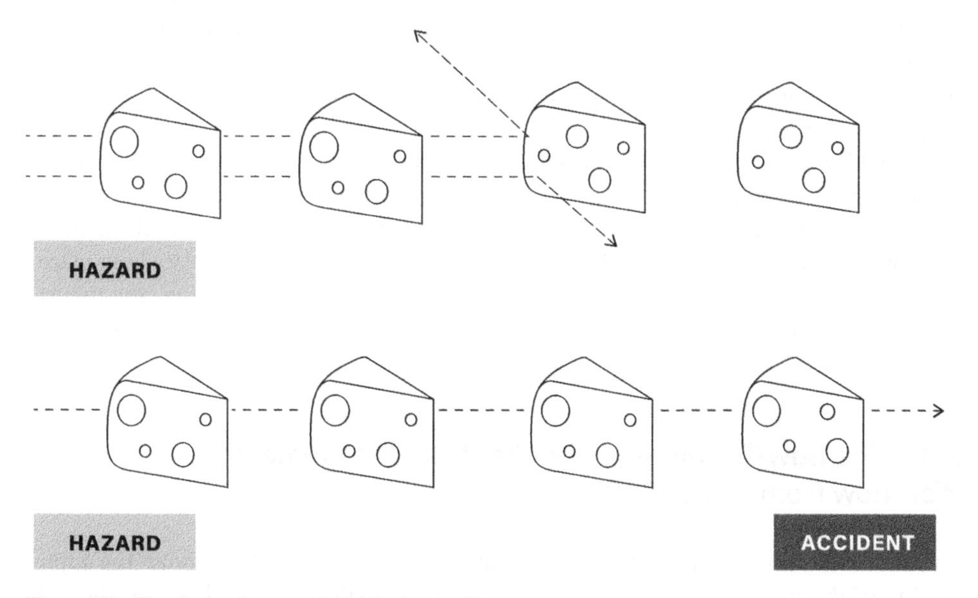

Figure 77 The Swiss Cheese Model by James Reason.

The holes in Swiss cheese slices represent their never being 100% effective (or, if you prefer, reliable): some holes are due to latent conditions, others to active failures and in general to random failures. Generally, the barriers (at the design stage) are placed in such a way as to avoid the alignment of such weaknesses, thus ensuring that, although each barrier is not 100% effective, the overall system is still safe. However, under certain conditions (with a specific frequency), it may happen that the holes, i.e. the intrinsic weaknesses of each barrier, align with each other, thus allowing the danger to become a real accident.

As will be made clear in the course of reading this book, the Bow-Tie method is used:

- Reasoning in a structured way with respect to complex systems;
- Making risk-based decisions;
- Communicate and conduct risk awareness, information and training activities;
- Monitor barriers with periodic audits to record their status.

When compared to other risk analysis methodologies, the Bow-Tie method has a number of advantages that will be explored in more detail next in this book:

- It is easier to interpret due to effective and intuitive notation as well as strong graphics.
- It reduces the complexity of information.
- It makes barriers (preventive or mitigative) immediately visible.
- Shows the effectiveness of barriers.
- It is combined with incident analysis (reactively).
- Combines with audit plans (proactively).
- Provides, well implemented, a structure to the management system consistent with a risk management process developed according to ISO 31000 principles.

On the other hand, one of the disadvantages is the impossibility to hold against multiple simultaneous causes (normally connected with AND logical doors in a fault tree). It is, however, possible to overcome this limitation by grouping within the same Bow-Tie case all those causes that must occur simultaneously in order to reach the top event, as better explained in Section 2.10.

Given this evident limitation, it is possibile to say that in any case the Bow-Tie method is the most intuitive approach to risk assessment and to risk management for all those organizations, operating in different contexts and sectors, that have the goal to approach an initial risk-based thinking built around the evolution of the barriers they have already in place, initially to understand the need of an increase in number, quality, definition, and robustness of those. Subsequent further assessments will be then performed with more sophisticated methods.

2.4 The Bow-Tie Method and the Risk Management Workflow from ISO 31000

The Bow-Tie method, as also indicated in Table 1 of ISO 31010, which shows the applicability of the methods suggested by IEC/ISO 31010 to the various phases of the risk management process, is particularly suitable for the risk analysis phase. In particular, as will be illustrated below, the integration of the Bow-Tie with LOPA allows a frequency quantification of the same, thus obtaining one of the two input elements for the determination of the risk level: the probability (the other is the magnitude of the incident scenario).

The Bow-Tie diagram, whose construction follows the steps reported in Section 4.1, therefore offers itself as a tool for organizations wishing to manage risks in accordance with the dictates of the ISO 31000 technical standard, and in particular its principles, and its management process (see Figure 18).

In particular:

- The definition of the context and purpose of the work is undoubtedly the preliminary stage for conducting a Bow-Tie workshop. On this basis the level of abstraction will also be determined (see Section 2.4).
- Communication and consultation are vital elements for a truly productive Bow-Tie workshop, where profiles with different skills work synergistically to identify beforehand (e.g. through a HAZID) and analyze afterwards the risks.
- Risk identification can be carried out using any of the techniques suggested by IEC 31010, including brainstorming, checklists, HAZID, and so on. Taking the offshore Oil & Gas sector as an example, ISO 17776, *Petroleum and natural gas industries — Offshore production installations — Major accident hazard management during the design of new installations*, could be used to provide a list of hazards to be brought to the attention of the working group, identifying those that are actually present and are part of the scope of the work initially defined.

Risk analysis is the heart, and main goal, of the Bow-Tie session: the various elements making up a diagram are identified as best specified, i.e. top events, causes, consequences, primary barriers, escalation factors, and secondary barriers. Their precise definition may require further investigation, which must be postponed to a second stage when more information and evidence is made available to the team. In this phase the possible deviations from the control conditions of the identified hazards are identified, as well as the possible causes of these deviations and their possible consequences on the targets under consideration (safety, environment, quality, reputation, etc.). The control measures that can prevent certain causes or mitigate certain consequences are then identified, taking care to provide all the relevant information described in those: owners' responsibilities, any reference documents, performance criteria, and so on. Risk analysis using the Bow-Tie method, in line with what is suggested by the standard, can be both qualitative and quantitative. In the first case it is sufficient to express the data characterizing the risk (i.e. probability and magnitude) in qualitative form according to a scale of values that must be defined in the context definition phase. In the second case, it is possible to integrate the quantitative analysis thanks to the layer of Protection analysis (LOPA), of which more details are given later on; in this way, thanks to the Bow-Tie method, a risk assessment of the accidental scenarios (i.e. consequences) can be added. The risk values obtained must then be compared with a risk acceptability/tolerability criterion, also defined in the context definition phase.

The representation of the risk analysis in a single image (the Bow-Tie diagram) allows an immediate view of the benefits offered by the various risk treatment options. For example, you can see how the probability of a consequence occurring is changed if one or more control barriers are removed, or if an initiating cause is suppressed.

Finally, the Bow-Tie diagram, offering a graphical display of the dynamic concept that is risk, must be kept up to date, changing the causes, barriers, and consequences identified when required. This may, for example, occur downstream of audits, observations, or recommendations resulting from an analysis of incidents, near-misses, or non-conformities.

The points presented here are discussed in more detail in the following paragraphs. This summary is intended to demonstrate that the Bow-Tie is a method fully aligned with risk management under ISO 31000.

2.5 Application of Bow-Ties

The Bow-Tie method, as underlined by the collection of examples in Section 4.4, can be used to analyze risks of any kind, in any organizational context. Unlike some methods that have found their driving force and consolidation in certain areas of business, the Bow-Tie method has managed to preserve the universality of its field of application, thus offering itself as a method for the management of any risk in accordance with ISO 31000.

The purpose of this section is, therefore, even before defining the constituent elements of a Bow-Tie diagram, to make the reader appreciate the ease of consultation of the diagram, which, thanks to its powerful graphic notation, offers itself as a tool for information and knowledge of risks at different levels of the organization and for different stakeholders.

The following is an example of the management of a business process (the transfer of a data center), whose undoubted complexity is analyzed using the Bow-Tie. For this example, we would like to thank Domenico Castaldo from TIM S.p.A.

The Bow-Tie model (Figure 78) of risk management lends itself very well to managing the risk analysis of any project regardless of context (type, entity, size and complexity). To support this statement, the applicability of the Bow-Tie model to a facility management project is presented below: the decommissioning of a data center (DC). This example is summarized in Figure 79.

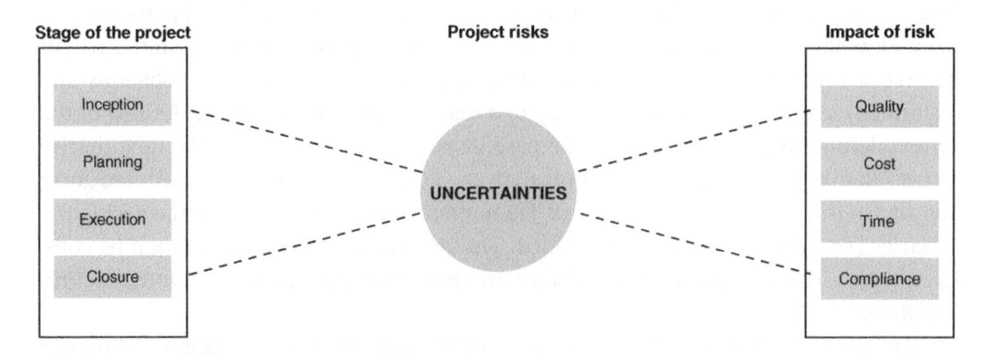

Figure 78 Bow-Tie project risk assessment.

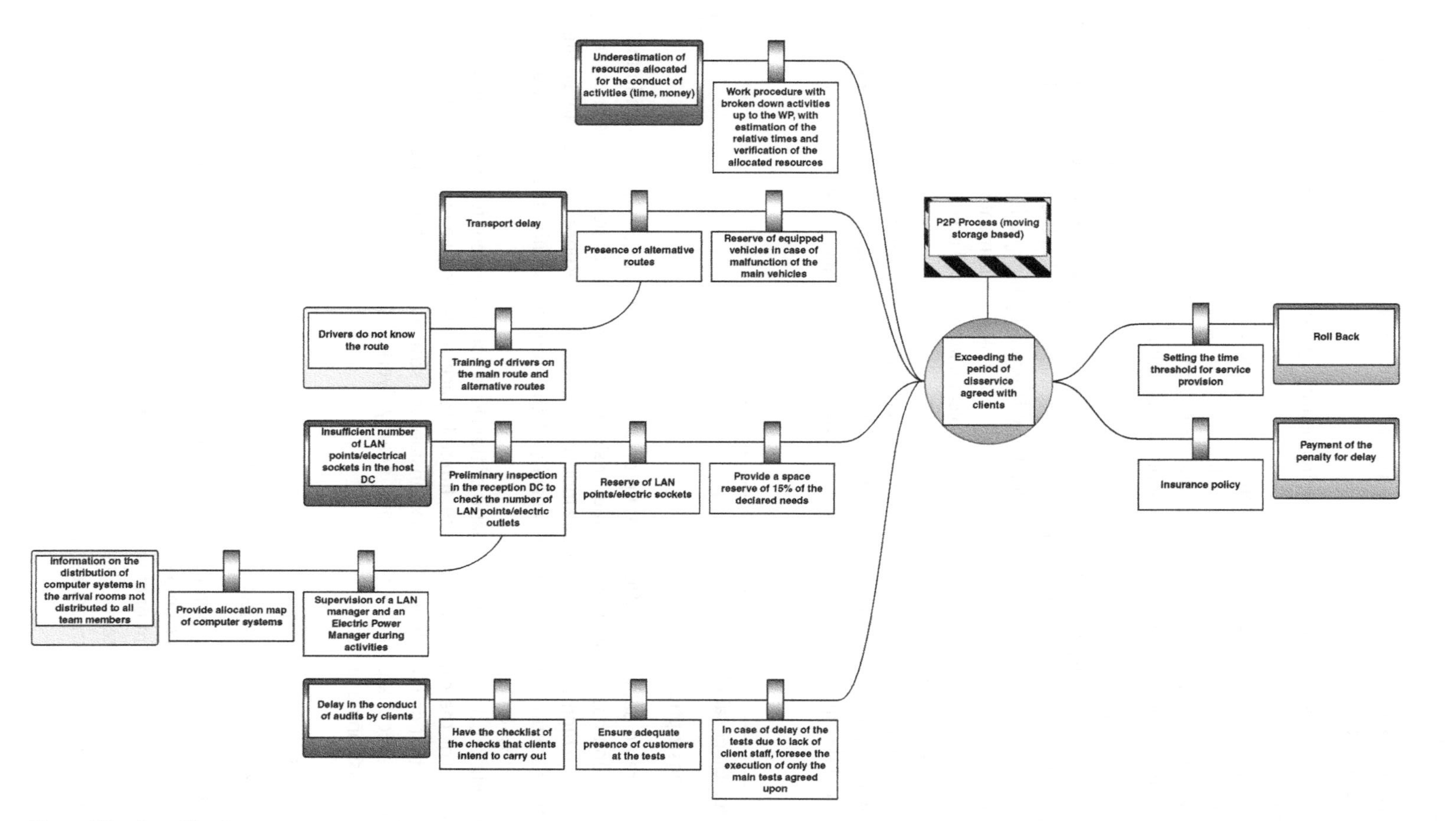

Figure 79 Bow-Tie diagram – transfer of a data center.

In a DC there is a high level of complexity; just to give an idea of the complexity there are four macro areas:

1) A diffuse electrical component (switchboards, generators, electrical backup systems, etc.);
2) An internal network component connected to the external one with redundancy to guarantee access to data and applications;
3) Hundreds of servers on which the applications run;
4) Storage areas for data with the corresponding backup areas.

Therefore, in the case of a project whose objective is to complete the decommissioning of a DC, including the displacement of its valuable content (applications and data) into other DCs, we can conclude that the project is complicated. If we add that there is no benchmark, but, at most, guidelines, because the specificity is given by the age level of the environments to be moved, we can conclude that the project risk management is (always, but now even more so) an activity that starts from the project planning and accompanies the project life cycle until the project closure.

The following is a presentation of the project with its description, objectives, limits, assumptions, and constraints; the operational modalities for the implementation of the project and the application of the Bow-Tie model to a given project process.

Project: Disposal of a Data Center

Project description

The migrated data center (DC) is set up on a site originally not designed as an industrial site and has a higher level of fault risk than the average of other DCs due essentially to the high obsolescence of the industrial plants, the presence of important single points of failure, and the limited capacity of the floor that limits the installation of enterprise systems.

In addition, the adaptation scenario of the DC is not convenient for several reasons: closeness of the expiry of the lease, limited space remaining compared to other DC available that do not allow further expansion, and structural constraints (location close to high-traffic roads and flooding of a nearby river). These considerations led to the decommissioning project.

Objectives

The objectives of the project concern three areas:

1) Data Center: moving non-obsolete equipment and maximizing the infrastructure of the destination DCs (compaction).
2) Facility: Replacement of obsolete equipment and implementation, in some cases, of solutions of absolute excellence in the target DCs.
3) Information Infrastructures: definition and revision of backup policies in the target DCs.

Project limits

The planned activities do not include functional increases in migrating applications. The project will develop all and only the actions and solutions related to the infrastructural improvement and evolution of the applications. The actions will be focused on the expected time and costs, binding everyone to favour all and only the solutions that will be compatible with these drivers.

Recruitment

The organizational responsibilities already present in the company that owns the DC have guided the roles in the organization of the project to ensure the highest level of effectiveness and efficiency in achieving the objectives.

The work of the project has been broken down into many "vertical" components (subprojects), as many homogeneous macro areas are present in the DC with the "horizontal" components Farm, Network, Data Migration, and Backup; this has been done to clearly identify roles and responsibilities while maintaining control of the interdependence between layers (dependent multilayer strategy).

Constraints

Respect for the expiry of the lease requires that the project end within three months from the date of delivery to the property in order to guarantee the restoration of the rooms.

The nature of the applications involved requires technical solutions that *minimize service unavailability times.*

Operating Modes

Migration operations concern the relocation of all the departing DC's IT systems and the equipping of the operating rooms for the receiving DCs.

In this context, the term *computer systems* refers to the whole made up of hardware (server and rack), storage (internal to the server or external to the server – e.g., SAN) and the network (LAN and electrical).

The migration methods have been divided into three different operational processes according to the nature of the type of server in the computer system to be migrated. The servers can be: virtual (V – i.e. partitions of physical machines) or physical (P) and can be migrated on virtual machines if they are V- or F-compatible or on physical machines if the servers are F and not portable on V machines.

Hence the three processes and their main characteristics:

1) P2P (physical to physical): uneven, variable and not repetitive
2) P2V (physical to virtual): homogeneous, standard and self-correcting
3) V2V (virtual to virtual): homogeneous, standard and self-correcting

In practice, P2V and V2V processes can be implemented by migrating applications (apps) running on computer systems and related data; once migration is complete, rooms can be freed from computer systems (*moving apps based*).

The critical process is P2P, which can have two different levels of criticality: high (H level) or low (L level).

The criticality is L level when servers have internal storage or external storage (SAN), but they are not too obsolete and it is possible to migrate storage (apps data) via LAN. In this case, in the operating rooms set up in the reception DC, the servers will physically move to the new storage that will be different from the initial ones, but, despite the technological upgrading of the storage, the servers will be able to acquire their data.

When, on the other hand, the servers are too obsolete, it is unlikely that they will be able to read their data acquired from new technological storage and therefore it is necessary to move the storage from the starting DC at the same time as their servers (H level criticality). In this case the migration methods are completely different because you have to move all the IT systems whose apps have the data on the same external storage so as not to block the apps by depriving them of their own data (*moving storage based*).

Since this is the most critical case, we will apply the Bow-Tie model to the P2P process in the *moving storage based* case with the aim of controlling it so that the duration of the suspension agreed with customers to ensure business continuity is not exceeded; in practice, the agreed period of disruption should not be exceeded.

In line with the Bow-Tie model we define the contents of the constituent elements.

Danger: P2P (moving storage based) process

Top event: Exceeding the period of disservice agreed with clients (respect of time)

Cause 1: Planning – the list of activities may not be detailed up to the elementary work package; the estimated timing of individual activities may be optimistic; allocation of resources for activities may be underestimated.

Preventive Barriers 1: Planning – check the list of activities, making sure that it is broken down to the elementary activities (WP); make estimates of the timing of individual activities with different methods (estimates by analogy, parametric estimates, PERT with the analysis of the three points – most likely, optimistic, pessimistic– etc.), considering the longest estimate as valid; check with the use of project management software products the estimate of resources allocated and submit it to expert colleagues for further verification.

Cause 2: Transport – alternative road routes may be missing; the number of equipped vehicles may be reduced.

Preventive Barriers 2: Transport – check with the mobility manager that the main route has alternative routes to use and estimate the journey times; make sure that the mobility manager makes the necessary number of equipped transport vehicles available and activates a "hot reserve," i.e. a sufficient number of equipped vehicles ready to leave in case of malfunction of the main vehicles.

Escalation factors: Drivers may not have a detailed knowledge of the route.

Secondary Barriers: The mobility manager ensures that drivers have detailed knowledge of the route by means of a meeting in which the manager will present the map of the entire route area (main and alternative routes) in detail and give a copy to the drivers' team.

Cause 3: Preparations – in the reception DC there may be a lack of LAN points/electrical sockets; in the arrival rooms, the areas prepared for the distribution of incoming computer systems may be smaller than necessary.

Preventive Barriers 3: The project manager (PM) ensures that the arrival DC's facility manager has set up the correct number of LAN points/electrical sockets to meet needs and a "cold reserve" of LAN points and electrical sockets has been set up as needed; rebook all the spaces in all the pre-allocated areas with the arrival DC's facility manager to ensure that there are no allocation errors and still leave a reserve of space equal to 15% of the declared needs.

Escalation factors: Information on the distribution of computer systems in the arrival rooms was not disseminated to all team members (delivery and acceptance); presence of a LAN manager and an electric power manager.

Secondary Barriers: The facility manager will ensure that all members of the arrival DC team have a map of the computer system allocation for each room and will activate the meeting where they will present the map for each room in detail and retrace the related activities. At the same meeting, the facility manager will set up the operational supervision for the duration of the entire operations of a LAN manager and an electric power manager.

Cause 4: Testing and validation – client departments may not have shared the app functionality checklist; the client team may not be present in sufficient numbers to parallelize the testing; the client team may be later than the agreed time.

Preventive Barriers 4: The PM must request the checklist of the tests that the client intends to carry out; the PM must know the matrix prepared by the clients between participating clients and the tests to be carried out to ensure full client-test coverage; the PM must agree with the client manager that in the event of delay in the start of the tests due to lack of their staff, only the main tests agreed upon will be carried out.

Consequences 1: Roll Back – ensure that the service is restored by returning the computer systems to the starting DC.

Mitigation Barriers 1: The PM, together with the mobility manager and facility managers (DC arrival and departure), must estimate the service delivery times (applications *up and running*) back in the DC of departure in order to set the time threshold beyond which one should not go in order to guarantee rollback.

Consequences 2: Penalties – activate the financial recognition of the delay penalty to the client for the restoration of the app service in the DC of arrival.

Mitigation Barriers 1: At the contractual level, the contract manager must provide penalties and SLAs for extra time suspension of business continuity in the event that exceeding the times within the extra time SLAs can avoid rollback; the contract manager will take out an insurance policy to mitigate the cost of the penalty.

The next example concerns the emerging risks of virtual classroom training. We thank authors Toncelli Luca and Francesco Fanelli of Human Factor Italia. The authors apply Bow-Tie methodology to a qualitative enhancement of risk assessment involved in the design and use of the training to integrate IATA guidelines on the available methods to improve the quality of training where classrooms are delivered via virtual online webinars. The resulting Bow-Tie diagram is shown in Figure 80.

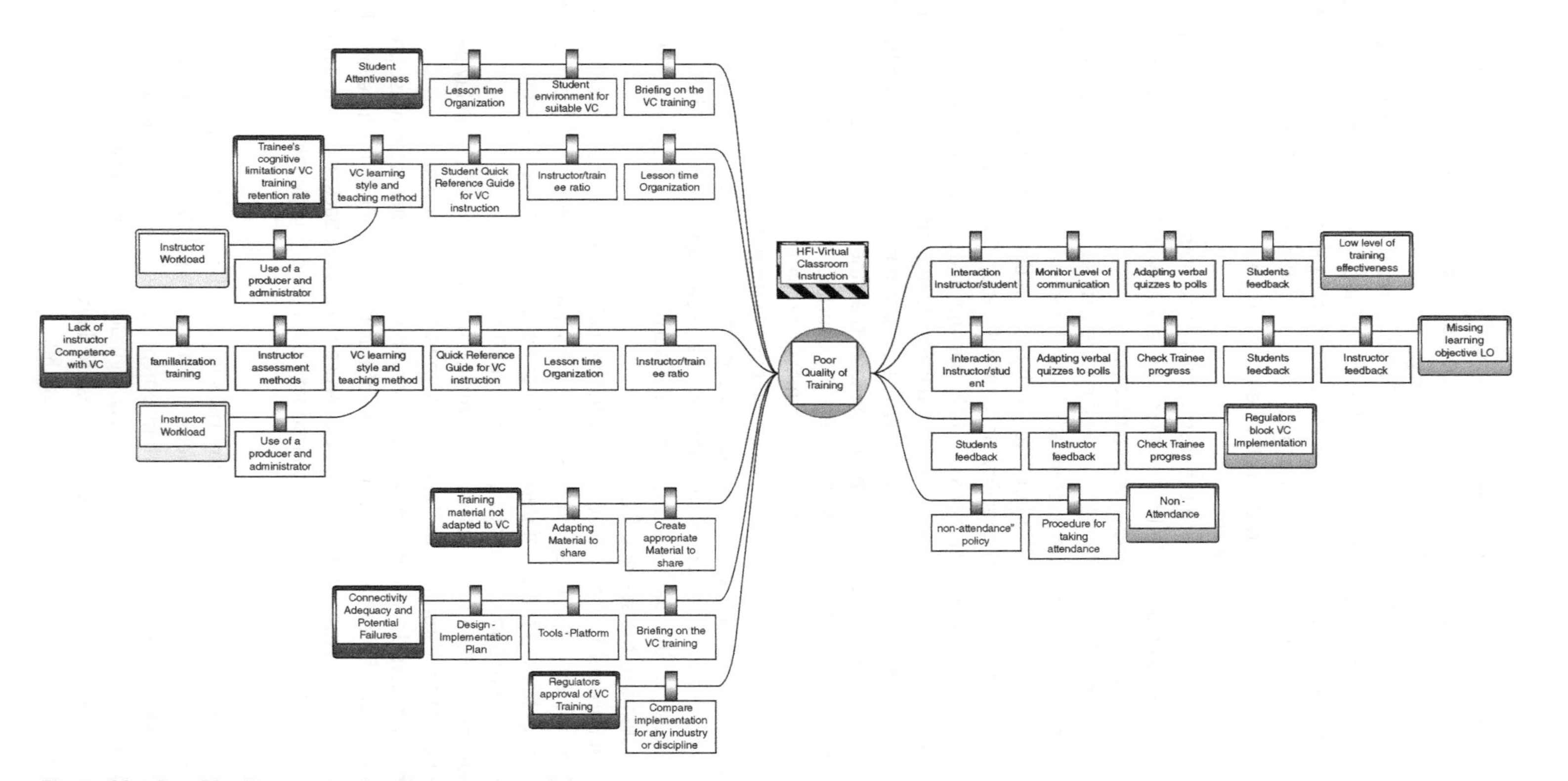

Figure 80 Bow-Tie diagram on virtual classroom training.

2.6 Level of Abstraction

The definition of the context and scope of work is, as already mentioned, the first step in the risk management process. In this delicate phase it is also necessary to define the level of abstraction that the Bow-Tie diagram should have. Any analysis of the risk, for example on the safety of the operators of an industrial plant working with dangerous materials, must give priority to establishing the general rules on which the entire assessment phase must be carried out. Taking the image in Figure 81 as a reference, the ideal Bow-Tie is a middle ground between being too specific and being too generic. During a Bow-Tie work session, there is often a tendency to be too general in defining causes, consequences and barriers. The risk of being too generic is to lose, perhaps because you take for granted, the key information that needs to be considered to ensure effective risk management. For example, one could omit indicating the figure responsible for the functionality and integrity of a control measure, or identifying in an unidentified operational error the cause of a possible top event, or indicating the consequence of pollution without specifying the environmental matrix involved or the quantities of pollutants released by the damaging event. On the other hand, however, one should not be too specific, otherwise one runs the risk of losing the general vision of the evaluation conduct by enriching it with technical information that is not necessary at this stage or by integrating details that limit the scope of the work carried out. Defining the level of abstraction is therefore a preliminary choice.

Usually people building a Bow-Tie diagram, especially in the initial period, decide to employ this method for risk assessment, and tend to be too generic, overly simplistic, and confused by the limited number of Bow-Tie elements and its very intuitive notation.

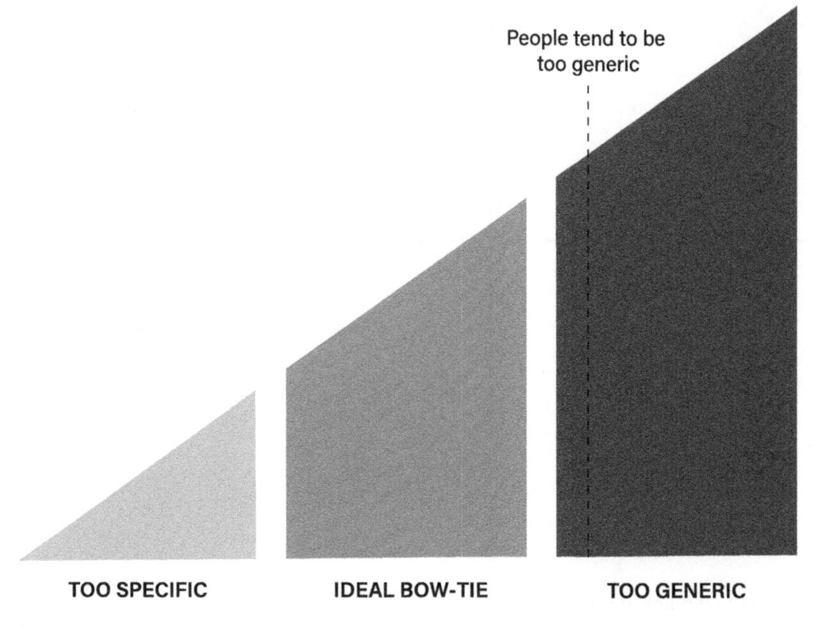

Figure 81 Level of abstraction.

Bow-Tie has been conceived to visually represent via an intuitive approach even very complex situations: it is a simple approach to complex problems, not a way to underestimate situations.

Diagramming should always consider the reality to be modelled and the objectives of the assessment. Since Bow-Ties can be used for both qualitative, semi-quantitative, or quantitative risk assessment, the detail level of the Bow-Tie should be defined accordingly, considering that, on the basis of the results, risk-based decisions should be made.

Both the "zoom level" of the risk assessment and the time scale should be defined (see Figure 82). For example, an analysis could take into consideration all the same plants of a multinational group or analyze in particular one of them and thus specify the additional or deficit control measures that may be present, or causes or consequences peculiar to that plant due to its specific hazard, vulnerability and exposure.

To understand the importance of fixing the point in time, consider Figure 83 and ask yourself what the top event to consider is:

- Residual water in tank;
- Corrosion;
- Loss of containment;

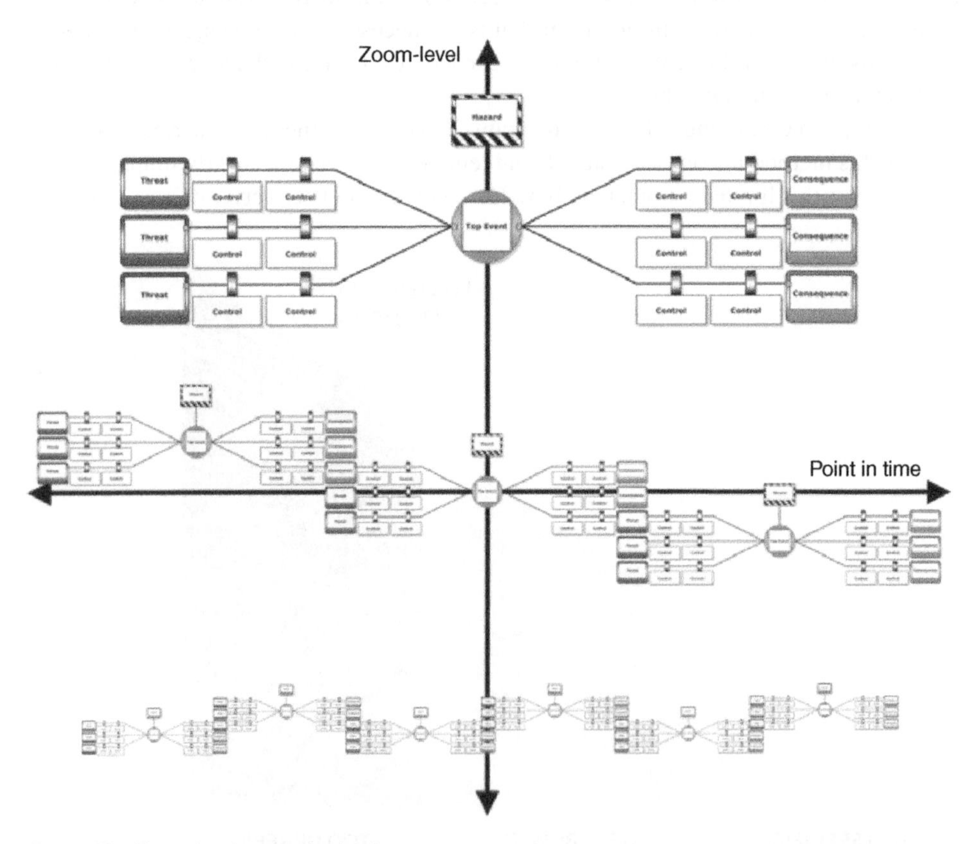

Figure 82 Zoom level and point in time.

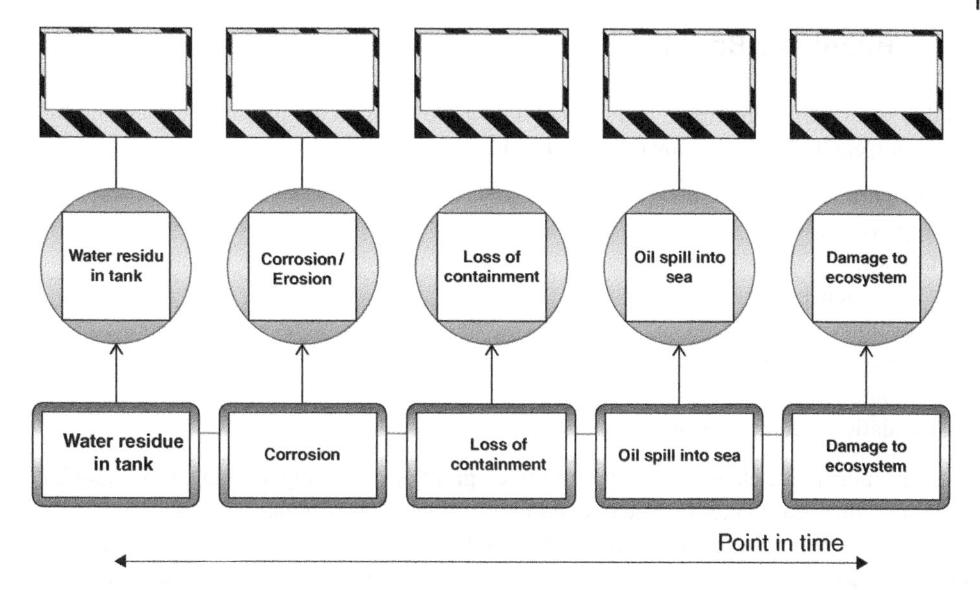

Figure 83 Example of point in time.

- Oil spill into the sea;
- Damage to the ecosystem.

The answer is: it depends. It depends on which point in time is being considered, i.e. who the party interested in the risk assessment is. In particular:

- For those dealing with detergents for hydrocarbon tanks, the top event is the residual water in tank.
- For the maintenance department, it is corrosion.
- For the operations department, it is the loss of containment.
- For the emergency response team, it is the release of hydrocarbons into the sea.
- For the fisheries government authority, it is damage to the ecosystem.

Zoom level is an important feature of the Bow-Tie's diagramming activities that can be joined with the quality, detail of data, and information associated with the single elements of the Bow-Tie.

If during the risk assessment process detailed and "zoomed" diagrams (eventually, quantitative ones) should be used, more abstract versions of the same diagrams could be used to communicate main issues and most important outcomes to the organization leadership and top management or to internal and external stakeholders; simplified and barrier-focused diagrams could be given to single managers having direct responsibilities for those; chained high-level diagrams can be shared with process owners to give an overview of the entire organization in terms of main risks, shared risks, common cause failures and other relationships and correlations; vulnerabilities and consequences on target-focused diagrams can be shared with specific in-place domains management systems owners; simplified Bow-Ties could be used for training and for information sharing on public boards.

2.7 Building a Bow-Tie

In this section the constituent elements of a Bow-Tie diagram are presented. These are the following eight elements, shown in Figure 84:

1) Danger;
2) Top events;
3) Causes;
4) Consequences;
5) Preventive barriers;
6) Mitigative barriers;
7) Escalation factors;
8) Escalation factors controls.

Each informative element making up the diagram is univocally associated to a graphic representation: this makes the elaborated model immediately readable and comprehensible.

It is clear from the outset that the names used are not the only ones that can be encountered in the various sectors of industry. The same elements in fact are also indicated through some synonyms, more or less widespread depending on the application context. It is worth remembering that the Bow-Tie method, born in the bedrock of the practices and methodologies for guaranteeing the safety characteristics of industrial risk in the oil & gas sector and in particular in that of offshore industrial assets, is now also widely used for completely different types of risk. Therefore, we often hear about proactive and reactive barriers, indicating those placed on the left (preventive) or right (mitigative) side of the Bow-Tie; or we could use the term "threats" instead of "causes," or "incidental scenarios" instead of "consequences," or "degrading factors" for escalation factors, or "secondary barriers" for "escalation factor controls." The same "barrier" may be referred to as "control measure," "risk reduction measure," "IPL," "individual protection layer," or "independent protection layer," and so on, depending on the organizational context and how certain terms have already been used (and established) in this context. What is important is to understand that these are synonyms and that, within the methodology illustrated in this book, they all assume the meaning that is clarified in the following paragraphs.

In any case, while fully respecting the principles set out in ISO 31000, the user of the method is free to integrate further elements and specific taxonomies aimed at better describing, with respect to the use of the basic system, the risk scenarios typical of the context in which it operates, in the same way the user can extend the basic information tools of the notation with the elements, also of a quantitative type (discrete or stochastic), as a result of the application of other techniques in parallel specialist in-depth studies.

It is therefore very important to underscore that, if ISO 31000 doesn't require the use of a specific risk assessment method (or even the barrier-based approach), multiple methodologies could be combined in order to achieve better and more useful results for the organization's goals. In this book we emphasize the combined use of the Bow-Tie method and LOPA for the risk quantitative assessment, as well as for the use of Bow-Tie and BFA methods to optimize the resources in the active and reactive phases of risk management; but there are other infinite combinations. FTA and ETA could be used to calculate the PFD of technical

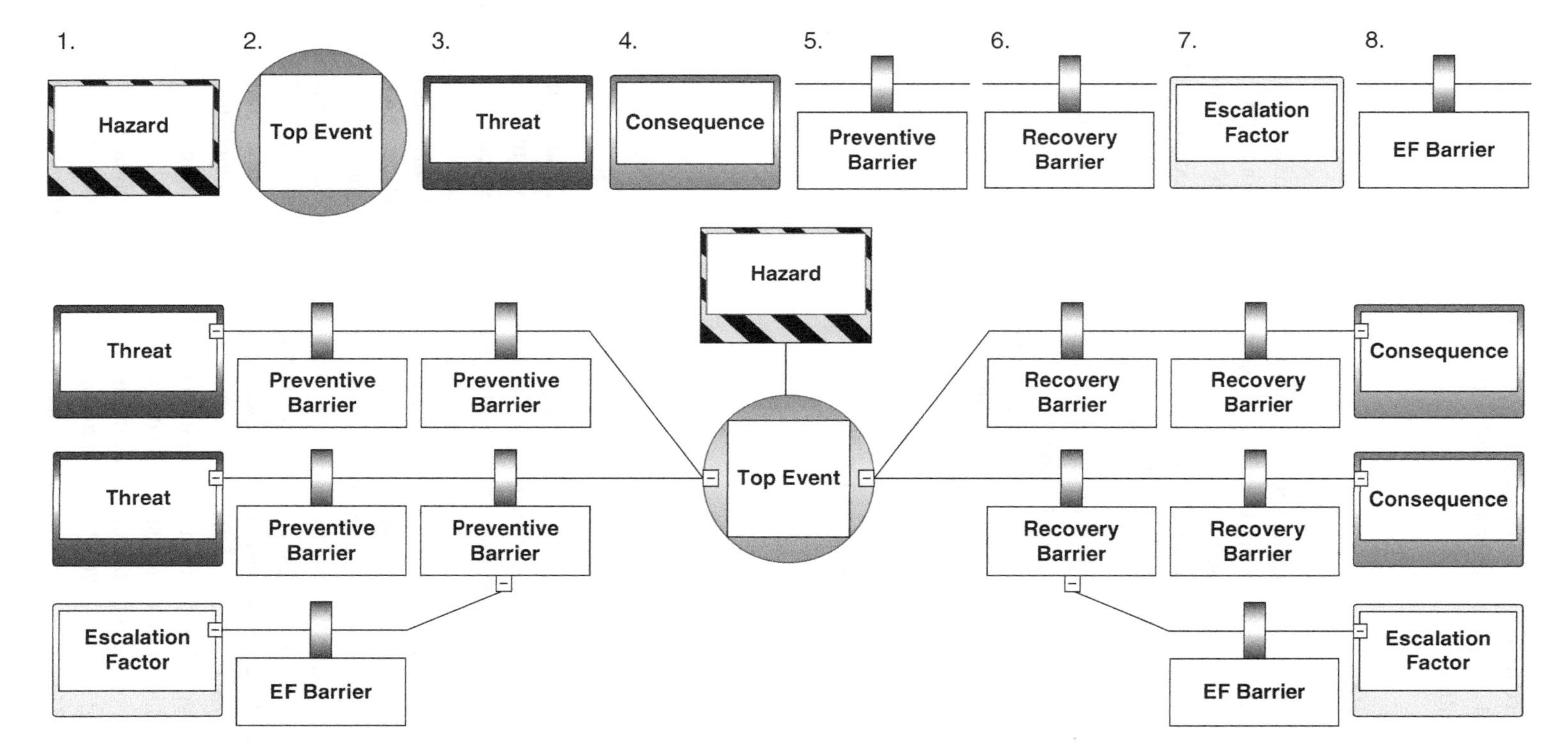

Figure 84 Basic elements of a Bow-Tie diagram.

barriers, HRA could be used to assess the probabilities associated with human factors (errors as threats, responses as behavioural mitigation barriers), and so on, with a number of new possibilities.

2.8 Hazards

The word "danger" might suggest something unintended, to be avoided. In the case of the Bow-Tie methodology this is not the case: "danger" is an operation, activity, or material that has the potential to cause damage, but is still a normal part of the business. It represents the opportunity that the organization takes to pursue its objectives (in the case of a company, profits). It could be said that without danger there would be no business. For example:

"Driving a car" is the danger for anyone going to work with their vehicle, but it is the opportunity to satisfy the need for transport.

"Working at height" is the danger for a labourer who has to carry out restoration work on the exterior of a building, but it is the opportunity to satisfy the need to carry out the work.

"Gasoline in a closed tank" is the danger for anyone in the proximity of that tank, but it is the opportunity for those who have stored it to be able to use or sell it later.

These are therefore activities or states that are part of normal business. Very often they involve different forms of energy, including:

- Chemical;
- Gravitational;
- Thermal;
- Biological;
- Kinetic;
- Nuclear;
- Pressure.

In a broad sense, considering the general applicability of the method to various contexts, it could also affect other forms of energy, such as social, financial, organizational, and so on.

It is clear that, in any case, the danger must be identified, understood, and managed: in fact, as long as it is kept under control, the fulfilment of the objectives is safely guaranteed.

However, identifying dangers is not as simple as it may seem. A correct definition of the hazard is a prerequisite for proper risk analysis with the Bow-Tie method, as it has the important function of defining the scope of the Bow-Tie from the outset. For example, "Driving the car" and "Driving the car in an urban context" are dangers that define completely different areas: the first is more general and identifies the danger of driving a car regardless of where you are driving it; the second describes a specific context, the urban one, within which the action of driving the car is carried out, thus highlighting threats and consequences typical of that area. Its definition therefore also identifies the zoom level that is to be adopted to analyze the risks associated with it.

In the world of the offshore oil & gas industry, reference is often made to the list of hazards defined by ISO 17776, *Petroleum and natural gas industries – Offshore production installations – Major accident hazard management during the design of new installations*, which is often adopted as a list of keywords to be adopted during HAZID sessions prior to the definition of Bow-Tie. Although this technical standard originated for offshore, it contains valid suggestions for onshore industry too, as well as a good example of taxonomy that can be used for the description of hazards attributable to a specific context (potential magnitude of damage, investigation priorities, areas of impact associated with the hazard).

2.9 Top Events

Given a well-defined danger, the next step is the identification of the top event positioned in the centre of the Bow-Tie diagram.

This is the precise moment when control over the danger is lost, allowing it to release its potential. Although the top event has occurred, it is not yet an accident (or impact), as there is still time for the reactive barriers to intervene in order to stop the sequence of events and prevent the consequence, or at least mitigate its magnitude. Examples of top events are:

- Loss of control of the car;
- Falling objects from above;
- Loss of primary containment.

It is absolutely possible to identify several top events for the same danger, as loss of control can occur in different ways. Therefore a single hazard can generate several Bow-Tie diagrams. For example, the danger "Performing work at height" can generate both the top event "falling objects from above" and "person falling from scaffolding at height."

The term "top event" is derived from the fault tree methodology, which has similarities with the left-hand side of the Bow-Tie model. It would also be the starting point of an event tree analysis (ETA), which would identify all possible consequences (superimposed on the right-hand side of the Bow-Tie model), starting from the variability of the states associated with the control barriers present.

Precisely because it represents the moment in which one loses control over the danger, the top event is very often identified in terms of "Loss of control over . . .," although this is obviously a suggestion that must be combined with the need to briefly communicate as much information as is considered useful for the purposes of conducting the analysis with respect to the set objectives.

The more or less detailed definition of the top event helps to define the level of "zoom" (i.e. the degree of depth) to be assigned to risk analysis using the Bow-Tie method. Generally we tend to be too general in defining the zoom level. It is also necessary to identify the point in time where the top event can be identified. Taking the example in Section 2.6, you might ask which of the following events, listed in ascending chronological order, is the top event:

- Residual water in the tank;
- Corrosion;
- Containment loss;
- Release of hydrocarbons into the sea;
- Damage to the marine ecosystem.

The answer is not so obvious, although many readers will have identified "Loss of containment" as the top event for the case in question, based on the considerations in the previous paragraphs. In fact, the correct answer would be "it depends," since all the events listed are potential top events for a specific context. In this case:

- The presence of residual water in a storage tank is a top event for tank cleaners.
- The triggering of the corrosion phenomenon is a top event for those responsible for inspection and maintenance.
- The loss of primary containment is a top event for the department in charge of operations.
- The release of hydrocarbons into the sea is a top event for the emergency response team and for those dealing with safety and the environment.
- Damage to the marine ecosystem is a top event for the authorities with jurisdiction over environmental protection.

It is clear that the level of detail and the way risk assessment is conducted should be a function of the organization's objectives, context and stakeholders (internal and possibly external).

2.10 Threats

Causes are the factors that could determine the top event. In other words, they are the potential reasons for the loss of control over the threat, thus resulting in the identified top event. For each top event there are normally multiple causes on the left-hand side of the Bow-Tie diagram. Each of these represents a single scenario that could directly and independently (i.e. independently of the other causes) lead to the top event. Therefore, if on a branch that links a cause to the top event there are no preventive barriers, then that cause should determine the top event with a direct match.

Cause independence anticipates an aspect of fundamental importance when we talk about Bow-Tie quantification with LOPA: causes are combined only through OR logical doors.

For example, if you analyze the top event "loss of car control" (Figure 85) it is not possible to say that "beer availability" and "propensity to drink" are valid causes for Bow-Tie analysis. In this case, a possible cause is the overlap of the two, i.e. "propensity to drink when beer is available" or "drunk driving." These constraints are typical of the method and it is important to highlight that in the selection of any methodology these peculiarities must be taken into account by the analyst. As an example: in an FTA analysis concomitant failures are excluded so that failure is due to the malfunctioning of a single component (and system(s) containing it) and in ETA the time factor is not considered in the implementation of the barriers and great importance is given to the order of intervention for the determination of the pool of potential outcomes. Simplifications are necessary for the representation of any complex system.

In the case of loss of containment of hazardous material, possible causes are:

- Overfilling;
- Overpressure;
- Corrosion;

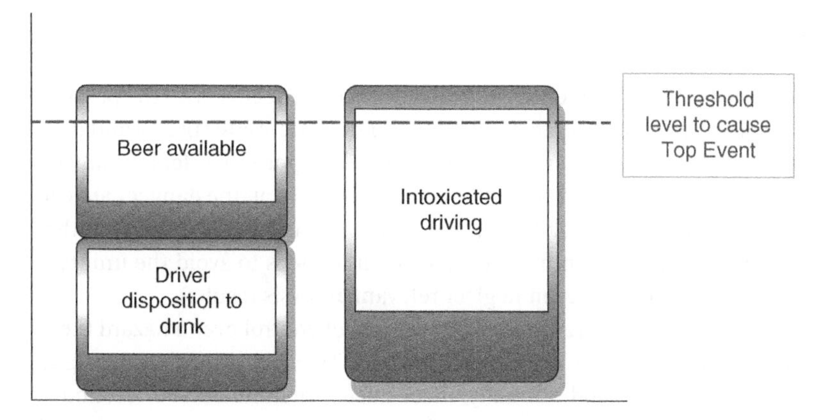

Figure 85 Determining the threshold level to cause the top event.

- Stress/fatigue;
- Infragilization;
- Erosion;
- Impact with physical object;
- Naval collisions;
- Helicopter collisions.

As a rule of thumb, causes are usually identified by exploring these three sets:

1) Primary equipment that does not operate within normal operating limits (e.g. mechanical failure);
2) Environmental influences (e.g. physico-chemical effects due to solar heating, earthquakes, etc.);
3) Operational problems (e.g. insufficient personnel for the required task).

The use of "human error" or a consideration of this (e.g. "lack of information, training and education") among the causes of a Bow-Tie is highly discouraged, because it commonly leads to "structural" errors, i.e. violating the rules of construction of a Bow-Tie diagram, as better detailed in the section on guidelines. At this stage it is sufficient to remember that human error is, on the basis of the author's experience, often attributable to a degradation factor.

If two causes share the same preventive barriers along the branches that lead them to the top event, then merging them under a single cause that includes both may be considered.

A frequent mistake is excluding those causes that rarely lead to the top event, arguing that there are already sufficient barriers to prevent them. In fact, all credible causes must be identified, noting that risk control measures have been implemented precisely to make it possible for these causes to be low risk. It is therefore essential to provide a comprehensive overview of possible credible causes, regardless of their likelihood to occur.

This is also in the light of the fact that very often some common factors can lead to the failure or partial effectiveness of several barriers.

2.11 Consequences

The next step is to identify the consequences of the undesired event effects on the possible vulnerable targets. Some analysts prefer to define them immediately after the formulation of the top event, going against the "natural" sequence, but this approach offers the advantage of helping the analysis team to define only the causes that act on the danger can lead to certain consequences. Both approaches are, however, valid and correct provided that they ensure an adequate degree of depth of analysis in all its steps to avoid the unintentional elimination of information (or even neglect relevant analysis results).

A consequence is a potential event resulting from the loss of control over a hazard (i.e. a top event), which implies a direct loss or damage (impact). They are placed on the right-hand side of the Bow-Tie diagram and normally affect safety, environmental or economic aspects (including reputational and asset damage), in relation to the specific objectives of the risk analysis and what was established in the preliminary context definition phase. Very often several consequences are identified for the same type.

They therefore include the most relevant events in terms of safety, environment and economic losses. However, minor consequences can also be identified if the objective of the analysis is to fully identify the barriers protecting all possible incident scenarios. Although a top event may of course have more consequences, only those with the most important consequences are generally developed to show the mitigation barriers present, thus excluding trivial consequences. Information on the level of risk can be displayed in order to show the severity of the damage associated with a specific consequence and its frequency of occurrence.

These are therefore those unintended events that an organization would like to avoid at all costs – or rather, at the costs justified by an ALARP study.

The category of consequences with impacts on the economic sphere is very wide. As a general principle, organizations should identify the consequences of accident scenarios in terms of (but not limited to):

Direct impacts

- Financial replacement value of the lost asset;
- Cost of acquisition, configuration and installation of new assets or backups;
- Cost of operations suspended until recovery conditions are met.

Indirect impacts

- Time and cost of investigation;
- Missed opportunities;
- Damage to reputation and image;
- Breach of legal or contractual obligations (as well as, in certain areas, legal requirements).

2.12 Barriers

2.12.1 Primary Barriers

There has been extensive discussion of how risk management is focused in its control. This is done through control measures, i.e. barriers. This book differentiates between primary and secondary barriers. The former are those found on the main branches of a Bow-Tie diagram, directly interposed between causes and top events or between top events and consequences. The latter, on the other hand, only appear if escalation factors are also defined and serve as measures to support primary barriers from the degrading threat of escalation factors. Therefore, secondary barriers do not directly prevent or mitigate the sequence of events: this is the task of primary barriers.

A barrier can be defined as any measure taken against an unwanted force in order to maintain a desired state. They can be both physical and intangible. For example, in the case of a Bow-Tie diagram having:

- Top event = loss of containment.
- Consequence = fire.

The "automatic deluge fire-fighting system" barrier clearly represents a mitigation barrier, i.e. to be placed on the right side of the Bow-Tie diagram, as its effectiveness is only expressed after the top event has occurred. The analyst then asks the following question for each barrier: "Does the barrier act before or after the top event has occurred?" If before, then it should be placed on the left side; if after, then it should be placed on the right side. By way of example, a containment basin intervenes after the accidental release of material has occurred: it is therefore a mitigating barrier. On the other hand, the execution of the activities foreseen in the inspection and maintenance plan can be clearly classified as a preventive barrier.

In reality, the placement of barriers in the Bow-Tie diagram is not always so immediate. In fact, barriers generally have one of the following five functions (Figure 86):

1) Eliminate or replace the danger;
2) Eliminate the cause;
3) Prevent the top event;
4) Separate the consequence;
5) Mitigate the consequence.

Since it is clear that the cause-effect nexus should be read from left to right and excluding the barriers that directly act on the danger (i.e. those that can be associated to the discipline known as "inherent safety" characterized by the privilege of actions on the danger rather

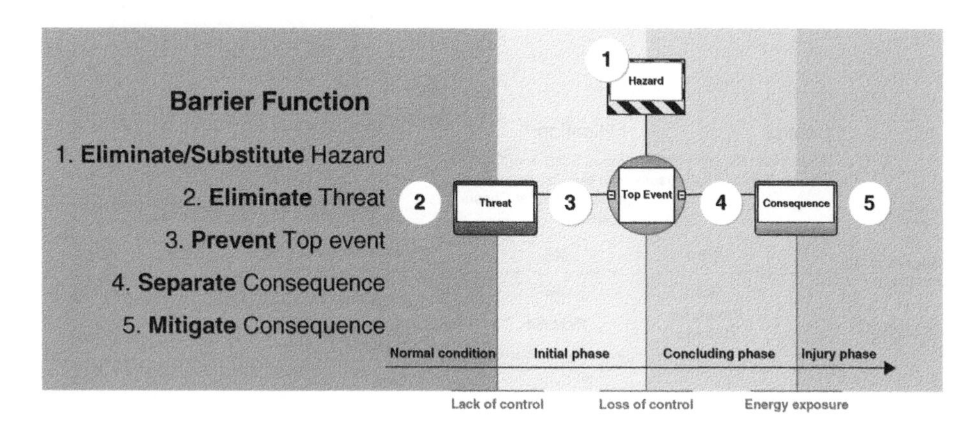

Figure 86 Barrier functions.

than on the prevention/mitigation of risks), for the remaining ones it could appear logical to identify the following positions:

- Elimination barriers: to the left of the cause;
- Barriers of prevention: between the cause and the top event;
- Barriers of separation: between the top event and the consequence;
- Mitigation barriers: to the right of the consequence.

However, in Bow-Tie analysis the positions to the left of the cause and to the right of the consequence are not permitted. For this reason, elimination barriers must be placed to the right of the cause (Figure 87), while mitigation barriers must be placed to the left of the consequence (Figure 88).

A typical mistake is to list as separate barriers the elements that actually refer to the same barrier, whose task is to detect the danger, decide how to act, and ultimately act as decided.

In general, a barrier performs its function by performing the following three actions:

1) Detecting the danger
2) Deciding how to act
3) Acting as decided

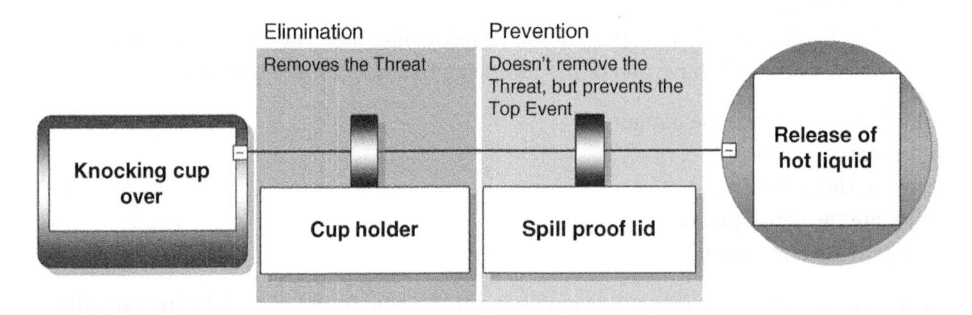

Figure 87 Location of elimination and prevention barriers.

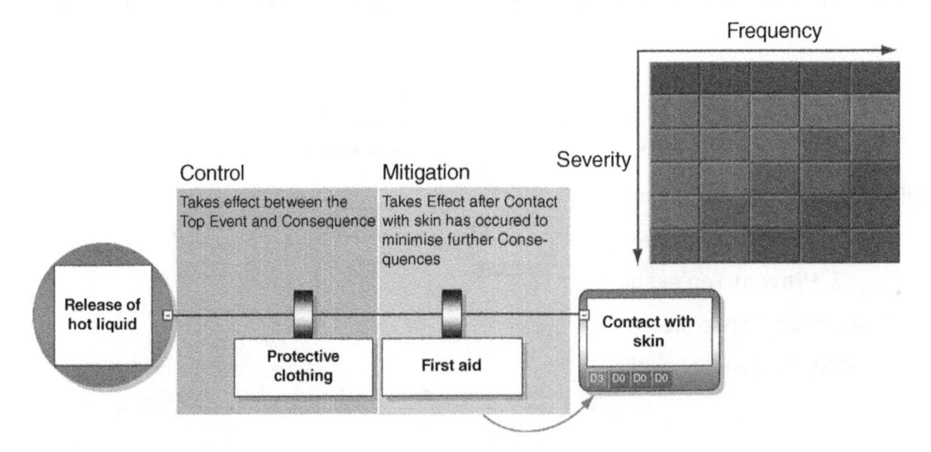

Figure 88 Location of control and mitigation barriers.

It is therefore essential to identify the subjects that carry out these actions and any subsystems of the barrier that can be traced back to them. In the example in Figure 89, it can be observed that the three functions have been differentiated and the subjects responsible for the various subsystems are also different. If, on the other hand, the responsible subject is common to all three subjects carrying out the three actions (detect, decide, act), or in any case one prefers to observe the problem in terms of systems and not functions, then it is recommended to use a single macro-barrier: the one at system level.

These and other properties of barriers, as well as the other characteristic elements of a Bow-Tie, are discussed in detail in Section 4.3.

Finally, it may be questioned whether a barrier may appear on either side of a Bow-Tie diagram. This is generally not recommended.

As shown in the example in Figure 90, the "Competent driving" barrier on either side of the Bow-Tie diagram indicates that good driver skills are a measure of control both in the preventive phase, preventing a slippery road from causing the vehicle to lose control, and in the reactive phase, by readily recovering its trajectory before crashing into an object. It is then preferable to differentiate these two barriers: the first will be "defensive driving," i.e. prudent driving; the second will be "competence to regain control of the vehicle after losing it", i.e. "slip recovery competence."

Given the centrality of barriers in the development of barrier-based risk management, it is necessary to offer the reader some insights that also recall appropriate scientific reference literature. Important references, among the others, are Liu (2020) and Sklet (2006).

Due to the origins and the widespread use of the Bow-Tie method in the chemical sector, most of the literature focuses on safety and process safety. But definitions given in those contexts are pretty useful in a number of difficult cases.

Safety barriers can be technical or organizational and perform one or more safety functions, thus determining their purpose (DNV GL, 2014). Generally, the function of a barrier is defined by a verb and a noun, but what is important is that it is clearly identified. Specific indications

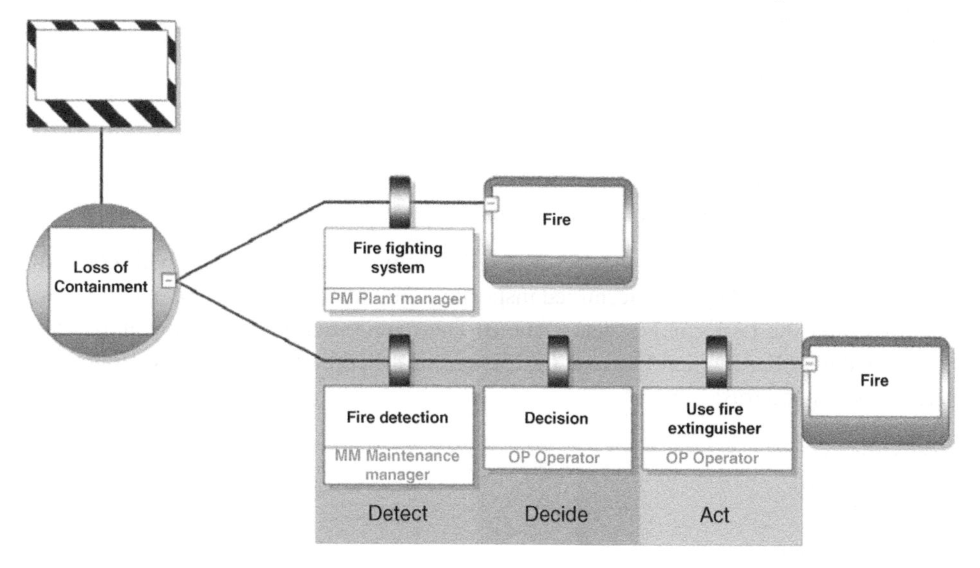

Figure 89 Barrier systems.

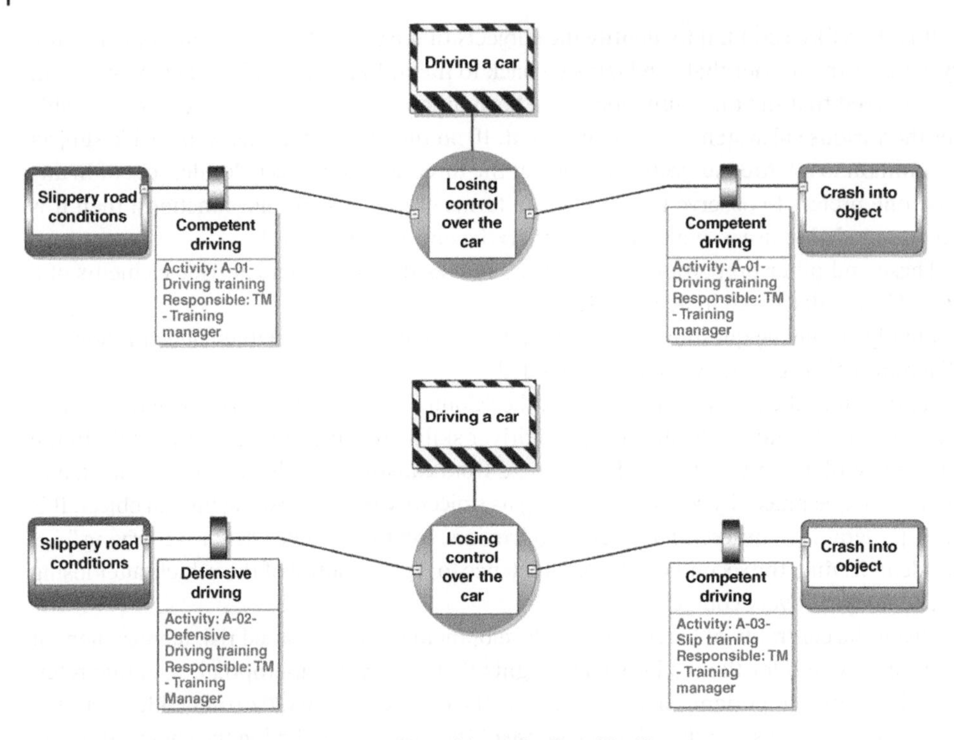

Figure 90 Using the same barrier on either side of the Bow-Tie diagram.

are provided by the AIChE CCPS guidelines, as discussed in the later section discussion them. In their system hazard identification, prediction and prevention (SHIPP) methodology, Rathnayaka, Khan and Amyotte (2011) classify safety barriers in the process industries as a:

- Human factor barrier;
- Management and organizational barrier;
- Release prevention barrier;
- Dispersion prevention barrier;
- Ignition prevention barrier;
- Escalation prevention barrier;
- Damage control emergency management barrier.

Pitblado et al. (2016) in their study distinguish between static barriers, with constant performance achieved by predetermined inspections and maintenances, and dynamic barriers that offer the performance degradations. The requirements for a barrier have been highlighted by Johansen and Rausand (2015), among others, who identified the following performance standards:

- Specificity;
- Functionality;
- Reliability;
- Response time;
- Capacity;

- Durability;
- Robustness;
- Auditability;
- Independence.

Prashanth et al. (2017) proposed seven groups of performance factors for safety barriers:

1) Performance;
2) Confidence;
3) Trust;
4) Limit;
5) Perception;
6) Dependability;
7) Robustness.

The first four groups are related to design and maintenances while the other three to management. It is important to highlight the role of management in assuring the barrier functionality and integrity, because:

- Barriers are a combination of technical and behavioural detection, decisions and acts (Pitblado et al., 2016).
- Barriers interacct with other components of a larger system (NRC-National Research Council, 2007; Xie, Lundteigen and Liu, 2018a and 2018b).
- Designing and implementing a barrier is a systematic work (Kjellén, 2007).

Indeed, as (Pitblado et al., 2016) have highlighted, the effectiveness of barriers decreases due also to management issues (people, documentation, procedures, and so on) and not only for technical reasons. The scientific bibliography offers different classifications between physical and non-physical barriers, as shown in Table 9.

Table 9 Different classification of barriers as physical or non-physical.

Terms		Reference
Physical	**Non-physical**	**Johnson (1980)**
Hard defence	Soft defence	(Reason (1997)
Physical – technical	Administrative	Wahlstrom and Gunsell (1998)
Physical – technical	Human factors/organizational	Svenson (1991)
Technical	Procedural/administrative – Human actions	US Department of Energy (1996)
Technical	Human/organizational – Human	Kecklund, Edland, Wedin and Svenson (1996)
Technical	Organizational – Operational	Bento (2003)
Physical	Management	US Department of Energy (1997)
Hardware	Behavioural	Hale (2003)

Another classification has been proposed by Sklet (2006), as shown in Figure 91.

As discussed in-depth in the Guidelines section of Section 2.14, the classification proposed by the AIChE CCPS guidelines is shown in Figure 92.

Adopting the barrier-based perspective, risk management is therefore barrier management and barrier management is is an integrated part of risk and safety management (Petroleum Safety Authority, 2011), where after identifying hazards, estimating frequency, and assessing consequences, barriers are analyzed too. According to the theory of safety management developed by Li and Guldenmund (2018), barriers are the input of the model, while safety performance is the output. Of course, barrier management includes managing processes, systems, solutions and measures (Petroleum Safety Authority, 2011).

Verification is an important step in barrier management, since the barrier's performance has to remain suitable over time, maintaining adequate conditions to meet the requirements (Petroleum Safety Authority, 2011).

The concept behind a safety barrier is usually related to the so-called energy model (Haddon, 1980), widely used as an accident model. According to it, a barrier is seen as a measure to protect the vulnerable target from the unwanted release of energy from a hazardous source (Figure 93).

The evolution of the barrier concept has been highlighted by many authors, including Hollnagel (2004) and Fleming and Silady (2002). Their reasonings converge on the definition of barriers as multiple lines of defence (including not only technical ones but also strategies and tactics) to protect the public health and safety.

Other authors proposed the generic safety functions related to a process model as shown in Figure 94.

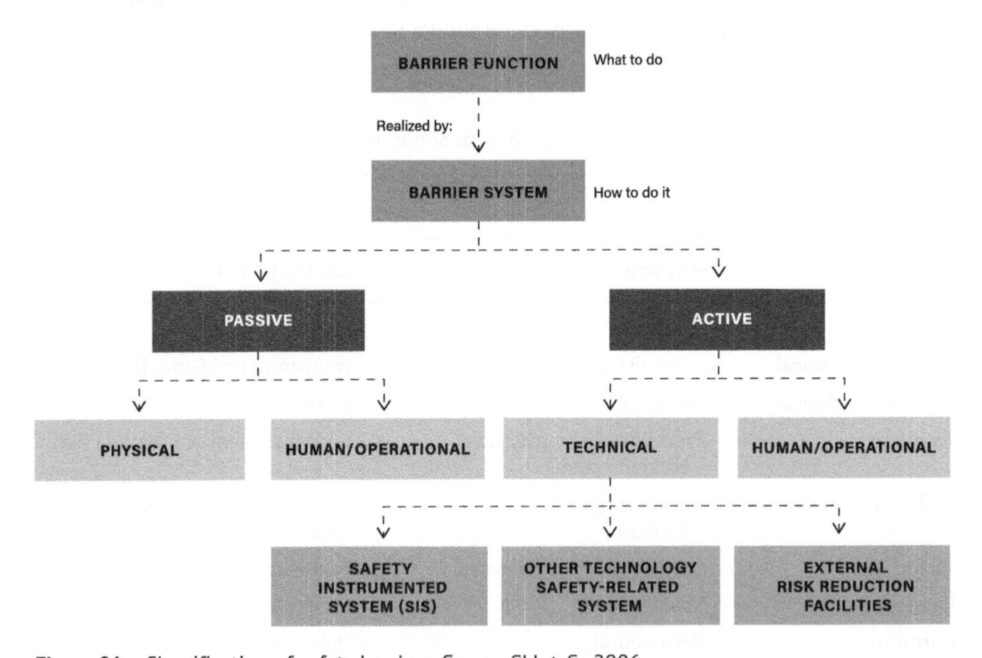

Figure 91 Classification of safety barriers. *Source:* Sklet, S., 2006.

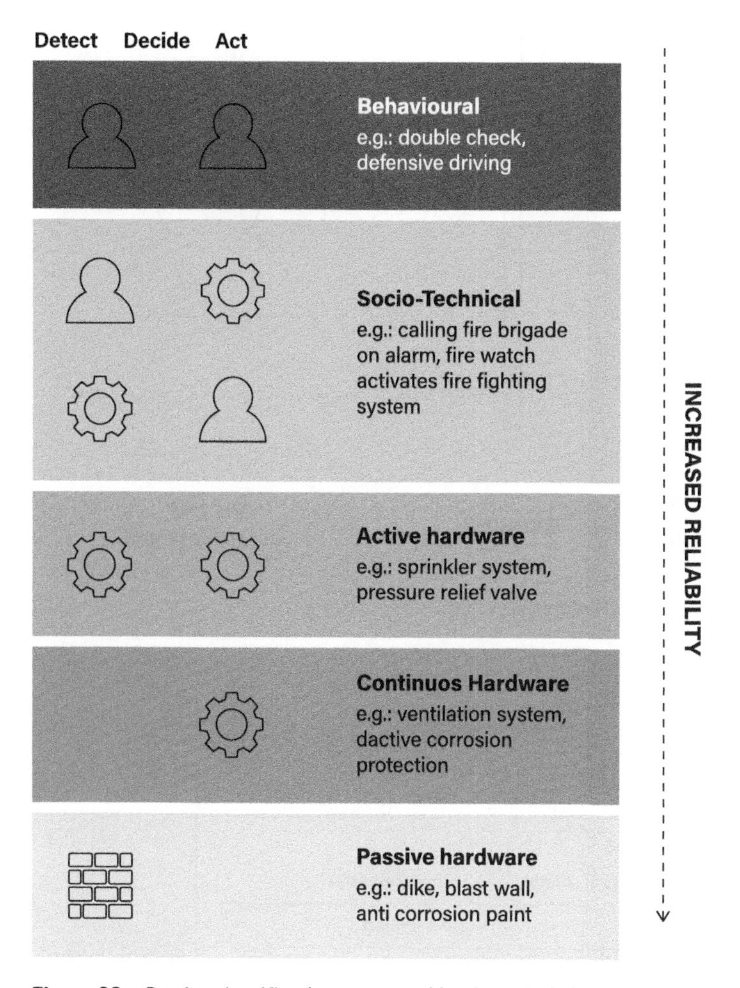

Figure 92 Barrier classification promoted by the AIChE CCPS Guidelines.

Hollnagel (2004), taking inspiration from Taylor (1988), proposed the requirements of barrier quality: adequacy, availability/reliability, robustness, and specificity. In Hollnagel (1995), another set of performance criteria was presented: efficiency or adequacy, resources required, robustness (reliability), delay in implementation, applicability to safety critical tasks, availability, and evaluation.

These are not the only performance standard for barriers: in a later section, the FARSI approach will be presented, in relation with the concept of criticality of a barrier.

2.13 Escalation Factors and Associated Barriers

Ideally, a properly identified, designed, installed, and tested barrier allows the stop of a flow of events, so that the top event or the consequences do not have the potential to manifest themselves, giving rise to an impact. We have already seen, thanks to the James Reason

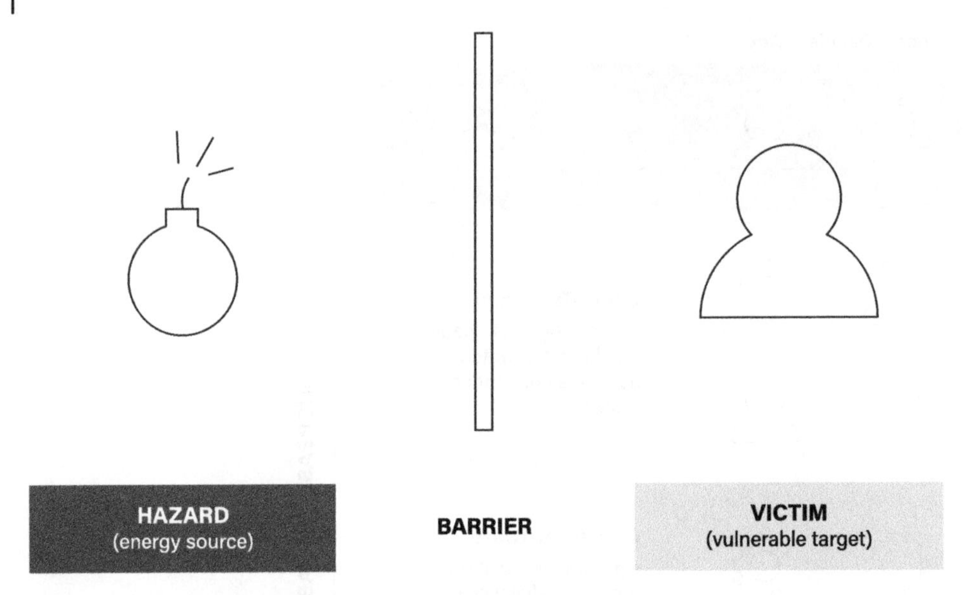

Figure 93 The energy model. *Source:* Haddon, W., 1980.

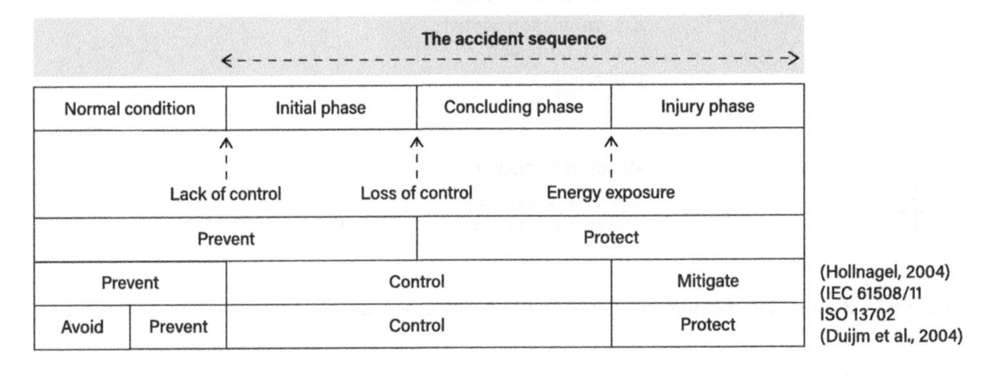

Figure 94 Generic safety functions related to a process model. *Sources:* Hollnagel, E., 2004. Barrier And Accident Prevention. Hampshire, IK: Ashgate; Duijm et al., 2004.

model, that this assumption is only true in the ideal world, where the reliability of a barrier is always 100%. In reality this is not the case and the definition of a probability of failure on demand is the most important proof of this. But beyond the intrinsic vulnerability of any barrier expressed through its PFD, there are also conditions that do not directly cause a top event or consequence, but can still reduce the effectiveness of a barrier; these conditions are known as escalation factors.

Within a Bow-Tie diagram, escalation factors are placed below the barrier to which they apply. They are like the causes of the Bow-Tie diagram, only the threat is no longer the top event or consequence, but the barrier.

Four categories of escalation factors can be identified:

1) Human factors;
2) Mechanical failures;
3) Abnormal conditions;
4) Loss of critical services (e.g. lack of electricity or instrument air) and more generally common causes of failure.

Examples of escalation factors are:

- Lack of electrical energy (which frustrates the effectiveness of a barrier working with this form of energy);
- Adverse weather conditions (which reduce the effectiveness of an organizational barrier such as rescue activities at sea);
- Lack of operator training (which reduces the effectiveness of a behavioural barrier that requires a period of information/training to perform a particular task). In other words, a well-designed operator response procedure given an emergency condition is made to fail in terms of risk reduction if the procedure is not known or if it is not experienced.

2.13.1 Secondary Barriers

To ensure that an escalation factor does not threaten a primary barrier, so-called secondary barriers – i.e. escalation factor controls – can be identified. Often, such barriers do not meet the standard requirements for a primary barrier (effectiveness, independence and measurability, as will be described later in this book), but it is accepted that they are also identified as barriers.

Their position within a Bow-Tie diagram is between an escalation factor and the primary barrier on which it is acting.

Examples of secondary barriers are as follows:

- Diesel emergency generator (against a possible lack of electricity);
- Emergency fire pump (against a possible failure of the main one);
- Refresher course with training to be carried out every three months (against a possible threat of lack of training) with verification of skills maintenance.

2.14 Layer of Protection Analysis (LOPA): A Quantified Bow-Tie to Measure Risks

LOPA, the acronym for layer of protection analysis, is a standard method typically used as a risk assessment tool that can also be used for incident investigation. Developed to assess industrial risks related to process safety, different protection layers are in place in chemical plants to reduce risks related to undesired events (Figure 95). Sorted from the inner layer to the outer layer, they are:

- Process design (inherently safety culture and plant design);
- Basic process control system (BPCS);
- Critical alarm and human intervention as response to the alarm system;

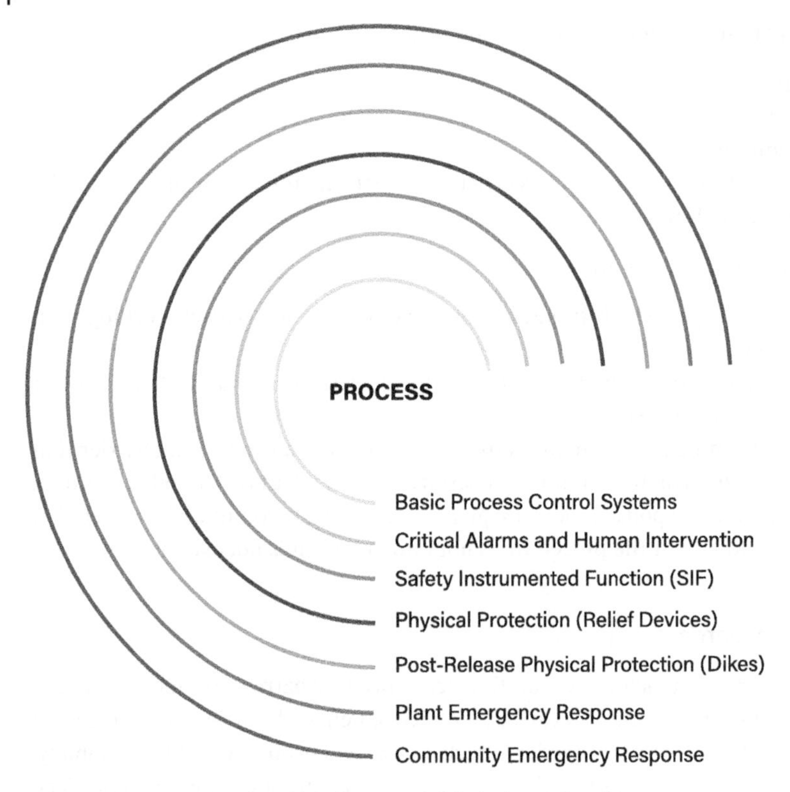

Figure 95 Layers of defence against a possible industrial accident.

- Safety insrumented systems, automatic interlocks, emergency shutdown systems;
- Physical protection (relief valves);
- Post-release physical protection (like dykes, walls);
- Plant emergency response plan;
- Community emergency response plan.

The aim of LOPA is to analyze the effectiveness of the proposed protection layers, comparing their combined effects with the risk tolerance criteria (Franks, 2003). Indeed, LOPA uses orders of magnitude for both initial event frequencies, consequence severities, and the probability of failure of IPLs, approximating the risk scenario and determining if the existing protection layers are sufficient to mitigate risk below the tolerability limit.

Safeguards can be classified as active or passive and preventive (pre-release) or mitigating (post-release). All IPLs are safeguards, but not all safeguards are IPLs. A safeguard is any system, device, or action that can stop the chain of events following an initial event. In order to be an IPL, a safeguard must be effective (having the capacity to take action in time), independent (avoiding the common causes of failure), and auditable (to demonstrate that it meets the risk mitigation requirements). In particular, the EN/IEC 61511-3 establishes that:

- Each IPL must be independent from any other IPL.
- Each IPL must be different from any other IPL.
- Each IPL must be physically separated from any other IPL.
- Each IPL must not share common causes failure with any other IPL.
- Each IPL must be highly available (availability > 90%).
- Each IPL must be validated and auditable.

This technique can be used throughout the process of safety lifecycle (AIChE-CCPS, 2001). It is generally used to examine those scenarios coming from other PHA tools, like HAZOP, and to define the SIL targets to meet the risk acceptability criteria. However, it can also be used at the initial design stage, to evaluate alternative protections, or to identify the safety critical equipment (SCE), that is to say, the equipment used as a protection layer maintaining risk in the tolerable region. This often resulted in a decrease in the number of SCE (AIChE-CCPS, 2001), still maintaining safe conditions, in contrast with the old idea that adding equipment equals increasing safety. LOPA can also be used to identify the critical administrative control (CAC), i.e. those operator actions or responses that are critical to keeping risk inside the tolerable region. The method is also used to identify the ALARP risk scenarios.

It is clear that, in theory, a single layer of protection is itself sufficient to stop the incident sequence and prevent the risk scenario. However, no layer is 100% reliable, and no one is perfectly effective. This is why a set of protection layers is generally identified to provide the required risk mitigation. If the risk is not tolerable, additional IPLs should be prescribed.

To perform a LOPA, the consequence categories, component failure data, and human error rates must be available. Obviously, in order to have consistent results, the risk tolerability and acceptability criteria must be defined and shared among the participants before performing LOPA.

LOPA is not a fully quantitative method, but a simplified numerical approach to evaluating the effectiveness of the protection layers for a precise incident scenario and, eventually, the need for further risk reduction measures (new IPLs or more robust ones). The fact that LOPA uses numbers does not mean that it provides precise risk measurement: it only gives an approximation that could be useful to make comparisons. However, its methodology can be seen in parallel with other QRA methods, like the ETA, as shown in Figure 96. The thickness

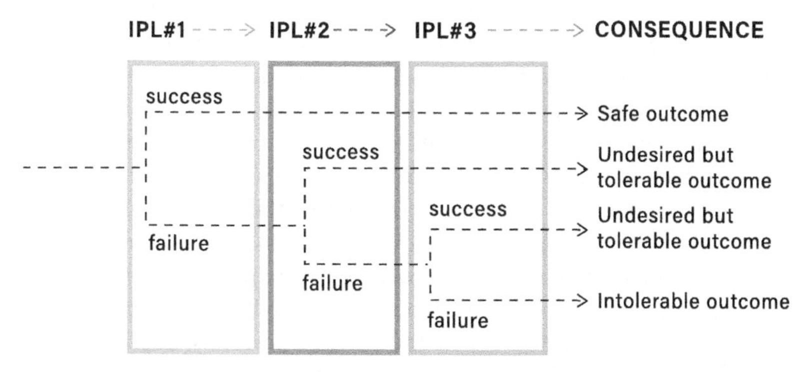

Figure 96 A comparison between ETA and LOPA's methodology.

of the arrow represents the frequency of the scenario that is reduced by effective safeguards. It is clear how they share the same concepts at the base of common reasoning.

Basically, LOPA consists of the following steps:

- Collect scenarios developed in prior studies, like HAZOP.
- Select an incident scenario and evaluate its consequence.
- Identify the related initial events (IEs) and their frequency.
- Identify the related IPLs and their probability of failure.
- Estimate the risk, combining IEs' frequency, consequence severity, and IPL data.
- Evaluate the risk and make risk decisions (evaluate if further risk reduction is required).

LOPA should not be confused with HAZOP: they are different techniques, with different goals. HAZOP is used to brainstorm the possible hazards and identify incident scenarios, whose risk can be evaluated only from a qualitative point of view. Instead, during LOPA, the analyst uses a predefined scenario and estimates its risk in a quantitative way, even if approximate. From this point of view, LOPA is a complement of a HAZOP analysis.

The consequence analysis should take into account the nature of the scenario (LOPC, the release of toxic substances, fires, explosions) estimating injuries, fatalities, environmental damages or business losses. To do so, qualitative or quantitative approaches can be used. The former uses a predefined categorization, identifying higher-severity classes depending on the quantity released, the number of economic losses, or the qualitative evaluation of injuries and fatalities. The quantitative approach for consequence analysis, instead, usually requires complex computer models to evaluate the scenario, like the dispersion of a toxic cloud or the extension of a jet fire, using mathematical models. Great attention to hypotheses and surrounding conditions should be paid when using fully quantitative approaches for consequence analysis. When assessing the IE frequency, the analyst should take into account the different mode of the intervention of an IPL: in continuous mode or in demand mode. Moreover, the failure probability must consider the time to risk, i.e. the adjustment to correct failure probabilities (expressed as occurrence per year) infraction of years when the component is operating.

LOPA is one of the tools used to establish the SIL targets in the functional safety life cycle. SIL targets are the quantitative measure of the required risk mitigation to meet the risk tolerability criteria. The SIL level is related to the PFD of the SIS performing a specified SIF. LOPA analysis allows establishing whether one or more SIFs are required and, if any, to allocate the SIL level to the SIF, on a not-generalistic approach, so to achieve the required risk mitigation with smaller capital expenditure, if compared with risk mitigation based on a generalistic approach (like assigning a SIL 3 reduction everywhere).

In conclusion, LOPA can also be used in incident investigation, being a powerful analysis and communication tool. It can be used to show how additional IPLs could prevent the incident occurrence, or to identify those scenarios sharing the same failed IPL and to show how to reduce the scenario frequency by adding new IPLs. Identifying IPLs is important to support the management system because if they are not maintained properly over time, the estimated risk can be wrong and the unwanted scenarios may occur more frequently than expected. To be sure that the barriers performs as wanted, an effective management system should be put in place, to test, verify, check, inspect and maintain those barriers. Procedures, training/experience, auditing, inspection, testing, and preventive maintenance, leadership and culture, incident investigation, management of change, and human factors are all elements to be taken into account from a management system perspective.

2.14.1 How to Build a LOPA Assessment

Detailed guidance on how to conduct a LOPA assessment is given in Chapter 4, to which reference should be made in full. When conducting a LOPA assessment integrated with the Bow-Tie method, the suggestions offered by the AIChe guidelines, shown next, should be taken into consideration.

2.14.2 AIChE CCPS Guidelines

The Center for Chemical Process Safety (CCPS) of the American Institute of Chemical Engineers (AIChE) has issued, in collaboration with the Energy Institute, some important guidelines to be adopted for the correct and standardized use of the Bow-Tie methodology.

The purpose of this section is to present the main guidelines, without, however, taking the place of the work of the above-mentioned primary companies that have dedicated specific publications on the subject.

The need to produce guidelines is derived from extensive and unregulated use that has led to a large number of incorrect applications of the Bow-Tie methodology. Needless to say, in the majority of cases of failure to implement a risk management process, the primary cause lies in the lack of knowledge of the general principles (of the reference standard) and in the belief that the method or, even worse, the instrument selected is in itself a guarantee of obtaining the result.

The main application guidelines concern:

- The specificity of the elements;
- The independence of the causes;
- The differentiation between causes and failed barriers;
- The definition of barrier types;
- The definition of criteria for the validity of a barrier;
- The correct use of escalation factors and secondary barriers;
- The use of Bow-Tie as a barrier management tool.

Specificity of the Elements

The constituent elements of a Bow-Tie must be defined using the right words, capable of providing synthetically all the information necessary to characterize its peculiarities.

Danger The danger must be described in its controlled state: we will say "transport of fuel in tanker truck from A to B"; we will not say "explosion of tanker truck containing fuel."

In addition, the necessary details will be added to determine the scope and level of zoom. Consider the examples in Table 10. It is understood that the responsibility of defining the level of detail is an essential activity from the earliest stages of the study. An unintentional error in this attribution could determine a severe potential loss of useful information in the evaluation phases with a consequent impact in the even more critical decision-making phases.

Table 10 Comparison of defined hazards with insufficient detail and optimal degree for evaluation.

Danger not properly defined	Danger suitable for an assessment
Electricity	Voltage electric machine > 440 V
Storage tank	Gasoline in the storage tank
Oil	Oil under pressure
Helicopter	Helicopter personnel transport from A to B
Hydrochloric acid	Hydrochloric acid (10 t) under pressure in the storage tank

Top Event The top event should describe in its formulation which control over the danger is lost and how. If possible it should also give an indication of the scale of the problem: Is a limited or significant loss of flammable substance from a tank being analyzed? Furthermore, as is well known, the top event should not be confused with a consequence: the analyst should then ask whether what has been identified really represents loss of control or is already a consequence. For example, the death of an employee is an event generally at the end of a chain of events with previous mitigation barriers, so it is not a good top event.

In general, defining the top event may not be as simple as the loss of containment of a tank of flammable substances. In fact, the identification of the top event is crucial to identify the real risks and causes that you are trying to control, but a top event can often be defined a little to the right or left of the study objective. A good way to verify that you have selected the best top event is to identify at least two causes and two consequences and to ask whether they are the responsibility of the organization (or department) that is responsible for the Bow-Tie. If the answer is positive, then the right event has been selected, both in time and in the business process, to be identified as the top event. If the answer is negative, it is better to stop the analysis and go back to the identification of the top event. As a general rule, if only a couple of causes and consequences can be defined, the top event has probably been defined in too limiting a way. If, on the other hand, multiple (let's say more than 10) causes and consequences have been defined, it is likely that the top event has been defined in an exclusively general way and without the details useful for the subsequent request for information, thus running the risk of entering causes and consequences that do not fall within the specific field of study.

Even for top events, it is therefore essential to take care of their definition, as the examples in Table 11 show.

Causes The causal link between the causes and the top event must be clear, without further clarification. In other words, the causes must be direct.

Table 12 shows a comparison of defined causes with insufficient detail and with an optimal degree of evaluation.

Consequences The consequences should be defined in the formula "<Damage> (or impact in some contexts) due to <event>" – for example, "Deaths due to fire" or "Environmental damage due to liquid loss." One should avoid being too specific, especially when it is possible to identify the same set of mitigation barriers between different consequences, for

Table 11 Comparison of defined top events with insufficient detail and with an optimal degree for evaluation.

The top event defined badly	Best defined top event
Fall	Fall from above
Super level	Gasoline comes out of the tank
Helicopter impact	Loss of helicopter control
Random break	Primary containment loss
Storm	Staff exposed to a catastrophic event

Table 12 Comparison of defined causes with insufficient detail and with an optimal degree for evaluation.

Cause defined poorly	Better defined cause
Wind during lifting operations	Strong winds (>80 km/h) during lifting operations
High pressure in the well	An unexpected increase in pressure in the well
Level not properly maintained	Overfilling the tank

which a fusion into a single consequence is recommended. Table 13 compares defined consequences with an insufficient degree of detail and with an optimal degree for evaluation purposes.

Independence of Causes

The causes must be able to cause the top event independently of each other (they are therefore in the OR logic between them). This is a fundamental assurance placed at the basis of the implementation of the method.

Differentiation between Causes and Failed Barriers

Confusing a cause with the failure of a barrier is a very common mistake that must be avoided. Identifying the lack of maintenance on the level switch as a cause is an error that, if not removed, cancels any possibility of improvement, because it hides an opportunity for the organization to pursue continuous improvement. Similarly, identifying ABS failure as a cause of a loss of control of a vehicle is a mistake: ABS is a preventive barrier and failure to intervene is a failed barrier. The cause is “Abruptly braking” or “Slippery road” or “Abruptly braking on slippery road,” depending on the context and scope of the Bow-Tie.

It is clear that a consistent use of the methodology cannot ignore the knowledge of terms, references and definitions, otherwise the benefits of applying the method will not be exploited and, more strictly, significant errors of assessment will be made.

Definition of Barrier Types

Three verbs are used to describe the functioning of any barrier present in the system under consideration: “detect,” “decide,” and “act” (Figure 97).

Table 13 Comparison of defined consequences with insufficient detail and with an optimal degree of evaluation.

Misjudged consequence	Better defined consequence
Environmental pollution	Severe damage to marine wildlife due to oil pollution
Delay	Business process stopped for three months
Damage	Injury or death of the driver due to the impact of the vehicle against an object

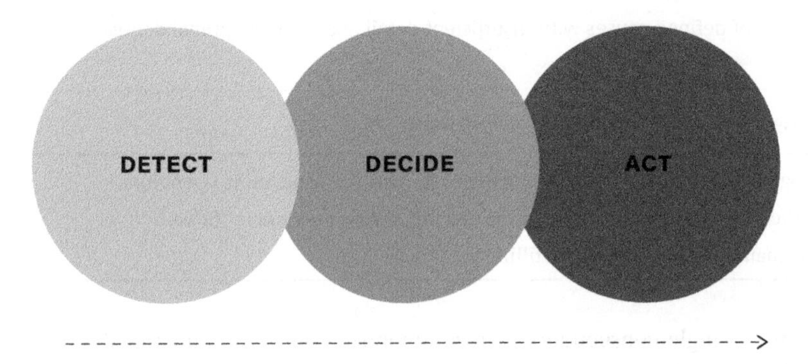

Figure 97 Actions of a barrier.

In relation to specifying the subject who detects, decides, or acts, it is possible to identify the five types of barriers described in Table 14.

Passive hardware barriers are those whose task is to absorb energy when it is released, but do not intervene to do so: they simply passively wait for it to happen. Examples of passive hardware are containment dams or fire doors. In this case, no one detects, decides, or acts.

In active hardware barriers, technology is the main subject that detects the danger, decides autonomously how to intervene, and then acts. An example of active hardware is the safety instrumented system (SIS): when a deviation from the operating parameters of a process is detected by a sensor above a certain threshold, the programmable logic unit decides how to intervene by sending a signal to an actuator which, for example, will intervene by sectioning a line.

Behavioural barriers are instead characterized by human intervention in all three phases. A human detects (e.g. a fire principle through a field inspection performed by procedure), decides how to intervene, and acts (e.g. using fire extinguishers).

In socio-technical barriers at least one of the subjects who detects, decides, or acts is different from the others. For example, fire detection systems could include the activation of an alarm in the control room via a sensor. In the control room, however, it is the operator who decides how to intervene, sending an emergency team to the site or pressing a button to activate the flood system.

Table 14 Barrier Types.

Barrier type	Detect	Decide	Act
Passive hardware	N.A.	N.A.	N.A.
Active hardware	Technology	Technology	Technology
Behavioural	Human	Human	Human
Socio-technical	Technology/Human	Technology/Human	Technology/Human
Continuous hardware	N.A.	N.A.	Technology

Finally, in continuous hardware barriers the technology is always in operation, regardless of the external stimulus that is not even read. Examples of continuous hardware are forced ventilation systems: they are always in operation, regardless of the concentrations of flammable or toxic gases inside the compartment.

One of the strengths of the method lies in the possibility to describe various types barriers and evaluate them.

Definition of Criteria for the Validity of a Barrier

A barrier, which from the previous paragraphs represents both the fulcrum of the risk assessment and the key element on whose operation the process of ensuring the maintenance of the level of risk over time is grafted, to be considered valid within the Bow-Tie methodology, must meet at least the following three requirements:

1) *Effectiveness*. The barrier must be able to prevent the top event or mitigate a consequence by acting as and when expected. "Training" or "competence" are not effective barriers; they are secondary barriers. Likewise, the "Fire & gas detection system" is not an effective barrier because although it is an important system, it alone is not capable of mitigating the expected consequence, merely of detecting a danger (i.e. "decide" and "act" actions are not envisaged). An effective barrier is instead: "Operating intervention on fire & gas system alarm with emergency shutdown activation."
2) *Independence*. The barrier must have a direct and independent impact on the cause, top event or consequence. This criterion therefore excludes those systems that share common causes of failure or failure modes.
3) *Auditability*. The barrier must be capable of being evaluated for efficiency and effectiveness to verify expected performance. Formally, this criterion is met by assigning requirements or performance standards to the functionality of the barrier, which can be periodically verified by comparison with minimum performance criteria judged to be acceptable (e.g. the technical barrier consisting of an active fire protection system operating with foam can be considered suitable and functioning not only if present, not only if when activated it supplies the extinguishing agent, but only if tests carried out at a suitable frequency, if necessary in accordance with specific regulatory requirements, show compliance with the performance defined in the design activities of the barrier itself with respect to the risk scenario for which it was designed, such as the expansion ratio and the specific discharge density).

Correct Use of Escalation Factors and Secondary Barriers

Escalation factors, which are not mandatory within a Bow-Tie diagram, must be used with the utmost care in order to avoid improper use. Firstly, it should be made clear that an escalation factor does not simply deny the barrier. For example, if the barrier is "automatic high-level lockout," the escalation factor cannot be simply "automatic high-level lockout fails"; it is best to use a formulation that enriches the analysis with additional data and useful elements, such as "incorrectly calibrated level meter." You also need to be clear about what can cause the primary barrier failure. For example, instead of "procedure not performed," it would be better to say "suppliers not aware of the procedure," so as to better focus attention on the specific problem that needs to be analyzed.

As already seen for the other components of a Bow-Tie diagram, if on the one hand the extremely generic graphic notation allows the analyst ample freedom to go into detail, on the other hand it is essential to specify and detail each constituent element.

The use of escalation factors, if not controlled, can lead to a nesting of escalation factors and secondary barriers, if these are used recursively (see Figure 98).

Think for example of the barrier "Manual activation of the fire extinguishing system on signalling of the smoke detection system." A possible escalation factor that could threaten the functionality of the barrier is "Lack of electricity," having assumed that the system mentioned is equipped with an electric fire pump. To protect the barrier from its escalation factor, there is the secondary barrier "Supply of electricity with emergency diesel generator" (it could also be an emergency diesel pump). However, the diesel generator can also be "threatened" by an escalation factor: "Failure to fill the diesel generator tank." In this case, therefore, a probable secondary barrier could be the "Generator diesel tank filling procedure," which describes how and frequency of the checks aimed at keeping the fuel level under control. The procedure may then not be followed for lack of training refreshers, in order to cope with which the organization provides itself with an annual training plan. In short, it is clear that such a structure of continuous escalation factors and secondary barriers could negate one of the key advantages of using Bow-Tie, namely their readability. As they cannot give up their graphical notation, the alternative is to transform the nested

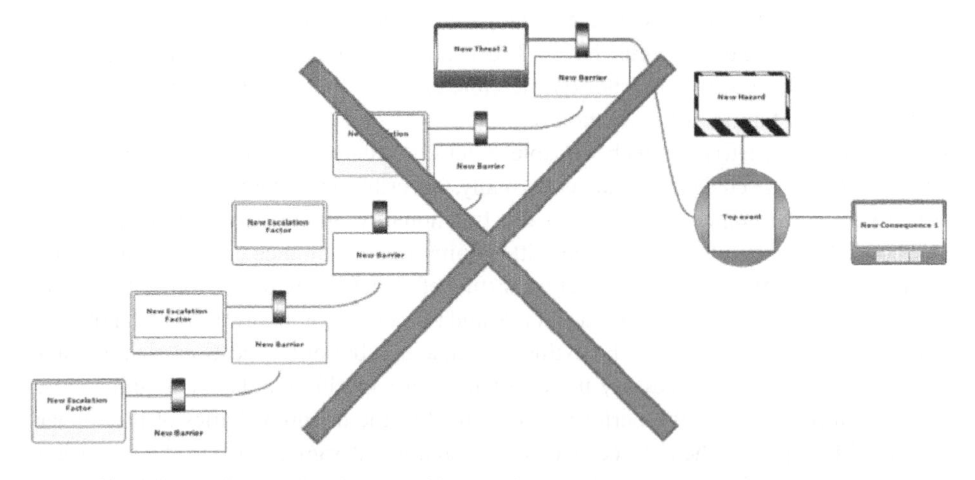

Figure 98 Misuse of escalation factors, with nested structure.

structure of secondary barriers into "activities." Therefore, the "Manual activation of the fire extinguishing system on smoke signalling" barrier will be accompanied by a series of activities that the organization must perform, according to a pre-established schedule, in order to guarantee the integrity and functionality of the primary barrier. In the case under consideration, these activities are represented by what is described within the emergency diesel activation procedure in case of detachment of the primary energy source, the control and refuelling procedure of the diesel level of the diesel generator, the training refresher procedures and so on.

Using Bow-Tie as a Barrier Management Tool

The Bow-Tie method, selected for this introductory volume to the principles of ISO 31000 compliant approaches to risk management across different industry sectors, is not only the approach to initial context definition and risk assessment, but also a valuable aid for the periodic verification of the state of barriers in a systemic perspective. In fact, there is clearly a direct correlation between the level of risk of an organization and the instantaneous state of the barriers, i.e. all those preventive and mitigative controls designed and implemented to obtain a degree of risk reduction congruent with the criteria of acceptability and tolerability adopted by the organization.

Thanks to the introduction of the "activities," the chance offered by the Bow-Tie method to support the organization in barrier management activities is even clearer. In fact, not only will it be possible to identify a responsible person within the organization for each barrier to ensure its integrity and functionality, but also the individual activities in support of a barrier will have a responsible person, a certain frequency, a priority, one or more supporting documents, and so on. In fact, the risk management system is being implemented in a barrier-based perspective, as indicated by the use of specific barrier metadata in Figure 99.

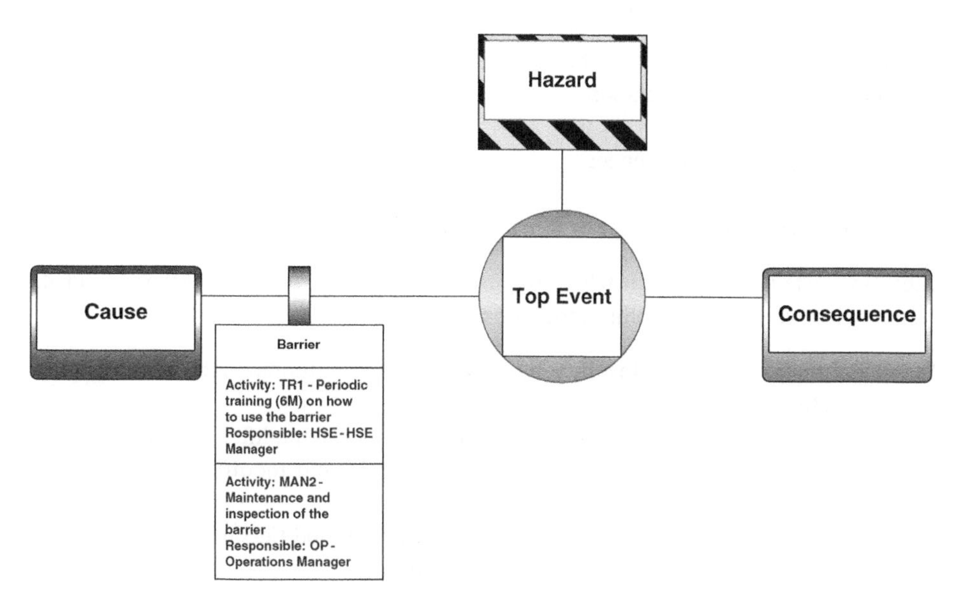

Figure 99 Defining "activities" for a barrier.

2.14.3 Conditional Modifiers and Enabling Factors

A conditional modifier can be defined as one of several possible probabilities included in scenario risk calculations, generally when risk criteria endpoints are expressed in impact terms (e.g., fatalities) instead of in primary loss event terms (e.g., release, vessel rupture). This definition, as is much of the material in this section, is taken from a CCPS project developing Guidelines for Conditional Modifiers and Enabling Conditions (AIChE-CCPS, 2013). Conditional modifiers are sometimes called by other names, such as mitigating events. As was the case for enabling conditions, these are often not events but rather system states at the time a scenario occurs.

Each conditional modifier is expressed as a dimensionless probability. They are not used in every LOPA.

One distinction between IPLs and conditional modifiers has to be made:

- IPLs are engineered or set up to respond or be in place when a process deviation or loss event occurs, to bring the system to a safe (safer) state.
- Conditional modifiers are factors not specific to the scenario being evaluated.

Conditional modifiers are used to obtain a more accurate estimate of the frequency of getting to an impact expressed in ultimate consequences, such as fatalities, and to obtain a more accurate overall risk estimate by considering probabilities typically included in a QRA. Typically, there are three common conditional modifiers: probability of ignition/ release; probability of person(s) in effect area/loss event; probability of injury or fatality/ person(s) in area. These are risk reduction factors but not IPLs. These conditional modifier probabilities get plugged into our scenario risk equation right along with the initial event frequency and the IPL PFDs. They should be used carefully since this risk reduction factor can be awarded if they are maintained over time.

Other conditional modifiers include:

- Probability of a hazardous atmosphere;
- Probability of ignition or initiation;
- Probability of explosion;
- Probability of personnel presence;
- Probability of injury or fatality;
- Probability of equipment damage or other financial impact.

There are certain situations in which the use of conditional modifiers is best avoided, similar to what we saw for enabling conditions. These include:

- The LOPA analyst(s) has(ve) insufficient knowledge of conditional modifiers to employ them correctly.
- Insufficient information is available to assess the conditional modifier probability.
- The established LOPA procedure indicates that conditional modifiers are not to be used.
- The established LOPA procedure is to not use conditional modifiers unless they provide a full order-of-magnitude effect on the risk calculation.

- The severity part of the LOPA risk evaluation is based on a loss event (hazardous material release, vessel rupture explosion, etc.) rather than on the assessed possible impact(s) of the loss event (injuries/fatalities, property damage, environmental damage, etc.).
- Management systems are not in place to ensure the validity of conditional modifier probabilities over time.

It needs to be emphasized that many companies choose not to use conditional modifiers in their LOPAs, for various reasons. Factors like the probability of personnel presence are often taken into account when estimating the severity level of a given scenario, rather than explicitly including them as conditional modifiers. Likewise, the probability of ignition is often taken as 100% in all situations. Reasons to exclude them are:

- A company prefers an implicit consideration of probability of personnel presence, and so on in choosing a severity level, rather than explicit consideration.
- A company doesn't see them as being comparable to IPLs; they consider them "luck factors."
- A company wants to avoid multiplying too many probabilities.
- The probability of ignition is always taken as 100%.
- And so on.

Citing explicitly (AIChE-CCPS, 2013), the following definitions are provided:

- *Human Error Probability (HEP):* The ratio between the number of human errors of a specific type and the number of opportunities for human errors on a particular task or within a defined time period. Synonyms: human failure probability and task failure probability.
- *Independent Protection Layer (IPL):* A device, system, or action that is capable of preventing a scenario from proceeding to the undesired consequence without being adversely affected by the initiating event or by the action of any other protection layer associated with the scenario.
- *Independent Protection Layer Response Time (IRT):* The IPL response time is the time necessary for the IPL to detect the out-of-limit condition and complete the actions necessary to stop progression of the process away from the safe state.
- *Initiating Event (IE):* A device failure, system failure, external event, or wrong action (or inaction) that begins a sequence of events leading to a consequence of concern.
- *Initiating Event Frequency (IEF):* How often the IE is expected to occur; in LOPA, the IEF is typically expressed in terms of occurrences per year.

If in the LOPA analysis, human error is considered to be initiating event, then those data should be verified using site-specific human performance validation. Specific conditional modifiers can be adopted to evaluate the PFD for those IPLs that are influenced by human factors. It is an extensive topic that is out of the scope of this book. The interested reader can find additional information in the *Handbook of Human Reliability Analysis with Emphasis on Nuclear Power Plant Applications* (Swain and Guttmann, 1983) and *Human Reliability and Safety Analysis Data Handbook* (Gertman and Blackman, 1994).

2.15 Bow-Tie as a Quantitative Method to Measure Risks and Develop a Dynamic Quantified Risk Register

The Bow-Tie method can be used for both qualitative and quantitative risk assessment, overcoming in fact the use of the method for the only immediate and intuitive graphic notation that, given the knowledge of the fundamental elements, can be used in a short time in real applications. In this second case, through some very simple rules, the Bow-Tie allows the expected frequency of consequences to be calculated once the following input data are known:

- Frequency of all independent causes;
- Probability of failure on demand (PFD) of all barriers.

Obviously, other frequency reduction/amplification parameters will also have to be assessed, which the analyst will have to evaluate, based on experience and objective data defined by the context and scope, in relation to time to risk, exposure and other factors, such as the degree of vulnerability of the targets to a given potential impact. These have been discussed in Section 2.14.3.

In the following paragraphs the operations to be performed for the frequency quantification of Bow-Tie will be presented, one step at a time. Once the frequency of the consequences is known and combined with the magnitude level that the analyst will obtain on the basis of other methods that are not dealt with here, then the level of risk associated with the consequences of the Bow-Tie diagram, for example through the use of a risk matrix, will be immediately determined.

2.15.1 LOPA Analysis in Bow-Tie

The tool that allows the frequency quantification of Bow-Tie is the LOPA (layer of protection analysis), already discussed in Section 2.1.4 and which represents, according to ISO 31010, a proper methodology.

Thanks to this tool, deliberately illustrated here in simplified form, it is possible to combine the frequencies of the causes with the probability of failure of the barriers, in order to obtain the frequency of occurrence of the consequences.

First of all, the difference between frequency and probability must be clear to the reader. While frequency is a dimensional quantity, measured in occasions per year [occ/year], probability is instead an adimensional measure (expressed from 0 to 1) used here as the possibility that a barrier does not perform its function satisfactorily.

Therefore, a barrier with PFD = 0 is obviously an ideal barrier, since it never fails, i.e. it has 100% reliability and availability. In reality, no barrier has a PFD = 0, because there are always latent failure mechanisms (in the case of hardware barriers) or human factors (in the case of software barriers) that make the PFD, however small, different from zero. To admit, at the opposite extreme, that PFD = 1 means that the barrier in question always fails whenever it is called upon to intervene, so in fact it offers no reduction in risk.

The LOPA analysis therefore foresees that the frequency of each independent cause is multiplied by the PFDs of the barriers placed along that specific "cause-top event" branch.

This multiplication is carried out for all independent causes. If the contribution of all independent causes is added together, the frequency of the top event is obtained. In mathematical terms:

$$f_{\text{top event}} = \sum_{i=1}^{N} f_i \cdot \prod_{j=1}^{M(i)} PFD\left(j(i)\right)$$

where

- $f_{\text{top event}}$ is the frequency of the top event;
- $i = 1 \ldots N$ indicates the ith cause;
- f_i is the frequency of occurrence of the ith cause;
- $j = 1 \ldots M(i)$ indicates the jth preventive barrier (whose maximum number M depends on the branch under consideration, i.e. the ith case under examination);
- $PFD(j(i))$ indicates the PFD of the jth preventive barrier placed on the ith cause.

At this point, the frequency of the consequences is obtained by multiplying the frequency of the top event by the PFD of the barriers placed on the branch of the consequence considered. In mathematical terms:

$$f_k = f_{\text{top event}} \cdot \prod_{j=1}^{M(k)} PFD\left(j(k)\right)$$

- f_k is the frequency of occurrence of the kth consequence;
- $j = 1 \ldots M(k)$ indicates the jth mitigation barrier (whose maximum number M depends on the branch under consideration, i.e. the kth consequence under consideration);
- $PFD(j(k))$ indicates the PFD of the jth mitigation barrier placed on the kth consequence.

This method can be enriched, as mentioned above, by conditional modifiers, i.e. multipliers that aim to also include, within the quantitative analysis carried out, evaluations on the time of exposure to risk, or other factors such as temporal or spatial constraints that are necessary to satisfy in order to be exposed to risk.

The application of LOPA analysis to the Bow-Tie method used preliminarily to construct the risk model to be assessed, with the mathematical formulas presented above, is based on two important hypotheses:

1) Independence of causes (the frequency of occurrence of a cause is independent of whether or not all other causes occur);
2) Independence of barriers (there are no common causes of failure between barriers identified on the same branch).

Summing up:

- Causes are quantified by assigning each cause an occurrence frequency (occ/year), which can be obtained either through technical-scientific literature data (database) or through more detailed evaluations such as a fault tree or human factor assessments;
- The quantification of preventive and mitigative primary barriers is done by assigning to each of them a probability of failure on demand, which can be obtained either through

technical-scientific literature data (database) or through more detailed assessments such as a failure tree or human factor assessments.

- The quantification of the top event is done by combining the frequencies of the causes with the PFD of the preventive barriers, according to the logic discussed above;
- The quantification of the consequences is done by combining the frequency of the top event with the PFD of the mitigation barriers, according to the logic discussed above;
- There is no explicit quantification of escalation factors and secondary barriers. Any reduction in the reliability or availability of the primary barrier by the escalation factor and secondary barriers must be considered when determining the PFD of the primary barrier to which the escalation factor and secondary barriers refer. Generally, this results in an increase in the PFD of the primary barrier.

Finally, it may also be useful to distinguish three types of occurrence frequencies:

1) Those unmitigated calculated without barriers, i.e. assuming that PFD = 1 for all barriers. In this way, the risk of incidental scenarios is assessed if the organization has not installed any barriers.
2) The current ones calculated assuming the PFD of preventive and mitigative barriers installed by the organization to lower the risk.
3) The mitigated ones calculated considering the contribution of any additional barriers that the organization should implement to further mitigate the level of risk.

This distinction makes it possible to immediately appreciate the contribution made by barriers (whether they are actually present or only assumed) to the reduction of the level of risk.

For the quantification of a Bow-Tie, since it is not always easy to carry out calculations by hand or on a spreadsheet, special plug-ins available for the main software solutions on the market can be used, leaving the burden of the calculations entirely to the machine.

Consider the simplified Bow-Tie in Figure 100.

The input data are as follows:

- Frequency Cause #1 = 0.0005 occ/year.
- Frequency Cause #2 = 0.0001 occ/year
- PFD Barrier #1 = 1E-1.
- PFD Barrier #2 = 1E-1

With this data, the current frequencies are:

$$f_{\text{top event},cu} = 0.0005 \cdot \left(1 \cdot 10^{-1}\right) + 0.0001 = 0.00015 \text{ occ / year}$$

$$f_{1,cu} = 0.00015 \cdot \left(1 \cdot 10^{-1}\right) = 1.5 \cdot 10^{-5} \text{ occ / year}$$

In the absence of barriers, however, unmitigated frequencies are:

$$f_{\text{top event},un} = 0.0005 \cdot 0.0001 = 0{,}0006 \text{ occ / year}$$

$$f_{1,un} = 0{,}0006 \text{ occ / year}$$

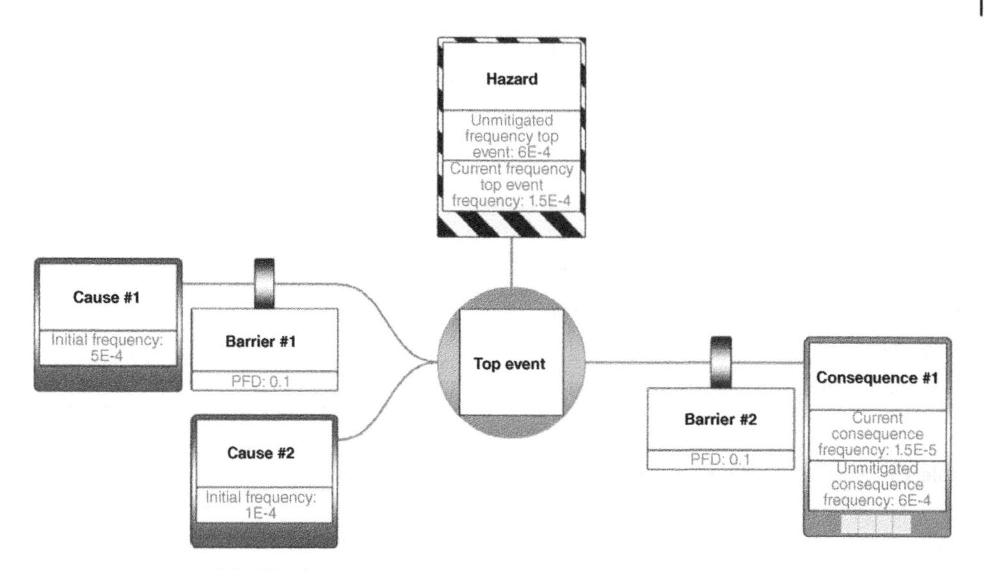

Figure 100 Quantifying a simplified Bow-Tie.

Therefore, a significant increase in the unmitigated frequency is observed compared to the current one. This difference, depending on the sensitivity of the risk assessment criterion used, could be decisive in triggering one or more levels of risk from one case to another.

The quantification of a Bow-Tie on the basis of a LOPA approach (respecting the formulation criteria of both methodologies) also makes it possible to make assessments in this regard:

- On the degree of risk reduction associated with a control measure characterized by a certain PFD compared to the reduction factor obtained by the set of controls, i.e. the strategy implemented by the organization;
- On the reduction factor associated with alternative control strategies.

These assessments are particularly useful for the identification of any judgement of importance or criticality of the barriers, for the conduct of cost-benefit analyses, or for the prioritization of risk-oriented adaptation or improvement measures.

Finally, it is sometimes useful to define a correlation factor between the PFD of a barrier and its effectiveness, in order to correct the value of the PFD according to a set of performance factors recognized as driving factors for the integrity and functionality of that specific barrier. This is extremely useful when it is necessary to perform a relative risk ranking on a pool of similar assets, for which it is not important to know the "absolute" risk level, but the relative risk level of one asset over the others, in order to properly prioritize possible barrier improvement actions on the basis of risk-based decisions.

For example, the effectiveness of a barrier can be measured on a normalized scale from 0% (zero effectiveness) to 100% (maximum effectiveness). The numerical scores obtained can also be associated with a qualitative judgement, summarized in Table 15.

The criterion for assigning a numerical value to the effectiveness of a barrier is as follows:

$$\text{EFFECTIVENESS} = \text{EXISTENCE} * \text{AVERAGE}(\text{PS})$$

Table 15 Quality scores and judgments on the effectiveness of barriers.

	Range	
Effectiveness	**Min**	**Max**
Very poor	0%	29%
Poor	30%	59%
Good	60%	79%
Very good	80%	100%
Not applicable	NA	NA

Table 16 Standard Performance Scores (PS)

PS	Score	Description
++	1	Present and effective
+	0.66	Present and medium effective
–	0.33	Present and poorly effective
––	0	Absent
IM	0.33	Missing information

In other words, first the existence of each individual barrier is assessed, assigning, for each individual asset:

- Yes, if the barrier exists (EXISTENCE = 1);
- No, if the barrier does not exist but should exist (EXISTENCE = 0, which forces the effectiveness to 0%);
- NA, if the barrier does not exist and it is not necessary to foresee it.

In the last two cases, no further evaluation is necessary.

In the first case (existing barrier) it is instead necessary to assign qualitative scores to the performance standards identified for each individual barrier, such as the existence of an updated procedure, the execution and frequency of maintenance and inspection activities, and so on.

Table 16 can be used to assign scores to the performance standards.

Finally, the effectiveness of the barriers, evaluated in step 3, and therefore the peculiarities of the barriers plant by plant, is taken into account, using the following formula:

$$\text{PFD}(\text{corrected}) = \left[\left(\text{PFD}(\text{theoretical})\ \ 1\right) * \text{EFFECTIVENESS} / 100\right] + 1$$

In other words, this expression makes it possible to correct the EFP of barriers by modifying it within a range ranging from PFD(theoretical), in the case of 100% effective barrier, to 1 in the case of 0% effective barrier. It has therefore been hypothesized that the variability within this range is directly proportional to the percentage value of effectiveness evaluated in the previous phase, as represented in Figure 101 and Figure 102.

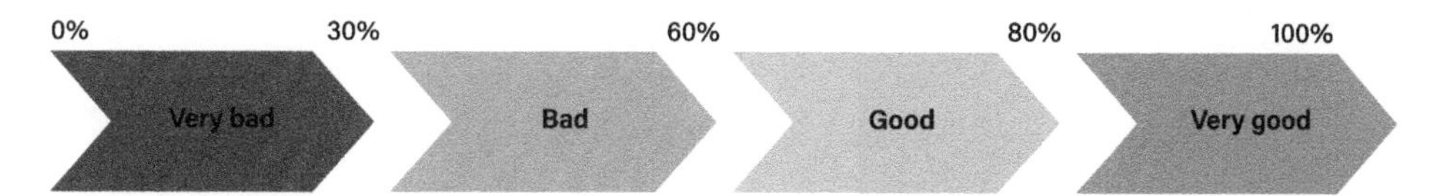

Figure 101 Scale of the effectiveness of a barrier and the relationship between effectiveness and PFD (correct).

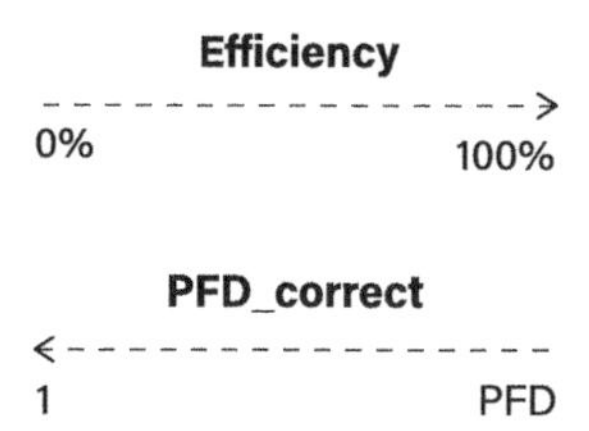

Figure 102 Relationship between effectiveness and PFD (correct).

2.16 Advanced Bow-Ties: Chaining and Combination

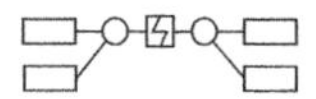

In some cases, the need to analyze the consequences of the consequences of a Bow-Tie diagram, or the causes of the causes, may arise. In other words, it may be necessary to further explore the temporal-causal link between causes and consequences, thus considering the consequences of the Bow-Tie diagram as the starting point (i.e. the causes) of another top event, and thus of another Bow-Tie. At the same time, it may be necessary to analyze the initiating events of a given cause, which will therefore also be the consequence of another top event, and therefore of another Bow-Tie.

This need can be addressed by concatenating Bow-Tie events (Figure 103). This is an advanced use of the Bow-Tie methodology, which should only be used when really necessary. In most cases, once the scope and perimeter within which the risk analysis is to be carried out has been defined, developing a single Bow-Tie is sufficient.

Other times, however, the need to highlight intermediate states in the logical flow that leads from causes to consequences, or the need to separate hazards as they impact different organizations, or the need to highlight the presence of certain control measures impacting intermediate stages, may require the concatenation of Bow-Tie events. This can take place between cause/consequences and top events. Therefore, a cause or consequence in diagram X will be a top event in diagram Y.

The relationships between causes/consequences and top events of various Bow-Tie events can also be very complex. It is therefore recommended to use the concatenation of Bow-Ties only when actually useful and in any case always sparingly, in order to always have a simple and immediate overview (which we would like to point out once again to be among the unquestionable advantages of the Bow-Tie method that have established its widespread use in various sectors).

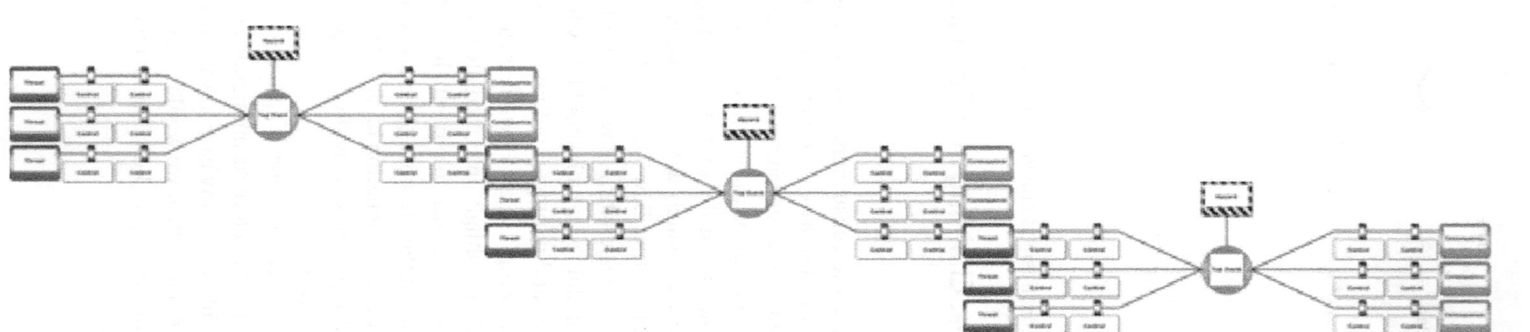

Figure 103 Bow-Tie concatenation example

3

Barrier Failure Analysis

3.1 Accidents, Near-Misses, and Non-Conformities in Risk Management

By "operational experience of organizations" we mean all deviations from expected results (or even process deviations) which, leading to an undesired condition, may be accidents, near-misses or non-conformities (anomalies).

Before explaining its meaning, it should be made clear that the intention of this chapter is to provide the reader with some basic information for the analysis of such events, not for their investigation. In other words, although the analysis of incidents necessarily starts from their investigation, for the purposes of this chapter all the information, data and evidence necessary for their analysis is considered already available to the organization so that it can be used to derive other information and lessons.

There is no unambiguous definition of "accident" in the literature, as different versions have been given over time. According to one of the most widespread, an accident is an event that can cause injury or other health problems. However, this definition limits the impact of an accident to the sphere of human health and could be very limited.

Other authors define an accident as an unwanted event that causes damage to health or property. A broader and more complete definition describes it as the final event of an unplanned process that causes injury or illness to employees and possible damage to property. In even more general terms, an incident can be described as an unplanned event or sequence of events that leads to negative impacts on the organization; this definition finally covers all possible targets that can be damaged by an incident, such as health, safety, environment, economic damage – generally speaking, "negative impacts."

Following the barrier-based approach, an incident occurs when all control barriers fail in their attempt to stop, with their own probability of success (assessable in 1-PFD), the sequence of events that from one or more causes lead to a top event first and then a consequence. These events occur because barrier failures exist. The aim of this statement is therefore to exclude from the definition of "accident" all those events that involve uncontrollability, bad luck, or surprise, sometimes defined in literature as an "act of God," for which the English term *accident* is used instead of *incident*. The analysis of accidents reveals that they are generated by so-called immediate causes, that is, technical failures or

Bow-Tie Industrial Risk Management Across Sectors: A Barrier-Based Approach,
First Edition. Luca Fiorentini.

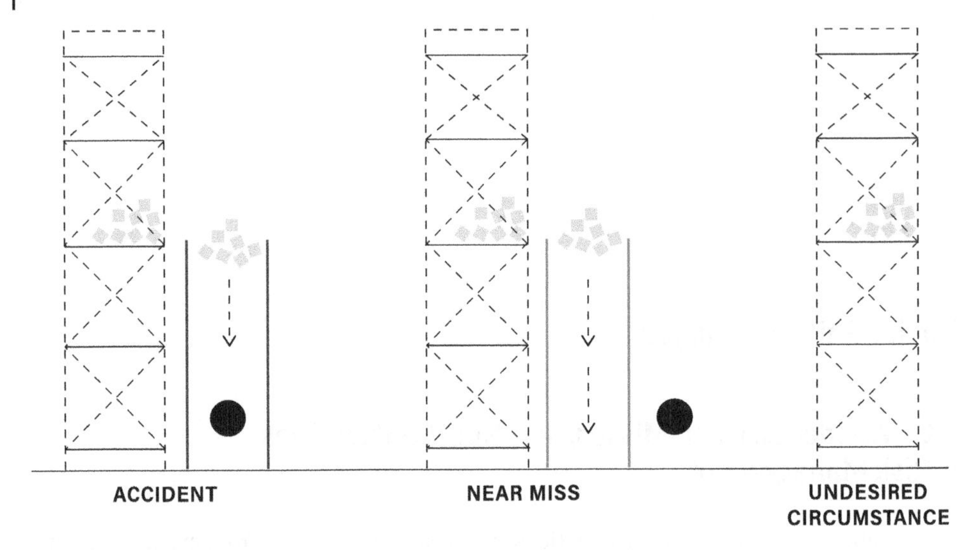

Figure 104 Difference between accident, near-accident and unintended circumstance.

human errors that have led to the failure of one or multiple barriers. However, they are induced, facilitated or accelerated by latent organizational factors known as *root causes*.

A near-miss is an incident without adverse consequences but with the potential to generate a loss. Some authors summarize this concept by stating that the difference between an accident and a near-miss is simply "luck." In fact, while an incident results in actual loss or damage, a near-miss is an event where an incident (i.e. loss or damage) could have occurred if the circumstances had been slightly different. Consider the example in Figure 104. Materials are stored on a scaffold at height. If, when falling, the materials hit a worker, then an accident has occurred. If, on the other hand, when falling, the materials stop their fall thanks to a safety net (so in the barrier-based approach only one barrier remained intact), then we speak of a high-potential incident. But what happens if the materials fall from the scaffolding, reach the ground, and do not hurt anyone just because no operator was in the line of fire at the time? In this case we speak of a near-miss, because although the potential energy that could have caused damage was released, no target was exposed to it and therefore no loss or damage was recorded.

The undesirable circumstance, i.e. the set of conditions that have the potential to cause damage or loss, still needs to be defined. For example, storing material at height on a scaffold without other safety precautions is an undesirable circumstance, i.e. a non-conformity, that has the same importance as accidents and near-accidents to be investigated and analyzed, since it has its own root causes and, in another occasion, could evolve in an accident.

3.2 The Importance of Operational Experience

Considering safety-related emergencies as an easy-to-understand example, when an accident occurs, the emergency response machine (set up by the local institutions and/or the same organization that has suffered an accident) intervenes to stop the cause of the accidental event – or at least to stop its propagation, such as the domino effect of the accident to adjacent areas, facilities or units – and to provide the necessary assistance where required. Once this phase is over, all that is left are significant losses, such as injuries to workers, spills of pollutants that cause environmental damage, and major economic damage.

However, the occurrence of an accident becomes, as with any threats, an opportunity to learn some lessons, starting with the analysis of what went wrong. This allows a better understanding of unresolved questions and unknown aspects whose impact (on safety, the environment, or business economics) is considerable, as evidenced by the occurrence of the accident itself. In order to avoid the recurrence of accidents, an investigation into them is therefore carried out with the aim of identifying the causes that led to the event. This is why it is important to analyze the organisations' operational experience.

The Learning From Experience (LFE) process provides opportunities to learn from incidents, audits or other events, leading to improvements in the organisation's Operational Excellence Management System and requests for corrective action through LFE directives or notices. It is important to understand that the analysis of an incident, near-miss or non-conformity is beneficial only if:

- Research is conducted on the facts, not on guilt;
- You get to the root causes;
- Reporting, sharing and preserving activities are put in place.

The importance of accident analysis is, in fact, based on three simple principles, illustrated in Figure 105: what is not reported cannot be investigated, what is not investigated cannot be modified, and what is not modified cannot be improved. However, the final improvement is the aim of an employer, as Figure 106 reminds us. In fact, the investigation of an accident is of primary importance also for third parties, stakeholders and, in some cases, institutions, whose real interest, however, is not the actual improvement of safety levels from risk (whether related to health, environmental or economic aspects), but respectively to protect private and public interests.

At the end of this section, it is reiterated that this chapter, while presenting an overview of the entire process indicated in Figure 107, intends to focus on the analysis phase.

In conclusion, the organization should learn from actual outcomes in order to improve its performance, preventing similar unwanted events reoccurrence. These should include losses, near-misses, non-conformities and opportunities that were identified in advance, occurred, and yet were not acted on. Points that may be considered in such a review include:

- What happened;
- How and why the outcome came about;
- Whether any assumptions need to be reviewed as a result of the outcome;

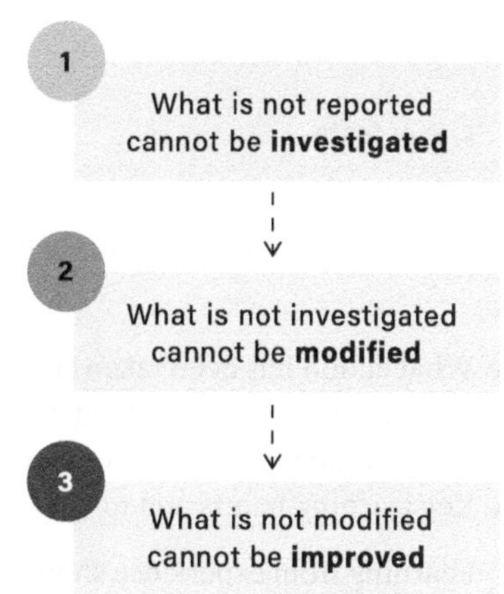

Figure 105 Principles of incident analysis.

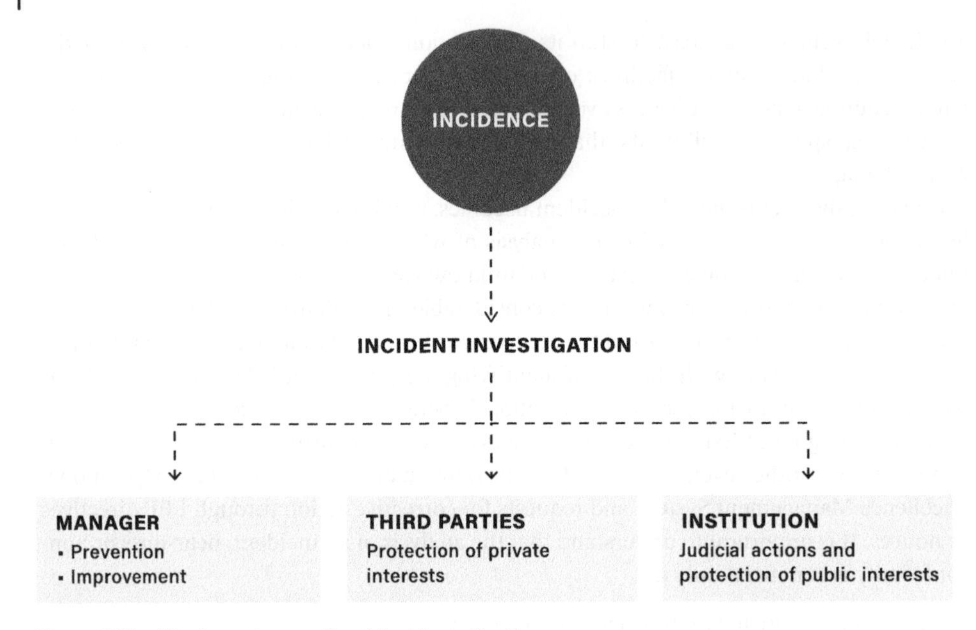

Figure 106 The importance of accident investigations.

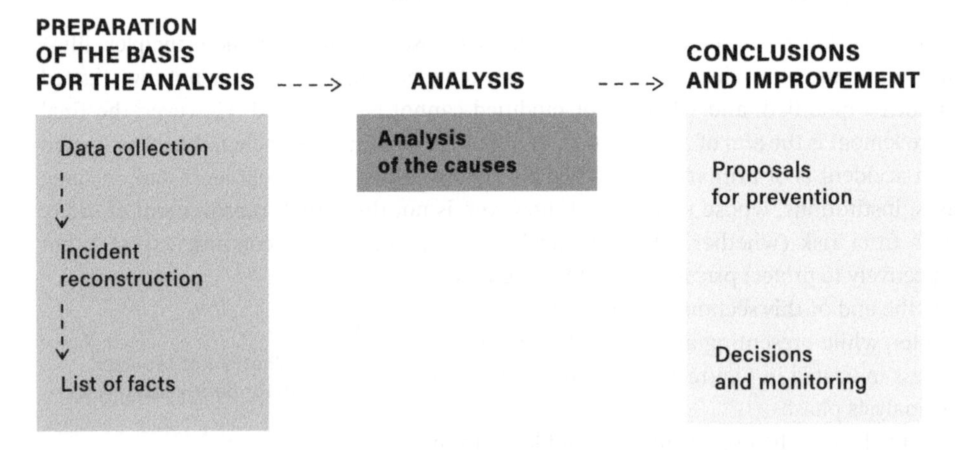

Figure 107 Steps in the analysis of the operational experience of organizations.

- What action has been taken (if any) in response;
- The likelihood of the outcome happening again;
- Any additional responses or steps to be taken;
- Key learning points and to whom they need to be communicated.

Learning from experience should be considered a periodic process, to be adapted also (in terms of frequences, tools and methods, and outcomes presentation) on the basis of results, in order to guarantee its compliance with the complexity of the organization and its strategic goals.

3.3 Principles of Accident Investigation

The search for incidents and near-misses, together with related investigation and analysis activities, is one of the most valid methods to improve the safety and reliability of business processes (both hardware and software). The risk assessment discussed in the first chapters of this book shares the same objective as the analysis of operational experience (i.e. improving risk management), but it is a predictive method. This characteristic of risk assessment implies the following limitations:

- The analyses are speculative. Therefore, it is highly possible that all plausible events are not identified.
- It is difficult to predict the real level of risk, because its estimation is usually based on approximate assessments of the likely consequences.
- It is difficult to identify events resulting from multiple causes, such as accidents.
- It is difficult to predict human error and take it into account in risk assessment.

On the other hand, the investigation of accidents, near-misses and non-conformities provides useful information, even if difficult to obtain, reducing prejudices, ignorance and misunderstandings that may influence the theoretical preventive analysis (i.e. risk assessment). The investigation of accidents is a fundamental element of any management system (safety, environment, financial losses) whose main objective is to prevent the damaging occurrence. In fact, one of the fundamental assumptions of accident investigation concerns the possibility of finding, as root causes, a malfunction in the management system. In other words, it is always possible to find certain aspects of the management system that, if correctly organized and applied, could have prevented the accident from happening. This malfunction may be related to a lack of planning, organization, updating or control of the management system. According to this, when developing, evaluating or improving the incident management system, it is essential to:

- Have a strong commitment at the leadership level.
- Involve relevant staff, define an appropriate scope, implement the programme consistently across the company and monitor the effectiveness of incident investigation to maintain reliable investigative practices.
- Identify potential incidents to investigate, monitor all possible sources and ensure reporting activities.
- Adopt appropriate methods to investigate, collect appropriate data, be rigorous, and provide expertise and tools for investigative staff.
- Enhance the results of accident investigation, with a clear link between identified causes and developed recommendations.
- Monitor survey results, implement recommendations, and share findings both externally and internally within the company perimeter.
- Analyze the data to identify any trends in the occurrence of similar incidents.

The execution of an accident investigation may benefit from the use of a structured method, such as those presented in this chapter. However, an effective accident investigation is not simply limited to the application of a method, but also requires the investigator to:

- Establish trust and safety, thereby creating a favourable environment to discuss the incident;
- Be inclined to listen to what people say and base all findings on verifiable facts;
- Establish a clear cause-and-effect link, based on solid evidence;
- Be assisted by technical experts when dealing with specialized issues;
- Understand the root causes identified and eventually link them to management system elements and organization processes;
- Manage the investigation and analysis of incidents as any other ordinary project would require (activity planning, budgeting).

The scientific method is the fundamental basis for investigating accidents, near-misses and non-conformities. Its rigorous approach is the key to conducting effective investigations that look beyond the immediate and trivial causes. But it is also a systematic, comprehensive and intellectually honest method.

The accident investigation process consists of many activities. In the context of a simplified approach (useful in many cases and affordable also by small organizations as well as those willing to implement a structured risk management process), it is possible to identify the three phases in Figure 108: data collection, evidence analysis (which can guide the collection of evidence towards new objectives, generating a new hypothesis or rejecting others) and developing effective recommendations.

The investigation of an accident is a process that may be required for different purposes by different entities (among them the organization leadership, its stakeholders, authorities,

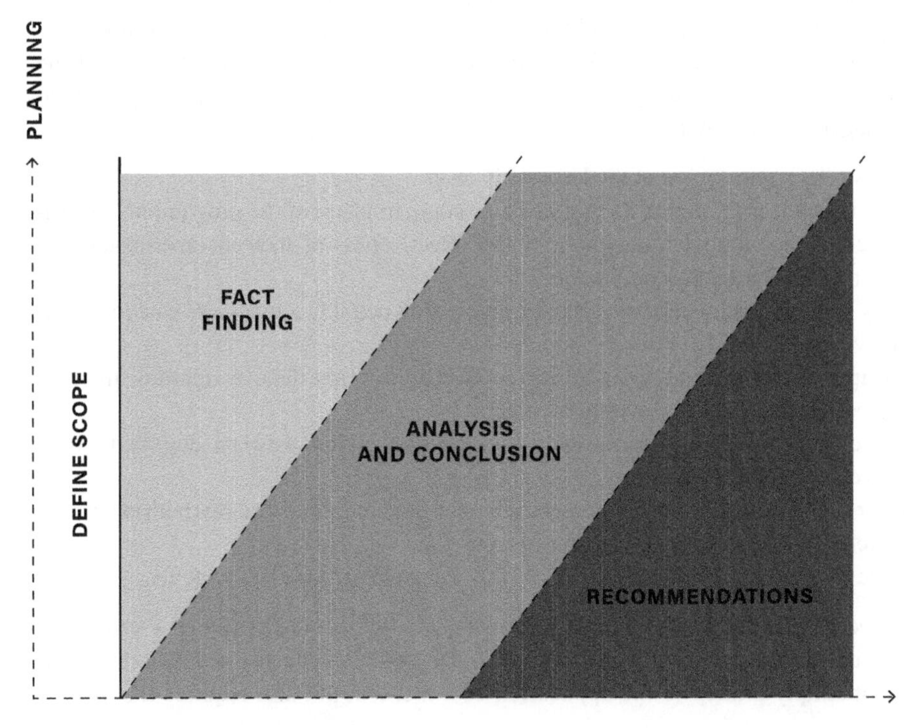

Figure 108 Steps in accident investigations.

etc.). The main purpose of the investigation is to determine the cause of the incident, exploring both immediate and root causes, and to develop recommendations to prevent its recurrence. Its purpose is never to assign the blame for the damaging event to anyone. However, there may be other objectives in conducting an accident investigation, such as verifying compliance with laws and standards or resolving problems relating to insurance liability for damages. For example, after an industrial accident, an authority having jurisdiction may assume that a crime has been committed and then decide to conduct an investigation. In this case, it is carried out to assess the basis for potential criminal prosecution, so it is legitimate to go and find the "faults" as well. Very often, the investigation is commissioned by the same company that suffered the accident. This is done not only to comply with the internal rules established at the corporate level, but also to ensure the successful outcome of the learning from experience process, thus avoiding similar events in the future.

Investigating accidents is also a good way to demonstrate how positive the company's attitude towards health and safety is, regardless of whether a real dispute arises. It is common to investigate the reasons why a part, component, material, procedure or management system fails. In theory, the reasons they were successful before an accident occurred should also be investigated, but the reasons that produce positive results are generally taken for granted and attention is understandably focused only on undesirable outcomes.

On this basis it is possible to define the investigation as the management process through which the root causes of undesirable events are discovered and measures are taken to prevent similar events.

Investigations usually start at the end of the story: once the incident has occurred, people wonder how it could have happened, in order to go "beyond the widjet" and understand the root causes rather than the observed immediate ones.

Starting from the chronologically final point, the investigator starts his work, in an attempt to determine who, what, when, where, why and how the accident happened. Only when the event has been explained in its entirety, the sequence reconstructed and the main causes identified, can the investigation be said to be over. The investigative analysis is based on physical evidence and verifiable facts. The investigator then uses selected scientific principles and methodologies to collect, recognize, organize and analyze the evidence. Reasoning by similarity, the investigation of the accident is structured like a pyramid (Figure 109). The facts collected and the physical evidence are the basis of the investigative pyramid. They are therefore the input element for the analysis, carried out in accordance with scientific principles. Finally, the analysis is the basis to support a small number of conclusions (the apex of the pyramid), which must not be based on other conclusions or hypotheses; otherwise the investigative pyramid collapses.

The peculiarity of each incident, its complexity, the subjectivity given by the personal contribution of the investigator, and the lack of all the data useful for the resolution of the case make it impossible to find universal guidelines to be adopted for the investigation and analysis of each incident.

However, it is always possible to identify three levels of causes:

1) *Immediate cause*. This is the most obvious reason why an adverse event occurs (e.g. incorrect positioning of a valve). A single incident can be related to several immediate causes.

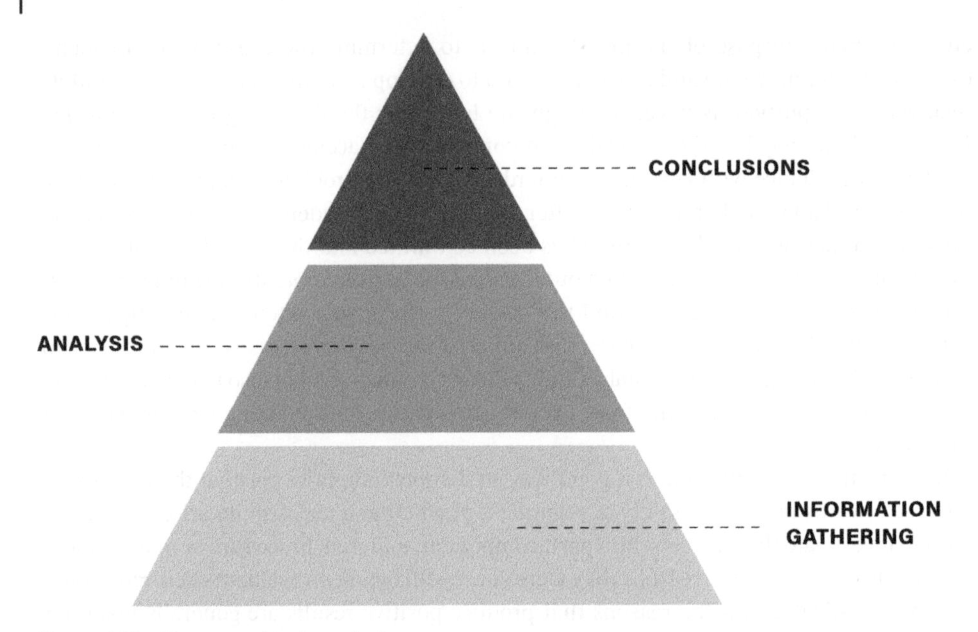

Figure 109 The pyramid of conclusions.

2) *Latent cause*. This is the least obvious reason found at the end of the investigation and often concerns the condition of the surrounding environment, e.g. consolidation of bad habits, high pressure to finish a job in time, poor ergonomics and so on.
3) *Root cause*. This is the initiating event from which all other causes originate. They are generally related to problems of management, planning or organization.

In general, recommendations are also developed on these three different levels, reflecting the distinction presented for causes. They will address different aspects of the management system. It is clear that if a recommendation aimed to solve immediate causes could prevent the same event reccurrence in a short period with limited resources, a root cause–related recommendation will require more effort but will address the avoidance in the medium and long terms of events in a similar class.

How thoroughly should an accident be investigated? A good criterion is to stop with the analysis of the causes when those identified are no longer controllable, in relation to the scope of investigation, by the subjects who have suffered and investigated the accident, taking care never to confuse an event (e.g., the malfunctioning of a control system or a spill of hydrocarbons) with a cause.

According to the scientific method, once the causes have been preliminarly hypothesized, on the basis of the available evidence, the credible ones are selected and in the end the so-called "falsification" is carried out, i.e. a logical test on the selected hypothesis that aims to eliminate the confirmation bias, that is, that error that occurs when the investigator tends to rely on a hypothesis only because "there can be no other explanation."

This book does not cover the forensic aspects of accident investigations whose legal, ethical and insurance repercussions deserve separate attention.

Before presenting the BFA, selected here as the most intuitive barrier-based incident analysis method, the Tripod Beta method is also briefly explained, an example of which is also given in Chapter 4 on case studies.

Tripod Beta is an incident investigation methodology developed in the early 1990s. It was explicitly created, in line with the human behaviour model, to help accident investigators model incidents to understand how the environment influenced the sequence of events and to discover the root organizational deficiencies that triggered the incident. Indeed, the idea behind Tripod is that organizational failures are the main factors in accident causation (Sklet, 2002).

Using Tripod Beta, incident investigators model incidents in terms of:

- Objects (something acted upon, such as a flammable substance or a piece of equipment);
- Agent of change (something – often an energy – that acts upon objects, such as a person or fire);
- Events (the result of an agent acting upon an object, such as an explosion).

Working back from the top event (the incident) allows a full understanding of what happened and how. A set of shapes consisting of an agent, an object and an event is called a trio and is the basic building block of the Tripod Beta method. Events themselves can also be objects or agents, allowing the investigator to chain these trios into a large diagram.

To understand why the incident happened, the next step is to determine what barriers were in place to prevent those objects and agents acting in the way they did and why they failed. Tripod Beta teaches looking at the immediate causes of the act that led to the incident, the psychological precursors to that, and ultimately the underlying organizational deficiencies that allowed those precursors to exist. An example of a Tripod Beta diagram is shown in Figure 110.

Performing a Tripod Beta analysis means following these steps:

1) Collection and preserving evidence.
2) Creating a timeline/storyboard. This is a listing of the main important events and the relevant factors sorted in a temporal sequence. This is especially suggested for complex incidents, involving many people and systems (i.e. an evident complexity), helping to understand how latent issues (design aspects, unrevealed failures) affect the outcome.

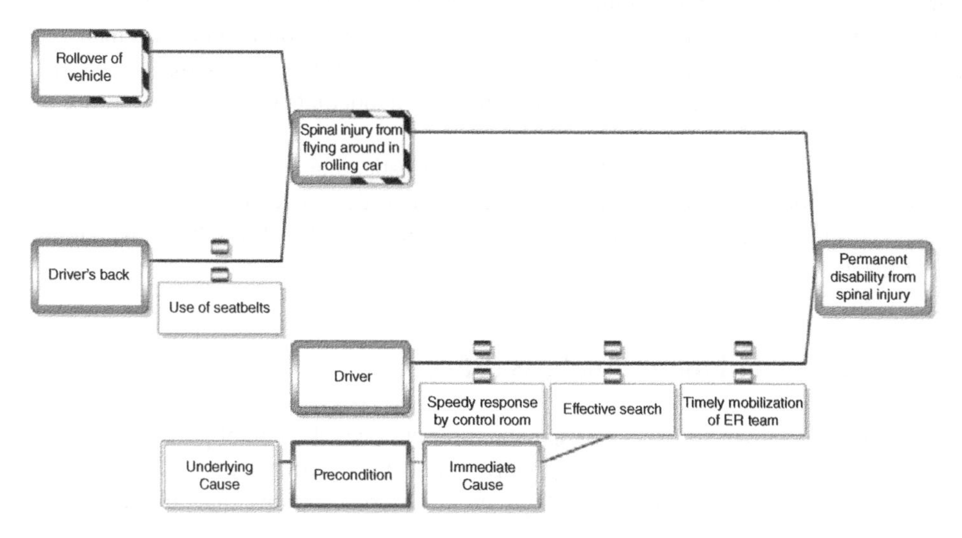

Figure 110 Example a Tripod Beta diagram.

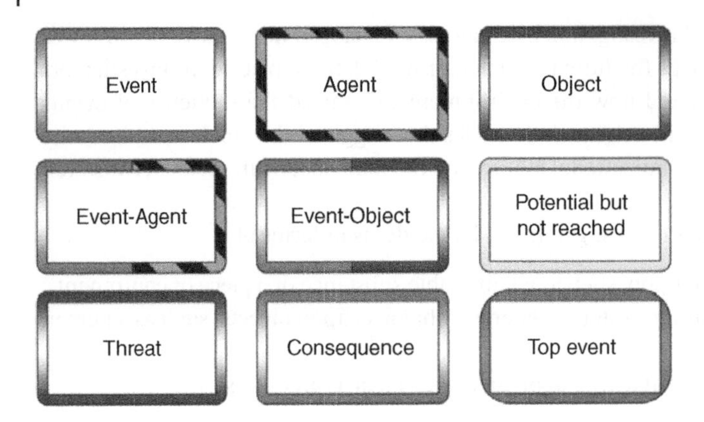

Figure 111 Possible Tripod Beta appearances.

3) Identifying the trios (agents, objects, and events) and linking them to each other, creating an event flow diagram. The combined nodes "event-agent" and "event-object" are the outcome of the linking process of an event to an agent or an event to an object respectively.
4) Identifying the barriers between the event and agent or object.
5) Analyzing the barriers (What has gone wrong? How? Why?).
6) Performing a Tripod Beta causation assessment on each barrier.
7) Generating the final report.

The possible Tripod Beta appearances are in Figure 111.

The idea behind Tripod is that substandard situations do not just occur; indeed they are generated by mechanisms coming from decisions made at the management level. These underlying mechanisms are the basic risk factors (BRF). They cover a broad range of factors (human, organizational and technical), including psychological precursors like time pressure, poor motivation, and so on. Prevention of incidents is therefore performed by removing these latent factors. Examples of possible BRFs are shown in Table 17.

The Tripod Beta method merges two different models: the hazard and effects management process (HEMP) model and the original Tripod model. The results are merged in a computer-based instrument that exploits a menu-driven tool that guides the investigator in representing the accident dynamics.

The Tripod Beta method follows the new way of investigating incidents; indeed after the BRFs are identified, recommendations are developed in order to decrease or eliminate their impact. This means that the real source of the problem is faced, not simply the symptoms (Sklet, 2002).

3.4 The Barrier Failure Analysis (BFA)

Barrier failure analysis (BFA) is a pragmatic, un-opinionated, general-purpose incident analysis method. It has no affiliation with any particular organization. It can be seen as a good companion for the Bow-Tie method in those organizations that are willing to

Table 17 Definition of BRFs in Tripod Beta.

Basic Risk Factor	Short Description
Design	User-unfriendly tools or equipment
Tools and equipment	Poor quality, suitability or availability
Maintenance management	Inadequate performance of maintenance tasks
Housekeeping	Insufficient attention to keep the floor cleaned
Error-enforcing conditions	Not-appropriate physical performance
Procedures	Insufficient quality or availability of procedures
Training	Insufficient competencies or experiences
Communications	Ineffective communications
Incompatible goals	Financial/production goals inconsistent with optimal working conditions
Organization	Inadequate or ineffective management
Defences	Insufficient protection of people, material and environment

implement a risk management framework compliant with the ISO 31000 standard and have found in the barrier-based approach a way to implement the standard principles. BFA is a way to structure an incident and to categorize parts of incident taxonomy. The structure has events, barriers and causation paths. In this, BFA shares a number of elements with the Bow-Tie method described throughout the book. These common elements facilitate knowledge sharing, methods applications, and results communication in the organization, especially those having limited resources or those new to risk management. Events are used to describe an unwanted causal sequence of events. This means that each event causes the next event. There can also be parallel events that together cause the next event.

Barriers are used to highlight certain parts of our environment as being primarily designed to stop a chain of events. They are not necessarily independent, or sufficient. Since the unwanted events still happened, causation paths are added to explain why the barriers did not perform their function. The causation path goes three levels deep. The levels are simply called primary, secondary and tertiary. These labels can be changed, but the idea is that a barrier can be analyzed in three causal steps. It does not specify whether the analysis should end on an organizational level or not, although this is what would happen most often.

Each level in the causation path can also be given a category or a classification. Because there is an infinite number of possible categorizations and a large number of different kinds of organizations, it is not possible to create a single definitive set of categories. Instead, users can create custom categories. This is why any categorization should go through iterations, to add and remove categories as they become wanted or obsolete. Optionally, any organization can make the categories specific to their context. This has a downside of not being able to make comparisons between different organizations, but that is not the primary goal of this methodology. It is better to use categories that are relevant than ones that are standardized.

Any organization should go through an initial period of testing and iterating categories. At some point a steady state should emerge that will capture most incidents. There will always be exceptions, but they should be exactly that – exceptions. Once exceptions happen more frequently, they stop being exceptions and should be integrated into the existing categorizations in a new iteration of the taxonomy. Taxonomies are a core element of a structured and systematic risk management framework, especially if they are shared among different processes or the entire organization. They also facilitate discussions on result, definitions of KPIs based on results over time, and decisions on root causes level recommendations.

Examples of BFA diagram are shown in Figure 112 and Figure 113.

The core elements of a BFA diagram are shown in Figure 114.

Once developed, a BFA diagram looks like the structure shown in Figure 115.

The BFA is carried out in seven steps:

1) Fact-finding (timeline);
2) Event chaining;
3) Identifying barriers;
4) Assessing barrier state;
5) Causation analysis and categories;
6) Recommendations;
7) Reporting/link to Bow-Tie diagram.

The first step is essential to get an overview by arranging the facts. It is suggested that the focus be on gathering as much info as possible, beware of assumptions, focus on the facts, and use a timeline tool to organize the facts. A common pitfall is making recommendations at this stage, but it must be avoided.

The second step is the "event chaining" (Figure 116). It is first necessary to define what an event is: "a happening or a change of state, in which the incident sequence changes."

Two points are certain: the normal mode of operations (operating condition) and the moment of incident. The zoom level will depend on the scope: complex incidents typically require more events. It is important to not confuse defeated barriers with BFA events (Figure 117).

The third step is the barrier identification (Figure 118). At this stage, the investigator usually asks in a brainstorming session, "Which barriers should have prevented the next event?" A barrier should be defined in the normal/wanted state, and should be able to prevent the event at its right. The barriers are put in the order of their effect.

Attention should be paid at this step, in order to avoid some common pitfalls, as shown in Figure 119.

The fourth step is the assessment of the barrier state, according to what discussed in Section 4.3.4.

The fifth step is the BFA analysis (Figure 120). The aim of this step is to understand what caused the barrier to fail.

For each barrier, the analysis tries to find:

- *Primary cause*. What exactly happened? (Operational).
- *Secondary cause*. Why did it happen? (Line management).
- *Tertiary cause*. How could the management prevent it? (Management).

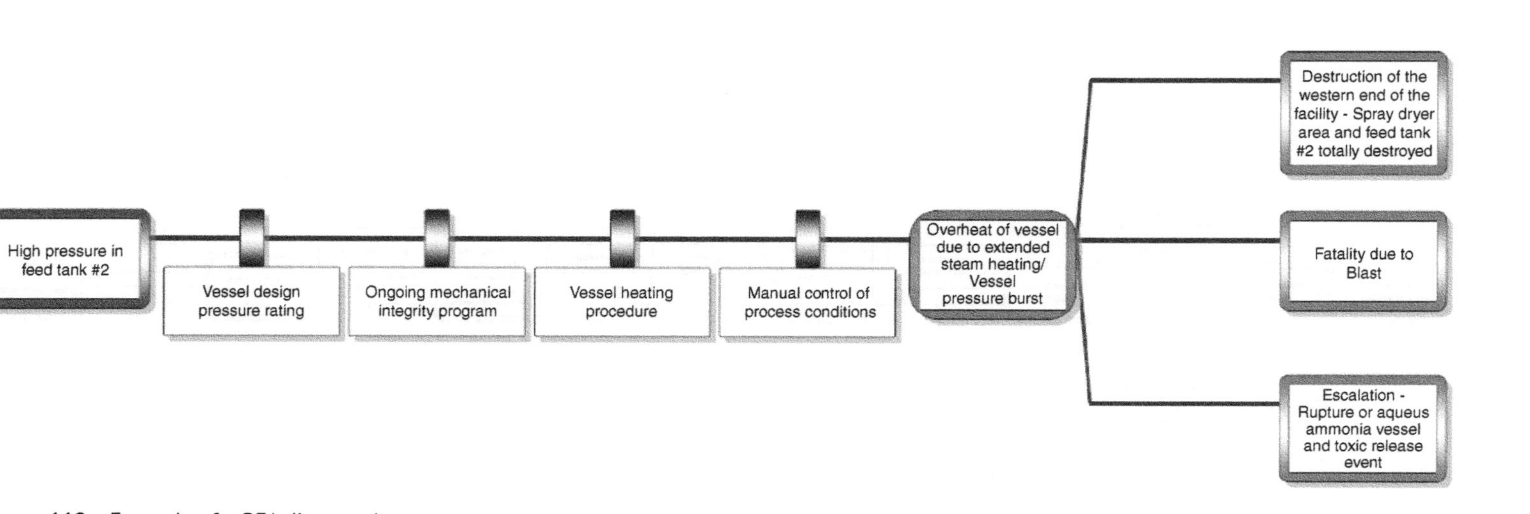

Figure 112 Example of a BFA diagram 1.

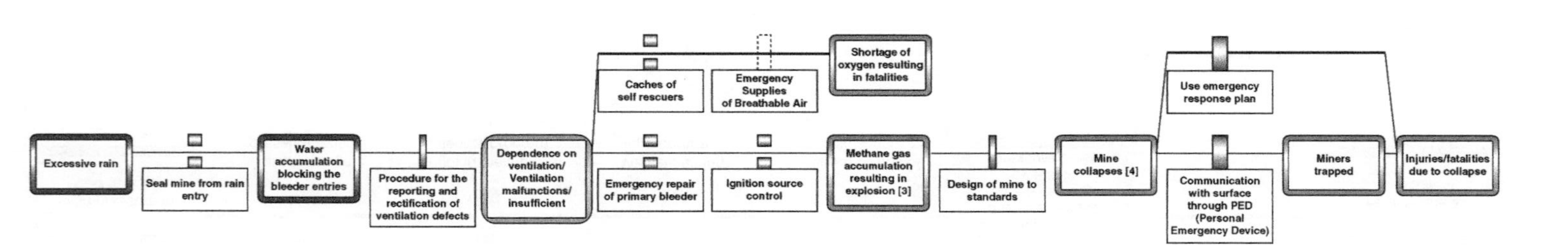

Figure 113 Example of a BFA diagram 2.

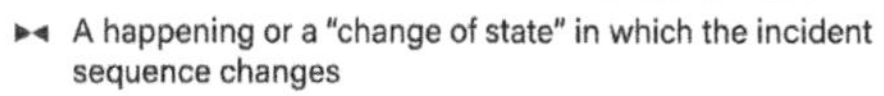

EVENT
A happening or a "change of state" in which the incident sequence changes
Place events in chronological order
Usually the first event is "normal" business activity or process
The amount of events depends on the scope of the analisys and complexity of the incident.
There can be parallel sequences for events
POTENTIAL EVENT
Events can have various appearances
Potential event appearance can be especially powerful to communicate about effective barriers and near misses
e.g.: Event appearances: event, threat event, top event, consequence event or potential event
BARRIER
A barrier is a measure which should prevent one event leading to another event
A barrier should be defined in the normal or wanted state
Place barriers in the order of their effect
Apply a barrier state to the barrier
e.g.: Defining a barrier in normal state: "Wearing a seatbelt" instead of seatbelt not worn
e.g.: Barrier state: effevctive, unreliable, inadequate, failed or missing
BARRIER
PRIMARY CAUSE
SECONDARY CAUSE
TERTIARY CAUSE
The causation assessment helps you to analyze why the barrier did not function as desired.
Primary cause is a direct act or omission from an actor.
Secondary cause shows the context of the work environment which influenced the primary cause
Tertiary cause shows the underlying organizational influences, this is where the improvement actions should be made
Link your incident analisys causation categories to the causation assesment to be able to perform trand analisys
e.g.: Actor: human or equipment
e.g.: Primary cause: Procedure Y not performed.
e.g.: Secondary cause: Contractor was not aware of need to perform procedure Y.
e.g.: Tertiary cause: Contractors are not in the communication system of the organization
ACTION
Recommandation can be placed anywhere in the diagram to improve the SMS.
If it was never consiered to be an industry standard or described in the SMS, it is not a missing barrier but recommendation

Figure 114 BFA core elements.

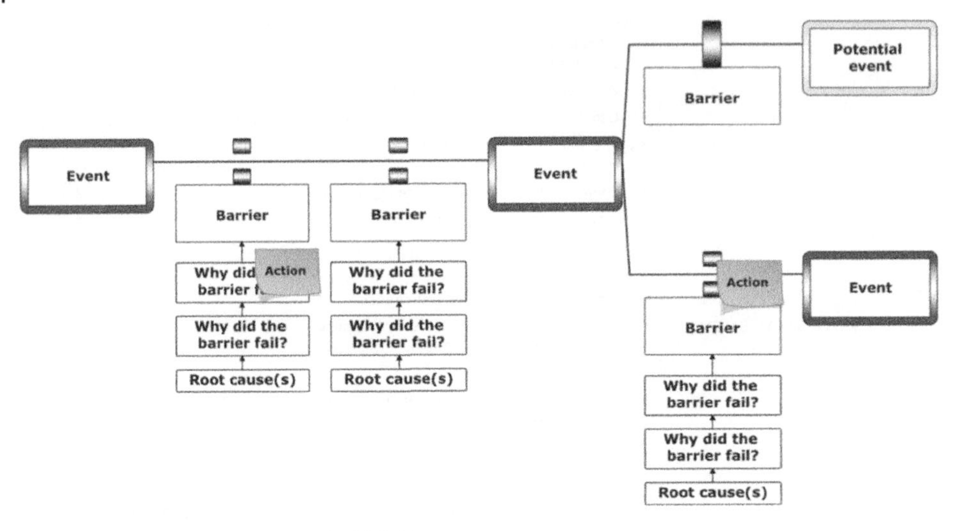

Figure 115 General structure of a BFA diagram.

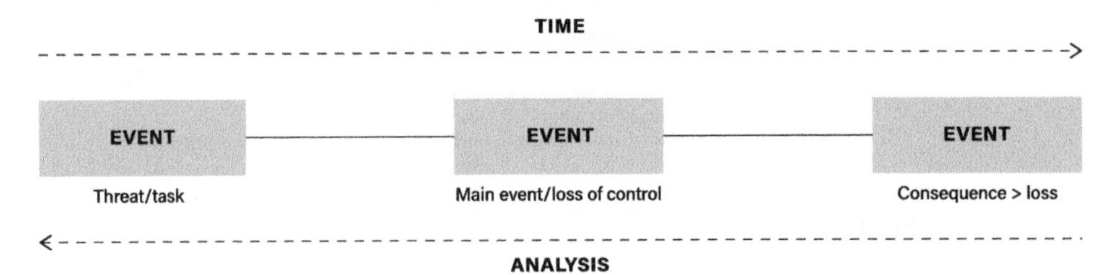

Figure 116 Event chaining in BFA.

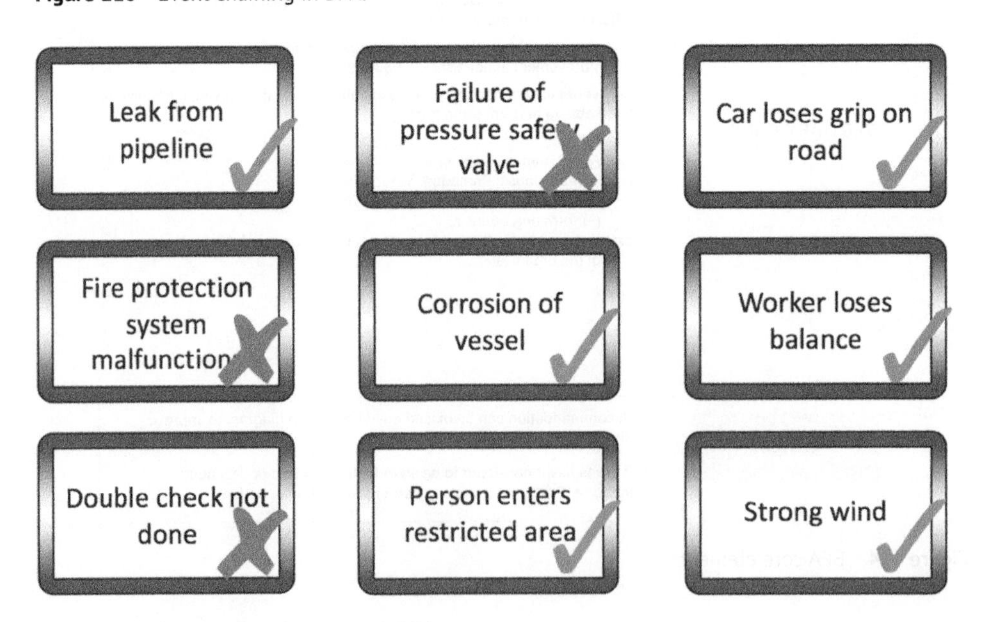

Figure 117 Defeated barriers are not BFA events.

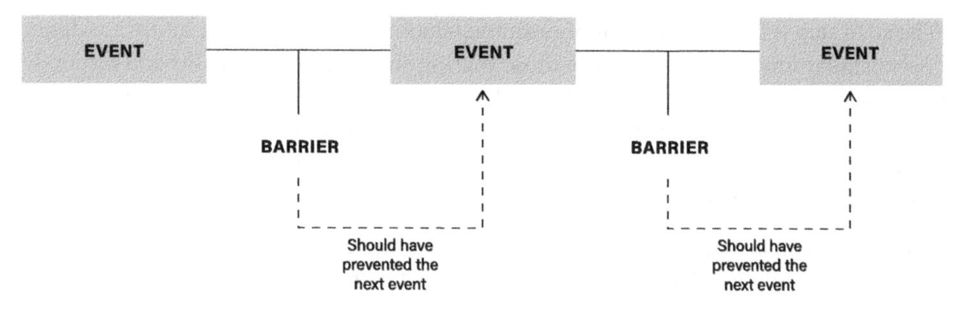

Figure 118 Barrier identification in BFA.

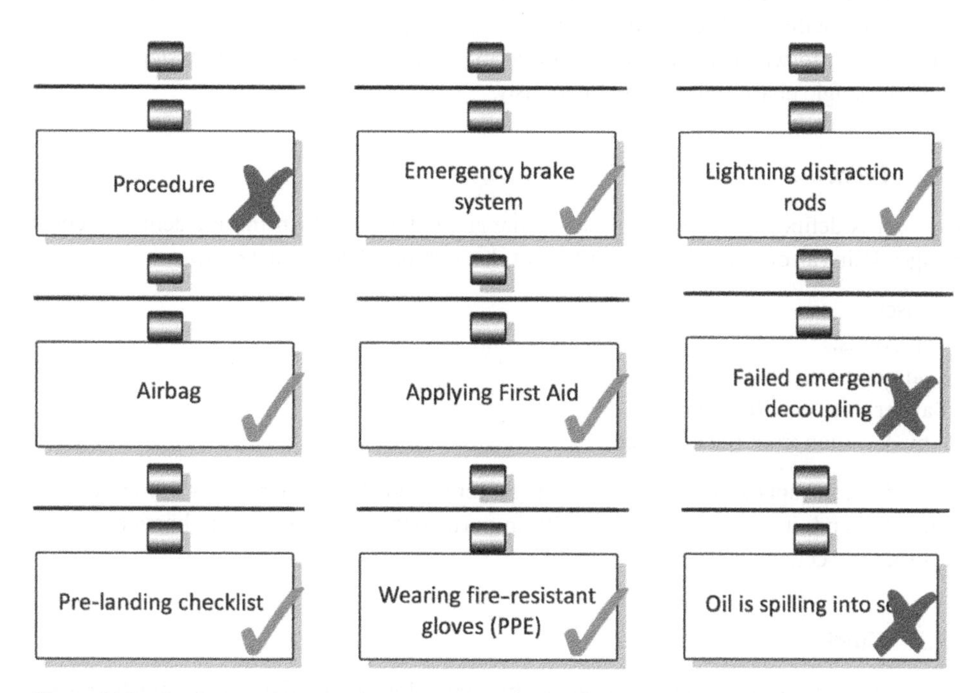

Figure 119 Correct and incorrect barrier identification in BFA.

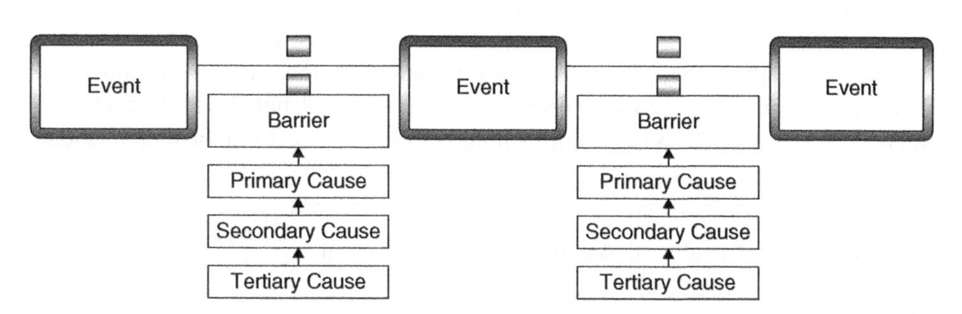

Figure 120 BFA analysis.

The sixth step is the development of recommendations. How to prevent future accidents or even similar unwanted events from those sharing similar causation paths and responding to the causes, in one or more ordinary levels? By improving the safety management system. Actions should be formulated as tasks, assigning job titles to actions, and defining a target date. The topic is discussed in deep in the next chapter. However, at this stage, it is useful to distinguish among:

- *Short-term solution:* Barrier level. Improve barrier quality before resuming operations again.
- *Medium-term solution:* Barrier level. Add barrier before resuming operations again.
- *Long-term solution:* Organizational level. Correct management system/underlying cause.

The last step is the reporting preparation about the outcomes of the assessment and the eventual link with Bow-Tie, as already discussed in Section 3.6.

Section 4.3, to which reference should be made in full, offers a step-by-step guide to tackle the single phases mentioned previously.

3.4.1 Event

An event is defined as a happening or a change of state in which the incident sequence changes. Using an overlap with the constituent elements of a Bow-Tie diagram, an event can be:

- Causes;
- Top events;
- Consequences;
- Barrier state modification;
- Any deviation from the expected.

The different appearance options for the events and the barrier states are shown in Figure 121 and, in Chapter 4, Figure 139 (in particular, how to assess the barrier state is discussed in Section 4.3.4).

3.4.2 Timeline

Cause-and-effect analysis is a very good technique to investigate the causal links that lead to an incident. However, it has one evident drawback: it does not provide information about the relative timing of the events.

A timeline is one of the most effective tools to organize and catalogue data. It is a chronological visual arrangement of the main events, data, and evidence associated with the incident being investigated (Noon, 2009). It helps the team to see the events in chronological order and it is very useful not only for being an investigative tool, but also for its capability in graphically displaying the relationships among the different facts and the final incident.

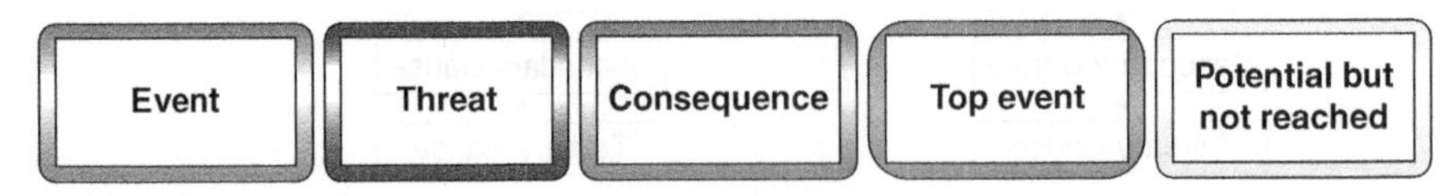

Figure 121 Events types in a BFA diagram.

In addition to the events, which are active items like "pipe failed" or "the pump started up," a timeline may also include conditions. Conditions are passive items, like "the pump was running" or "the pipe was corroded" and represent pre-existing elements in the context of the incident (they are presented using the words "was" or "were"). Obviously, the timeline can also include failures and omissions, if relevant to the incident (AIChE-CCPS, 2003). The development of a timeline covers the entire investigation, as data and information can be added throughout the investigation process to fill the gaps and solve the eventual inconsistencies.

In its simpler version, a timeline is a list of the events in columns, where it is easy to understand which came first, second, and so on. A tabular format is suggested to implement this type of timeline, as shown in Table 18.

More complicated formats of timelines are also available, to provide more information. For example, the Gantt chart format can be used. It enriches the previously discussed version providing the duration of every single event, which is therefore correlated in a general view together with all the events that occurred during the incident (note that a Gantt chart can be also used to plan the investigation activities). An example of the arrangement is shown in Table 19. Obviously, the suggested spreadsheets can be enriched with additional columns to specify additional information about the actors, the people or the equipment involved, and so on.

A very communicative way to arrange data in a timeline is shown in Figure 122. The example shows the first part of the timeline developed for the investigation into the *Norman Atlantic* fire.

Table 18 Example of spreadsheet event timeline.

Time	Remarks
03:24:02	Description of Event #1
03:24:09	Description of Event #2
03:24:44	Description of Event #3
03:25:58	Description of Event #4
03:26:01	Description of Event #5

Table 19 Example of Gantt chart investigation timeline.

		Time and Duration				
ID	What	Start	Finish	03:00:00	03:30:00	04:00:00
1	Event #1	03:00:02	03:30:07			
2	Event #2	03:00:09	04:00:45			
3	Event #3	03:29:44	04:01:29			

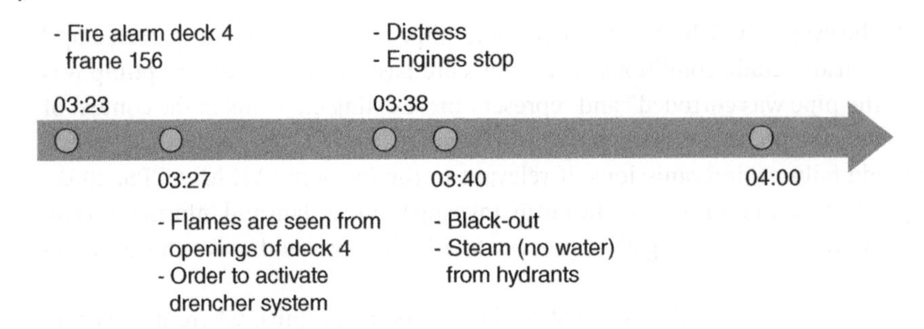

Figure 122 Example of timeline developed for the *Norman Atlantic* investigation.

Another way to arrange events in a time sequence consists of using a blackboard to draw the essential timeline and Post-it® notes to add information about the events. This alternative has the advantage of being extremely flexible (notes can be moved when needed) and has a considerable communicative impact. It is suggested that different colours of Post-it® notes be used for different types of data. Especially in the first stage of the timeline reconstruction, using software with a pre-defined approach could be limiting. It is preferable to use a simple and flexible format. Moreover, the level of detail should be kept manageable, avoiding adding everything that is known to the timeline. During the construction of the timeline, the timing of events and conditions may have different accuracy. For instance, data from BPCS have the accuracy of a tenth of a second while a field operator's observation is far from being so precise and can be very approximate (e.g. "more or less at noon"). This leads to timelines where data are used in combination with both precise and imprecise timing. Using this combination is a challenge for the investigator, since data need to be put in a chronological order at the end. However, a clear advantage of using this combination of data is that imprecise data can be detailed if coupled with more precise data. For instance, if the operator realized that during the intervention to close valve A, valve B was already automatically closed, then we can conclude that the intervention to close valve A was in a narrow window of time, more precise than the approximated value referred to by the operator. Timelines combined with computer simulations are powerful tools for analyzing the sequence of events and accurately re-creating the dynamics of the incident.

When detailing a single event, i.e. a single building block of the timeline, it is suggested that the following four rules be followed:

- Use complete sentences, avoiding fragmented information.
- Use only one idea per building block (concatenated phrases should not be adopted).
- Be as specific as possible (avoid qualitative terms and favour quantitative assessment).
- Document the source for each event and condition, to assess the validity of the data.

For some complex incidents, it is suggested that the investigator use parallel timelines showing the event sequences differentiated by location, actors, input or output variables, and so on. In any case, it is always recommended that two or more timelines be combined into a single, unitary chronological viewpoint of the incident. The timelines are important tools to assess the potential suspects in case of sabotage or malicious deliberative acts (which are not treated in this book, as clarified since the Introduction). To have a clear portrayal of how a person acts, the timeline can be supported by a plot showing the movements: the combination of the temporal and spatial information may be extremely valuable for those complex incidents where it is essential to know the exact position of a person within a certain time interval. Following the suggestions by Vanden Heuvel et al. (2008), timelines are usually constructed following this path:

- Identify the loss event. It needs to be defined specifically, according to what the investigation wants to focus on. If multiple loss events are selected, multiple timelines need to be created.
- Identify the key actors (like people, equipment, parameters).
- Develop building blocks for each actor, event, and condition. Then, add them to the timeline.
- Generate questions and identify data sources to fill in the eventual gaps.
- Gather data, according to what emerged from the previous step.
- Add additional building blocks to the timeline, according to the results of the previous step.
- Determine whether the sequence of events is complete.
- Identify the causal factors.

It is evident that some steps of the timeline construction are in common with the causal mechanisms analysis.

When creating a timeline, you can't always rely on precise time information (hours, minutes, seconds). Therefore, when the accuracy of the data is uncertain, but it is still possible to identify the relative position of one event in relation to others, then it may be sufficient to express the time in qualitative terms, such as "at about midnight" or "after lunch break" and so on. Using both quantitative and qualitative temporal data, it is often possible to give greater value to qualitative ones, since the determination of their chronological position immediately before, immediately after, or between two well-known events greatly narrows the temporal uncertainty.

When the event to be analyzed is complex, that is, the "actors" of an accident, quasi-accident or non-conformity are multiple, or the temporal development of the event is complicated, you can build more detailed timelines to integrate the temporal data with the actor of the action. This is accomplished by building matrices with:

- The temporal evolution in the column.
- The actors in line.

Events will be inserted at the specific time-actor intersections, following these simple rules:

- Use complete sentences, avoiding fragmentary information.
- Use only one idea for each event (do not concatenate multiple phrases).
- Be as specific as possible (favour quantitative valuations to qualitative terms).
- Document the source of each event to validate the data.

The actors are the subjects who perform the action described in the event: they can be people, but also objects or places whose status is changed because of the event.

An example of such an array is shown in Figure 123.

When developing a timeline, the investigator should always keep in mind what has already been stated about time reversibility and irreversibility (Dekker, Cilliers and

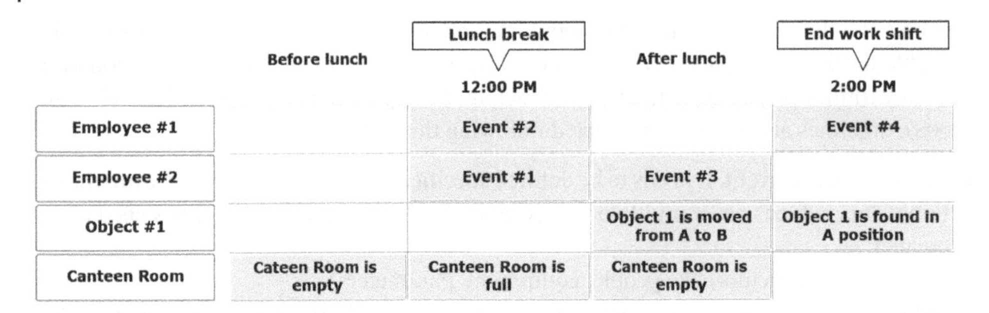

Figure 123 Timeline example.

Hofmeyr, 2011). We briefly repeat here that according to the Newtonian standpoint, the trajectory of the events can be drawn towards both the future and the past. This is an assumption that is based on the idea that the only limit in the reconstruction of an incident is the effectiveness of the used method, since the knowledge is always fully available. But the most recent approaches, taking inspiration from the complex theory, claim that the precise set of conditions that characterize a complex system (like an accident) cannot be exhaustively known. This happens because of the continuous changes and evolutions that affect the system and its relationships, following the adaptive nature of complexity. This second approach implies the loss of any effective predictive measures, since knowledge cannot be fully possessed. This is why an investigator uses both deductive and inductive methods, because they help him or her in moving in the time sequence of the events, looking for the causal link between them. Moreover, the simple selection of events to be arranged in a timeline is a transformation of the real story: it is not a description of what happened but a description of the most important events that happened. It is therefore crucial to be experienced in creating a timeline, since there is not an a priori objective method to establish what is important and what can be omitted from the incident timeline sequence.

For investigation purposes, it is relevant to determine the conditions at the time of failure. This activity is in the middle between the evidence collection and the root cause analysis, and can be handled thanks to a timeline. The searched condition can be short-term (i.e. immediately before the failure) or long-term (i.e. an existing latent condition), depending on the particular incident. Knowing those conditions and having evidence correspondence is a key part of validating failure hypothesis and the entire back-in-time reconstruction.

Timelines always comprise two sections: the events prior to the incident, and the incident itself. Sometimes, they may also focus on the events after the incident (Sutton, 2010).

In conclusion, the timeline tool takes into account all the required information to properly start the investigation, also suggesting where it is necessary to focus the efforts. Its creation is one of the very first steps in having a single manageable record of events and an introductory tool to the causal analysis and the root cause determination.

3.4.3 Barriers

BFA analysis is a barrier-based method, so the same considerations on barriers already presented in Chapter 2, dedicated to the Bow-Tie method, apply. Therefore, please refer to Section 2.12 in full.

3.4.4 Causation Path and Multi-Level Causes

Focusing on the system, rather than the individual, represents the proper way to conduct an investigation, at least for two reasons (Sutton, 2010). Firstly, if equipment and systems provided to persons are not effective, it is not the individual responsibility that has to be pointed out as the fault cause. Secondly, it is much easier to change a managerial choice than a person or his/her behaviour, which is susceptible to varying daily. Third, human errors may often be the consequence of insufficient training, motivation, or attention to safety, all being aspects that the management should promote and monitor. It is a matter of controllability and reliability, as they are the two most essential ingredients to ensure that the lesson learnt will guarantee an increase, or a restoration at least, of the safety level accepted in the industry at the corporate, field and line levels. Metaphorically speaking, an accident investigation is like peeling an onion: this concept, cited in Kletz (2001), gives us a live image of what we are called to solve. Technical problems and mechanical failures are the outer layers of the onion (Figure 124): they are the immediate causes. Only once you peel them you can find the inner layers (the underlying causes), like those involving the management weaknesses.

Older investigations were superficial, since they only identified obvious causes and developed poor recommendations. In the more modern layered approach, a deeper analysis is carried out and additional layers of recommendations are developed: immediate technical recommendations, recommendations to avoid the hazards, and recommendations to improve the management system.

To sum up, the research on the causes of an incident span over three different levels (Health and Safety Executive, 2004):

1) *Immediate cause.* This is the most obvious reason why an adverse event happens (e.g. the valve is in the incorrect position). A single adverse event may be correlated to several immediate causes identified.
2) *Underlying cause.* This is the less obvious reason found at the end of the investigation's outcome and it concerns the system. Examples are preliminary checks not carried out by supervisors; non-robust risk assessment, excessive production pressures, poor safety culture, and so on.
3) *Root cause.* This is the initiating event from which all other causes originate. Root causes are generally related to management, planning or organizational failings.

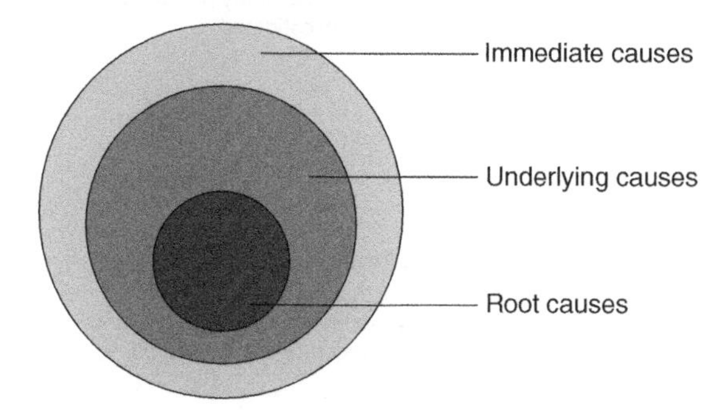

Figure 124 The onion-like structure between immediate causes and root causes.

This theoretical foundation is fully represented within a BFA analysis through the identification of three levels of causes: primary, secondary and tertiary.

The primary cause identifies the "practical" reason why the barrier failed. The secondary cause, on the other hand, indicates the "environmental" factors (see also human factor considerations) that defined the context in which the primary cause occurred. Finally, the tertiary cause represents the root cause itself, i.e. the system causes that generated those environmental factors that have (or at least could have) determined the establishment of the primary cause.

3.5 From Root Cause Analysis (RCA) to BFA

Root cause analysis (RCA) is among the most widely used incident investigation techniques (Figure 125). It is simple: it starts with the occurred incident and, continuously asking "why?," it delves into the chain of events to find the root causes. The basic idea of drilling down to the underlying cause is adopted by several proprietary methods that share the common basic idea of RCA: keep asking "why" to find root causes.

When developing an RCA analysis, the investigator needs to know when to stop. Indeed, if the method is used without such precaution, its result can be a useless jumble of elements, and the investigation team members could find themselves talking about theology or the two chief world systems by Galileo. To avoid this, categorizations and stopping criteria should be defined before starting with an RCA.

A powerful definition of cause analysis is given in (Forck and Noakes Fry, 2016), and it is cited in its entirety here:

> *Simply put, cause analysis is the process by which you discover the invisible thoughts – mental! Models, beliefs, values – that influence, and then produce the visible behaviours/actions. The reasons need to be discovered so actions to prevent recurrence can be initiated to prevent future incidents. Hue structured search is called root cause*

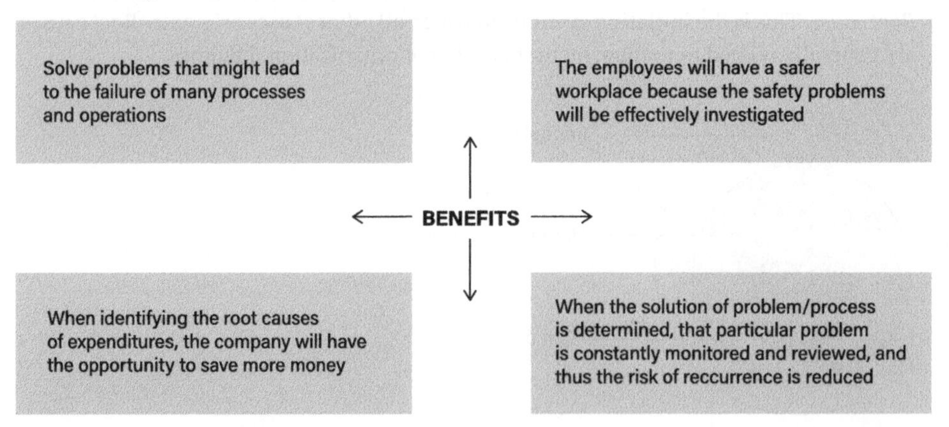

The following are considered to be the benefits of RCA:
Solve problems that might lead to the failure of many processes and operations
The employees will have a safer workplace because the safety problems will be effectively investigated
BENEFITS
When identifying the root causes of expenditures, the company will have the opportunity to save more money
When the solution of problem/process is determined, that particular problem is constantly monitored and reviewed, and thus the risk of reccurrence is reduced

Figure 125 Benefit of RCA.

analysis. The underlying drivers or reasons are called root causes. Once your investigation allows you to fully integrate cause analysis techniques into a systematic methodology for analyzing and solving problems, you will be able to take the lead in decision-making and quality control that will be reflected in better and more consistent results for your organization.

In the following introductory paragraph, the basic ideas about the RCA are presented, so to have a common knowledge at the base of the explanation of the proprietary method discussed in this book.

3.5.1 Introduction to RCA

An extensive bibliography is available on root cause analysis, being the most widely adopted approach to incident investigation. Taking inspiration from it, some basic concepts are presented in this section, highlighting the common hypotheses at the foundation of the technique, regardless of the specific adopted methodology (which might also be proprietary).

Basically, as underlined by Sklet (2002), the term *root cause analysis* refers to any techniques that identify root causes, i.e. those underlying weaknesses in the management system that eventually result in the incident and whose correction would prevent the occurrence of the same event and similar ones as well. The input data for the RCA are those coming from the previous investigation stages, when the investigator answered questions about what, when, where, who, and how the incident happened. With RCA, the investigation process adds new information: why the incident occurred. It is evident that accurate and comprehensive root causes derive from an exhaustive listing of causal factors. The RCA process is shown in Figure 126.

The differences between traditional problem solving and RCA is that the latter is structured, thus developing effective conclusions and recommendations, focusing not only on individuals but also on all those factors affecting the performance of the task, such as the environment, the equipment, and other external factors. Instead, traditional problem solving does not possess the rigorous approach that identifies solutions connected to the causes of the incident (Vanden Heuvel et al., 2008).

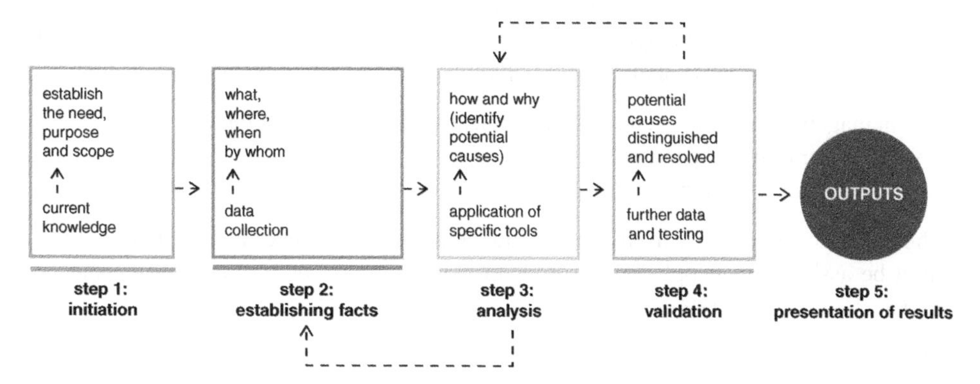

Figure 126 RCA Process.

The identification of root causes is undoubtedly the most challenging goal of an incident investigation. It requires the preliminary acquisition of all the causal factors. If the RCA is performed too early, poor conclusions will be reached, and ineffective recommendations will be developed. Actually, the identification of root causes is a double challenge for the investigator (ESReDA Working Group on Accident Investigation, 2009): firstly, it is necessary to identify the remote factors; secondly, causalities, i.e. their influence on the incident occurrence, must be proven. This is why performing an RCA requires additional competencies for the investigators; indeed, recalling the influence of human factors, knowledge of social and human science should be possessed by the RCA performer. It must be noted how rare are these skills in a technician.

One way to pass from immediate to root causes is to look for safety barriers, which did not prevent the incident. This approach relies on models for risk assessment, with their own pros and cons. In particular, the limit is the weak capability to highlight the rationale behind the described actions. Therefore, comprehensive approaches should be used to also take into account the human and social aspects of the complex system.

Following the path suggested in Vanden Heuvel et al. (2008), a root cause analysis consists of:

- Selection of a causal factor from the timeline, or cause-and-effect tree, or any other causal factor identifier;
- Brainstorming, to generate a list of potential management system weaknesses for each causal factor (investigation tools, such as the Root Cause Map™ in Vanden Heuvel et al., 2008, can be used to stimulate thinking about potential root causes);
- Documenting the results.

In RCA, all the possibilities within the mission statement should be considered, avoiding the a priori exclusion of some of them. Since the analysis is focused on root causes, the management systems should be thoroughly investigated, questioning those assumptions that are taken for granted during the proactive analysis just to save time. Indeed, every business is equipped with a management system, to ensure that the potential losses identified by the proactive analysis (like a PHA) would be low-frequency. But low-frequency is different from zero, and regardless of the efforts, incidents do occur. The output of the reactive analysis, like an RCA, are therefore essential to drive the proactive analyses and the management system towards continuous improvement. Having an effective incident investigation procedure is not enough if proactive analyses and management systems are not able to receive the outcome of an incident investigation. Therefore, RCA may question:

- The management of change, both technological solutions and procedures;
- The level of training of personnel;
- The accuracy of written procedures.

The method requires a certain level of judgement by the investigator: indeed, RCA cannot be assigned to neophytes. In order to perform an RCA, the investigator should (Vanden Heuvel et al., 2008):

- Think creatively, to identify new failure modes;
- Adopt a shared approach, to use knowledge of other people, inside and outside the company, experiencing the incident;

- Think inquisitively, to be curious about how things and people works;
- Be sceptical, to refuse poor explanations (i.e. those typically including the terms "everything, everybody, all, obvious," and so on);
- Think logically, to test the hypothesis with available data;
- Always remember the system, the macro, not only the micro, the details.

Apparent and root cause analyses differ in the level they reach. The apparent cause analysis stops at immediate causes, exploring the causal factors; only the root cause analysis goes deep into the underlying and root causes, as shown in Figure 127. Moreover, testing a root cause is not the same as testing an immediate cause, where tests or simulation can be sufficient. A root cause could be validated by testing the analyses on the actors at its base.

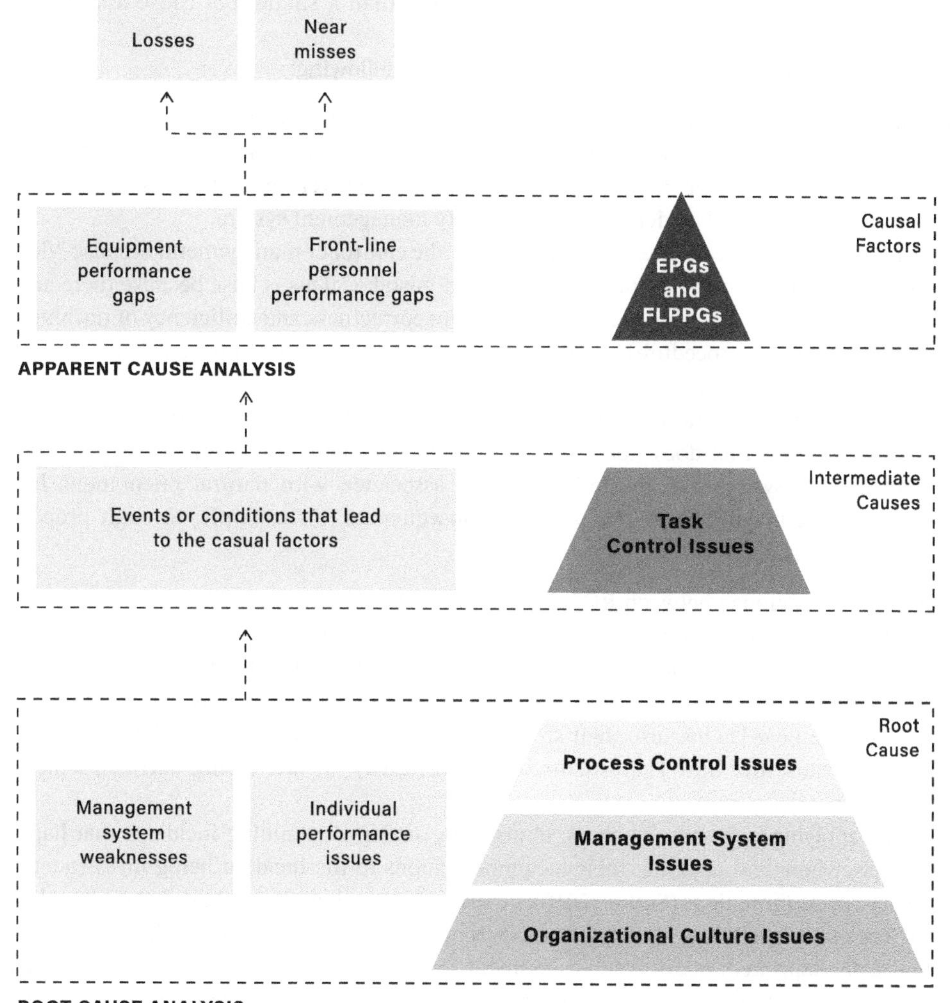

Figure 127 Levels of analysis.

Figure 127 also shows how the different levels actually affect different issues, on a scale from equipment performance gaps (EPGs) and front-line personnel performance gaps (FLPPGs) to organizational culture issues – increasing the depth of the analysis, increasing the level of learning, and increasing the scope of corrective actions.

It is essential to distinguish between fundamental causes and root causes. If an incident occurred because someone slipped and fell, gravity could be seen as a fundamental cause. However, it is out of management's control, and therefore it is not a root cause. Indeed, a root cause deals with the weaknesses of the management system. From this point of view, the identification of a fundamental cause will increase the completeness of the investigation, but it is irrelevant when the investigator develops the corrective actions to prevent further occurrences of the event. As cited in Vanden Heuvel et al. (2008), root causes are "intended to be as deep as can reasonably be addressed with practical and measurable recommendations." Moreover, as already anticipated, there is rarely one single cause for an incident, and a single causal factor may have more than a single root cause associated with it.

Some common traps when performing RCA are the following:

- Equipment failure is seen as an event out of the control of management because parts get old or simply work poorly. This is false, because equipment inspections, tests and maintenance are under the management system and prevent most failures. Moreover, defective parts should be detected by the quality management system.
- Human performance failures are seen as out of the control of management, because "the procedure was right, the employee just made a mistake." This is false because there are some correlations with the management system: correctness and sufficiency of training, the accuracy of procedures, commitment to error, already-happened failures but overlooked, and so on.
- External events, like natural phenomena, are obviously out of the control of management. However, this does not result in poor management involvement. Indeed, the management system can minimize the risk associated with natural phenomena by reducing the magnitude of the probable consequences, for example, through proper structural design.

The main difference between RCA and other incident analysis methods is that RCA is not barrier based. Everything in RCA is an event, including those things that would be considered barriers or barrier failures in BSCAT™, Tripod, or BFA. Therefore, whereas the barrier-based incident analysis methods like BSCAT™, Tripod, and BFA can be mapped back onto the Bow-Tie because their structure is similar, RCA cannot be linked back to a Bow-Tie because the Bow-Tie structure depends heavily on identifying barriers, which RCA does not do.

When applying root cause analysis, to argue by analogy, examining incidents that happened elsewhere and applying their recommendations to the incident being investigated, may have some limitations (Sutton, 2010). Firstly, attention should be paid to false extrapolation. Reasoning by analogy is useful, but it does not provide that thorough understanding of what really happened, because the assessment is no longer fact-based but analogy-based. Secondly, stories from past experience are generally linear: first, this happened, then this occurs, finally this results. The simplification of linearity may return poor reasoning by analogy; indeed, events in an industrial incident are usually in relation to each other, creating a complex system. Finally, a storyteller possesses his/her own worldview: what is good, obvious, a priority for someone is not necessarily for another. It is not about the inherent rightness or wrongness of statements; it is about their relative value, because of the person's point of view. Therefore, storytelling is encouraged, but it is important to be aware of its limitations.

For example, TIER diagrams (US Department of Energy, 1999) are one method to perform root cause analysis, discussed in Sklet (2002) and mentioned in Sklet (2004). They help the investigator in finding not only the root causes, but also the corresponding management level that has the power to promote, implement, and follow up with the corrective actions. Another example of RCA is the PROACT® RCA, discussed in Latino, Latino and Latino (2011), where a proactive method for using the RCA is presented, not limiting its application only to accident investigation.

In conclusion, the reader should note that RCA is a structured method to uncover the underlying factors, typically of undesired performance, even if it could also be used to investigate positive results. The interested reader may find additional information about RCA in Latino, Latino and Latino (2011).

3.6 BFA from Bow-Ties

The BFA technique is also used in combination with a Bow-Tie, to give a wider view of the incident investigation (Klein, 2016). The Bow-Tie methodology is used for risk assessment, risk management and risk communication. The method is designed to provide a better overview of the situation in which certain risks are present, to help people understand the relationship between the risks and organizational events. Risk in the Bow-Tie methodology is represented by the relationship between hazards, top events, threats and consequences. Barriers are used to display the measures an organization has in place to control the risk. All these are combined in an easy-to-read diagram, as shown in Figure 128.

The word *hazard* suggests that it is unwanted, but in fact, it is the opposite: it is precisely the thing you want or even need to do business. It is an entity with the potential to cause harm, but without it, there is no business. For example, in the oil industry, oil is a dangerous substance (and can cause a lot of injuries when treated without care), but it is the one

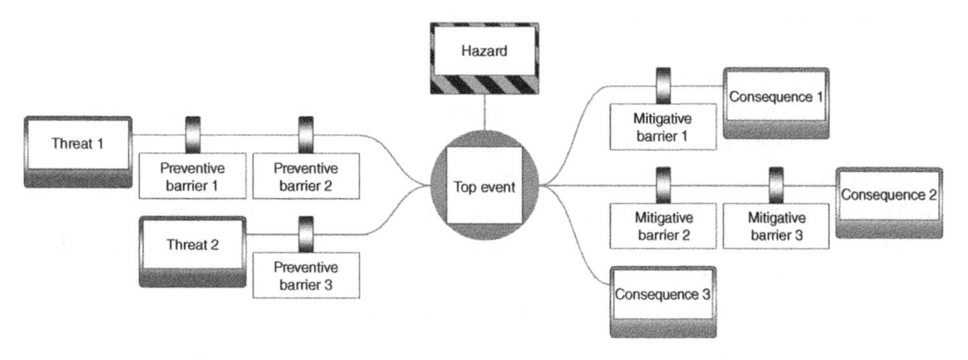

Figure 128 The Bow-Tie diagram.

the thing that keeps the oil industry in business! It needs to be managed because insofar as it is under control, it is of no harm.

Thus as long as a hazard is controlled, it is in its wanted state – for example, oil in a pipe on its way to shore. But certain events can cause the hazard to be released. In Bow-Tie methodology, such an event is called the top event. The top event is not a catastrophe yet, but the dangerous characteristics of the hazard are now in the open. This is the moment in which control over the hazard is lost. For example, oil is outside of the pipeline (loss of containment) – not a major disaster, but if not mitigated correctly, it can result in multiple undesired events (consequences).

Often several factors could cause the top event. In Bow-Tie methodology, these are called threats. These threats need to be sufficient or necessary: every threat itself should have the ability to cause the top event. For example, corrosion of the pipeline can lead to the loss of containment.

When a top event has occurred, it can result in certain consequences. A consequence is a potential event resulting from the release of the hazard which results directly in loss or damage. Consequences in Bow-Tie methodology are unwanted events that an organization by all means wants to avoid, for example, oil leaking into the environment.

Risk management is about controlling risks. This is done by placing barriers to prevent certain events from happening. A barrier (or control) can be any measure taken that acts against some undesirable force or intention, in order to maintain the desired state. In Bow-Tie methodology, there are proactive barriers (on the left side of the top event) that prevent the top event from happening, for example, regularly conducting corrosion inspections of the pipelines. There are also reactive barriers (on the right side of the top event) that prevent the top event resulting in unwanted consequences, for example, leak detection equipment or a concrete floor around the oil tank platform. Note that the terms *barrier* and *control* are the same constructs, and depending on the industry and company, one or the other is used. In this book, we use the term barrier.

In an ideal situation, a barrier will stop a threat from causing the top event. However, many barriers are not 100% effective. Certain conditions can cause a barrier to fail. In Bow-Tie methodology, these are called escalation factors. An escalation factor is a condition that leads to increased risk by defeating or reducing the effectiveness of a barrier, for example, an earthquake leading to cracks in the concrete floor around a pipeline.

Escalation factors are also known as defeating factors, or barrier decay mechanisms – which term is used depends on the industry and company. In this book, we use the term escalation factor.

After creating the basic Bow-Tie diagram, there are several ways to work out the barriers in more detail. One good way is to identify and link the underlying management system activities to the barriers. This will tell what should be done to keep the barriers working, like maintenance activities on hardware barriers. Mapping the management system onto a Bow-Tie also demonstrates in more detail how barriers are managed by a company. Furthermore, responsibilities could be attached to barriers, as well as a rating of their effectiveness and what type of barrier it is.

In conclusion, the following terms should now be familiar:

- The hazard, part of the normal business but with the potential to cause harm, can be released by:

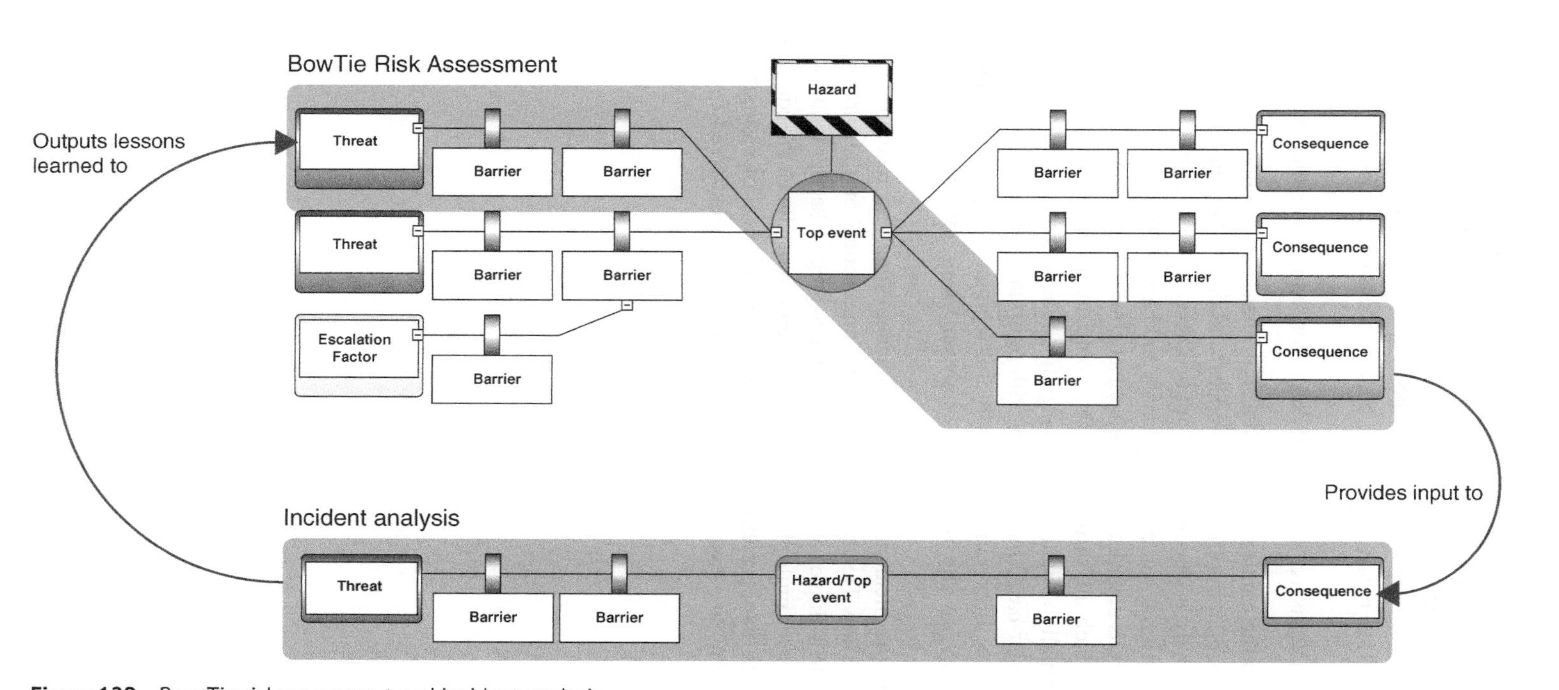

Figure 129 Bow-Tie risk assessment and incident analysis.

- A top event, no catastrophe yet but the first event in a chain of unwanted events;
- The top event can be caused by threats (sufficient or necessary causes);
- The top event has the potential to lead to unwanted consequences;
- Proactive barriers are measures taken to prevent threats from resulting in the top event;
- Reactive barriers are measures taken to prevent the top event from leading to unwanted consequences;
- An escalation factor is a condition that defeats or reduces the effectiveness of a barrier.

Coming back to the combination of BFA and Bow-Ties, it is possible to reuse and link existing risk assessment information (Bow-Ties) and do a full integration of incident investigation and risk analysis (Figure 129). If applicable Bow-Tie diagrams are available for use during the investigation, you can bring events and barriers from the Bow-Tie directly into your incident analysis diagram. This results in a better fit between incident and risk assessment analysis, which in turn allows the company to improve the risk assessment. Particularly for small incidents, this is a significant advantage – it allows staff to analyze incidents in a barrier-based methodology with minimal training. Creating a barrier-based incident diagram requires training, but with this templated approach, all incidents that fit onto existing Bow-Ties can be quickly analyzed using any incident analysis method, including BFA.

After finishing the incident analysis, the incident data and recommendations can be linked back to the Bow-Tie risk assessment and represented on the barriers. Bringing all your incidents into a single case file allows you to aggregate barrier failures and lets users do trend analysis over multiple incidents, and therefore it allows the company to see the weaker areas in its management system. This entire process allows gauging barrier effectiveness and availability based on real-world information extracted from the incident analyses.

4

Workflows and Case Studies

4.1 Bow-Tie Construction Workflow with a Step-by-Step Guide

Conducting a Bow-Tie risk assessment requires executing the phases shown in Figure 130. Following are descriptions of the detailed steps to perform a Bow-Tie assessment.

4.1.1 Case Study Anatomy

Analyzing risks with the Bow-Tie method may require, especially for complex and articulated cases characterized by multiple causes, consequences (perhaps on different aspects) and control measures, the preliminary identification of the "anatomy" of the case study to be analyzed. Making a Bow-Tie can in fact be extremely simple or complicated. For relatively linear and streamlined cases, as well as those of immediate conception, even a sheet of paper and a pen may be sufficient; follow the instructions in the previous chapter to construct the Bow-Tie diagram. Other times, however, the amount of data may be important, or the person to whom the Bow-Tie diagram is shown may be particularly demanding, or there is a need to supplement a basic Bow-Tie diagram with additional information. In all these cases, which in the experience of the authors are by far the majority, it is convenient to prepare for the construction of the diagram by defining the anatomy of the case study in advance. In this case the support of some commercial software can be used, but the reader can implement the following as desired. What is needed at this stage is to:

- Identify the attributes that will be integrated into Bow-Tie diagrams, such as categories and types of causes, consequences and barriers, roles of managers, frequency classes and any other information that can be used to customize Bow-Tie diagrams and make them fully compliant with company standards.
- Identify the attributes that need to be defined to enrich the (reactive) phase of incident analysis.

Bow-Tie Industrial Risk Management Across Sectors: A Barrier-Based Approach,
First Edition. Luca Fiorentini.

Figure 130 Bow-Tie preparation workflow.

- Define the structure of the documentary evidence, internal or external to the perimeter of the organization, which can be attached to the various elements of a Bow-Tie diagram (also referring to the elements constituting the document body of any organization's management system).
- Define the relationship between the elements of the diagram and the structure of any management system. For example, a particular barrier, in order for it to be integrated, requires periodic checks to be carried out by the system. These activities should be recalled (and integrated) in the Bow-Tie diagram (possibly enhancing them).
- Identify the applicable regulations and technical standards (with particular reference to all mandatory regulatory requirements in relation to elements to be considered, values to be used, and acceptability and tolerability criteria, for all those areas already standardized or for which there is a technical rule of reference for the purposes of conducting a risk assessment).
- Identify, in the case of complex applications, the groups of systems and subsystems that constitute a barrier, in order to assess their status (online or offline), also thanks to integrations with SAP applications;
- Define any in-depth information that can be linked to the elements of a Bow-Tie diagram in order to provide further details about assumptions, elements, and so on.

4.1.2 Case Study Taxonomy

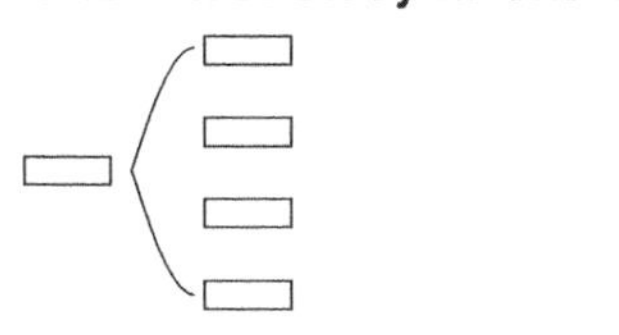

After defining the structure of the database, all the taxonomies that can be applied to the various elements of a Bow-Tie must be determined. We often arrive at an exhaustive and complete taxonomy only after having completed the creation of the diagrams, i.e. when the structure, order and presence of the main constituent elements of a Bow-Tie are clear – when the nuances and differences that we want to highlight through a proper categorization in classes and subclasses appear evident.

For example, a barrier, initially indicated qualitatively, may be labelled according to the type of response offered (e.g. active or passive, but more detailed classifications are possible), or according to who is responsible for it, or according to the elements of a predetermined list that may distinguish between some standard categories of barriers (e.g. separating those for hazard prevention, risk identification and control, and those for evacuation, escape and rescue).

There are no limits to the extent of the taxonomy that can be applied to the various elements of a Bow-Tie. Evidently, a reasoned choice must be made – even using the experience of risk analysts – so as not to burden the diagram with overwhelming information that would make it no longer so easily readable. A good compromise must therefore be found between the amount of data and the readability of the diagram, a feature that has in any case determined the success of Bow-Tie in recent years. Based on the experience of the authors, it is suggested that the following taxonomy be applied, reserving further categorization only in cases where it is strictly necessary to introduce it. It is possible to apply different taxonomies to the same diagram, so that it can be used with different perspectives in relation to the use, the degree of detail, and the target users.

For the causes, we recommend identifying the different types. It may be useful to distinguish human factor causes from those that categorically exclude them, such as environmental factors external to the organization. It may also be useful to differentiate causes by qualitative ranges of frequency of occurrence; however, this taxonomy should only be defined if it is not intended to quantify the diagram, e.g. by LOPA. In relation to the purpose of the diagram, it could also be useful to differentiate whether a cause linked to an operational error can be assumed in the normal execution phase or only in the maintenance phase, since different subjects could be involved (sometimes even outside the organization, although in any case under its own direction and coordination).

Concerning barriers, we recommend to distinguishing at least type, level of criticality, and responsible figure. Thanks to the typology it will be possible to filter passive barriers from active ones, or those activated also by human intervention from purely automatic ones. The definition of a criticality level, on the other hand, makes it possible to assign a high priority for maintenance and inspection activities to be carried out on barriers considered critical in the area of interest (e.g. safety, environment, business continuity, corporate reputation). A barrier labelled as critical should therefore be managed differently from other barriers, since its integrity and its correct and effective functioning are essential requirements to guarantee the maintenance of the risk level within the previously defined acceptance threshold. Although there is no unambiguous criterion for defining when a barrier can be defined as critical, extending what is now established in the offshore process industry, a barrier can be considered critical:

- Whose failure could cause a significant accident;
- Whose failure could lead to a significant accident;
- Whose purpose is to prevent a significant accident;
- Whose purpose is to mitigate a significant accident.

The abovementioned criterion therefore shifts the problem to the definition of "significant accident." Every organization should have a criterion to provide such a definition, in accordance with its policy and its sensitivity to the issues covered by the risk analysis. In the process industry, this definition often coincides with the definition of "major accident" or "significant incident," which, simplifying in the safety-related risk assessment, especially those related with process safety and industrial risks, encloses fires, explosions, releases of flammable or toxic substances, and all the consequences associated with them. It is clear that, for an organization operating in the healthcare sector, a significant accident could be the reputational damage following an unsuccessful surgery, or the consequences associated with a sudden lack of utilities in the wards. The definition of "significant accident" therefore depends on the context of the organization, its policy and sensitivity to certain issues, i.e. its policy and sensitivity to risk, or specific rules if they exist for the scope of application.

Each barrier should then be associated with a manager, who can be considered the focal point for all aspects of integrity and effectiveness that must be guaranteed during all phases of its life cycle: design, commissioning and installation/application, normal execution, ordinary and extraordinary maintenance, decommissioning and removal. As shown in this book, Bow-Tie diagrams are very useful to manage organization process risks together with the information about roles and responsibilities. In this sense the method supports high-level (HL) management systems (HLS) and guarantees consistency in all the recordings and documented evidences (e.g. the recommendations and actions on barriers coming from the risk assessment process can be directly given an owner).

For the consequences of a Bow-Tie diagram, we recommend at least three taxonomic classes. The first taxonomic class is credibility, i.e. it discriminates whether or not the supposed accidental event is credible, in relation to predetermined thresholds that must be established before the Bow-Tie analysis is carried out. The second taxonomic class proposed is that of the addressed target. Although the risk analysis may be specifically aimed at knowing the risk to a predetermined target (e.g. security, or the environment, or business continuity, or corporate reputation), very often the same top event produces consequences that impact different targets. Thus, for example, an uncontrolled fire within the scope of a fire risk assessment may affect all four targets proposed in the example, while the consequences of a hacker attack, in the case of an assessment of cybersecurity aspects, may impact – among those mentioned – only business continuity and reputation. The third taxonomic class that could possibly be used is that of the type. Incidental scenarios with identical targets and credibility could, however, differ in type. For example, the credible safety consequences of loss of control over the driving of a car may only affect the driver or it could involve other people. Or the credible safety consequences of a loss of primary containment of a flammable substance could result in a fire or explosion (possibly with the associated specific probability being investigated using more appropriate methodologies, such as the event tree, or by direct calculation of the resulting effects using dedicated simulation packages).

There is really no limit to taxonomies and to the single taxonomy extension: it is therefore recommended that it be used sparingly and only when a differentiation of classes brings added value to the analysis or to the subsequent phases of communication, sharing and interpretation of the results as well as, obviously, to the final phase of the risk analysis, which is that of decisions.

Taxonomies, furthermore, could become the preferred way to organize and structure information in documentation during reporting activities with limitless possibilities (e.g. extracting actions per responsible people, organizing threats posed by human errors from different processes to prioritize training activities, developing KPIs and dashboards with specific queries, verifying the performance influence factors affecting one or more barriers, etc.).

4.1.3 Acceptability Criteria

Before any risk assessment, the general rules to be followed during this activity should be defined. These include the acceptability criteria to be used for typical Bow-Tie elements: causes, consequences, and barriers. This is a set of rules that must be respected in order for a cause, consequence or barrier to be used within the Bow-Tie diagram. It is therefore essential that these criteria are defined upstream. These acceptability criteria include those defined by the guidelines of the American Institute of Chemical Engineers – Center for Chemical Process Safety, reported in Section 2.14.2 of this book, to which reference should be made. As a further example, consider the company's need to ensure that each barrier has an associated responsible figure, or at least an associated management system activity. Or consider the need, for a specific category of causes, to have two different types of barriers associated with it, while for other causes only one type of barrier may be sufficient. Therefore, a preliminary definition of acceptability criteria is indispensable for the continuity of the activity and represents an integral phase of the workflow, unfortunately often forgotten.

The risk acceptability criteria could derive from a regulatory requirement specific to the scope of application considered or from generally accepted values possibly codified in a recognized guideline (AIChE-CCPS, 2009).

4.1.4 Definition of Risk Matrices

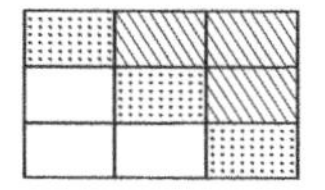

The consequences (and top events) are assessed for their potential to cause damage, estimated to fall within a given category, and their probability of occurrence. In other words, the risk matrices, already presented in Section 1.3 of this book, are used. It is possible to define as many risk matrices as there are addressed targets whose risk is to be assessed: there will then be a matrix for safety, one for the environment, one for business continuity, and so on.

Each risk matrix can be used twice on the same consequence: one for the evaluation of the unmitigated risk (i.e. evaluated in the absence of prevention and protection barriers), the other for the evaluation of the residual risk (i.e. evaluated in the presence of prevention and protection barriers). This use makes it possible to have an immediate count of the risk reduction capacity due to the barriers established.

Obviously, the detailed definition of risk matrices is in the hands of the organization, which, on the basis of its own risk management policies, may decide to create them internally or to refer to some risk matrices already used in other organizations operating in the same sector or recognized standards and guidelines.

In constructing risk matrices, once risk acceptability criteria have been defined, it is recommended that no less than three risk categories be defined:

- *Acceptable risk*. Groups with the lowest risks within the risk matrix. No further risk reduction is required.
- *Tolerable risk*. ALARP assessments may be required. If convenient, further risk reduction measures are required.
- *Intolerable risk*. In this category the highest risks fall within the risk matrix. They are an expression of an unacceptable situation at which immediate and urgent corrective action must be taken to achieve at least a tolerable level.

Obviously, depending on the reference company context, it is possible to have risk matrices with more than three risk levels, where for example the tolerable risk region is further subdivided into several sub-regions, to differentiate when any ALARP studies are only recommended or mandatory.

It is worth remembering, however, that given the uncertainty that inevitably characterizes a risk assessment at each stage (model construction, allocation of failure rates, frequency of occurrence, PFD, identification of assumption), it is completely useless and often misleading to aim at an extensive fragmentation of the resulting risk categories, for each of which the analyst must always be able to associate a justified evaluation and judgement; it is indeed very useful to limit the extensions of risk matrices in a way that it is easy to associate categories to the input parameters and that is possible to define regions coming from groups of cells sharing the same risk category to be developed further for a better insight with other risk assessment methods.

4.1.5 Diagram Construction

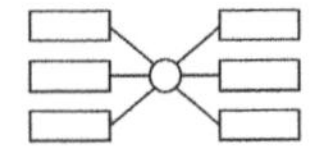

The construction of a Bow-Tie diagram follows the steps detailed in Chapter 2 of this book and summarized below:

1) Definition of danger and top event;
2) Definition of causes;
3) Definition of consequences;
4) Definition of primary barriers;
5) Definition of "escalation factors";
6) Definition of secondary barriers ("escalation factor barriers").

These steps can be considered the standard actions for constructing a Bow-Tie diagram, according to the proposed logical sequence and allowing one to be created with the minimum set of necessary information, although sometimes it is not even necessary to define the escalation factors and related barriers.

However, it is possible to create “enriched” Bow-Tie diagrams, i.e. integrated with data and information resulting from the anatomy of the case study and the identified taxonomic classes. As an example, the following detailed information can be added to a barrier:

- Satisfaction of acceptability criteria;
- Responsibilities;
- Activities to be performed to ensure the integrity of the barrier (e.g. by entering a reference to an activity foreseen in the safety management system);
- Category and type of barrier;
- Unique identification code;
- Criticality (degree of);
- Description;
- Reference document;
- Efficacy (degree of);
- Reference family (other taxonomic class useful to group barriers belonging to the same family);
- Name;
- Objective;
- Expected performance.

It may be useful to build thematic perspectives on this additional information.

As the reader may notice, one important advantage of the Bow-Tie method lies in its “scalability” from a simple illustration of the mail elements to a complete and detailed, quantitative map of chained risks across multiple organization processes.

This scalability allows informed decisions on the basis of the available data.

It is reiterated that the graphic power of the Bow-Tie is the element that has contributed most to its success, so we recommend adopting the graphic solutions that may be necessary for greater readability and transparency of the diagram. For example, it may be convenient to show limited portions of the diagram in order to better understand the logical flow that determined its final configuration. One can therefore think of using “masks” (possibly based on theme-based perspectives) to show:

- Only top events and dangers;
- Top event, danger, causes and consequences;
- Top events, danger, causes, consequences and primary barriers;
- Top events, danger, causes, consequences, primary barriers and escalation factors;
- Top events, danger, causes, consequences, primary barriers, escalation factors and secondary barriers.

Graphics management is an essential element of the Bow-Tie method, so we recommend using software that can easily filter, expand and narrow the diagram according to the audience it is shown to. Some software tools, for example, highlight the elements of a Bow-Tie that satisfy certain filters, and set up transparently those that do not. In this way, if you are interested in showing where the barriers that are considered critical within a complex Bow-Tie diagram are located, it will be easy and immediate to do so, just as it will be possible to create customized views without excessive rework for the different stages of the risk management life cycle, or for the different users of the diagram, or for the different documentation needs of the system elements.

4.1.6 Versioning Activities and Track Changes

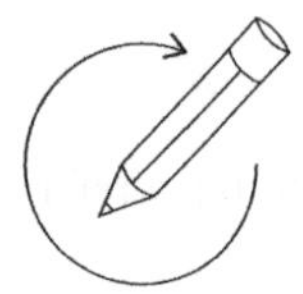

Risk is a dynamic concept. It evolves (very easily it alters over time towards lower levels of security) and what is photographed today may not coincide with what will be in the future. Bow-Tie diagrams, like any technical documentation, must therefore contain information about the revision in order to establish a time of creation and track the various changes that may become necessary as a result of actions arising from observations and recommendations downstream of audit plans or analysis of accidents, near-misses or non-conformities.

The information that should be included, perhaps within a cartouche, is as follows:

- Author;
- Revision number;
- Revision date;
- Documentary reference;
- Notes, if any.

The ability to identify the context and responsibility for creation is clearly fundamental for the assessment documents relating to specific areas (e.g. occupational safety, major accidents, food, etc.).

This part, very simple to understand, is a very important concept: Bow-Ties support risk management systems and could also be selected to deal with risks underling HL management systems. Since the management of change (MOC) process is a fundamental activity to be implemented, operated and reviewed (e.g. the MOC process in the chemical industry dealing with plant modifications that should be managed from the initial conceptual and design phases to the latest pre-start-up safety review to be assured that no new major risks are introduced given the industrial risk level developed in the plant safety report).

4.1.7 Terminology

Without prejudice to the content of the concepts expressed so far, it often happens that some recipients of a Bow-Tie diagram have already gained experience with the same concepts, but from using a different terminology than the one proposed.

It is therefore a good idea to set up from the outset which set of words you intend to use and therefore always adhere to this choice during the creation, modification and display phases of a Bow-Tie diagram.

For example:

- A barrier could be called "control" or "control measure";
- A "preventive" barrier could be called "proactive";
- A "mitigative" barrier could be defined as "reactive";
- A cause could be called a "threat" or "initiating event";
- A consequence could be called an "outcome" or "scenario" or "incident scenario";
- An escalation factor could be called a "defeating factor" or a "barrier decay mechanism" or a "degradation factor."

There is no limit to the choice of terminology to be adopted, drawing from a set of words already in use in a given organization or in any case from a set of synonyms validated by use in a given industrial sector. The important thing is to be consistent with the choice made and use the same terms as much as possible.

The terminology used is in fact an inalienable element of the project and, therefore, must be, like the assumptions used, documented and highlighted to every reader by using a specific glossary. This glossary should be used during brainstorming sessions but also during training activities and leadership/management review of the results coming from risk assessment.

4.1.8 Using Colors

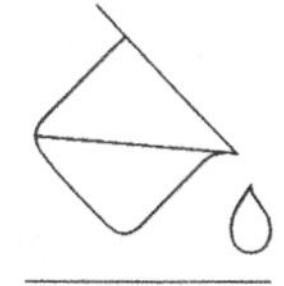

The choice of colours is crucial when making a Bow-Tie diagram. If on the one hand colours mitigate the portability of a diagram and increase the complexity of preparation (also suggesting the use of supporting information tools such as software packages with a cost), on the other hand they further enhance the comprehensibility of the graphic notation that the Bow-Tie relies on for the effective reading and dissemination of information. First of all, different colours help to identify different elements of the Bow-Tie diagram. Thus, for example, the causes will be described in blue boxes, while the consequences in red boxes, the escalation factors are coloured yellow, the top event in orange and the danger in yellow and black stripes. But that's not all. You may also decide to colour the individual elements in relation to the value of a taxonomic class assigned to them. You can then decide to colour the barriers according to the criticality: blue for the critical ones and yellow for the non-critical ones (for example); or on the basis of effectiveness, with colours varying gradually from red to green depending on whether the taxonomic class "effectiveness" has been assigned a value closer to "very poor" or "very good." The consequences can instead be coloured according to whether the credibility threshold of the scenario has been reached: the consequences considered credible in red, those not credible in green. Or again, if the interest is to differentiate the consequences according to the target audience, then green could be adopted for consequences with environmental impacts, red for those on security, blue for business continuity and purple for reputation.

If colouring the perimeter of the boxes of the elements may disturb matching the colour choices of other documentation accompanying the Bow-Tie analysis and therefore a more discreet use of colour is preferred, a fixed set of colours could be chosen for the elements (blue for the causes, red for the consequences, orange for the top event and black for the barriers), and the colour table could be used to enrich only the individual information inside the boxes, to better identify different taxonomic classes and different values.

Obviously, these proposed are only examples and readers are free to choose the colours they like best. However, the aesthetics of the Bow-Tie diagram, its linearity and ease of reading that could be compromised by an unregulated use of colours must always be preserved. There are therefore no universal guidelines to be adopted on the use of colours, only common sense to follow.

4.1.9 Attach Documentation

Much of the data contained in a Bow-Tie diagram relates to information that is already documented somewhere, and on many occasions you will want to refer to it. For example, when assigning the effectiveness value to a barrier, you would want to refer to a specific analysis for that barrier (possibly using more in-depth techniques).

To address this need, some commercial software packages allow you to insert a pointer within the Bow-Tie elements to link the required document to the specific Bow-Tie diagram element.

If you are not using software, which we have seen to be a viable choice only for extremely simple Bow-Tie and devoid of much information, you could at least write the code of the reference document inside the box of the element in question. However, the ability to use pointers to files of interest is far more appealing, so we recommend software solutions for anyone who wants to enrich their Bow-Tie with this kind of information.

It is convenient to first define the hierarchical structure of the document links you are going to use when defining the anatomy of the case study. It is then possible to define document sets, grouped by origin (internal or external to the organization), then by company function, then by company substructure or topic and so on. Each link will then have assigned a code, a name, a description and a pointer to an external file (when not directly attached to the Bow-Tie creation software). At this point, when you are going to define the properties of a Bow-Tie element, such as a barrier, for example, you will be able to specify that the data that distinguishes it is the 12345 document link that points directly, for example, to the management system document coded to 12345, which contains information on how to maintain a critical piece of equipment.

Cross-references are a rather useful tool in cases where the evidence is made up of a documental element of the reference management system or in cases where the Bow-Tie is itself the synoptic link between management systems pertaining to different areas.

4.1.10 Definition of Organizational Structure and Responsible Stakeholders

Many companies are multi-site entities, i.e. their business is distributed in several locations, even possibly distributed around the globe. All these sites may want to make their own Bow-Tie or, in some cases, share them if they are dealing with similar risks. To keep track of what data is relevant to which site, the risk analyst can create a specific anatomical structure for the case study: the organizational units. These are often linked to geographically distinct sites, although it is also possible to define organizational units in terms of departments, thus offering a more abstract definition. The advantages of providing such a distinction in risk analysis with the Bow-Tie method are different:

- It is possible to define the effectiveness of a barrier, distinguishing its level of performance for each organizational unit.
- It is possible to link organizational units to individual Bow-Tie diagrams, to keep track of which Bow-Ties are relevant to a specific organizational unit.
- Organizational units can be used as filter parameters, for example to produce the list of critical barriers at site X, or the list of credible incidental scenarios at site Y, and so on.

Regardless of whether it is a multi-site entity or not, it may be convenient to define business functions for those people who have certain responsibilities over the elements of a Bow-Tie. Thus identifying a person responsible for a particular barrier helps to identify that focal point on which all management activities for that specific barrier should be focused, including activities arising from action plans that have emerged following audits or analyses of incidents, near-misses, or non-conformities. A manager can then be associated with both the activities used to describe the regular performance of certain tasks such as maintenance and training (part of the safety management system) and improvement actions.

The possibility of standardizing the approach in complex organizations allows, especially in the case of company functions shared at a transversal level as well as top management figures, to:

- Have a horizontal perspective by function;
- Have synthetic performance indicators by site, function, or risk, for the purposes of an overall central review or function based in turn on site-specific insights in which detailed evidence and justification can be sought.

4.1.11 Defining Tasks

By adopting a barrier-based perspective, risk management has been shown to coincide substantially with barrier management. No barrier is exempt from the need to be managed: it is always necessary to provide a set of activities to be carried out regularly, at stable intervals, in order to preserve the performance standards of the barrier and ensure its integrity and functionality over time. For example, if reference is made to a barrier consisting of a mechanical technical system, greasing rotating parts every six months may be required, or in the case of a barrier of an organizational-management type, carrying out refresher training on an annual basis may be reequired. These activities, which it is logical to expect to be already defined within a safety or maintenance management system, can then be called up within the Bow-Tie. It could also happen that, thanks to the perspective focused on barriers, the need to carry out activities not foreseen by the current management system emerges, since it is set from a different perspective unable to grasp the management needs of a particular barrier. In fact, the non-respect of activities related to ensuring the performance of a barrier will lead to an increase in probability of failure on demand (PFD).

Bow-Tie is therefore not only a risk analysis tool, but also a method, i.e. a logical and structured approach to the definition of an effective barrier-based security management system.

For example, a cause that is controlled by a number of barriers and supported by a number of activities, all relying on the same single person, could be considered more "threatening" than a cause with barriers and activities that are not all dependent on one person. The identification of activities may therefore highlight common causes of failure.

This phase is also important because it makes it more evident why certain activities are critical, helping responsible people to understand their importance and thus ensuring better execution.

Generally, the definition of an activity requires that the anatomy of the management system within which the activity is included be defined in advance. Then, for each individual activity, at least the following data should be indicated:

- Activity code;
- Name of the activity;
- Description of the activity;
- Category (generic taxonomic class that the analyst can use at will);
- Periodicity of the activity;
- Responsible subject/function;
- Required skills (this class may be used to filter personnel who meet certain competence requirements and who could then be appointed to be responsible for the integrity of the barrier).

From this point of view, it is clear that the Bow-Tie is not only a powerful graphic notation but also the tool with which to observe and subsequently criticize the organization's articulation with respect to processes, roles and responsibilities, including aspects related to information, training and education, skills, inspection and maintenance.

4.1.12 LOPA Analysis

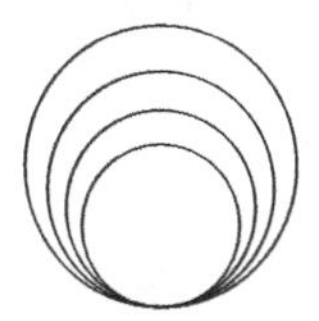

The need to carry out a quantitative risk assessment requires the adoption of ad hoc methods that can be integrated with the Bow-Tie methodology. The most commonly used is layers of protection analysis (LOPA).

The LOPA method, already introduced in Section 2.14 of this book, is widely used in the Oil & Gas sector and in other industrial sectors, such as automotive and manufacturing, which have already adopted the concepts related to functional safety provided by the technical standard IEC 61508/IEC 61511. It is often used in combination with event trees, offering both a qualitative and quantitative approach.

The integration of LOPA with the Bow-Tie method therefore provides a powerful risk analysis tool that is now enriched with an interesting functionality.

First of all, it is clarified, for non-experts, which objective is allowed to reach the LOPA: to determine the frequencies of occurrence of the accidental scenarios (the consequences of the Bow-Tie diagram), once the frequencies of occurrence of the causes and the PFD of the barriers are known. The LOPA assessment therefore makes it possible to evaluate one of the two inputs of the risk matrix: the one relating to frequency or the one relating to the severity of the consequences.

Without going into detail, the input data for a LOPA assessment can be extracted, to cite a few examples, from internationally recognized databases of proven reliability, from ad hoc fault trees, or from human factor assessments (HRAs) in the case of behavioural barriers (e.g. operator response to alarms, periodic housekeeping of workplaces, emergency response).

These data must then be appropriately combined to estimate the frequency of occurrence of the consequences. The method is simple: for simplicity consider a Bow-Tie diagram that has only one branch of causes and one branch of consequences, both with primary barriers. The frequency of occurrence (expressed in occasions/year) of the cause is noted in the corresponding box. This frequency should be multiplied by the PFDs of the barriers on the cause–top event branch. In this way the frequency of occurrence of the top event will be a fraction (i.e. less than or equal to) of the frequency of the cause, since PFD, like all probabilities, are between 0 and 1. Let's analyze two extreme cases: if PFDs are all equal to 1 it means admitting that all barriers actually fail and therefore it is logical to expect the top event to occur with the same frequency as the cause. If, on the other hand, PFDs are 0 (ideally), then whatever the frequency of the cause, the top event will have zero frequency, i.e. it will never occur because the barriers are infallible. In reality, even the most effective barriers have a PFD that, however low, will never be 0. This means that the frequency of occurrence of a top event, because of the PFD barriers, is certainly lower than the frequency of occurrence of the cause.

Now suppose you have a Bow-Tie diagram with two causes leading to the same top event, and the two branches are defined as barriers A and B on the first, and C and D on the second. Since the causes of a Bow-Tie diagram are independent of each other, the frequency of the top event is given by the sum (logical OR port) of the frequencies of occurrence of the two causes, each reduced by the barriers at its end. Therefore:

$$\text{f\_top event} = \text{f\_cause1} \times \text{PFD\_A} \times \text{PFD\_B} + \text{f\_cause2} \times \text{PFD\_C} \times \text{PFD\_D}$$

Obviously, in the case of several causes, we will always reason in the same way, adding new terms to the summation.

The frequency of consequences is therefore further reduced by the PFD of the barriers (which we will call E and F) that are in place in the single branch from the top event to the specific consequence:

$$\text{f\_consequence} = \text{f\_top event} \times \text{PFD\_E} \times \text{PFD\_F}$$

The LOPA assessment provides immediate feedback on the effectiveness of the barriers. Assuming the force of the PFD of all barriers to 1, the so-called unmitigated frequency is obtained, i.e. the frequency evaluated assuming the absence of prevention and mitigation barriers. If, on the other hand, the calculations are repeated using the actual PFD of all barriers, then the current frequency is obtained. It is almost always observed that the risk of the consequence assessed with the unmitigated frequency (unmitigated risk) falls within the region of the risk matrix defined as "intolerable," while the risk assessed with the current frequency (current risk) is significantly reduced and at least tolerable, when not directly acceptable. The analyst is therefore able to evaluate, also thanks to an expert and trained judgement, the causes, the preventive and mitigative barriers that have the greatest weight in determining the frequency of occurrence of an accidental scenario.

Obviously the LOPA assessment can also use some conditional modifiers, where required. These are multipliers (generally between 0 and 1) that are further defined for causes, barriers and consequences, in order to take into account in the analysis other factors impacting the assessment of the frequency of occurrence, such as the time to risk that is used when not exposed to danger 365 days a year, 7 days a week, 24 hours a day (e.g. batch operations in the process industry, or certain causes related to planned maintenance interventions).

It is important to note that the LOPA method combined with the Bow-Tie diagram structure allows the calculation (or estimation in risk matrix or indices semi-quantitative approaches) of the frequency of specific effects on targets given the PFD of the single barriers on the accident path having those effects. PFD values can be obtained from:

- Internal statics (given the results of periodic inspections on the barriers over time);
- Data banks;
- The application of further quantitative methodologies (e.g. FTA calculates the unavailability of systems of components given failure rates, maintenance policies and human probabilities).

4.1.13 Other Data

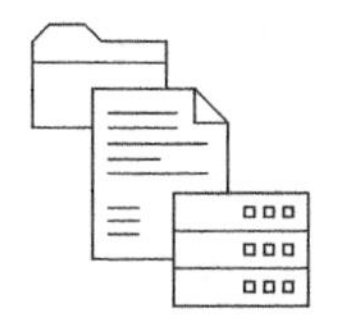

In relation to the peculiarities of the analyzed case, the analyst may need to specify further data in addition to those proposed in this chapter, thus defining further taxonomic classes. It is necessary to reiterate this possibility, to underline that a Bow-Tie diagram provides wide customization, including in the data contained.

For example, it may be useful to specify to which area of the system a given Bow-Tie diagram refers, defining the "area" field for the "danger" element. It can also be equally useful to make the last inspection date of a barrier immediately known by defining the "last inspection date" field for the "barrier" element. Again, in the case that preventive and mitigative control measures are inspected by a third-party entity (such as a certifying body or authority having jurisdiction over the specific barrier), you can define the "third-party entity" field for the barrier element, where the name of the third party inspecting the barrier will be specified.

There is no limit to the analyst's definition of data. The only limit is imposed by the need to keep the Bow-Tie diagram clearly legible, which requires additional taxonomic classes to be defined only when their adoption brings real added value.

4.1.14 Extraction of Critical Barriers and Performance Standard Register

The taxonomic class "criticality" of barriers has already been discussed. The labeling of critical barriers is essential to be able to extract from the generic list of all barriers the control measures that must be managed with priority, i.e. the critical ones.

Having this list can be requested by the authorities for the granting of clearances or authorizations in general in all the contexts where the risk assessment process and its results should be documented to competent third parties, including authorities having jurisdiction, that in some cases are obliged to conduct specific verification audits. But it is the organization itself that benefits most from this list.

Given that the organization's resources, in terms of time and money, are not unlimited, identifying critical barriers makes it possible to start an optimization process aimed at keeping the performance of the organization's own activities safe, investing primarily where the greatest need has been identified, thanks to the barrier-based perspective, in order to always guarantee their integrity and functionality.

For critical barriers, the performance standards required can therefore be defined, i.e. those performance requirements that must be guaranteed over time. There are several approaches to defining performance standards; one of the most widely used is the

Functionality, Availability, Reliability, Survivability and Interactions (FARSI) approach (Fiorentini et al., 2018). In essence, the performance required at a critical barrier is described in terms of five main characteristics:

- *Functionality*. This performance standard identifies what the barrier is required to do: interrupt an unsafe process, extinguish a fire, give an alarm, activate an instrumented safety logic, contain a loss of containment of hazardous or flammable liquid substance, and so on.
- *Availability and Reliability*. Should the critical barrier under examination always be available, 7 days a week, 24 hours a day? Or should/may it only be available for a defined and limited subset of time, perhaps only during a scheduled extraordinary maintenance operation? Furthermore, when called upon to intervene, with what reliability can the barrier perform what it is called upon to do without failure? In other words, what is your PFD? Within these two performance standards, all routine maintenance and inspection activities are of course carried out to ensure that the barrier remains available and reliable over time, as required.
- *Survival*. This performance standard is intended to answer the question "In the event of an accident, is the barrier required to survive?" Obviously, if the incident in question is a fire and the barrier is a firefighting system, the answer will be yes, at least for as long as needed to keep the flames under control (when it is not possible to extinguish them directly) until external help arrives. Similarly, if it is an automatic shut-off valve that has to intercept a line of flammable liquid because there is a fire, it is obviously required to be fire resistant. In other cases, survival in the event of an accident may not be required at all: this is the case, for example, for fire sensors, whose functionality is exhausted when the fire itself is detected.
- *Interactions and independence*. A critical barrier should not share common causes of failure with other barriers. Furthermore, appropriate interactions should be taken into account. If, therefore, the failure of an emergency diesel generator means that the flood fire extinguishing system (manual or automatic or otherwise) cannot be used, then there is an interaction between the "emergency power generation" and "flood system" barriers that needs to be highlighted. Clearly, if the manual and automatic flood system are powered by the same pump, then they cannot be considered as separate barriers, as they are not independent; in fact, a failure on the pump would result in the cascade failure of both systems.

These performance standards may be subject to certain areas of control by the authorities, as is already the case in the context of the process industry and as, for example, also established by the European Directive on major hazards of oil and gas extraction platforms at sea (Directive 2013/30/EU), the Major Accident Prevention Directive, and so on.

The safety critical equipment (SCE) is linked to performance assurance actions and the activities are composed of tasks. different levels, also of responsibility, for maintaining the related level of performance, as shown in Figure 131.

Barrier Criticality Assessment

As seen in various sections, Bow-Tie diagrams can be used as the central point of information and data gathering for risk-based processes. They can be completed and integrated with other methodologies that allow covering specific aspects. Given the identification of critical elements a further judgement about their relative importance in the overall map of risks can be obtained with a barrier criticality assessment.

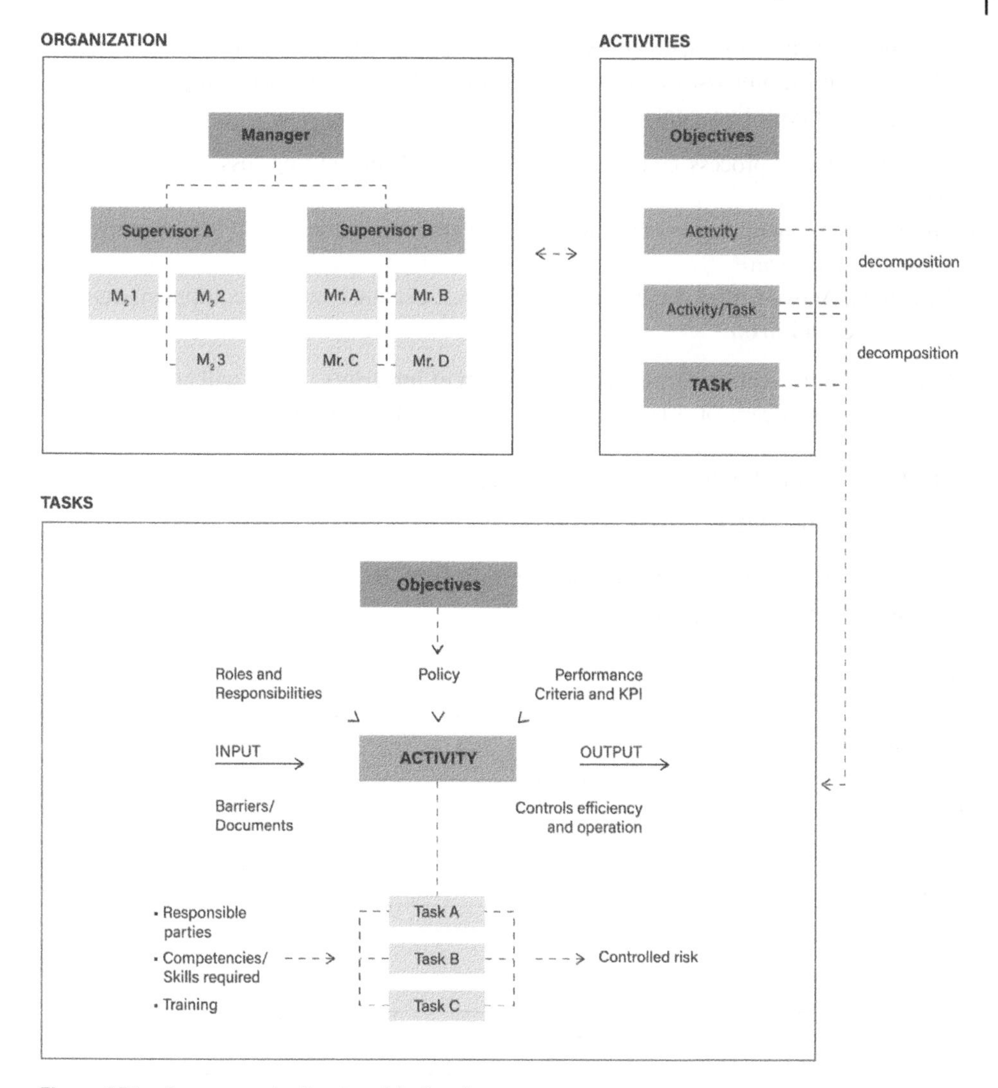

Figure 131 From organization to critical tasks.

This step can be performed once SCE has been defined at the equipment level.

Potentially each identified barrier can be treated as SCE. However, in order to focus on the ones playing major roles in preventing the occurrence of the major scenario or in controlling/mitigating its potential consequences, each identified barrier shall be criticality-ranked according to:

- Its role in managing the risk – function score (indicated with FS);
- Its impact on the event consequences in case the barrier fails – consequence of failure score (indicated with CS);
- The extent to which an alternative barrier can take over the function of the barrier failure – redundancy score (indicated with RS).

The possible values for FS, CS and RS are reported respectively in Table 20, Table 21, and Table 22. The criticality assessment shall be performed by a multidisciplinary team.

Examples of preventive, control, and mitigating barriers are:

- *Preventive:* PSVs, process containment integrity, SIF preventing loss of containment events, and so on
- *Control:* ESD system, depressurization system, detection systems, HIPPS, secondary containment, and so on
- *Mitigating:* Active fire protection, rescue facilities, personal survival equipment, temporary refuge, and so on

Note that for preventive barriers (e.g. containment integrity), the CS will be set equal to 3 by default, as in case of a barrier failure the hazardous event will occur (e.g. loss of containment).

The Barrier Criticality Index is then calculated as:

$$\text{Barrier Criticality Index} = \text{FS} \times \text{CS} \times \text{RS}$$

The final criticality ranking of each barrier is determined based on the ranges of criticality reported in the following table:

Table 20 Barrier function score (FS).

Barrier Function – FS	Score
Preventive	3
Control	2
Mitigation/emergency response	1

Table 21 Barrier consequence of failure score (CS).

Consequences in Case of Barrier Failure – CS	Score
Severity of the event is significantly affected by the failure of the barrier	3
Severity of the event is affected by the failure of the barrier	2
Severity of the event is not increased by the failure of the barrier	1

Table 22 Barrier redundancy score (RS).

Barrier Redundancy – RS	Score
There is no barrier that can fully duplicate the function of the barrier subject to failure or unavailability	3
The analyzed barrier is redundant by design	2
There is an independent alternative barrier that can fully assume the functionality of the failed barrier	1

A simplified example of barrier criticality assessment is shown in Figure 132. In this specific case, the initiating event (overpressure in vessel A) is represented by a spurious opening of a pressure control valve upstream vessel A, which can potentially develop into a loss of containment scenario.

The criticality assessment of the barrier 1 (high-pressure safety loop) is carried to define a barrier criticality ranking (Table 23) as in the workflow described in Table 24.

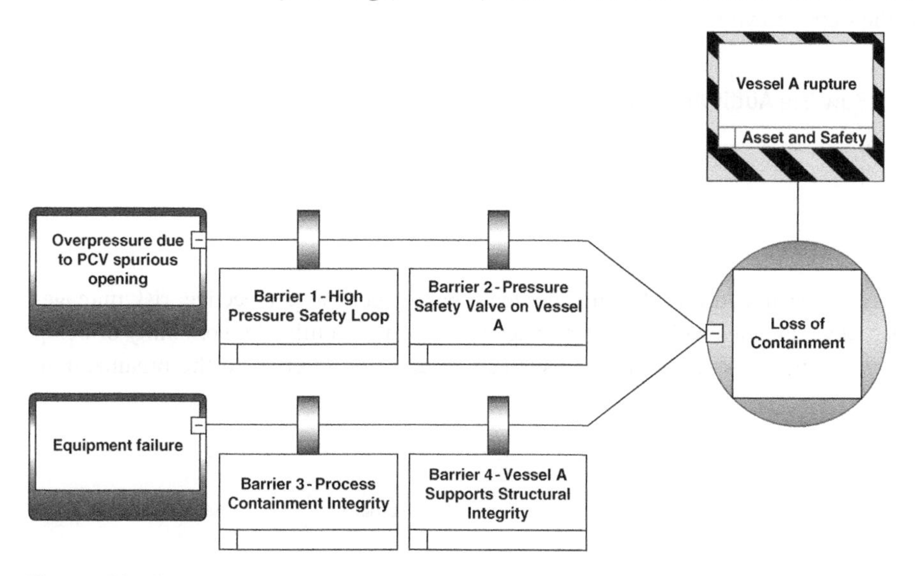

Figure 132 Example of Barrier Criticality Assessment.

Table 23 Barrier criticality ranking.

Barrier Criticality Index	Barrier Criticality Ranking
Index ranging from 13 to 27	High Criticality
Index ranging from 6 to 12	Medium Criticality
Index ranging from 1 to 5	Low Criticality

Table 24 Barrier criticality assessment example.

Barrier 1 – Function	**FS Score**
Preventive barrier as its intervention averts the occurrence LOC of containment event	3
Barrier 1 – Consequence of Failure	**CS Score**
Score set equal to 3 as the hazardous event will occur in case of failure of Barrier 1	3
Barrier 1 – Redundancy	**RS Score**
Barrier 2 (PSV) acts as independent alternative barrier that can fully assume the functionality of the failed Barrier 1	1
Barrier 1 – Criticality index	**Criticality**
Criticality Index = FS (3) × CS (3) × RS (1) = 9	Medium

The method presented is quite often used in the oil & gas field to define the criticality of the barriers that prevent or mitigate major accidents coming from process safety issues. Each organization could define its own method, qualitative or quantitative, if it is useful to have a measured estimation of the importance of a single control in relation with the consequences.

In conclusion, the identification of critical barriers starting from Bow-Tie diagrams can follow the steps shown in Figure 133.

4.1.15 Bow-Tie Audit Activity

It has already been said that, having adopted a barrier-centric perspective, risk management actually coincides with barrier management. Consequently, the possibility of being able to carry out controls (audits), by subjects internal or external to the organization,

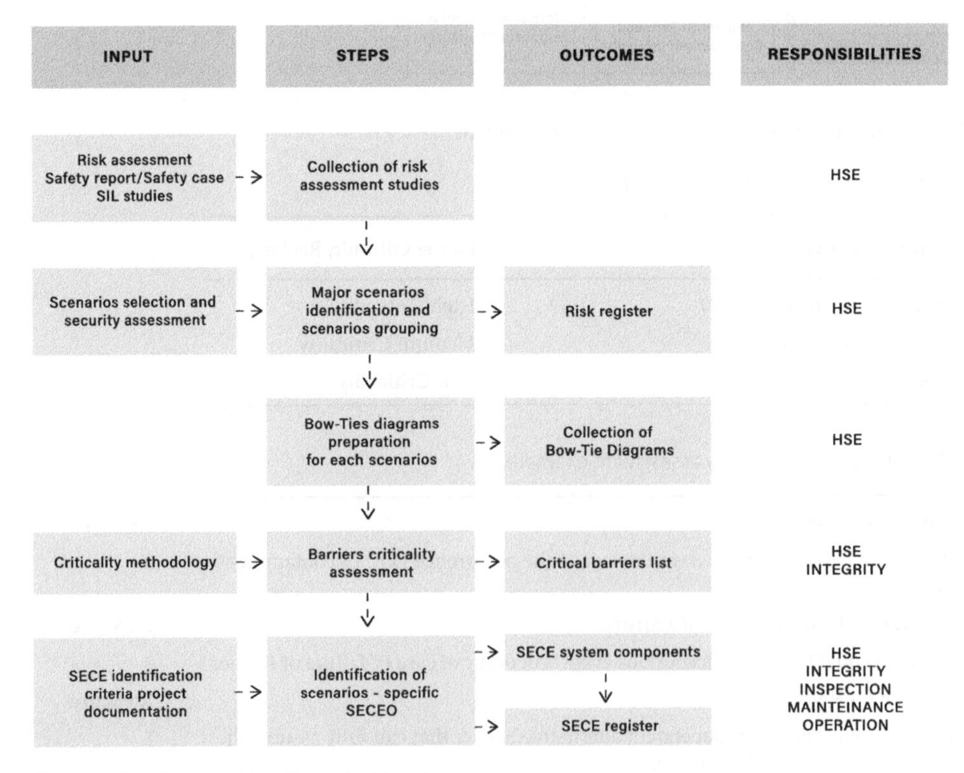

Figure 133 Steps to identify critical barriers.

represents an important step to close the Deming cycle applied to risk management. After having defined the barriers within the management system (plan) and after having implemented them (do), a check is required to ensure their quality (check). Furthermore, a thorough analysis of the barriers during this phase facilitates the development of effective recommendations in the next phase (do).

The audit is therefore an indispensable tool to have an in-depth analysis of the state of health of the barriers, their quality and effectiveness, and in order to have premonitory signals before an incident occurs, as relying on an ineffective barrier is extremely dangerous.

Auditing the barriers of a Bow-Tie diagram is possible. As usual, it is better to rely on software tools that already provide such functionality, but it is still possible to make do with do-it-yourself solutions, at least as long as the reduced complexity and small size of the company allow it.

The advantages of auditing with Bow-Tie are many:

- *The audit is also risk-based.* The scope of the audit is created by looking at a risk scenario and the results are rendered from a risk-based perspective.
- *Analysis of dependency effects.* In fact, a poor management system can have extensive effects throughout the Bow-Tie diagram.
- *Visualization of results.* The results of an audit can be shown directly on the Bow-Tie diagram, using intuitive charts that allow you to instantly highlight problem areas. The graphical display also facilitates expert judgement of the quality of a barrier and provides up-to-date feedback on the effectiveness of a barrier.

Questions for an audit applied to the Bow-Tie may involve different levels of the Bow-Tie audit:

1) Barriers (of any type);
2) The activities and documents of the management system;
3) General questions (e.g. on the organizational structure).

With regard to barriers, the auditor may ask whether the integrity of the barriers meets certain requirements (Is the barrier in place? Is it working properly?).

At the management system level, applications may relate to the areas of training, maintenance, availability and updating of manuals, procedures, instructions and operating manuals.

Finally, the third level covers those questions that cannot be linked to a specific Bow-Tie element, affecting a higher or more general level. Typically these questions question the existence of complete systems (Is a housekeeping system defined?) or more general issues (How does staff communicate? What is the status of housekeeping?).

The workflow for auditing Bow-Tie diagrams is as follows:

1) Define the questions that allow a consistent judgement to be made.
2) Associate questions with barriers and/or activities.
3) Create a "survey," i.e. a grouping of some questions (e.g. by barrier category, or by business function, or by any other criteria).
4) Create questionnaires with the questions of the survey and administer them to those who have to answer.
5) Collect the answers.

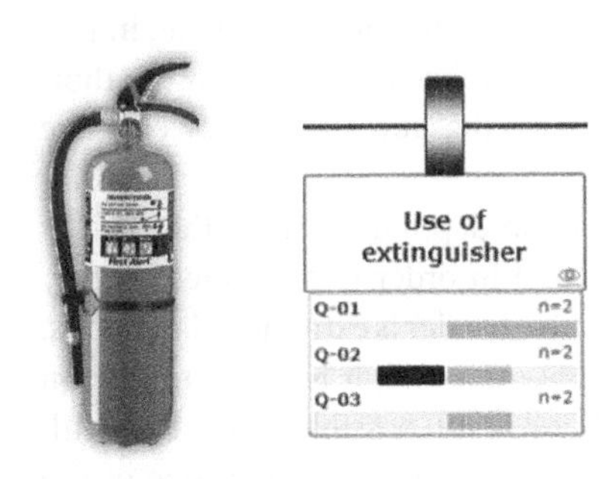

- Q01: (Maintenance)

"Are the extinguishers well maintained and filled?"

- Q02: (Training)

"Is the personnel capable of using the equipment?"

- Q03: (Procedures)

"Is the personnel aware of where to find the device?"

Figure 134 Example of a barrier audit.

6) Visualize and analyze the results.
7) Verify that the completion of the survey given allows, as it was designed, to produce an overall judgement.

A convenient way to view the results of an audit on the Bow-Tie diagram is to use histograms (Figure 134). Each question is associated with a histogram, whose value (and colour) changes from red to green depending on the answer given. This histogram can be reported just below the barrier to which the specific question is applied. The questions must therefore be of the multiple-choice type, and each answer is associated with a predefined position of the histogram bar, i.e. the most positive answer (bar entirely green) and the most negative answer (bar entirely red) among those possible. The intermediate answers will have an intermediate weight.

In order to avoid having innumerable histograms under the same barrier, the questions (and therefore also the answers and their graphical display by histograms) can be aggregated into one question. Not only that, the same survey can be administered at regular time intervals, or to several people. It is therefore necessary to have a system to collect the different answers given at different times and/or by different people, so that we can also have a wider database to analyze.

The results displayed in histograms must therefore be interpreted correctly, as suggested in Table 25.

One of the advantages of carrying out audit activities on barriers is abandoning the traditional approach of analyzing one element of the management system at a time (Figure 135), which involves costs that are often important for the organization that undergoes audits and control cycles so frequently that they are sometimes considered incompatible with operational activities.

This approach is therefore abandoned in favour of an analysis of all the elements of the management system related to the specific barrier audited (Figure 136). In other words, the subject of the audit is now the barrier, which is observed from all perspectives offered by the management system, while ensuring that all disciplinary areas are covered by the barrier-based audit.

Table 25 Interpretation of the barrier-based audit response histograms.

Example	Description	Meaning
Q2 What did you think about lunch? Q2 Somewhat okay ...	Single answer Somewhat positive answer	Single answer
Q10 How knowledgeable would you rate the trainer? Q10 Expert	Single answer Most positive answer	Single answer
Q1 Was the room temperature okay? Q1 n=8	Eight answers (n=8) Even distribution between negative and positive answers	Uncertainty
Q3 Were there enough breaks? Q3 n=8	Representing eight answers (n=8) Mostly negative answers	Negative tendency
Q7 Was there a cohesive story that ran through the course? Q7 n=8	Representing eight answers (n=8) Answered all most positively	Fully positive
Q10 How knowledgeable would you rate the trainer? Q10 n=8	Representing eight answers (n=8) Mostly neutral answers	Neutral tendency
Q10 How knowledgeable would you rate the trainer? Q10 n=8	Representing eight answers (n=8) All answers are all centered/fully neutral	Fully neutral
Q9 Were the sheets informative? Q9 <No Data>	No answer data available	Not answered

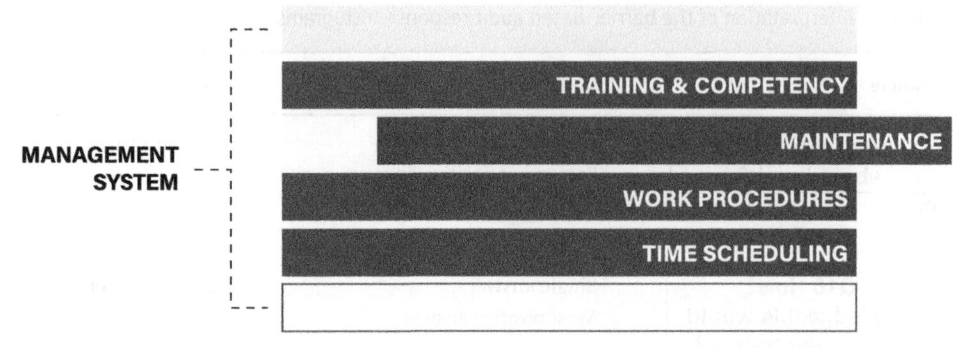

Figure 135 Traditional audit: one element of the management system is analyzed at a time.

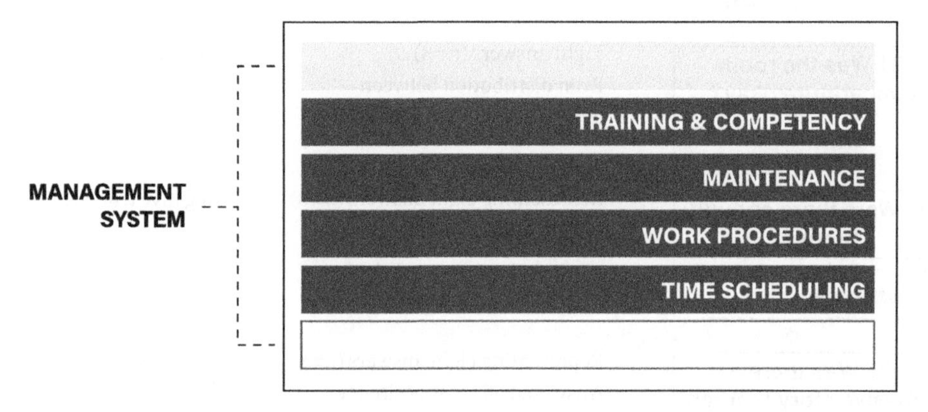

Figure 136 Audit barrier-based: all elements of the management system identified as relevant to a specific barrier are analyzed.

4.1.16 Work Organizational Schemes for Multi-Site Operations

There are different ways to structure your workflow when dealing with multi-site operations, perhaps equipping yourself with a server that can handle centralized information. The most common ones are discussed next.

"Top-Down" Approach

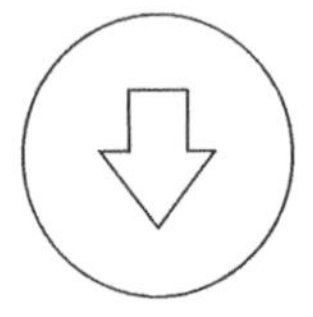

This approach is generally best suited for organizations with multiple-site operations, but with a high level of similarity. The following example is provided for a direct and more effective understanding of the approach.

Consider a company that builds and operates bio-gas plants (all similar) in similar areas. The headquarters are located in Paris, but the company has operations throughout Europe.

Although the physical location may differ, the process of building and running bio-gas plants is very similar. The company's corporate offices have therefore created a set of safety rules to which each site must adhere, regardless of its location. Thus, the central offices have created a set of Bow-Ties that are relevant to the process of building and running a bio-gas plant and intend to distribute these Bow-Ties to each individual site. They then want to audit each site to see if the barriers identified on the diagrams are implemented correctly. As the audits will be based on the barriers from the Bow-Tie diagrams themselves, the results from the different sites can be easily compared. This approach is called "top down" because the corporate offices create the Bow-Tie charts and then submit them to individual operational sites. For their part, sites only need to ensure that the Bow-Tie barriers defined by corporate are implemented correctly.

For this approach, it is suggested that only a few Bow-Tie diagrams be created. For example, the two Bow-Tie groups "Construction of a plant" and "Operation of a plant" could be created. The first group could include the Bow-Tie "Working at height," "Working with electricity," and "Entering confined spaces," while the second group could include "Containment loss" or "Filling a pressure tank."

Since corporate offices need to create and manage Bow-Tie diagrams over time, corporate users will be able to read and modify these diagrams. Users of individual sites should only be given the power to read, as they need only to consult the information in the Bow-Tie, not change it.

The advantage of this approach is that it achieves a high level of standardization and congruence of data, which makes it much easier to compare the performance of different sites. But this approach only works well if a high level of similarity in the processes performed at the various sites is guaranteed.

"Bottom-Up" Approach

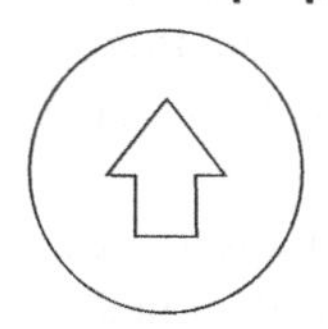

While the top-down approach is most useful for organizations with a high level of similarity between sites, the bottom-up approach is the best option for those companies with very different processes. For example, consider a mining company that has many different sites, each with its own very specific characteristics. One site may have underground mines, while other mines are open pits. Geological differences between different sites can also result in causes that are highly relevant to one site, while definitely not being applicable to another site. These individual differences between sites make it difficult to create a single generic Bow-Tie to which all sites must adhere, as the top-down approach would suggest. The generic Bow-Tie would simply not be an adequate representation of the peculiarities of individual sites. Instead, it may be better to allow different sites to create their own set of Bow-Ties, to ensure a good match between the operational reality and the Bow-Tie diagrams. With this approach, Bow-Ties are created from the lowest operational regions of the organization, hence the name "bottom up."

With this approach, the Bow-Tie of each site is stored within groupings managed by the individual site. Operational realities therefore have write-and-read access to their own Bow-Tie diagrams. In order to share knowledge between sites, read-only access to the diagrams of other sites should be allowed.

Opting for this approach, the role of corporate offices is totally different compared to the top-down approach. Rather than using Bow-Tie to dictate what barriers should be used at different sites, Bow-Ties are now used by individual operations to convince and prove to corporate offices that they effectively control their risks.

The advantage of using the bottom-up approach is that Bow-Tie diagrams are very much in tune with reality; on the other hand, the disadvantage is that it is much more difficult to compare the performance of individual sites and more resources are used to create all Bow-Ties.

Hybrid Approach

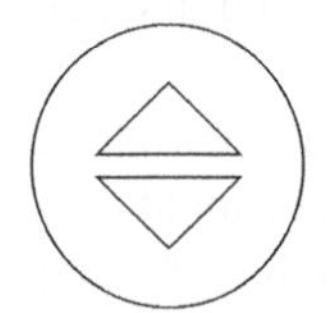

Obviously, not all companies could be recognized in the two approaches described above. There are hybrid forms to try to compensate for the excessive rigidity of the top-down approach and the lack of similarity of the bottom-up approach. The hybrid approach is therefore suitable for those companies that operate with a lot of common ground, but with slight differences from site to site.

For example, an airline may operate in several states. Although the company's main operation is "flying an aircraft," differences in legislation or facilities provided by airports may result in some barriers or causes being present in one place but absent in another. So even if there are some differences, a large part of the Bow-Tie content will still be the same for all sites. In this case, a hybrid approach may be the most appropriate.

In a typical hybrid approach, corporate offices create a set of Bow-Tie diagrams to use as templates, on a generic level. The barriers, causes and consequences that are identified on these diagrams should be seen as the minimum requirement to be met. Each individual site will then make a copy of the Bow-Tie diagram templates and add or remove certain content to make it as close to reality as possible. In this way, the overall structure of the Bow-Tie diagrams remains congruent, but minor adjustments dictated by the peculiarities of the individual sites are still allowed.

The approach described here may be more rigid or flexible, depending on how detailed the Bow-Tie set used as a template is. If corporate templates already provide a high level of detail, then they will ensure a high level of congruence between sites, but it is very likely that this detailed model will be difficult to adapt to all sites. Building a very flexible template (e.g. identifying only causes and consequences) ensures that each site can use its own barriers, but this would result in Bow-Tie sets no longer matching each other. Therefore, depending on the company's predominant need, a balance must be struck between rigidity and flexibility in the content of the Bow-Tie model when using a hybrid approach.

4.2 LOPA Construction Workflow with a Step-by-Step Guide

4.2.1 Identification of Objectives

The layers of protection analysis (LOPA) method, i.e. the analysis of independent protection levels, is a semi-quantitative risk analysis methodology based on the verification of the availability of control and mitigation measures for the identified risk level. This approach allows the degree of risk reduction associated with each control and mitigation measure to be verified, making the methodology particularly suitable for the identification, during the design phase, of alternative protection strategies with respect to the identified risks.

LOPA analysis can be used:

- In an exclusively qualitative way, for a review of the levels of protection existing between a hazard and the expected consequences with respect to the occurrence of an undesirable event;
- In a quantitative way, downstream of a HAZOP/HAZID or preliminary hazard analysis, in order to deepen the criticality that emerged during the execution of such analysis.

The choice to focus attention on the protection barriers allows the verification of the appropriateness of the risk reduction made by a barrier with respect to the risk value to be mitigated, subject to the definition of a risk acceptability criteria, defined, for example, from the requirements deriving from applicable norms, codes and standards. In case of an HSE/process safety LOPA, domains where this methodology is being applied most, layers of protections (or IPLs) are those safety barriers that come between the hazards of a process and vulnerable targets (usually people, environment and assets) they could affect. They are the safety measures providing so-called "defense" in depth against probable incidents. This LOPA approach could be easily described with a qualitative Bow-Tie: LOPA becomes the quantitative extension of the Bow-Tie basic methodology. It can be referred to a simplified method of risk assessment, intermediate between a qualitative hazard analysis (e.g. Bow-Tie) and a fully quantitative risk analysis. It can provide an order of magnitude risk estimate for each assessed scenario. Starting ingredients for a LOPA are a scenario to evaluate, a risk goal to attain, a set of rules to follow and a format to document the inputs and outputs of the analysis.

The general workflow of the LOPA analysis is given in Figure 137.

LOPA can be used at different stages of the plant/process life cycle: process initial development, process basic and detailed design, operation/maintenance and modifications, and decommissioning. Risk criteria and objectives should be related to the phase LOPA is applied to, since they may differ from one phase to the other. LOPA should be applied to prior analyses such as HAZOP/HAZID, qualitative Bow-Ties, and so on. LOPA gives a better, quantitative (order of magnitude) view of identified scenarios in order to make better

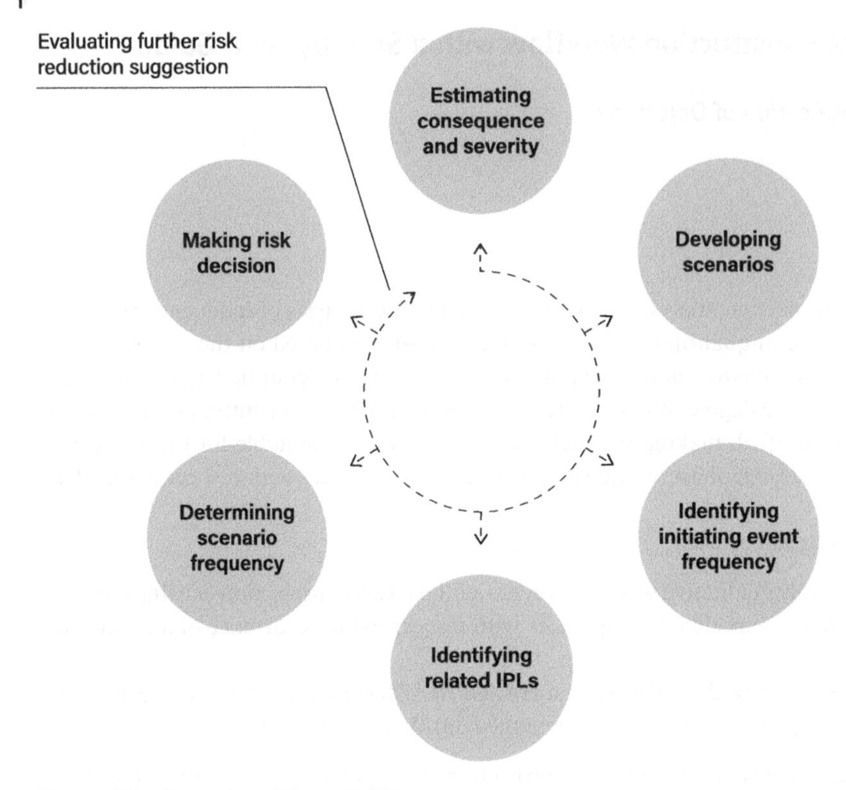

Figure 137 General workflow of LOPA.

informed, and risk-based decisions, on whether the frequency of the consequence for a given scenario meets the tolerance criteria for a particular organization. It is important to remember that LOPA is not intended to be used as a hazard identification tool.

4.2.2 Identify the Consequence to Screen the Scenarios

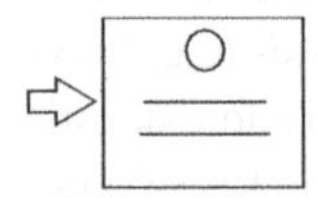

A prior assessment should be available since LOPA evaluates scenarios that have been already developed in a qualitative way via a different approach. LOPA could take advantage of a prior Bow-Tie study, already conceived in terms of barriers. Since prior studies often describe more scenarios it is fundamental to operating a scenarios screening: the most common screening method is based on consequence level of severity and on vulnerability category. Usually initial levels of severity are assigned to multiple scenarios using matrices and/or indexes. Consequence estimation could be refined during LOPA, even taking advantage of quantified parallel studies.

4.2.3 Select an Accident Scenario

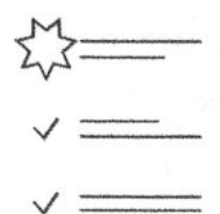

LOPA is applied to one scenario at a time that is intended as an unplanned event or incident sequence that results in a loss event and its associated impacts, including the success or failure of safeguards (barriers) involved in its incident sequence. A scenario should be defined as a well-described pair of cause and related consequence towards a specific single vulnerable target. When a single scenario involves more than one loss events, LOPA usually suggests focusing on the worst one. In any case, LOPA does not identify scenarios (and associated loss events); they should be evaluated from a different source and they should be available at the beginning of the LOPA assessment. Other sources of scenarios are SIL studies, FMEA assessments, root cause analysis, and so on. In LOPA, consequences are estimated to an order of magnitude of severity, and guarantee the ability to compare risk from different scenarios. Usually LOPA is conducted for 15–20% of identified hazardous scenarios while 1% of total scenarios are too complex for a LOPA and they should be directly investigated with quantitative risk assessment, supported by more precise mathematical modelling of the consequences.

4.2.4 Identify the Initiating Event of the Scenario and Determine the Initiating Event Frequency

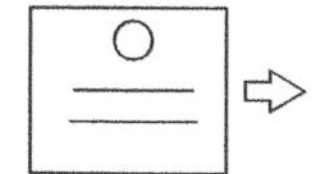

Consequences are raised towards one or multiple vulnerable targets along a specific path considering failures, escalation factors and conditional modifiers that could be easily described with a Bow-Tie diagram. The path starts from a threat, also known as an initiating event. The initiating event should be associated to a frequency of occurrence (in occasions per year). An initiating event is an equipment or human or external event (not combined) that starts a sequence of events leading to one or more undesirable consequences. The initiating event is referred to as a threat in a Bow-Tie diagram. Generic initiating event frequencies are available in data tables but they can also be determined given the analysis of real experience. Individual organizations should choose their own values, consistent with the degree of conservatism of the company's risk tolerance criteria. Failure rates can also be derived from internal experience in terms of "as-found conditions" and inspections and maintenance activities results. Using a set of standard data can guarantee consistency across different LOPA assessments (as well as those used during management of change assessment of potential new or modified hazards).

For procedure-based operations where the initiating event frequency is connected with a human error, frequency could be estimated by combining the frequency of performing the specific operation and the probability of error per operation (considering all the performance-influencing factors like stress, fatigue, training level, availability of written procedures, and complexity of the procedure to be followed). This assessment is particularly important in specific operations like tests, start-up, shut-down, decommissioning, emergency situation immediate response, batch operations, and so on.

LOPA also allows considering enabling conditions: a condition that cannot be identified in a failure, error or protection layer but makes it possible for an incident sequence to proceed to a concerned consequence. It consists of an operating phase or condition that doesn't directly cause the scenario but that must be active or present in order to achieve a loss. This condition is usually expressed as a probability and it usually describes the probability of a specific state of the system, or a campaign-related risk or even a seasonal risk. Its use is not compulsory but it allows for a more detailed description of the situation and for a more accurate estimation of the frequency of the considered abnormal situation that could drive towards a loss. During the initiating event frequency estimation process, state enabling conditions and seasonal conditions that could be combined in so-called time-at-risk enabling conditions.

4.2.5 Identify the Independent Protection Layers

The path from an initiating event to its associated consequences is described in terms of:

- Independent protection layers (barriers);
- C=Conditional modifiers.

Each scenario could require for one or more IPLs.

An IPL is a control/barrier placed between a hazard and a vulnerable receptor. IPLs in LOPA are equal to controls in Bow-Tie. A specific principle underlined in LOPA references is that IPLs, according to the James Reason approach (also known as the Swiss Cheese Model), reduce the frequency of the consequence for the selected scenario. If all the layers fail, the original consequence occurs at the initial estimated loss but at a reduced frequency of occurrence. The likelihood of a loss event is determined by the initiating event frequency and the combined PFD of the barriers. A Bow-Tie diagram is one of the best ways to represent the path from a threat to a loss and add the quantitative simple approach of LOPA. During the selection of barriers it is important to notice that not all the HAZOP/HAZID safeguards can be claimed as IPLs; furthermore, safeguards in those initial studies are not recorded without avoiding common cause failures that play a fundamental role in judging the independence of a barrier in a Bow-Tie/LOPA and that often should be better considered as escalation factors. Causes of dependence could be identified in design issues, construction issues, procedural issues and even environmental issues. In general if a component or human response is shared across barriers, those barriers couldn't be claimed as independent (e.g. a single operator required to respond to multiple different alarms: the operator is the common cause of failure of the alarm system, which represents the barrier to be considered in the LOPA).

Simple rules could be used in order to verify the suitability of a barrier. These are similar to the requirements seen for the controls in Bow-Tie preparation. Among these rules are the "three Ds": detect, decide and deflect (seen as Act in Bow-Tie), in order to respectively indicate the ability of each IPL to detect that a cause has occurred, the ability to decide to take some action (also raise an alarm), and the ability to take an action to deflect or to prevent the associated consequence to occur; and the "four Enoughs": each barrier must be "big enough" to do the intended job, "fast enough" to respond, "strong enough" to prevent the associated scenario, and "smart enough" to accomplish the requirement as best as it can.

The concept of IPLs is shared among a number of different barrier-based methods for assessing risk in conformity with the ISO 31000 standard on risk management: Bow-Tie, based on IPLs, and event tree analysis (ETA) visually describe scenarios as outcomes of the potential failures of barriers that are the gates of the event tree itself.

4.2.6 Characterize the IPLs in Terms of Probability of Failure on Demand

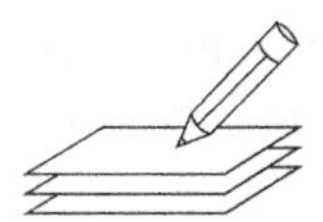

Each barrier may fail and the failure of a barrier would not avoid consequences. Partial failure may lead to different (usually lower) consequences. In general terms, risk reduction is achieved by increasing the number of IPLs and increasing their reliability or, better, lowering the associated probability of failure on demand (PFD).

IPLs could be given a credit (e.g. 1 × 10-1 aka 10X, 1 × 10-2 aka 100X reduction in event frequency) or PFD could be estimated with different techniques (such as fault tree analysis) or by the use of technical literature and data banks where results of FMEAs on similar components/systems are shown.

Given the fundamental characteristics of the IPL (the three Ds rule and four Enoughs rule), a PFD can be assigned to each barrier: this probability can be defined using data tables available in the technical literature or using other techniques like fault trees where the total probability for a system is calculated from the failure rates of single components coupled with repair/inspection/test interval data over the mission time considered. It is quite common to consider at least three parts: sensors, logic solvers and final elements.

LOPA provides the opportunity to determine the required PFD (the maximum one) for an IPL in order to get the tolerable/acceptable risk. This is generally done in the process safety or in the machine safety fields to identify the reliability to be assigned during design phase to specific categories of controls (in the specific case, those named as "safety intrumented functions" that should be described in terms of "safety integrity level" seen as 1/PFD).

4.2.7 Estimate the Risk

Risk associated with a pair (cause and consequence) is given by the frequency of the initial event (threat) and the combined probability of failure on demand of identified barriers. During the calculation of the risk, conditional modifiers could be taken into consideration to better describe the incident dynamics (e.g. presence of people, probability of avoiding the consequence under certain circumstances, probability of escape, probability of ignition, etc.). Risk is calculated as the combination of the expected frequency and severity of a scenario. Risk could be calculated as unmitigated (not considering the IPLs) and mitigated (considering one or more IPLs able to reduce the severity of the outcomes or the probability of the occurrence).

Risk values are given as a combination of the scenario frequency (defined as initiating event frequency and the combination of the PFDs of the selected IPLs) times the scenario impact.

Risk calculation could take into account conditional modifiers defined as one of several possible probabilities included in scenario risk calculations, generally when risk criteria endpoints are expressed in impact terms instead of in primary loss event terms. Often these modifiers are referred to as mitigating events. Three common conditional modifiers in the domain of process safety and industrial risk are probability of ignition given a release of flammable material, probability of people impacted in the area of the loss event, and probability of fatality. Conditional modifiers can be used when it is realistic to expect that the probability can be considered valid over time.

4.2.8 Evaluate the Risk and Make Risk-Based Decisions

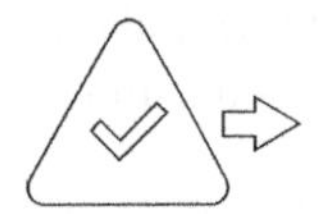

The resulting risk for each pair should be compared with tolerability and acceptability criteria, which are often defined by regulations, standards, and internal documents. The first result defines the need for additional/different barriers. Risk-based decisions are often supported, even if not strongly recommended, by expert judgement while they could take advantage of specific more detailed studies such as cost-benefit analyses and relative risk reduction of competing alternatives assessment (this could be an approach to deal with a portfolio of similar assets presenting the same scenarios but different levels of control). Through an ALARP approach, risk-based decisions could refer to three different situations: no risk modification is required (current level of risk is proper), calculated risk should be reduced (tolerable risk not achieved), or risk should be abandoned (risk is not even tolerable). LOPA results should help in determining how much risk reduction is needed. It is important to note that risk tolerance and acceptability criteria are often given as target values (e.g. 10^{-6} occasions/year) while in some companies there are specific "per scenario" criteria, "sum of all scenarios" criteria towards a specific vulnerable target, and so on. These approaches should be clear when initiating a LOPA on scenarios as well as in the consideration of results final phase.

4.2.9 Consideration of Results

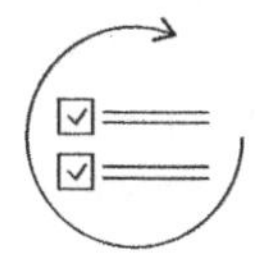

Given the results of the LOPA it is important to note that it is a simplified approach and it should not be applied to all scenarios identified during initial and supporting studies. Special consideration should be given to the results of a LOPA; in some cases the effort to cover all the pairs may be excessive for some risk-based decisions and in some other cases it is overly simplistic for other decisions that require a more detailed and quantitative approach. In some cases, when multiple scenarios should be considered, it is often advisable to directly include LOPA consideration into HAZOP/HAZID worksheets with specific columns supported by matrix approaches to probabilities and severities and develop template Bow-Ties to be customized for each pair.

Results should be properly documented and the conclusion document should mention the rules used to apply the methodology and those assumptions selected to guarantee consistent outcomes. LOPA should be organized in such a way to have, for each considered selected scenario, the following information: scenario description, risk target, factors considered in risk calculation, results, discussion about risk-based decisions (together with any eventual cost-benefit and or ALAR analysis), and documentation and justification of each parameter included in the calculation. LOPA worksheets can be presented as a single sheet depicting the workflow or in a table-like format (especially in the case of extended HAZOP/HAZID). A fully documented LOPA may also include supporting information for initiating event frequencies, PFDs, scenario outcomes, work sessions and involved people, recommendations, human factor associated probabilities, expected follow-up with roles, responsibilities and deadlines, and criteria for LOPA updating and revalidating over time.

LOPA actions should be managed as PHA actions:

- Review with top management and stakeholders;
- Commitment of resources;
- Defined deadlines;
- Assigned responsibilities;
- Document resolution;
- Follow-up with top management.

4.3 BFA Construction Workflow with a Step-by-Step Guide

Providing a universal workflow for the investigation of an accident is frankly impossible, as the organization of activities depends closely on the particular accident. However, it is possible to describe a general path, highlighting the essential steps that define a complete investigation.

The investigative workflow can be divided into six phases, as shown in Figure 138. Notification is one of the preliminary processes: it is necessary to inform all necessary personnel about the incident, activate the first emergency activities, and alert external authorities and the public. Once the emergency response activities have been carried out, the initial investigation begins. This phase aims to provide immediate feedback to a small number of people interested in knowing what happened and what immediate corrective action should be taken to restore the original conditions and ensure that similar events do not recur. During emergency response activities, evidence and data may change. However, the main objective of these activities is to prevent further damage and loss, so potential loss or alteration of evidence must be considered. Only if the investigation is initiated in parallel with emergency response activities can data be retained and collected before it is lost or altered.

Any investigation should start immediately after an accident, near-accident, or non-conformity, in order to counteract as much as possible the physiological deterioration of evidence, especially the most sensitive (e.g. data that risk being overwritten within a limited

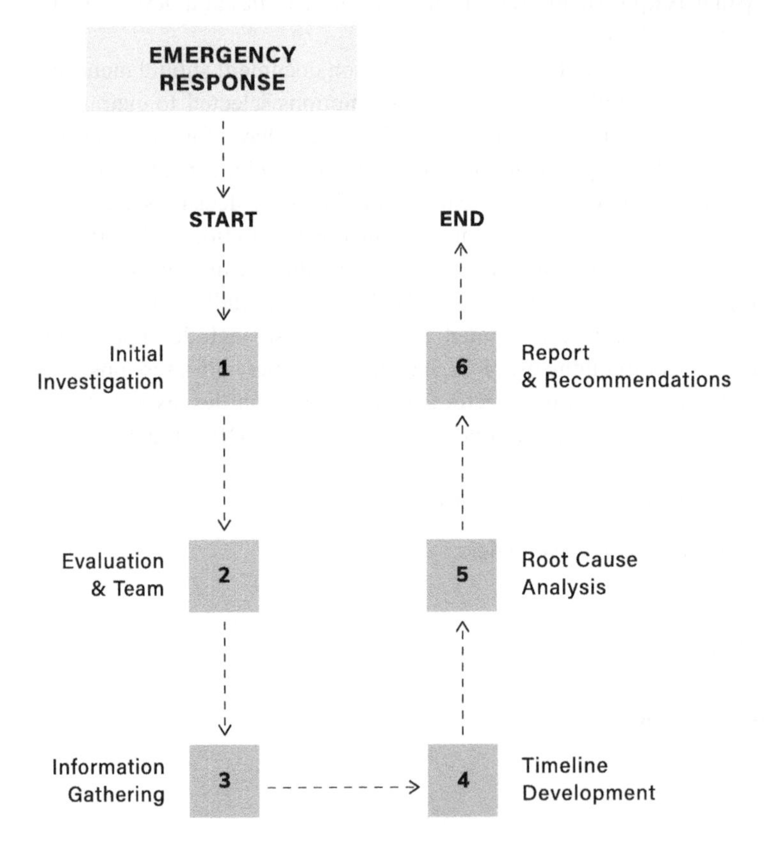

Figure 138 The general workflow of a survey.

memory, external scene exposed to atmospheric agents, oxidation of chemical residues and so on). Obviously, starting the investigation as soon as possible implies some challenges: first, the team must already be selected and assembled. It may also need to be trained and equipped and team members may need to go to the site, access to which, in the most serious cases, may be barred by external bodies or authorities.

It becomes important to clarify the investigative needs before the start of the investigation. Planning, training, and preparation for investigations require strong organizational skills in choosing the right people, at the right times, in the right places, with the right tools, following the right procedures. In other words, a high-quality incident investigation programme starts with the support of management.

Flexibility in team composition is a key factor demonstrating a well-designed management system. The training of the team normally takes place in two phases. A first formal training is conducted prior to the event in order to have a group of potential investigation team members who know the expectations, methods, definitions and objectives of an incident investigation. Then, when the investigation is initiated, a second refresher training is provided for team members, designed in relation to the nature of the event.

For simple incidents it may be sufficient to work individually, but for more complex cases working in groups, using a multidisciplinary composition, is strongly recommended. However, this second approach is preferable because, in general, it must be honestly acknowledged that it is almost impossible for a single person to correctly investigate, collect, analyze and extract conclusions from such a complex issue as the analysis of an accident. The team structure provides many other benefits, such as:

- More solid final conclusions and recommendations thanks to the redundancy offered by the different backgrounds and skills deployed by the various team members;
- Increased objectivity by eliminating subjective bias, such as confirmation bias;
- Overall improvement in the quality of the survey, thanks to peer reviews that provide constructive criticism;
- Division of investigative activities among the various team members, appropriately distributing the workload without excesses for anyone.

There is no preconfigured team composition: a specific incident requires a specific investigation team, depending on the type, severity and complexity of the event analyzed. It can certainly be said that senior management, although not directly involved in the investigation, plays an important role, reviewing and commenting informally on the activities progressively conducted by the team. There must also be a team leader specifically trained in the investigative process.

The leader is a kind of project manager, who manages the planning, the budget and the final report. He or she selects team members and assigns them responsibilities and tasks. He or she is also the intermediary with national authorities, the press and the public, as well as the person responsible for coordination with legal representatives, should any consequences on this plan arise. Also, he or she secures the chain of custody of evidence, preserving potential evidence, especially during the early stages when emergency activities could (voluntarily or involuntarily) erase it. He or she writes the final report and assigns assignments to external experts when extra help is required.

It is crucial for the team leader to establish the terms of reference of the investigation. They usually relate to the objective of the investigation, the identification of team members, the purpose of the investigation, the methodology to be used, the priority, the proposed timeframe for delivering the final report, the estimated time and cost required, and the required depth of analysis of the main causes.

A possible checklist for the development of an investigation plan is given below:

- Setting priorities;
- Rescue activities and medical care;
- Protecting the site and preserving the evidence;
- Environmental issues;
- Collection of evidence;
- Planning interviews with witnesses;
- Reconstruction/restart of affected business processes;
- Team leader selection;
- Selection, training and organization of team members;
- Initial visit and photographic inspection;
- Planning the recognition, collection and organization of evidence;
- Establishing a communication protocol;
- Identifying the necessary equipment and tools and planning their procurement;
- Planning special or refresher training.

Before investigations begin, a group meeting is usually held to present team members and their skills, establish communication protocols, manage evidence secure the chain of custody, and assign tasks and responsibilities to team members.

Each member of the investigation team must have both hard and soft skills, be objective, scrupulous, use logical thinking, avoid jumping to premature conclusions, not be haughty and show empathy. Other team members may be involved in a part-time consultancy role, depending on particular needs, such as a chemist, structural engineer, equipment specialist, process control engineer, environmental engineer, human resources representative and other specialists. External investigators, who are free from management involvement, are perceived as more independent and therefore are sometimes preferable. It is not a question of objectivity, as an internal team member can be as objective as an external investigator; in fact, credibility is a matter of perception.

Table 26 shows what members of the investigation team should and should not do.

Table 26 Survey team members should and should not.

Members of the survey team	
...should	**......shouldn't**
Have an open mind, with an independent perspective; Work well as a team; Have analytical, writing and communication skills; Be experts on a particular topic of the investigation.	Have a priori opinions; Identify the causes before the investigation begins; Be emotionally involved; Have tasks or work priorities that conflict with the workload required by the investigation; Impose restrictions that are not compatible with the timing of the investigation.

Finally, it is necessary to schedule regular group meetings in order to update all team members on the availability of new information, share interim results, report on activities carried out and so on.

4.3.1 Fact-Finding

Fact-finding starts with evidence collection. The term “evidence” refers to all data, of different natures, that are collected at the site of the accident or are generically related to it, allowing the reconstruction of the dynamics of the event.

Some evidence may help instantly when collected; some may require more in-depth analysis before providing useful information. Evidence only becomes evidence when it is correctly placed within the general context. Typically, it is not necessary for all evidence collected to support a specific deduction. However, it is undoubtedly essential that none of the evidence supports the supported investigative hypothesis.

In order for an investigation to be defined as well done, special care must be taken to preserve the evidence to ensure that the data is not modified, contaminated or lost. To prevent this risk, which affects all types of evidence in different ways, it is necessary to ensure that the evidence is separated from any possible external contaminating source. Working in this way will make it easier to demonstrate and maintain the integrity of the investigation, thus validating any conclusions drawn from it. In this perspective, cleaning activities that are not part of first aid activities should only be allowed after gathering the physical evidence of interest, for which a secure chain of custody is also essential. The chain therefore includes all aspects of evidence management, such as:

- Collection;
- Custody;
- Control;
- Transfer;
- Analysis;
- Conservation.

If the chain of custody is interrupted, then the evidence may be considered unreliable, especially when operating in a litigation context.

To ensure a broad, comprehensive and rigorous collection, an important factor to consider is time. The priority immediately after an accident is obviously to provide first aid to people who may be injured, while at the same time restoring the safety of the site. It is clear that these operations can alter the scene, but one can only take note of this eventuality. Some evidence is particularly sensitive to the passage of time that can cause its loss or, more generally, alteration (think of a fading sign, a witness forgetting something, an interruption in the energy supply that causes the loss of data stored in the temporary memory of a machine, and so on). It can also happen that a witness is influenced by external or internal pressures within the company's perimeter, so his testimony can no longer be considered reliable. As a general principle, the more fragile the evidence is as time goes by, the higher the priority of its collection. Although it is not possible to provide a universal list, there is no doubt that paper evidence, electronic files, materials subject to rapid decomposition, and metal objects that can oxidize and lose relevance from a metallurgical point of view are sensitive to the passage of time. But there is another equally fragile datum: the position of things and people. In fact, almost every piece of evidence becomes significant only when information about its position is known, which is why it is always advisable to produce abundant video-photographic documentation before the evidence is collected and moved elsewhere.

In conclusion, the evidence collection phase can be summarized in three key words:

1) Speed, because data is usually sensitive to the flow of time;
2) Accuracy in creating the collection, in compliance with standards and laws to certify and guarantee evidence;
3) Traceability of everything that is collected, through a secure chain of custody.

It is understood that this is not a single-cycle process but iterative, since the analysis of evidence can generate new hypotheses that in turn require research and collection of new evidence.

The information available through the evidence is like pieces of a puzzle found scattered on the floor. This similarity provides an immediate understanding of how important it is to organize the evidence. Pieces of evidence are generally grouped according to a shared characteristic or type, or simply because of their position within the "picture," just as we do when we organize the pieces of the puzzle before and during the attempt to solve it. In fact, in the absence of an initial systematic organization of the evidence, the number of attempts to solve the puzzle (i.e. solve the investigation) increase significantly.

It is therefore suggested that we ignore those pieces of evidence that do not belong to the puzzle that we intend to solve, and work in groups to create islands of puzzle pieces that progressively grow and merge with each other, i.e. classify the evidence according to its relevance, the place of collection, and any other criteria that may help in an efficient organization, depending on the particular incident. When all the pieces of the puzzle are arranged, it is easier to guess the image of the whole picture even when there are still a few pieces left to arrange. The same is true for accident investigations, near-misses, and non-conformities: if one piece of evidence is missing, but all the others fit together perfectly, the investigator can equally enjoy the overall picture of the reconstruction of the accident (although from a forensic point of view the lack of a piece of evidence can change the consequences in terms of civil or criminal action).

4.3.2 Event Chaining

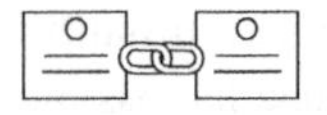

The collection of evidence and its punctual organization allows the salient events of the accident, near-accident and nonconformities under investigation to be identified. Conducting an incident analysis requires that such events (which in a Bow-Tie diagram can be overlapped with threats, top events or consequences) must be related to each other. There are two types of relationships: temporal and causal. The temporal relationship is concretized through the definition of a timeline, as already mentioned in Section 3.4.2. However, the cause-and-effect relationships do not always respect the chronological order of events, although it is far more frequent when the two relationships are superimposable, as shown in Figure 106. This is why it becomes important to study this relationship as well and this is what needs to be done at this stage. Indeed, in an incident investigation, it is crucial to establish the right time sequence of an incident to reconstruct the real dynamics in a proper chronology. To complete the discussion here, it is important to distinguish among the concepts of coincidence, correlation, and causation (Noon, 2009). It can happen that two events occur closely in time: it becomes fundamental to establish if their chronology reveals a cause-and-effect relationship or not. In other words, if event A occurred just before event B, can we conclude that A caused B? The answer is no, since an apparently ordered time sequence (i.e. a coincidence) does not automatically involve a cause-and-effect evidence. For instance, if you eat pizza the day before you sit an exam and pass it, this does not mean that the success you have is related to the pizza you ate. Coincidence is a random effect involving independent events. In the example, eating a pizza and passing an exam are events that occur independently. Very often, because of the improbability of the coincidence, people are prone to think that a cause-and-effect relationship must be present, because, according to them, the two events happening in the sequence are too improbable to manifest by coincidence. This argumentation is only good for sophistries, not for forensic engineers. Even according to the law of large numbers, coincidence may exists: extremely-low-probability events can occur if we consider a large number of possibilities to occur (i.e. if the set of events, whichever is their final state, is big enough). If something is not just a coincidence, then a causal link must be found between the two events: in other words, it must be demonstrated that the first event triggers, encourages, sets up, and causes the occurrence of the second event.

Therefore, the second concept needs to be introduced: correlation. A correlation exists when two events are linked with a demonstrable relationship. It implies repeatability of the chronological order, and provides a useful tool to test the time and event sequence. Correlation is the first step to indicate the existence of a direct cause-and-effect link between two events, but it may be not sufficient. Indeed correlation also exists when a common factor is shared between two events, regardless a direct relationship between them. For example, the increase in the number of car accidents is regularly followed by a similar trend in the collapse of agricultural outbuildings. Obviously this does not mean that the first event causes the second one. A correlation exists only because the events share a common factor, which is the snow: snow makes the road wetter and slippery, increasing car accidents, and increasing the structural loads on the agricultural outbuildings, which are not generally designed to endure extreme conditions.

It is clear that coincidence and correlation should not be confused with causation, which is what an investigator looks for. The Latin *post hoc ergo propter hoc* fallacy synthesizes this wrong approach to causation. Indeed, as discussed previously, it is not true that if B comes after A, then A causes B.

4.3.3 Identifying Barriers

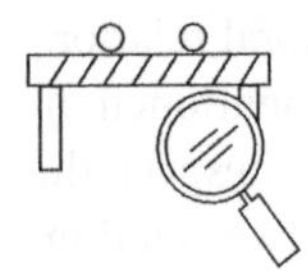

Once the events have been defined, next is the identification of the barriers (Figure 118), i.e. those control measures aimed at preventing or mitigating the event immediately following them. For the identification of barriers, the definition of their performance standards, the applicable taxonomy, the general rules to be followed for their correct application in a BFA diagram and to avoid the most common errors (Figure 119), it is necessary to refer in full to what has already been indicated in earlier sections (2.12, 2.14.2, and 4.1) of this book.

4.3.4 Assessing Barrier State

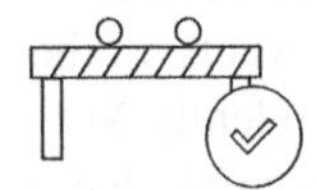

Once the evidence has been gathered, the investigator's task is to convert it into useful information. In fact, the investigator does not simply identify what was present or absent before the accident – the analysis is conducted to determine how the behaviour, conditions and latent system weaknesses contributed to the incident. Performing an analysis means dividing the incident into its individual events and then looking for those conditions that contributed first to each individual event, and then to the whole incident. To do this, basic assumptions must be made about what caused or contributed to the incident.

The analysis phase requires the achievement of two main objectives:

1) Validate what happened and how it happened;
2) Answer "why did it happen?".

The first objective involves a study to assess the plausibility of the hypotheses put forward, generated on the basis of the evidence gathered; the second requires the identification of the root causes.

The analysis is therefore at the heart of the investigative process: it lies between the research of evidence and the development of recommendations. Although distinct from the evidence collection phase, the analysis phase may overlap with the latter, due to the iterativity of the process, as already reported.

Thanks to the analysis of the evidence the investigator is able to give a structure to what he knows and what he does not know. There are no prescriptive rules, based on an informed judgement in conditions of uncertainty. For this reason, the analysis of the evidence requires a cross-check between the different data collected, which must not present incompatibilities or time inconsistencies.

In summary, there are two generic approaches: the deductive approach and the inductive approach. The deductive approach implies reasoning from general to particular while inductive reasoning goes from particular to general. The first is often used to chronologically retrace the sequence of events that led to the accident starting from its causes, while the second (speculative approach) is used to chronologically reconstruct the sequence of events in the opposite direction, i.e. moving from the accident and formulating hypotheses on possible logically predictable causes.

Therefore, there are different techniques of analysis of operational experience, including BFA, Tripod, RCA and others.

Their common objective is to identify:

- Events in time sequence;
- The cause-and-effect links;
- The actors of such events;
- The control measures present;
- The state of the control measures present at the time of the accident;
- The reasons the control barriers were in the identified state (often pushing the analysis to the three levels of immediate, latent, and root causes already discussed).

When analyzing the data, one should avoid so-called "fixation." People tend to see the world around them from a personal perspective, based on their own experiences and opinions. In simple terms, root cause analysis can be affected by the investigator's prejudice. This is a concept that goes beyond accident analysis alone. An example is the following: imagine being an engineer in the 1990s and having to design a lighting system for a spaceship to be used on the Moon. You probably turned immediately to standard bulbs (those with the tungsten bulb and filament) and you would end up engineering the solution to install them. However, in this process you have probably forgotten why bulbs have the bulb: to protect the tungsten from oxygen. But since there is oxygen on the Moon, the solution is probably far from optimal. To have considered bulbs as the "obvious" solution to provide light, based on daily experience, is an example of fixation.

Finding all the root causes is undoubtedly a great challenge: in fact, a common mistake is stopping investigations before all the causes are discovered. It is not unusual to work with investigators oriented towards finding "the" root cause. However, this approach can lead to poor investigation, where ineffective recommendations are developed. Indeed, incidents, especially complex ones, are rarely attributable to a single cause but are the result of multiple causes contributing in different ways to the occurrence and development of the event. Each primary cause is associated with a level of risk (combination of frequency and magnitude); from this point of view, it makes sense to say that some primary causes are more significant than others. But it is not possible to conclude that finding the most risky one is tantamount to solving the investigation of the accident: all root causes must be discovered, otherwise it becomes practically impossible to correct those weaknesses in the management system that allowed the accident to occur.

When assessing the state of each barrier, the icons in Figure 139 can be adapted to have a clearer view about them. The relation between barrier state and barrier life cycle is shown in the same figure. In order to determine the barrier state, the following logic should be followed:

- Was the barrier described in the company's management system, or was it considered an industry standard? If no, take action; if yes, ask the following:
- Was the barrier implemented, or could the barrier at one point perform according to its specification? If no, it is a "missing" barrier; if yes, ask the following:
- Did the barrier function according to its intended design (envelope)? If no, it is a "failed" barrier; if yes, ask the following:
- Did the barrier stop the next event in the incident sequence? If no, it is an "inadequate" barrier; if yes, ask the following:
- Are you confident the barrier will stop the next event in the incident sequence in the future? If no, it is an "unreliable" barrier; if yes, it is an "effective barrier."

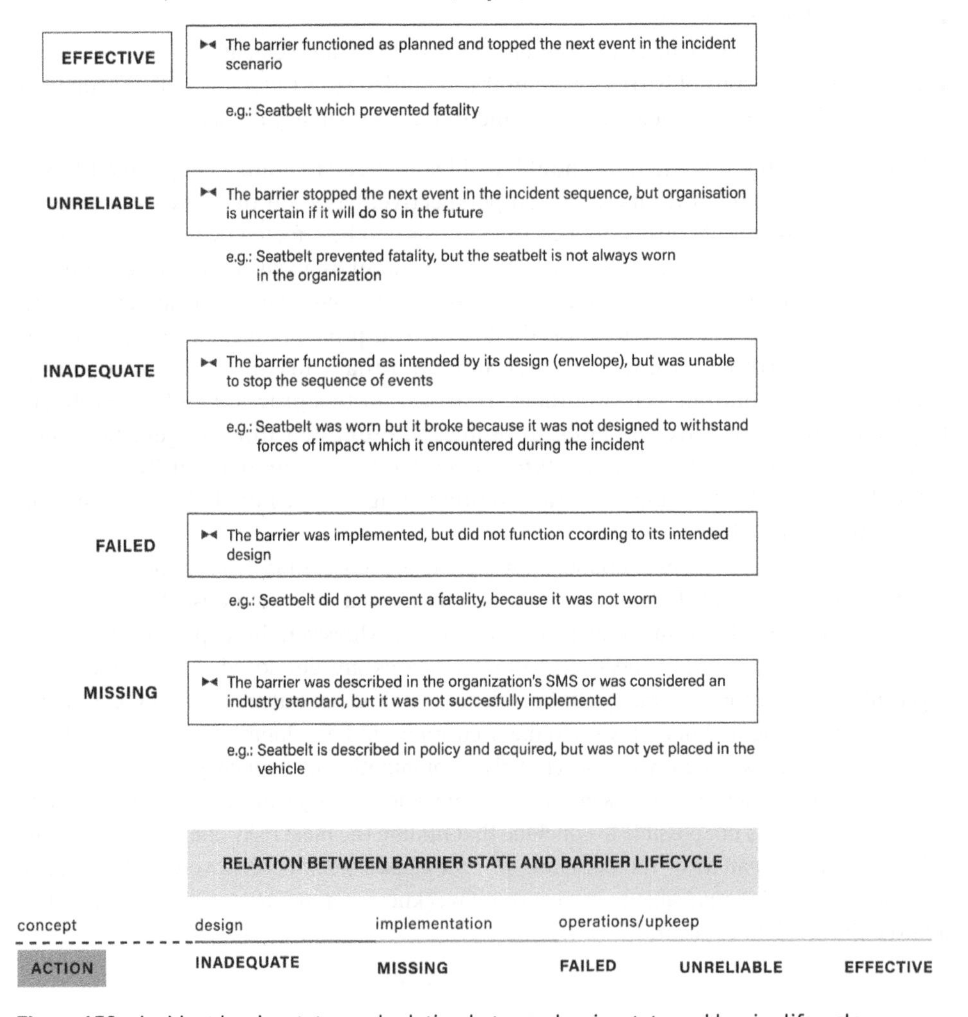

EFFECTIVE
The barrier functioned as planned and topped the next event in the incident scenario
e.g.: Seatbelt which prevented fatality
UNRELIABLE
The barrier stopped the next event in the incident sequence, but organisation is uncertain if it will do so in the future
e.g.: Seatbelt prevented fatality, but the seatbelt is not always worn in the organization
INADEQUATE
The barrier functioned as intended by its design (envelope), but was unable to stop the sequence of events
e.g.: Seatbelt was worn but it broke because it was not designed to withstand forces of impact which it encountered during the incident
FAILED
The barrier was implemented, but did not function ccording to its intended design
e.g.: Seatbelt did not prevent a fatality, because it was not worn
MISSING
The barrier was described in the organization's SMS or was considered an industry standard, but it was not succesfully implemented
e.g.: Seatbelt is described in policy and acquired, but was not yet placed in the vehicle
RELATION BETWEEN BARRIER STATE AND BARRIER LIFECYCLE
concept
design
implementation
operations/upkeep
ACTION
INADEQUATE
MISSING
FAILED
UNRELIABLE
EFFECTIVE

Figure 139 Incident barrier states and relation between barrier state and barrier lifecycle.

4.3.5 Causation Analysis and Categories

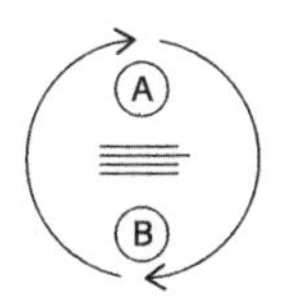

This step is the focus of the BFA analysis (see Figure 120). Its aim is to understand what caused the barrier to fail.

The analysis involves the identification of three levels of causes, in line with the most common theories in the field of accident analysis, including those of other methods such as Tripod Beta. For further details, see Sections 3.4 and 3.5 in full, where the topic has been already discussed in depth.

4.3.6 Recommendations

When an accident occurs, the highest price has already been paid. It is therefore essential to try to learn valuable lessons from this to improve the performance of the "organization" system.

The survey shows the areas of risk assessments that need to be improved: the investigative team, once the root causes are identified, develops those recommendations that can reduce the likelihood (or magnitude) of an accident. When recommendations are shared with designated stakeholders (usually management), responsibility for these actions is transferred from the investigative team to the management of the organization, which must evaluate, accept, reject, modify and implement the proposed actions. In fact, the responsibilities of the investigative team stop at the development of practical recommendations, subjecting them to management. It is therefore up to management to approve (or not) the recommendations, allocating the staff and economic resources needed for their implementation and following up on the resulting actions, to ensure that the measures are implemented as planned. It is clear that until these corrective actions are implemented, the risk profile will remain unchanged. Sometimes, immediate recommendations are developed even before the completion of the investigation, in order to immediately address those dangerous factors that can be mitigated in a very short time.

Turning the results of the investigation into recommendations to improve risk management is undoubtedly a real challenge. Recommendations are the most important product of accident investigations: they are developed only after the analysis and discovery phase of root causes. Corrective actions can be preventive or mitigating measures and have a different socio-technical impact. During this process, in-depth knowledge of the system under consideration is required in order to develop effective recommendations, involving stakeholders and also developing a communication strategy to share the lessons learned with them. Recommendations should be made to achieve the following objectives:

- Prevent the same or similar incidents from happening again;
- Mitigate the consequences if such an event were to happen in the future;
- Solve knowledge deficiencies uncovered by the investigation;
- Identify weaknesses in the system and, above all, in its interfaces (each combination between the technical, human and managerial sides of the system), which could be the weakest parts;
- Strengthen these weaknesses;
- Propose a rapid alert system.

It is generally recommended that a specific time limit to respond to a recommendation be established. It should not be described prescriptively; it's a good idea to let the technical details be defined by who will be responsible for implementing a recommendation. Of course, this is no longer true if the recommendations are identified by external authorities or authorities (such as the fire brigade, law enforcement, etc.).

In general, there are two strategies for making recommendations:

1) Restore the initial level of performance, which has deviated from the regulatory level, by addressing the "resiliency" of the system;
2) Address system deficiencies and enable change in the operating environment.

Similarities in these two approaches can be seen with two different strategies for structural design. Basically, there are two ways to ensure that a structure does not collapse: create more robust structures, with solid materials and thick geometry, or use more flexible and lightweight structures, leaving them free to deform in load conditions, without collapsing. The same approach applies to making recommendations.

The application of a recommendation plan should follow the following guidelines:

- Those responsible for the activities covered by the recommendations must take this into account and take appropriate corrective action.
- Before responding (accepting or rejecting) recommendations, the primary responsible party (PRP) must consider all relevant information to manage the risks involved.
- Responses to recommendations should be recorded. If the recommendation is rejected, a justification should be provided; if it is accepted, the relevant action plan should be attached.
- Actions should be tracked by their proposal until they are completed.
- Lessons learned should be stored in company memory, using a database for results and actions taken (in this regard it is good to remember that organizations have no memory; people do).
- Lessons learned should be shared with other organizations in the same sector.
- The lessons learned must not be possessed by individuals, but by the system; otherwise, they will be lost.
- Recommendations should be used proactively to improve hazard analyses and risk assessments. This can be done using a dedicated database, which should not be used passively, but actively, to develop continuous refinements.

Communication plays a key role. In the process industry, as well as in other industries, companies with excellent safety performance not only share lessons, but take actions to document and respond to these learning opportunities, identifying lessons and being aware of the value of sharing them with others; use an efficient system for sharing them; and incorporate lessons into business procedures and standards, checking whether changes to existing equipment, processes, and procedures are required. Sometimes, in order to carry out this practice, some companies analyze the causes and recommendations of past accidents relating to other companies operating in the same industrial sector.

As noted several times in this chapter, in order to develop effective recommendations, the lessons extracted from the learning from experience process must be based on root causes. If the lessons arise from immediate causes or preconditions, the actions developed will be ineffective, that is, unable to prevent future recurrences. From this point of view, recommendations can be developed on four different levels:

1) *Short term*. These are immediate corrective actions, generally related to immediate causes. For example, if an accident occurred because, as an immediate cause, a lock valve was left in a closed position when it should be opened, the short-term recommendation is to open and mark that valve, even in all other plants of the company.
2) *Medium-term*. This refers to those actions that take months to implement, which are addressed by the management of the structure, and which do not require a substantial change in company policy. For example, in the case of the incident previously used as an example, the facility manager may decide to conduct a hazard analysis to look for similar problems, ensure that contractors are aware of the latest procedures, provide them with formal training, and so on.
3) *Long-term*. Recommendations at this level are related to the root causes discovered. They relate to the system. In the example, assuming that root cause analysis revealed some weaknesses in communication between owner, contractors and subcontractors, the recommendation developed could be to evaluate and update the entire contractor management system.
4) *Industry-wide*. Major incidents can lead to the development of recommendations affecting the entire industry. For example, lessons learned from major oil and gas accidents (Bhopal, Seveso, Buncefield, *Deepwater Horizon* and others) have produced recommendations that the entire industry has enjoyed.

Evaluating recommendations is a key step in examining whether proposed actions can actually reduce the risk profile. In fact, some developed recommendations may actually create a new risky scenario, increase the existing level of risk, or are actually irrelevant to reduce the level of risk (in the authors' experience, the last option is not unusual when it comes to functional safety, since those who develop corrective actions may sometimes not be aware of the requirements required by international standards IEC-61508 and, for the process industry, IEC 61511). Therefore, risk assessment tools should be used to assess potential risks in implementing a given recommendation (for example, the use of nitrogen is a solution to create an inert atmosphere, but increases the risks associated with asphyxiation).

In the development of the recommendations, one should keep in mind the risk treatment options discussed in Section 2.5.1, preferring solutions that remove the source of risk to those that change its frequency and magnitude.

In drafting the recommendations, information on the benefits of implementation or the potential consequences of rejection should be included. Of course, this type of analysis is not carried out if the implementation of the suggested corrective action is required by law or other benchmarks, or when the cost of cost/benefit analysis exceeds that of implementing the corrective action itself.

In general, the person who will implement a recommendation is not the same person who wrote it: therefore, it is essential that the elements of the action are written clearly, without offering any interpretative doubt. The text of the recommendation must not be extended in order to avoid ambiguity and misunderstanding. As a general rule, the recommendations should include the reasons why they were developed. One possible format is: "To avoid X, then Y should be done."

When you want to give more flexibility when implementing a recommendation, you often use the terms "Evaluate" or "Consider if" and so on.

Sometimes the investigative team that writes the recommendations does not have all the elements to offer a full assessment. In this case, the recommendation may take the following format: "Confirms that the water mixture X is Y-soluble. If confirmed, take action No. 1; otherwise, take action No. 2." In the case of legal proceedings, the recommendations developed should be reviewed by the company's legal department in order to minimize exposures in the event of litigation.

The recommendations developed are often classified by risk, to guide management in prioritizing the proposed solutions. In other words, their mitigating effect is assessed, namely their ability to reduce the probability or magnitude of the incidental scenario.

Possible reasons for rejecting the implementation of a recommendation include the following:

- A detailed analysis shows that the implementation of the action is not as beneficial as initially thought.
- Further information, not originally available to the investigative team, reveals that the problem is not as critical as initially assumed.
- Something has changed, and the recommendation is no longer valid.
- One recommendation already implemented covers the objectives of another, which is no longer needed.
- The suggested recommendation is indeed beneficial, but not as much as required to mitigate the risk in the tolerable region.

In fact, an incident investigation is a reactive process, since it starts only after an unwanted event has occurred. However, the proposed effective recommendations, ranging from immediate corrective actions to management system reviews, allows the best lessons to be learned from the incident investigation, turning it into a proactive process. Technical and administrative controls are generally easy to develop. The real challenge is to convince management to make changes at the directional level: in fact, management generally recognizes the importance of corrective actions taken, but if they do not understand the benefits, their implementation becomes unlikely.

It is important to provide quality information when developing recommendations: quality in, quality out (QIQO). Otherwise, if not enough useful information is provided, you may run into the garbage in, garbage out (GIGO) principle. To do this, it is suggested that you answer the following six questions that help develop quality recommendations:

- What exactly is the problem?
- What's the story about the problem?
- What are the solutions that could correct the problem?
- Who's the decision-maker?
- What motivates the decision-maker?
- What will be the cost-benefit ratio of corrective actions and system improvements?

A good strategy is to provide the decision-maker with more alternatives, to increase the likelihood that he or she will choose at least one. For example, one option could be formulated independently of economic constraints, while a second option might consider limited funds.

Recommendations should be made to create measurable completion criteria: for example, it can be difficult to track the status of the "Provide a solution to mitigate risk" recommendation; instead, it is easy to determine whether the recommendation "Implement an interlock with a certified SIL2 to stop the compressor when the low-level alarm is activated in the tank" is completed or not, although such a rigid recommendation may appear more like a prescription, risking discouraging the implementation officer from following it.

The objectives of an action plan should be specific, measurable, agreed, realistic and timely (i.e. SMART, an acronym for the English translation of the adjectives).

Effective recommendations, therefore root cause agents, are able to eliminate the multiple causes of the system-related accident. In fact, the recommendation process consists of the steps shown in Figure 140.

In conclusion, it should be noted that sometimes no recommendation is developed. This is often the case with risks that are assessed through an ALARP study while remaining tolerable according to company policy.

To summarize, successful recommendations from the so-called learning process:

- Address the root causes;
- Are clearly expressed;
- Are practical, feasible and achievable;
- Add or apply a level of protection;
- Eliminate or reduce risks by acting on probability, consequences, or both.

4.3.7 Reporting

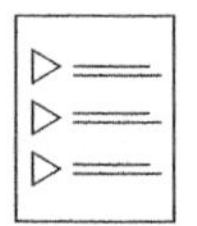

The results of the investigation should be presented in the form of a final report. To determine the format to be used, it is essential to know who will read the report, i.e. who the "public" is. For example:

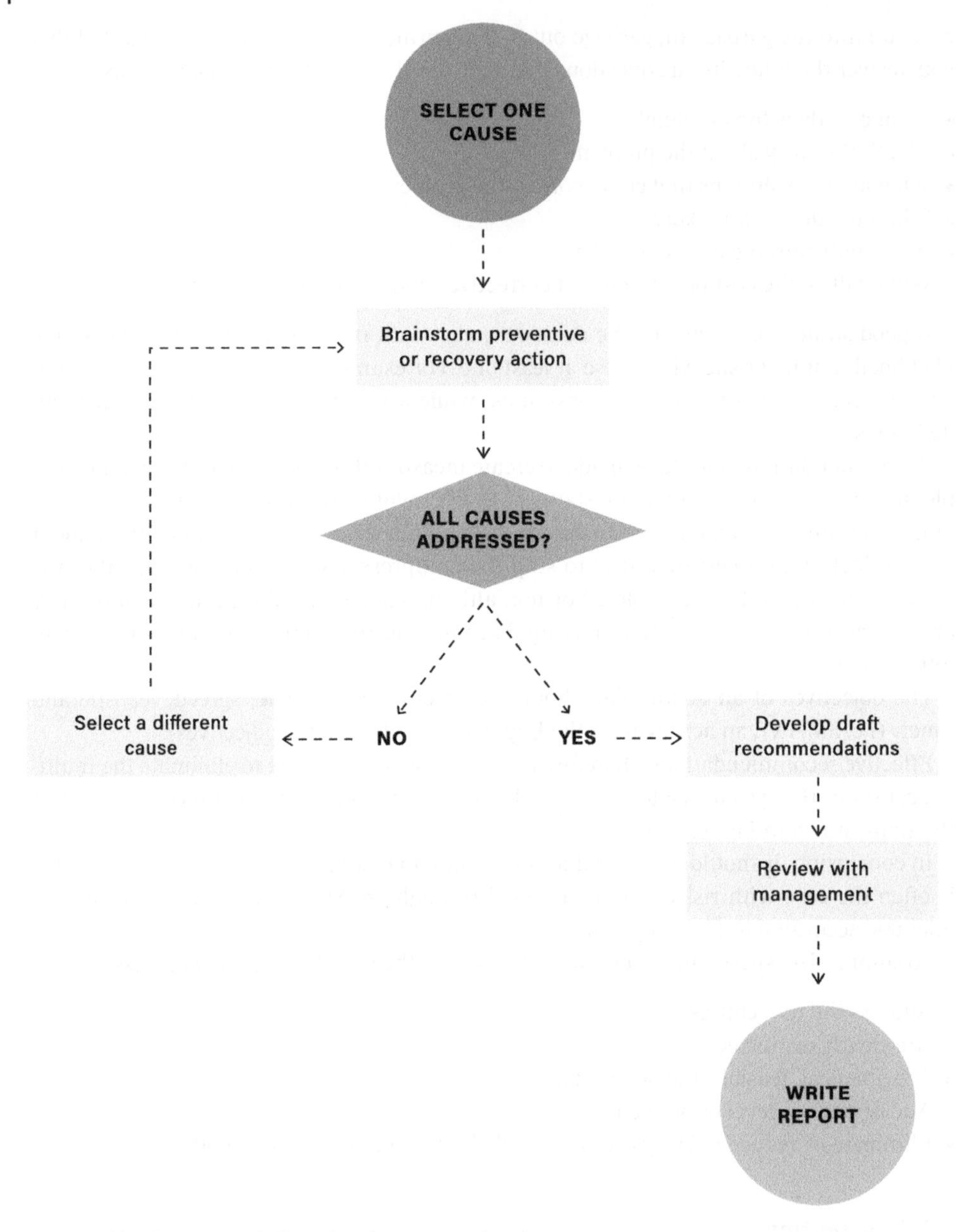

Figure 140 Recommendations development and review.

- Insurers interested in the report to determine whether or not the insurance company is jointly and severally liable for the payment of compensation.
- Law enforcement agencies, whose interest is to understand whether a crime or specific violations of the law have been committed.
- Lawyers, who will read the report word for word in order to gain legal advantage from possible interpretations of sentences or words.

- Technical experts, whose interest is in analyzing the technical causes identified in the report.
- The author himself, who is often asked to testify about the investigation several years after its conclusion, completely forgetting the content of the report.
- The judging authority, if the report is accepted as evidence in legal proceedings. Considering the limited technical knowledge of the judge, the report should be understandable, avoiding equations or statistical data as much as possible, if not strictly necessary.
- Management, interested in understanding what is wrong with the organisation, also suggesting corrective measures to prevent similar incidents. Depending on the level of maturity of the company's culture, the results of the report could lead to the dismissal of a person or could provide the basis for sound investment.

Writing a report, it is best to:

- Be brief, if possible;
- Stick to the facts;
- State whether an observation is an opinion or a fact that actually happened;
- Determine the causes, for each result;
- Pay particular attention to the formulation of recommendations;
- Avoid getting carried away with emotion (e.g. using superlatives);
- Avoid anticipating or mixing different conclusions when presenting results.

Objectivity and accuracy, pursued during the investigation phases, must also be transferred to the final report. In fact, poorly written reports may prove ineffective in preventing similar incidents from occurring.

4.4 Worked Examples

This section brings together some examples of the Bow-Tie approach applied to various business areas. It is the result of the kind cooperation of the following people, whom I personally thank for their availability and for having made an important contribution to this volume with their fully developed examples from their professional activity:

1) Prof. Chiaia from Politecnico di Torino and Prof. De Biagi from PoliTo for their Bow-Tie risk assessment on the structural damages to bridges;
2) CGE Risk and Ian Travers from Ian Travers Limited for the Bow-Tie risk assessment on the COVID-19 infection;
3) Jasper Smit from Slice for the Bow-Tie assessment about fire in flight;
4) Jasper Smit from Slice for the BFA about food contamination;
5) Tor Inge Saetre from AdeptSolutions for his Bow-Tie on web-based software development;
6) Tor Inge Saetre from AdeptSolutions for his Bow-Tie on IT operations;
7) Salvatore Tafaro from the Italian National Fire Brigade for his Bow-Tie assessment on crowd management in public events in open spaces;
8) Emily Harbottle from Harbottle Hughes Risk Management for her Bow-Tie about military helicopter operations;

9) Ed Janseen and CGE Risk for their patient safety Bow-Ties;
10) Arthur Groot, Process Safety Expert at Royal Haskoning DHV (NL), for his Bow-Tie about managing risks and compliance in the process safety field;
11) David Hatch, process safey expert, for his BFA diagrams on some of most famous incidents in the process industry;
12) Orazio Cassiani, risk manager in the healthcare sector, for his Bow-Ties about drug administration;
13) Prof. Luca Marmo from Politecnico di Torino for his reconstruction of the ThyssenKrupp fire investigation;
14) Paul Heimplaetzer for his Tripod Beta analysis of the Twente stadium roof collapse;
15) Annalisa Contos from Atom Consulting for her Bow-Tie about water treatment;
16) Rosario Sicari from MFCforensic for his BFA about a deadly explosion, here anonymized.

Worked examples have been selected from different industry sectors and they range from Bow-Tie risk assessment diagrams to near-misses and real accident investigations. Examples are, in most cases, reduced versions of the original ones assessing real cases, but it is quite evident that they all share a barrier-based approach and that this method allows the assessment of a variety of situations (with their risks coming from threats, hazards and consequences) from various domains. Given the principles underlined in this book the reader is able to understand the message of the diagrams even from a different domain with limited knowledge on a single case. Powerful notation underlines main elements and facilitates the identification of the relationships among them.

4.4.1 Local Reduction of the Resisting Capacity of a Bridge due to Ageing – Bow-Tie Risk Assessment

Bridges are fundamental parts of a transportation network. Briefly, they consist of horizontal elements, namely the girders, supporting the deck and supported by vertical members, i.e., the piers. Ageing is a natural process that results in the degradation of concrete and represents a hazard since a large part of our transport infrastructural heritage dates back to the second half of the twentieth century, i.e., it has approximately reached the end of its expected working life. This hazard implies the reduction of the resisting capacity of the structural members, which can fail. The local damage is the top event, whose consequences can entail the total collapse of the structure. Multiple causes of local damage can be highlighted: on one side there are the aspects related to the design, the construction materials and quality; on the other, all the loading scenarios and actions that occurred throughout the life of the infrastructure, traffic loads that increased, environmental actions, and accidental events like earthquakes or settlements that may have damaged the structure have to be remembered. The water on the bridge deck and its discharge play a major role in the speed of material degradation, thus attention must be paid to avoid water stagnation through correct road lateral slope, maintained road joints and appropriate downpipes. Action of downpipes can be seen in Figure 141A where the reader can notice the wet concrete, while in Figure 141B it can be observed that the concrete is in good condition due to the safe position of the downpipe. To prevent local damage due to ageing and humidity

Figure 141 On the left: pier with a damaged downpipe; the concrete is wet and deteriorated. On the right: a similar pier with a safe downpipe; the concrete is in good condition.

(Figure 142) and spalling due to corrosion (Figure 143 and Figure 144), various actions can be undertaken: first, periodic visual inspection on the structure is an effective task to be implemented for assessing the presence of defects. Further analyses encompass the structural safety assessment through calculations. The engineer must consider that the loads on the bridge might have varied throughout the years and reduction of cross-section sizes due to adverse environmental conditions may have occurred. Maintenance and monitoring are the lasting prevention measures to avoid local damage.

As soon as local damage occurs, propagation to the whole bridge through a progressive collapse mechanism can occur if the scheme is not robust and redundancy is not present.

Figure 142 Effects of ageing and humidity on the concrete. The reinforcement bars are corroded and there are signs of rust on the beams.

Figure 143 Concrete spalling on a Gerber support with a consequent capacity reduction. The cause of the damage has to be searched for on a damaged downpipe on the road joint (recently substituted).

Figure 144 The spalling of concrete caused the corrosion to progress. The reinforcement bars broken due to the limited cross-section are causing a reduction of the capacity of the girder.

Structural robustness can stop the propagation mechanism in such a way that the damage stays local. The choice on whether or not to repair the damage depends on various aspects: the possibility of downgrading the infrastructure (traffic limitations for cars and light trucks, only) or the availability of alternative and effective roads for rerouting the traffic. Although the former point depends on the structural characteristics of the bridge, in the latter the resilience of the transportation network is involved. Bow-Tie (Figure 145) could become a proper method to describe threats affecting the resisting capacity along with the consequences and the prevention and protection measures.

4.4.2 COVID-19 infection – Bow-Tie Risk Assessment

The pandemic spread of COVID-19 has imposed major changes in the lives of people and organizations. Figure 146 shows a Bow-Tie diagram relating to the management of the risk of virus contamination.

4.4.3 Fire in Flight – Bow-Tie Risk Assessment

The Bow-Tie shown in Figure 147 is the result of a short study on fire that can occur during flight. As specified in the hazard/top event, the diagram focuses on fire that originates in the cabin where the passengers and their luggage reside. It should be noticed that this excludes fire related to the engines, equipment and avionics outside the cabin area.

4.4.4 Food Contamination – Barrier Failure Analysis

The following BFA incident analysis diagram (Figure 148) was created to analyze a near-miss that occurred in a food processing plant. The diagram is based on actual events, although it has been altered for demonstration purposes.

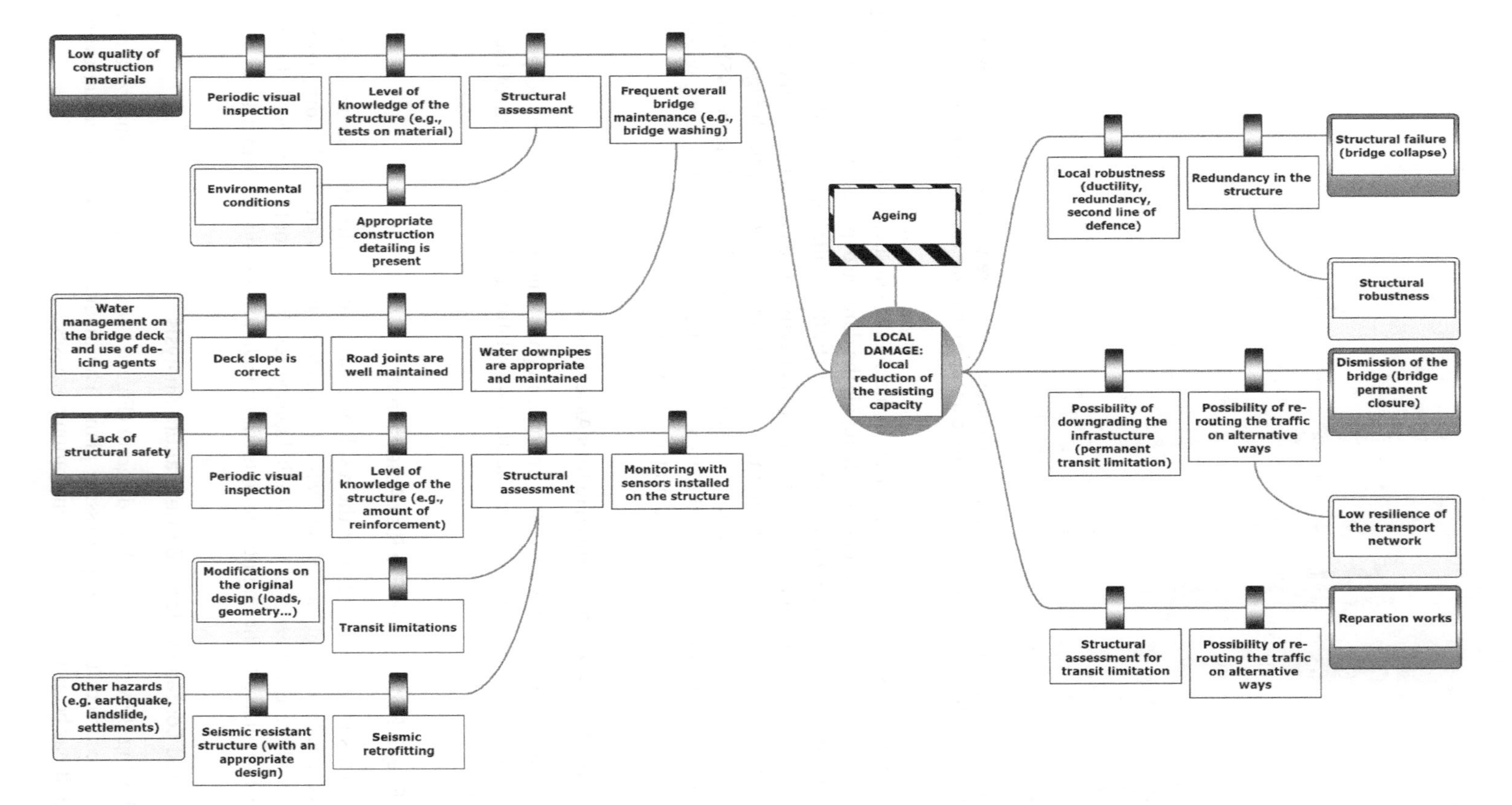

Figure 145 Bow-Tie diagram for "Local reduction of the resisting capacity of a bridge due to ageing".

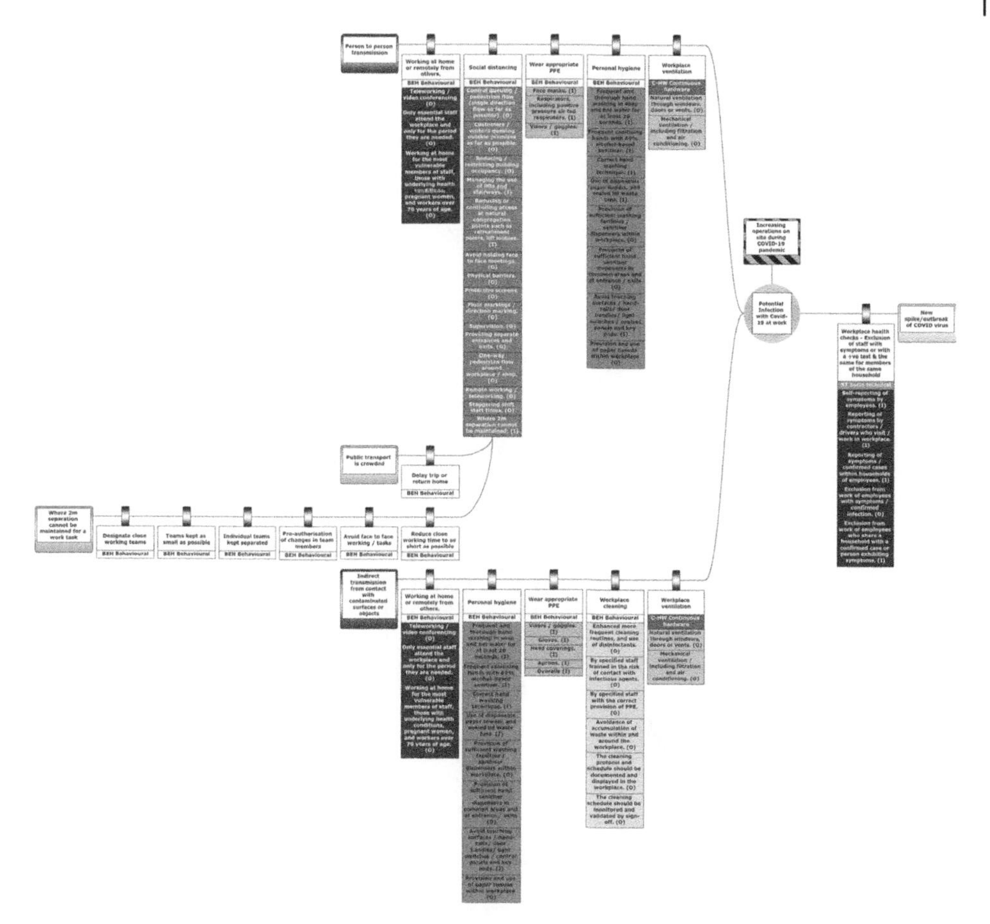

Figure 146 Employee infected with COVID-19 virus.

The diagram describes an unfortunate course of events that resulted in a restaurant guest feeling what turned out to be glass particles between her teeth. Even though no actual injury was sustained, it was a serious breach of hygienic standards and it devastated the reputation of the food processing company. The glass was traced back to originate from a broken light bulb that was positioned above a mixing machine that was used in the production process.

The diagram describes how the glass remained undetected until it reached the customer. Additionally, the effective barrier at the end demonstrates that further harm was avoided through an effective recall procedure.

4.4.5 Web-Based Software Development – Bow-Tie Risk Assessment

Following, some key elements of the Bow-Tie are described. This Bow-Tie, shown in Figure 149, is inspired by PCI DSS, a standard for the payment card industry.

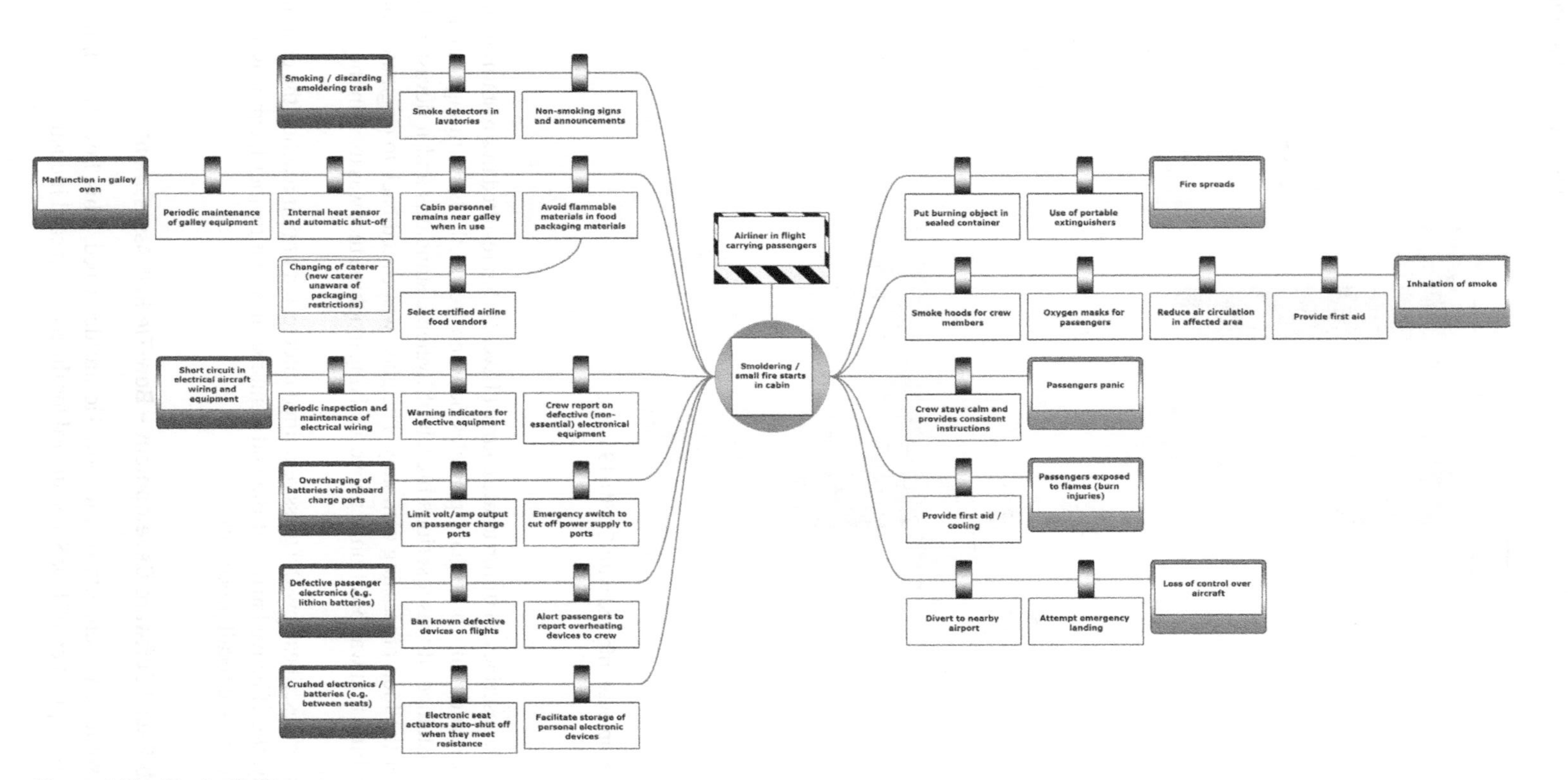

Figure 147 Fire in flight.

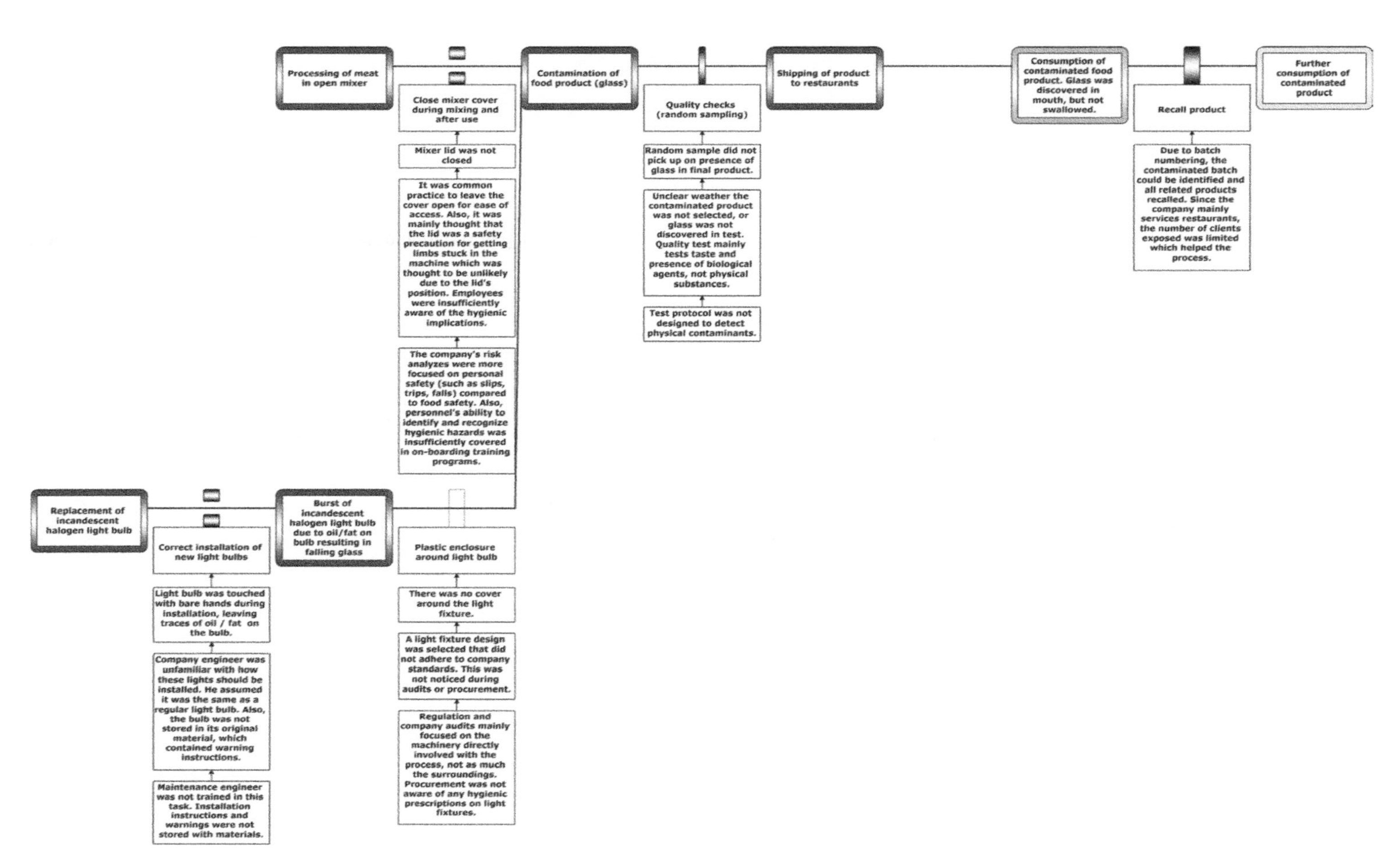

Figure 148 BFA on food contamination (near miss).

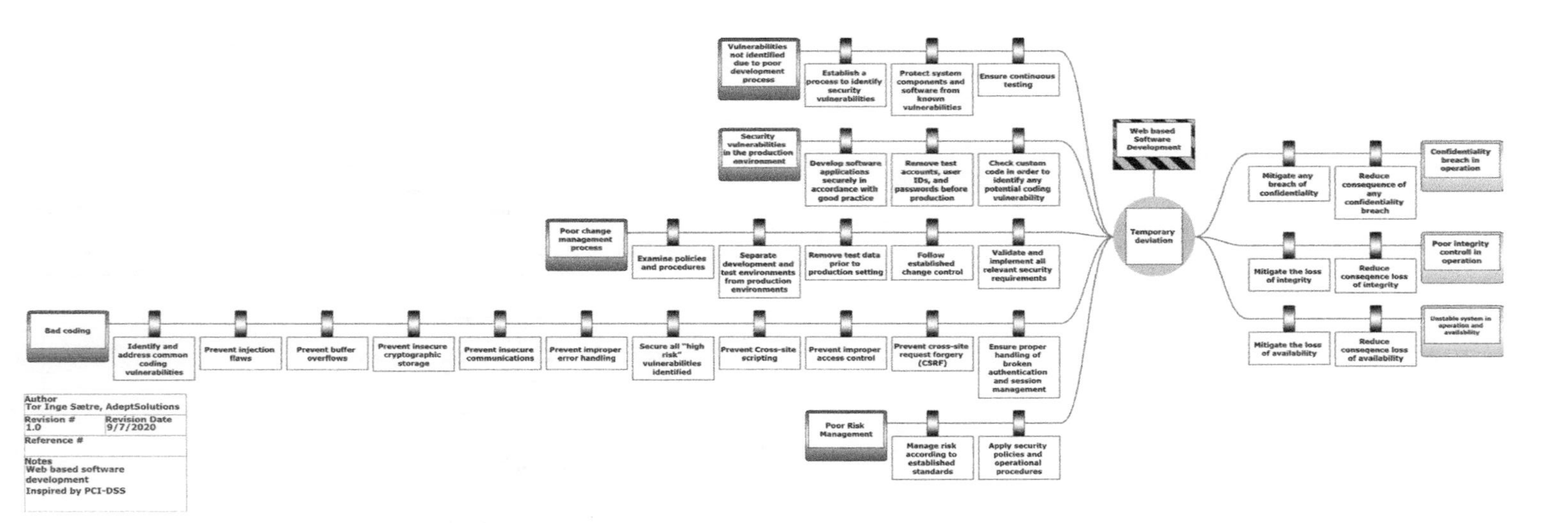

Figure 149 Web-based software development – Bow-Tie.

Threats

Security vulnerabilities in the production environment

Without the inclusion of security during the requirements definition, design, analysis, and testing phases of software development, security vulnerabilities can be inadvertently or maliciously introduced into the production environment.

Understanding how sensitive data is handled by the application—including when stored, transmitted, and when in memory—can help identify where data needs to be protected.

Poor change management process

Without properly documented and implemented change controls, security features could be inadvertently or deliberately omitted or rendered inoperable, processing irregularities could occur, or malicious code could be introduced.

Due to the constantly changing state of development and test environments, they tend to be less secure than the production environment. Without adequate separation between environments, it may be possible for the production environment, and cardholder data, to be compromised due to less-stringent security configurations and possible vulnerabilities in a test or development environment.

Bad coding

Application developers should be properly trained to identify and resolve issues related to these (and other) common coding vulnerabilities. Having staff be knowledgeable of secure coding guidelines should minimize the number of security vulnerabilities introduced through poor coding practices. Training for developers may be provided in-house or by third parties and should be applicable for technology used.

Preventive Barriers

Establish a process to identify security vulnerabilities

Establish a process to identify security vulnerabilities, using reputable outside sources for security vulnerability information, and assign a risk ranking (for example, "high," "medium," or "low") to newly discovered security vulnerabilities.

Note: Risk rankings should be based on industry best practices as well as consideration of potential impact. For example, criteria for ranking vulnerabilities may include consideration of the Common Vulnerability Scoring System base score, the classification by the vendor, and/or the type of systems affected.

Methods for evaluating vulnerabilities and assigning risk ratings will vary based on an organization's environment and risk-assessment strategy. Risk rankings should, at a minimum, identify all vulnerabilities considered to be a "high risk" to the environment. In addition to the risk ranking, vulnerabilities may be considered "critical" if they pose an imminent threat to the environment, impact critical systems, and/or would result in a potential compromise if not addressed. Examples of critical systems may include security systems, public-facing devices and systems, databases, and other systems that store, process, or transmit data.

Ensure continuous testing

Examine policies and procedures to verify that processes are defined for the following:

- To identify new security vulnerabilities.
- To assign a risk ranking to vulnerabilities that includes identification of all "high risk" and "critical" vulnerabilities.
- To use reputable outside sources for security vulnerability information.

Interview responsible personnel and observe processes to verify that:

- New security vulnerabilities are identified.
- A risk ranking is assigned to vulnerabilities that includes identification of all "high risk" and "critical" vulnerabilities.
- Processes to identify new security vulnerabilities include using reputable outside sources for security vulnerability information.

Develop software applications securely in accordance with good practice

- In accordance with security standards (for example, secure authentication and logging).
- Based on industry standards and/or best practices.
- Incorporating information security throughout the software-development life cycle.

Examine written software-development processes to verify that the processes are based on industry standards and/or best practices.

Without the inclusion of security during the requirements definition, design, analysis, and testing phases of software development, security vulnerabilities can be inadvertently or maliciously introduced into the production environment.

Understanding how sensitive data is handled by the application – including when stored, transmitted, and when in memory – can help identify where data needs to be protected.

Examine written software-development processes to verify that information security is included throughout the life cycle.

Examine written software-development processes to verify that software applications are developed in accordance with security standards.

Interview software developers to verify that written software-development processes are implemented.

Check custom code in order to identify any potential coding vulnerability

Security vulnerabilities in custom code are commonly exploited by malicious individuals to gain access to a network and compromise data.

An individual who is knowledgeable and experienced in code-review techniques should be involved in the review process. Code reviews should be performed by someone other than the developer of the code to allow for an independent, objective review. Automated tools or processes may also be used in lieu of manual reviews, but keep in mind that it may be difficult or even impossible for an automated tool to identify some coding issues.

Correcting coding errors before the code is deployed into a production environment or released to customers prevents the code exposing the environments to potential exploit. Faulty code is also far more difficult and expensive to address after it has been deployed or released into production environments.

Including a formal review and signoff by management prior to release helps to ensure that code is approved and has been developed in accordance with policies and procedures.

Examine policies and procedures

Examine policies and procedures to verify that the following are defined:

- Development/test environments are separate from production environments with access control in place to enforce separation.
- A separation of duties between personnel assigned to the development/test environments and those assigned to the production environment.

- Production data are not used for testing or development.
- Test data and accounts are removed before a production system becomes active.
- Change control procedures related to implementing security patches and software modifications are documented.

Separate development and test environments from production environments

Examine network documentation and network device configurations to verify that the development/test environments are separate from the production environment(s).

Examine access controls settings to verify that access controls are in place to enforce separation between the development/test environments and the production environment(s).

Remove test data prior to production setting

Test data and accounts should be removed before the system component becomes active (in production), since these items may give away information about the functioning of the application or system. Possession of such information could facilitate compromise of the system and related data.

Follow established change control

If not properly managed, the impact of system changes—such as hardware or software updates and installation of security patches—might not be fully realized and could have unintended consequences.

Validate and implement all relevant security requirements

For a sample of significant changes, examine change records, interview personnel, and observe the affected systems/networks to verify that applicable security requirements were implemented and documentation updated as part of the change.

Building this validation control into change management processes helps ensure that device inventories and configuration standards are kept up to date and that security controls are applied where needed.

Identify and address common coding vulnerabilities

Actions: Examine development policies and procedures to verify that up-to-date training in secure coding techniques is required for developers, based on industry best practices.

Examine records of training to verify that software developers receive up-to-date training on secure coding techniques, including how to avoid common coding vulnerabilities.

Prevent injection flaws

Injection flaws include SQL injection, OS Command Injection, LDAP and XPath injection flaws as well as other injection flaws.

Actions: Examine software-development policies and procedures and interview responsible personnel to verify that injection flaws are addressed by coding techniques that include:

- Validating input to verify that user data cannot modify the meaning of commands and queries.
- Utilizing parameterized queries.

Prevent buffer overflows

Buffer overflows occur when an application does not have appropriate bounds checking on its buffer space. This can cause the information in the buffer to be pushed out of the buffer's memory space and into executable memory space. When this occurs, the attacker has the ability to insert malicious code at the end of the buffer and then push that malicious code into executable memory space by overflowing the buffer. The malicious code is then executed and often enables the attacker remote access to the application and/or infected system.

Actions: Examine software-development policies and procedures and interview responsible personnel to verify that buffer overflows are addressed by coding techniques that include:

- Validating buffer boundaries.
- Truncating input strings.

Prevent insecure cryptographic storage

Actions: Examine development policies and procedures and interview those who are accountable/responsible to verify that insecure cryptographic storage is addressed by coding techniques that:

- Prevent cryptographic flaws.
- Use strong cryptographic algorithms and keys.

Prevent insecure communications

Applications that fail to adequately encrypt network traffic using strong cryptography are at increased risk of being compromised and exposing data. If an attacker is able to exploit weak cryptographic processes, they may be able to gain control of an application or even gain clear-text access to encrypted data.

Actions: Examine software-development policies and procedures and interview responsible personnel to verify that insecure communications are addressed by coding techniques that properly authenticate and encrypt all sensitive communications.

Prevent improper error handling

Actions: Examine software-development policies and procedures and interview responsible personnel to verify that improper error handling is addressed by coding techniques that do not leak information via error messages.

Secure all "high risk" vulnerabilities identified

Actions: Examine software-development policies and procedures and interview responsible personnel to verify that coding techniques address any "high risk" vulnerabilities that could affect the application.

Prevent cross-site scripting

Actions: Examine software-development policies and procedures and interview responsible personnel to verify that cross-site scripting is addressed by coding techniques that include:

- Validating all parameters before inclusion.
- Utilizing context-sensitive escaping.

Prevent improper access control

Actions: Apply coding techniques that include:

- Proper authentication of users.
- Sanitizing input.
- Not exposing internal object references to users.
- User interfaces that do not permit access to unauthorized functions.

Prevent cross-site request forgery (CSRF)

A CSRF attack forces a logged-on victim's browser to send a pre-authenticated request to a vulnerable web application, which then enables the attacker to perform any state-changing operations the victim is authorized to perform (such as updating account details, making purchases, or even authenticating to the application).

Actions: Examine software development policies and procedures and interview responsible personnel to verify that CSRF is addressed by coding techniques that ensure that applications do not rely on authorization credentials and tokens automatically submitted by browsers.

Ensure proper handling of broken authentication and session management

Secure authentication and session management prevents unauthorized individuals from compromising legitimate account credentials, keys, or session tokens that would otherwise enable the intruder to assume the identity of an authorized user.

Actions: Examine development policies and procedures and interview responsible personnel to verify that broken authentication and session management are addressed via coding techniques that commonly include:

- Flagging session tokens (for example cookies) as "secure."
- Not exposing session IDs in the URL.
- Incorporating appropriate time-outs and rotation of session IDs after a successful login.

Manage risk according to established standards

For public-facing web applications, address new threats and vulnerabilities on an ongoing basis and ensure that these applications are protected against known attacks by reviewing these applications via manual or automated application vulnerability security assessment tools or methods, at least annually and after any changes.

Apply security policies and operational procedures

Personnel need to be aware of and follow security policies and operational procedures to ensure that systems and applications are securely developed and protected from vulnerabilities on a continuous basis.

Actions: Examine documentation and interview personnel to verify that security policies and operational procedures for developing and maintaining secure systems and applications are:

- Documented
- In use, and.
- Known to all affected parties.

4.4.6 IT Operations – Bow-Tie Risk Assessment

Following, some key elements of the Bow-Tie in Figure 150 are described.

Threats

Threat (generic)

According to the Norwegian biannual *Mørketallsundersøkelsen*, around half of all information thefts may be attributed to employees or subcontractors. While this number may seem a tad high in the age of the data breach, statistically all large organizations may have a small percentage of ethically challenged individuals.

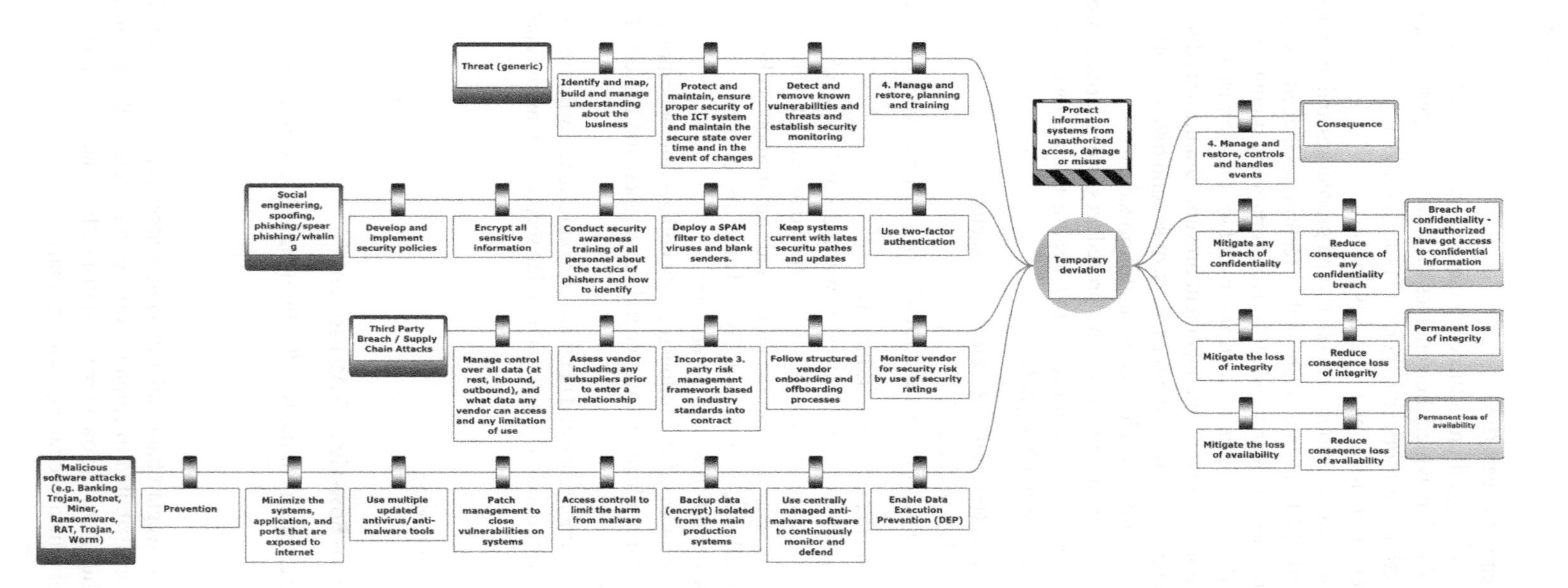

Figure 150 IT systems protection Bow-Tie.

Preventive Barriers

Identify and map, build and manage understanding about the business

The goal: The company identifies structures and processes for security and risk management to manage the work of securing the ICT systems. The company maps deliveries, information systems and supporting functions and assesses them against established tolerance limits for risk of establishing and adjusting security measures.

Lack of management structures and processes for risk assessment can lead to management not receiving sufficient information to prioritize and manage the company's safety work.

The company's information systems shall support the company's activities and deliveries so that these are carried out in accordance with the agreed quality. The company must identify, prioritize and protect its most important deliveries. In the absence of an overview, some less important parts of the ICT system can be well secured, while other vital parts are exposed and vulnerable to attack. The availability, integrity and confidentiality of information systems and data must all be considered in the company's security work.

Protect and maintain, ensure proper security of the ICT system and maintain the secure state over time and in the event of changes

The goal: Security is an integral part of the processes for procurement and development and the business minimizes the risk of new ICT products and ICT services introducing configuration and architectural vulnerabilities.

ICT security is important in all ICT products and ICT services, not just when procuring pure security products such as a firewall. If a company acquires ICT products and ICT services that have weak security or that do not integrate well with the company's other security architecture and existing ICT products, this can increase vulnerability and reduce the security level in the ICT system.

If the company lacks good processes for development, testing, verification and implementation, the probability will be high that the vulnerabilities will not be discovered. The cost of correcting this afterwards is often higher than the cost of good preparation.

Detect and remove known vulnerabilities and threats and establish security monitoring

Malicious code is a dangerous part of cyber threats and can be designed to affect systems, devices and data. Even the best products have flaws and vulnerabilities that can be exploited by attackers.

Malicious software can move fast, change as needed, and access end-user equipment, e-mail attachments, websites, cloud services, and removable media. The software is often activated by tricking users into taking action (opening, running, installing, etc.).

Modern malware can be developed to avoid certain security measures, or to attack or deactivate the measures. A business should have an overview of known vulnerabilities and known threats (such as malware) and protect itself against this.

Manage and restore, planning and training

The goal: The company must plan and implement efficient processes for incident handling so that incidents are detected quickly, controlled, the damage is minimized, and the cause of the incident is effectively removed. This includes restoring the integrity of systems and networks.

Recovery Barriers

Manage and restore, controls and handles events

Follow established procedures based on the classification of the incident and involve personnel with specialist expertise in the ICT system and day-to-day operations. Regardless of the type of incident, there will be some common principles that will apply. Examine the prevalence and potential for damage. Incident data and the situation picture should be kept as up to date as possible throughout the handling. One must be prepared for possible escalation and new, simultaneous events and ensure a good flow of communication between the parties involved.

Reactive measures should be initiated as soon as there is enough information. All activities and decisions should be logged so that the course of events can be subsequently addressed. Internal and external parties affected by the incident should, as far as possible, be kept up to date on the situation. When an incident has occurred, trust in services, systems and ICT infrastructure will naturally weaken. Effective event management will help restore this.

4.4.7 Crowding Bow-Tie Risk Assessment

Managing the crowd at large events in public spaces is a difficult task. Prescriptive approaches are not suitable while partial performance and risk-based regulations (at the national and local levels) revealed to be ineffective. Tragic recent events have demonstrated this (e.g. the Piazza San Carlo crowd incident in 2017, Turin – Italy). The reason of their ineffectiveness relies on the peculiarities of both each large event (a football match, a concert, a festival) and the space in which they are organized (a specific stadium, an old town centre, a public park), that may nullify the benefits of a rigid approach using checklists, even if yet performance-based.

Taking inspiration from the safety needs of the events for Matera 2019 (European Capital of Culture for 2019), the necessity to integrate art and safety when managing large events in public spaces also emerged during a dedicated conference held in Matera November 30, 2018. To satisfy this necessity means making the artistic and cultural exhibits, exhibited to large crowds during public events, safely accessible and exploitable, protecting both the people and itself from the risks of being hurt (the former) or damaged (the latter) in case of a loss of control on the crowd management.

Methodology

It is possible to highlight four key elements for safety and security of large events in public spaces:

1) Sharing information and accountabilities with stakeholders;
2) Risk assessment as the basis for defining the management strategy;
3) Standards and laws;
4) Usage of modern tools to verify if performance targets are reached, taking into account some key parameters when assessing complex situations (like the number of participants, their features, the layout and space configuration).

Of course, there is a need to identify the credible scenarios, representative and conservative, of what can go wrong when managing such complex and large events. This identification has to be formalized before the tools for advanced simulations are used, to define the scope and the "perimeter" of what needs to be analyzed and which are the safety targets to reach.

Moreover, there is a preliminary fundamental step before launching a software simulation: it's the knowledge of human behaviour, intended both as individuals and the crowd. The word "crowd" hides an intrinsic undetermination that is the new challenge for the risk analyst. Such problems, at both micro and macro levels, are the base of the emerging complexity and the need to use a shared, modern and pragmatic approach to analyze them from an engineering point of view.

Of course, Matera is a really good example of where to apply the proposed approach, because of its very complex layout, as shown in Figure 151.

Results

The first Italian regulations (2017), and their subsequent modifications about this topic, are addressed to:

- Evaluate the maximum crowding to avoid the loss of control of safety conditions and access monitoring and control (in one word, to avoid the overcrowding);
- Indicate the access and exit gates, with separate routes;
- Write an emergency and evacuation plan, indicating the evacuation routes;
- Divide the event area into sectors, leaving free spaces for emergency and medical aid;
- Use trained operators;
- Install a loudspeaker system for emergency communications;
- Evaluate the possibility of banning alcohol consumption during the event.

Figure 151 Satellite view of Matera.

Figure 152 Matera – Piazza Vittorio Veneto. On the right: steps. *Source:* Google LLC.

The definition of the access and outflowing routes can be challenging in a context like that of Matera. Let's take the example of Piazza Vittorio Veneto (Figure 152).

The walking surface of Piazza Vittorio Veneto is not uniform, because of the presence of some steps, like the ones shown in Figure 152. The square layout, with its lack of uniformity, might increase the likelihood of falling that, in the presence of high compressive forces due to the crowding, might cause a domino effect in the falling of several visitors. Moreover, the steps are themselves an obstacle to the evacuation of those people with physical disabilities; a compensatory measure could be, for instance, the usage of proper structures to break down those architectural barriers.

Communication is also extremely important to deliver a fast emergency response. In this sense, positioning, dimensions and design of the signage are critical factors that need to be properly evaluated. To achieve this goal, it is also fundamental to take into account the typology of visitors: age, cultural level, language, and vulnerabilities are key ingredients to determine the right communication strategy.

The Risk Assessment Phase Using the Bow-Tie method

The core parts of the proposed methodology are the in-depth risk assessment and the advanced crowd simulation. In particular, risk assessment has a fundamental role as a structured and systematic approach to identify the hazard, and analyze and evaluate the related risks. Even if it is a well-known workflow, structured according to ISO 31000, it is not very developed in crowding science. The first step is to identify the hazard of a large event in public spaces, to be then analyzed in the following phase of risk assessment.

Thanks to literature studies, former incidents analysis, and brainstorming, it is easy to identify that one of the principal causes of death in this context is asphyxia. It is related to chest crushing by the crowd, and it focuses the attention on the priority problem when managing large events: avoid the condition of overcrowding.

The "level of service" can summarize this condition: it is a quantitative measure of the concentration of people per unit of surface (pp/m^2). Establishing an acceptable level of this parameter is, therefore, the basic step to go deeper into the risk analysis. Using the "What-if" method, it is possible to analyze both the preventive and the recovery measures, taking into account the presence of hard and soft obstacles, different environmental layouts, procedures about flow management, distribution of visitors, and others. However, the Bow-Tie is one of the best methods to fully analyze the risks coming from an uncontrolled exodus in an immediately understandable way. This technique allows the scenario identification, taking into account both the preventive and mitigative measures, approaching the problem from a barrier-based perspective. A possible Bow-Tie is shown in Figure 153 and Figure 154.

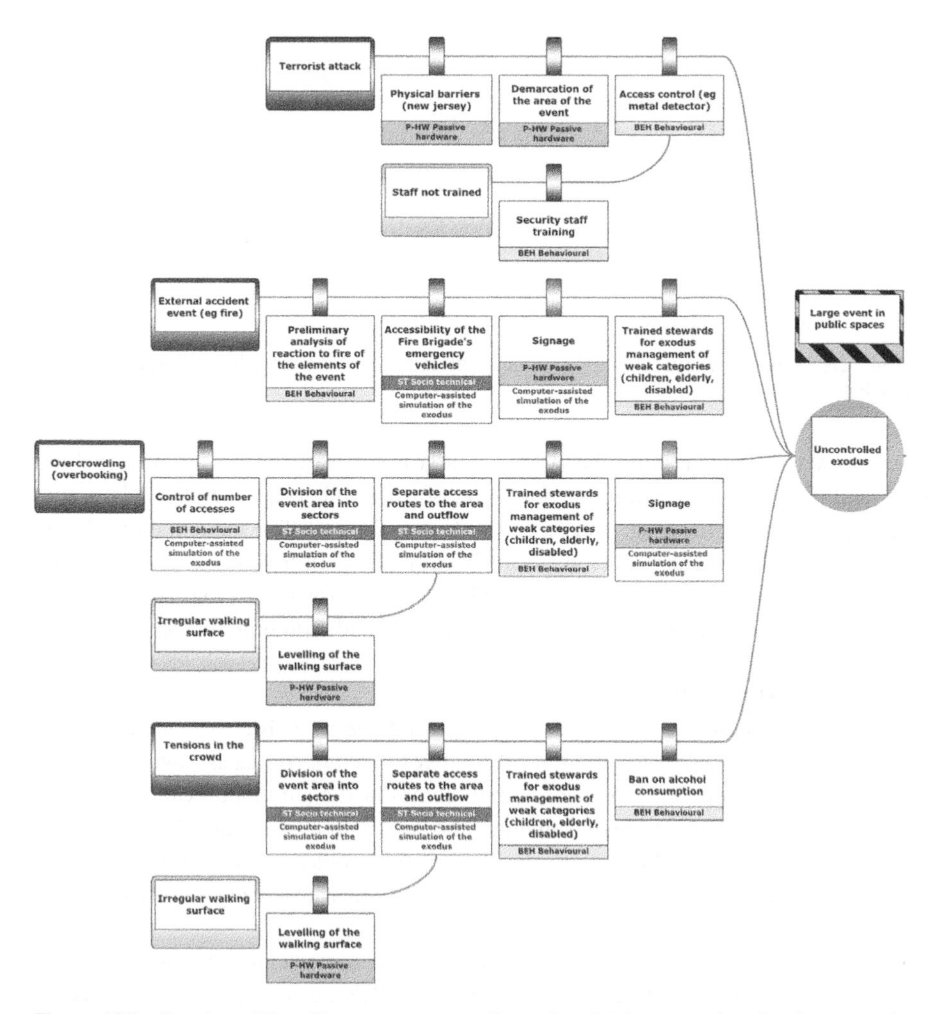

Figure 153 Developed Bow-Tie to assess crowding-related risks – zooming the threats and preventive barriers.

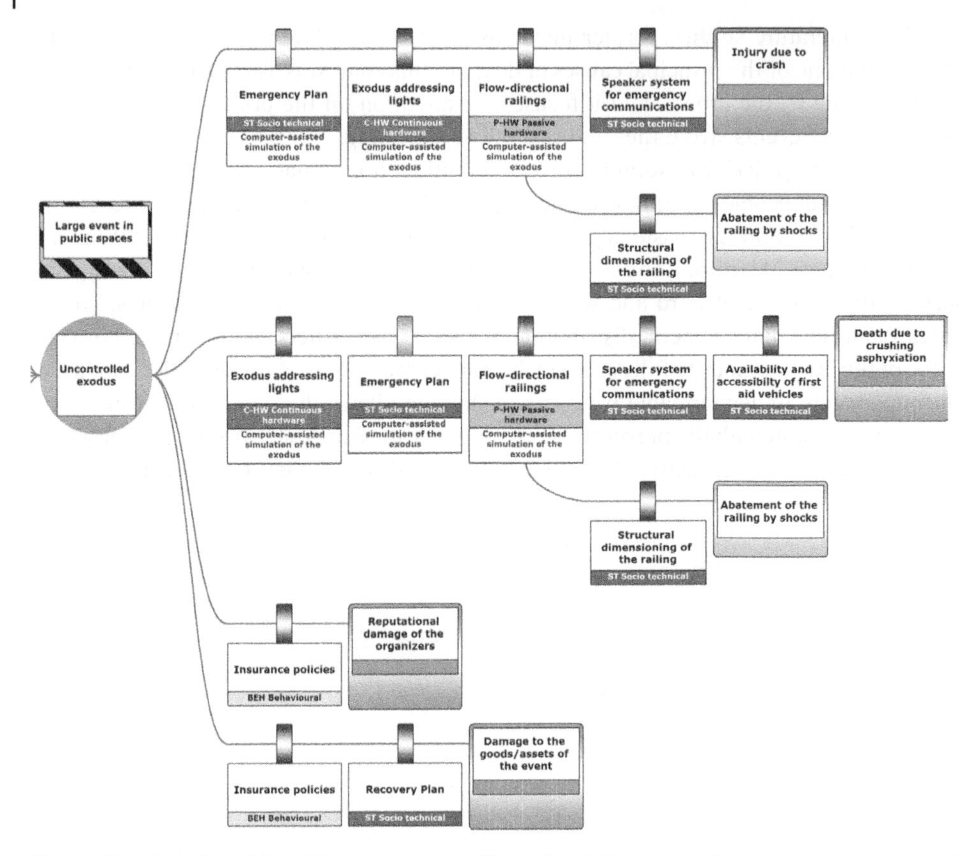

Figure 154 Developed Bow-Tie to assess crowding-related risks – zooming the consequences and mitigative barriers.

Those parameters that may affect the safety of a large event are taken into account, including safety culture, layout, staff training, definition of roles and accountabilities, plans and procedures to manage crowding, and level of communication. The combination of the "What-if" methodology, whose outcomes can be formalised in a Bow-Tie diagram, and the Bow-Tie itself satisfies the need to have a technical summary of the event, to be used in the next step of quantitative assessment for those credible and representative scenarios.

Computational Modelling of the Exodus

This last step is computer-assisted. It cannot be considered as a risk assessment tool by itself, but it is useful to analyze different alternatives (exodus strategies, layout) and to choose the best design option. Moreover, the outcomes of the simulation can be used to train the staff in emergency management. The simulation software (INCONTROL Simulation Solutions, 2019) offers a realistic view of the problems (like bottlenecks) and allows the designers to experiment with alternative solutions without a real impact on visitors and with a considerably cost reduction.

For instance, Figure 155 shows a map with a simulated scenario of an uncontrolled exodus from Piazza Vittorio Veneto in Matera. Of course, these models are just tools, and they need to be used critically; if improperly used, they may lead to incorrect and dangerous results, with

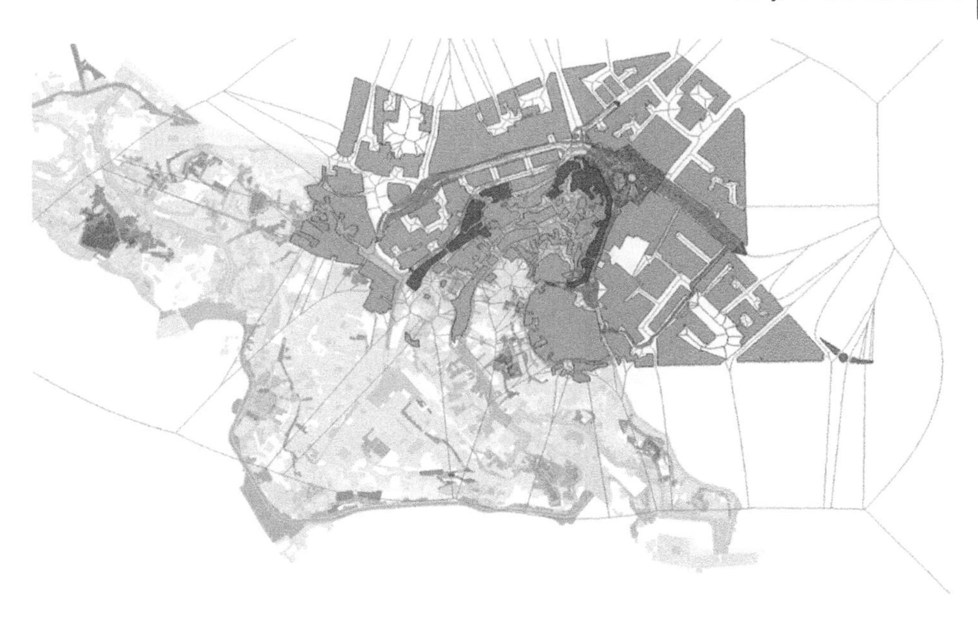

Figure 155 Map to develop simulated scenarios.

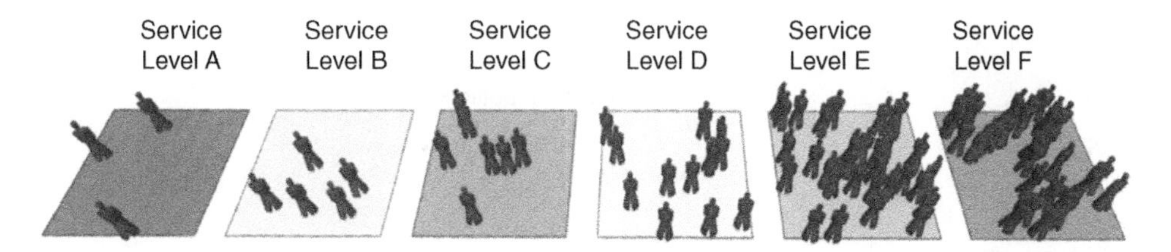

Figure 156 Different levels of service.

the risk of validating an ineffective emergency plan. Once the hypotheses on which these models are created has been accepted, the risk analyst has to identify those conditions that can cause a scenario: it is useful to identify different levels of service (from A to F, Figure 156), corresponding to a different concentration of people per unit of surface. In the international community, concentrations bigger than 6 pp/m^2 are considered to cause the flow to be congested: forces are transmitted through the crowd and individuals are literally "transported" by it.

Every simulation requires:

- Identifying scenarios;
- Defining input (people and environmental features);
- Evaluating routes and human behaviour;
- Developing the model and analyzing the results;
- Writing an emergency plan.

The user's features are defined via statistical parameters, rather than deterministic ones, always taking into account every particularity (e.g. presence of obstructions such as luggage). The output analysis allows identifying the bottlenecks (high congestion points), whose analysis may lead the organizers to make risk-based decisions. In the example of Piazza Vittorio Veneto, the simulation in three different configurations revealed the same

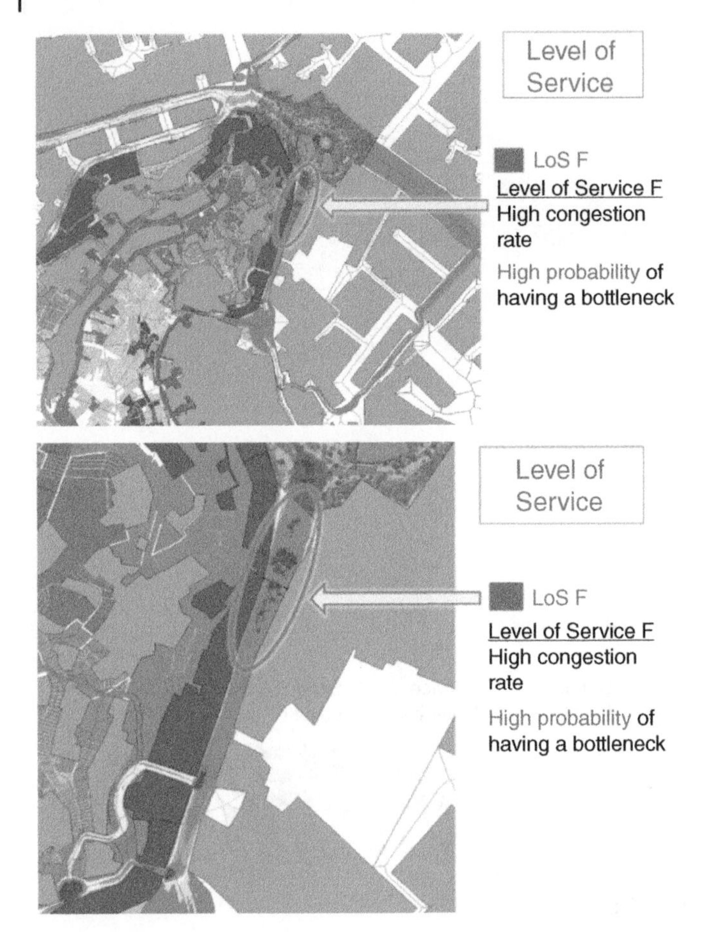

Figure 157 Piazza Vittorio Veneto and the bottleneck in Via San Biagio, Matera.

vulnerability: a bottleneck in Via San Biagio, as shown in Figure 157. Simulations also allow the evaluation of the impact of the soft obstacles, like parked cars, in the exodus (Figure 158).

Conclusion

In conclusion, only a shared and rigorous process (Figure 159) allows us to move from data to information, and from here to risk-based decisions by the stakeholders. Regardless of the tools, this is the real challenge for the insiders.

4.4.8 Military Helicopter Operations Bow-Tie Risk Assessment

Military helicopter operations can be complex and dangerous, owing to the environments in which they operate. Consider that they must be flown in poor weather, often close to the ground and in alien surroundings. While the precursor events you see on the left hand side

Figure 158 Impact of the soft obstacles on the pedestrian flow.

of the Bow-Tie shown in Figure 160 do occur quite frequently, we are very happy to say that loss of control is not a common phenomenon. We think this is because military helicopters are usually flown with at least two crewmembers and because the pilots are very skilled aviators. In the author's opinion, the barriers that mention two crewmembers are the critical ones.

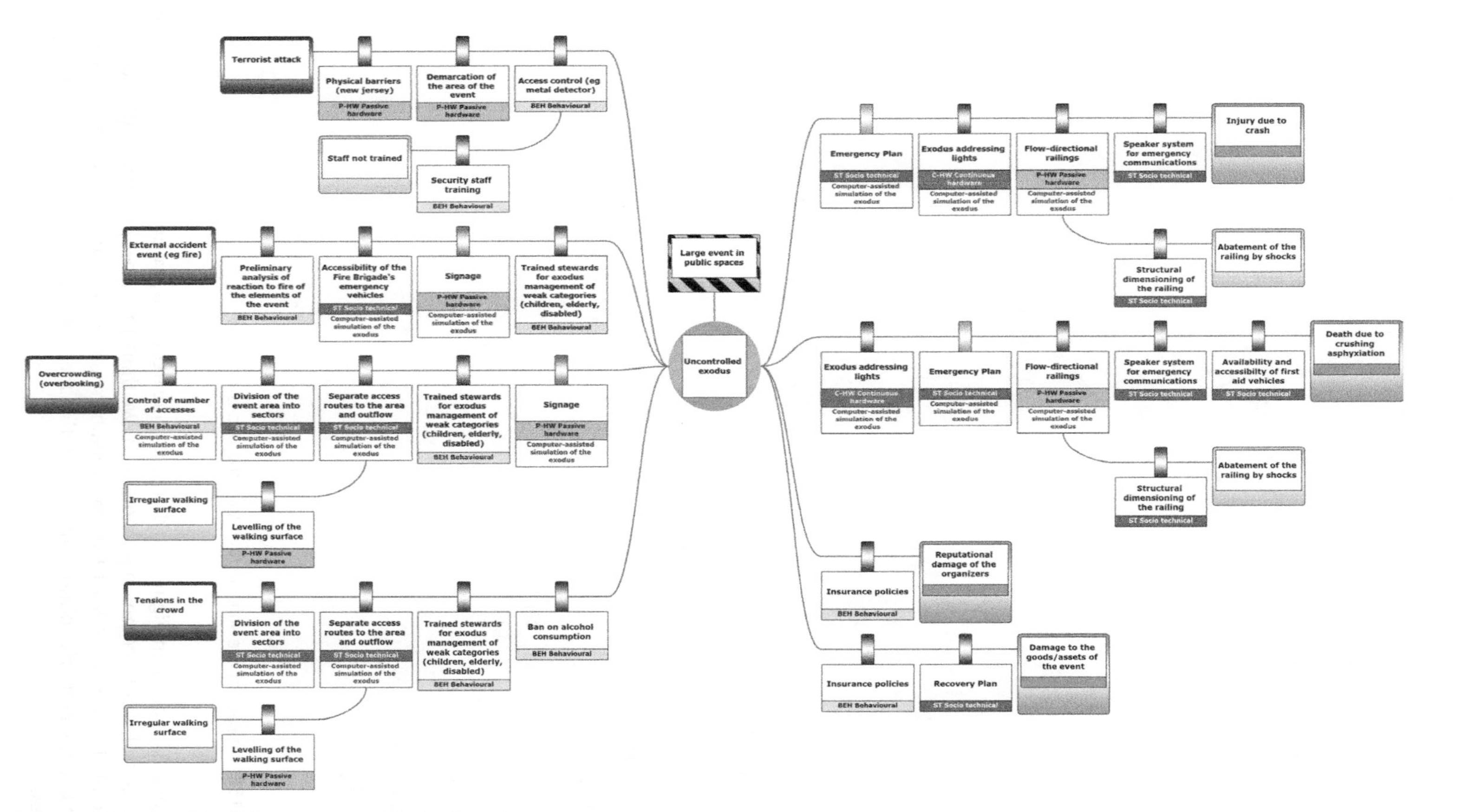

Figure 159 Bow-Tie Risk assessment (whole picture).

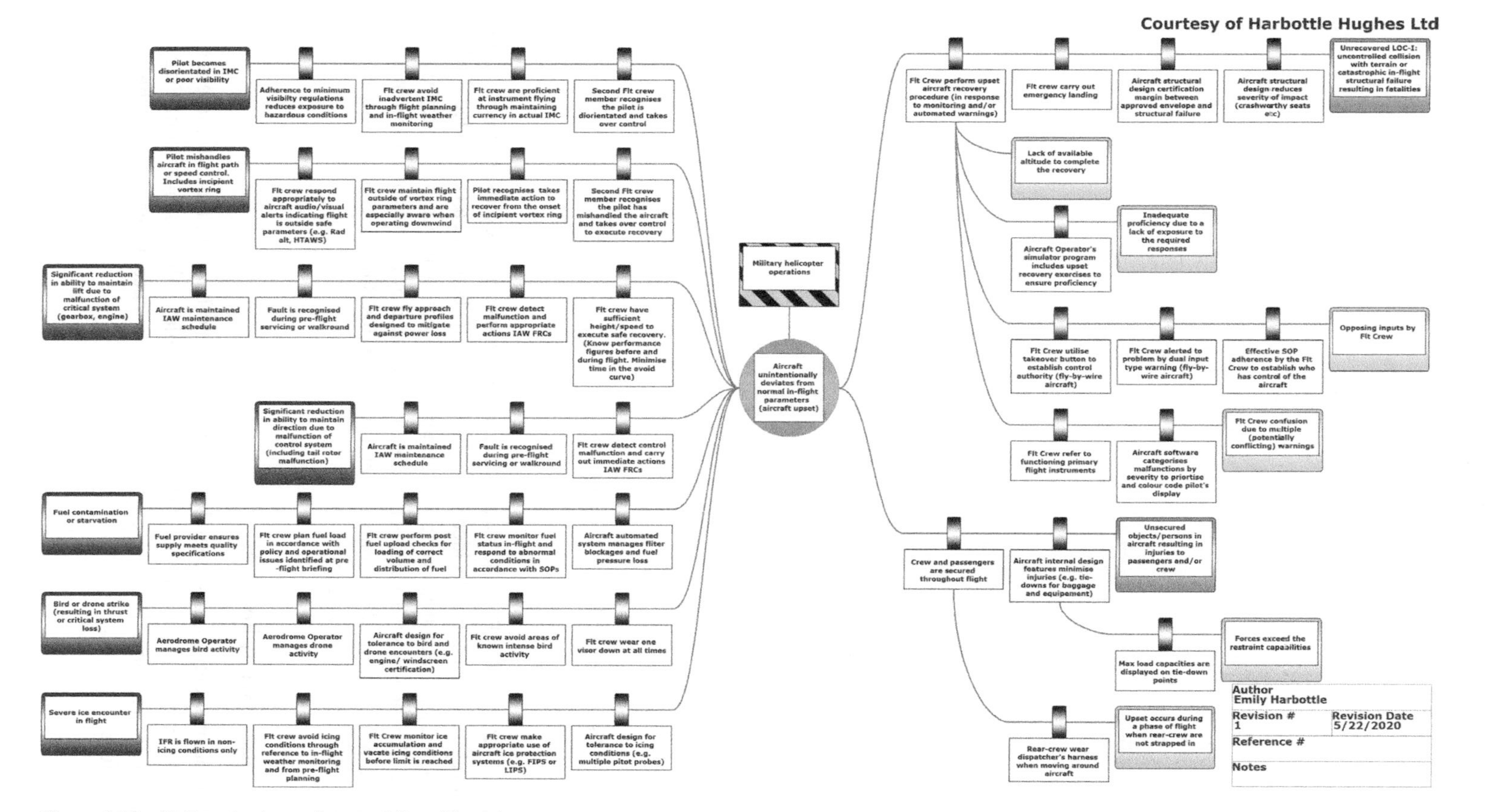

Figure 160 Helicopter loss of control Bow-Tie risk assessment.

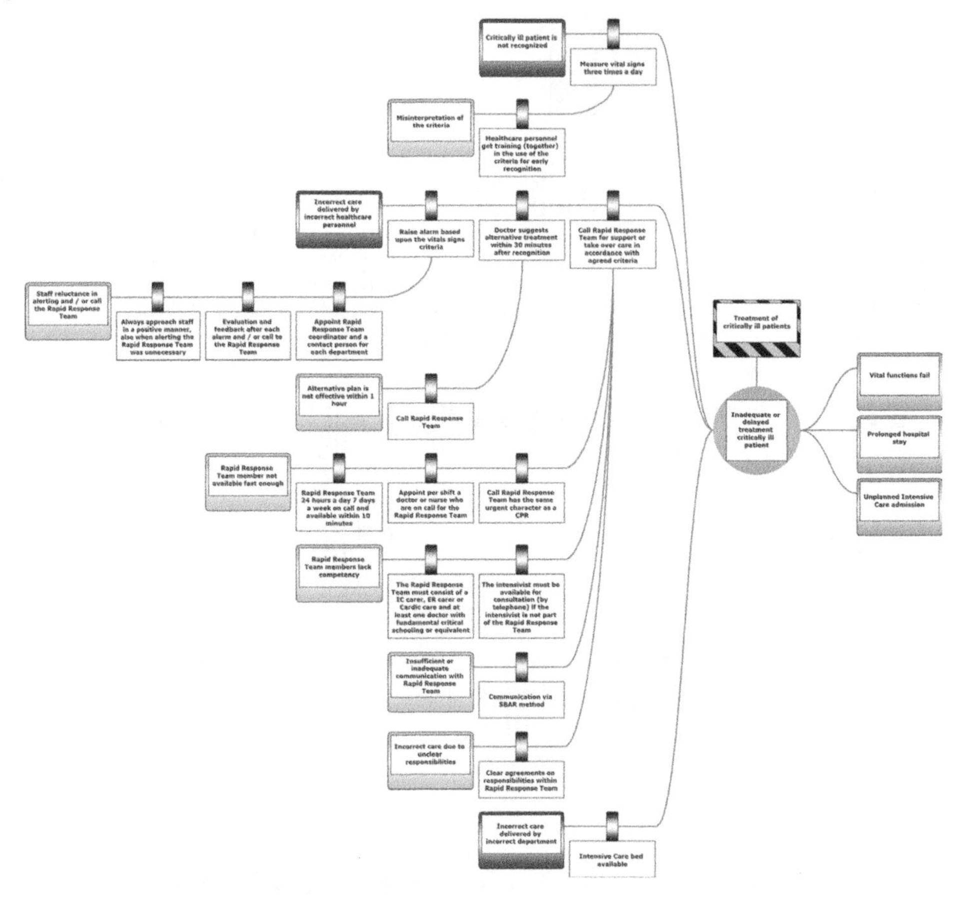

Figure 161 Treatment of critically ill patients.

When developing the Bow-Tie we considered the inclusion of "Collision with an obstacle at low level" as a threat but decided in the end that it fits better in a Controlled Flight into Terrain Bow-Tie, hence it is not included here. In recent years we have sadly witnessed several civilian helicopter accidents that have been caused by the loss of control in flight, so the author hopes many of the controls put forward in this Bow-Tie can be easily translated across to civil helicopter operations.

4.4.9 Patient Safety Bow-Ties

The healthcare sector has also found strength in the Bow-Tie diagrams to represent health-related risks in order to better inform, train and communicate their management. Figures 161 through 177 are some Bow-Tie diagrams relating to specific hazards in the area of patient safety.

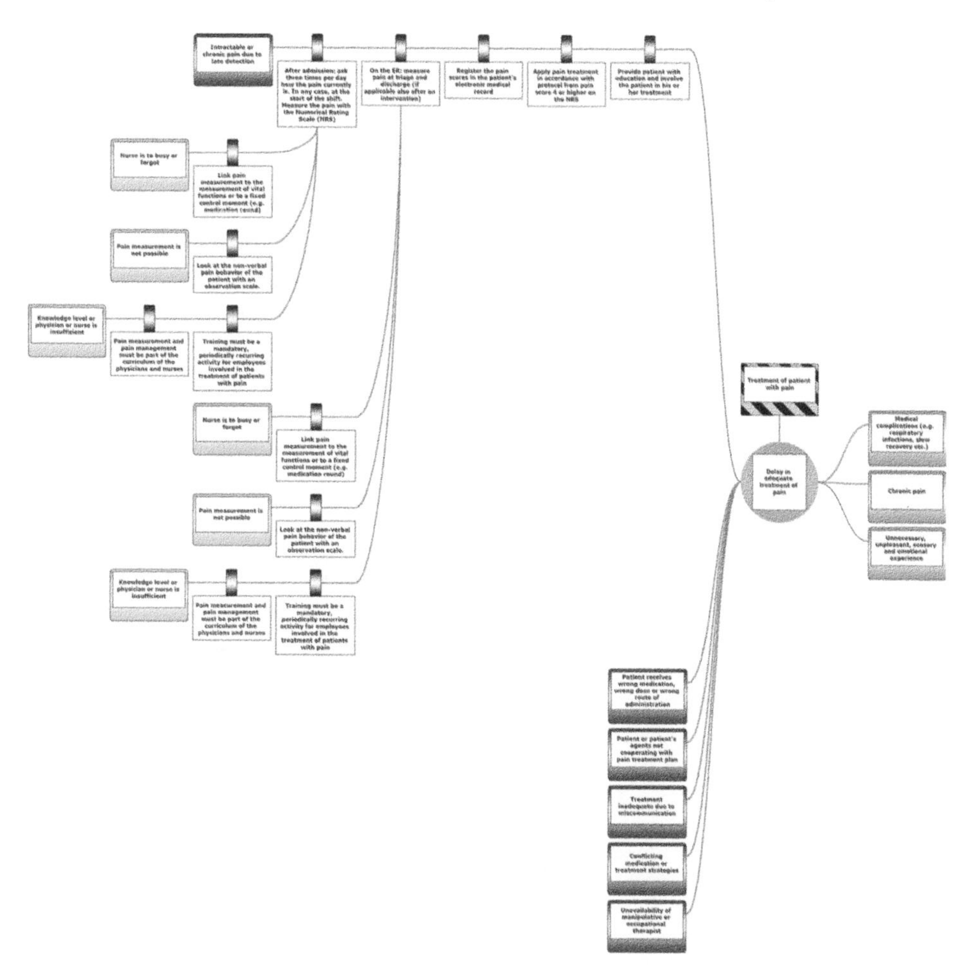

Figure 162 Treatment of patient with pain.

4.4.10 Process Safety Bow-Tie

High-risk industries, such as chemical production plants, face the need of managing safety risks as well as proving they comply with legal requirements regarding the management of these risks. In fact the risk management process itself and the compliance management process are two distinct processes with very similar process flows. Risk management is defined by general standards like ISO 31000 (*Risk Management – Principles and Guidelines*) as well as by industry specific standards like ISO 17776, *Guidelines on Tools and Techniques for Hazard Identification and Risk Assessment for Petroleum and Natural Gas Industries*. Compliance management is regulated by ISO 19600. Whereas the processes described in these standards are quite similar, their scope is different, the first two standards concentrating specifically on the management of safety risks, and the last standard focusing on the risk of not complying with rules and regulations in general.

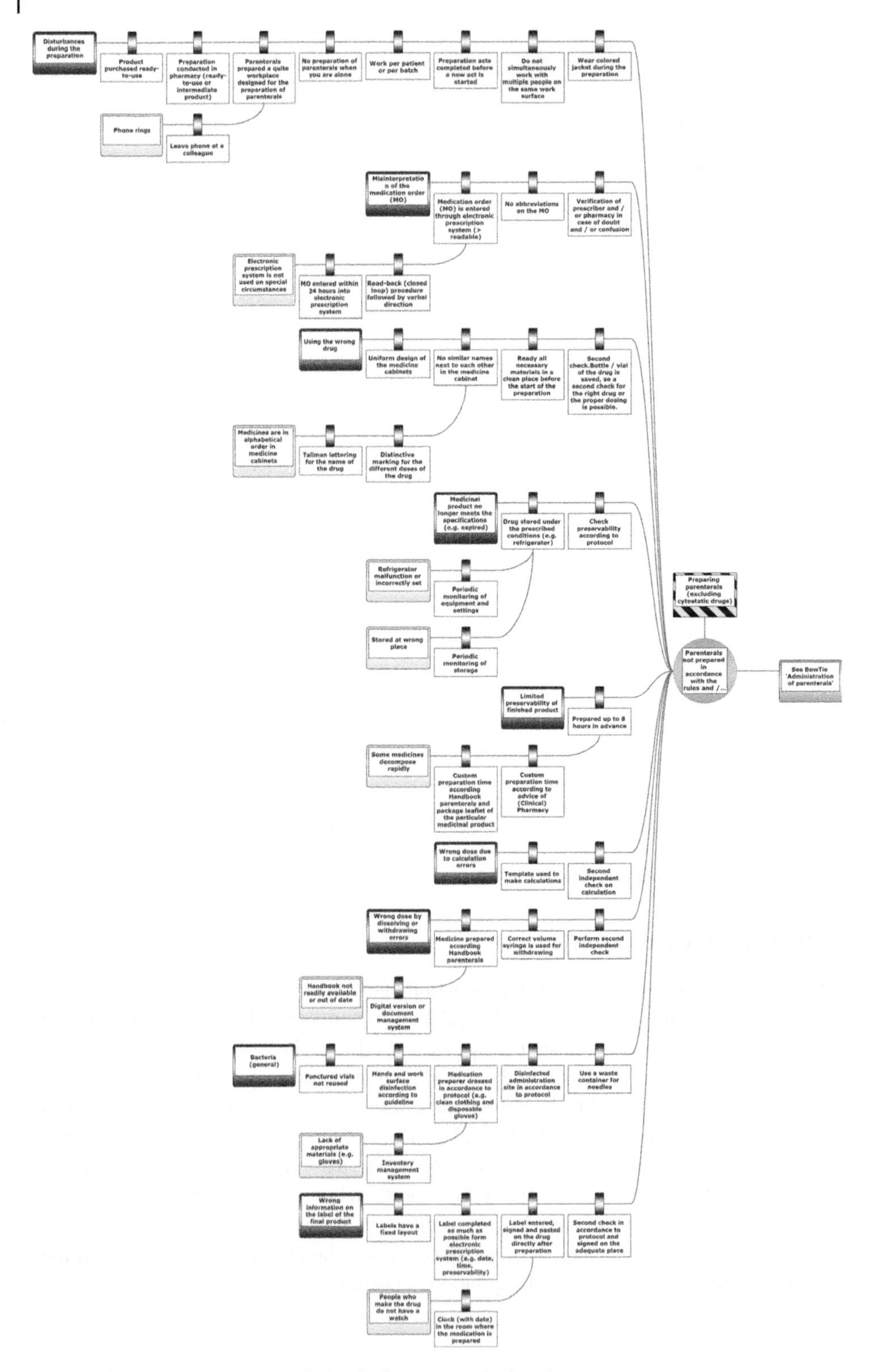

Figure 163 Preparing parenterals (excluding cytostatic drugs).

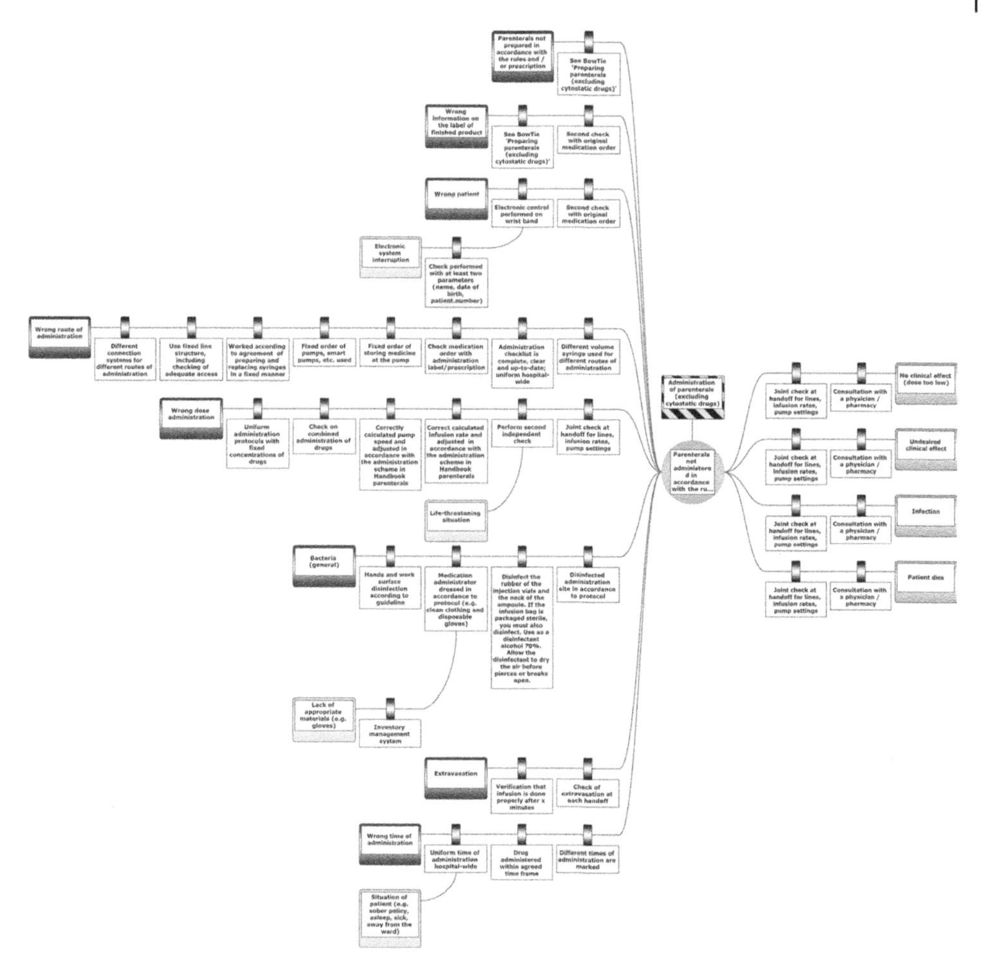

Figure 164 Administration of parenterals (excluding cytostatic drugs).

Industries therefore have to manage risks from a business continuity as well as from a compliance perspective. The Bow-Tie methodology serves as a tool to get a clear overview of the status of both these processes. It shows in a glance what threats are present, what preventive and mitigating controls are present and how these are linked to the business processes.

In brief, a Bow-Tie consists of a central node – the top event that is considered – and on the left side a number of threats that may cause the top event to happen. Each of these can be controlled by a number of threat barriers, which in its turn may be weakened by escalation factors, for which again a number of controls can be identified. On the right side possible consequences of the top event are identified that may be mitigated by recovery measures. Also these measures may be weakened by escalation factors, for which again controls can be identified.

This is illustrated by this case study, which regards an underground gas transport pipeline of more than 200 kilometers in the Netherlands. The central node, or top event, in this case is the leakage of product into the surrounding soil. Threats that may cause the leakage are both time-dependent and time-independent:

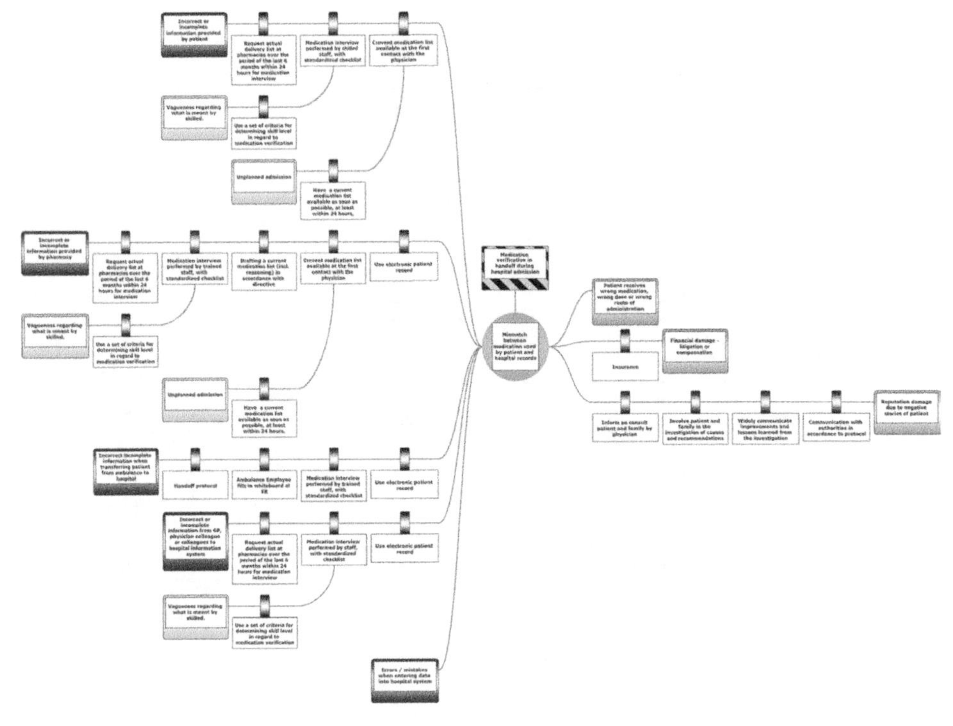

Figure 165 Medication verification in handoff during hospital admission.

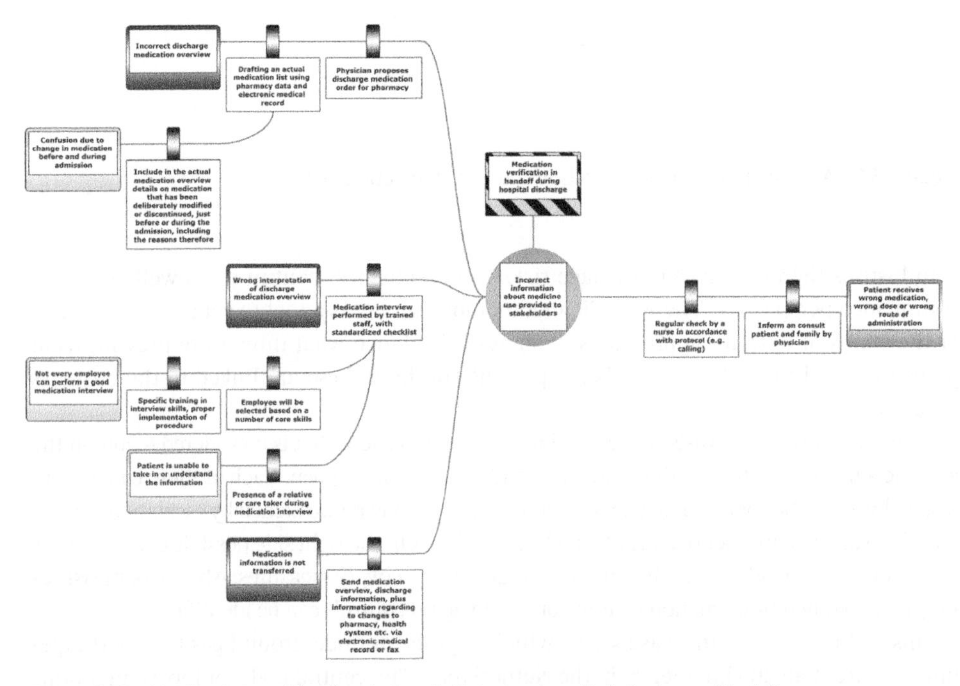

Figure 166 Medication verification in handoff during hospital discharge (1 of 2).

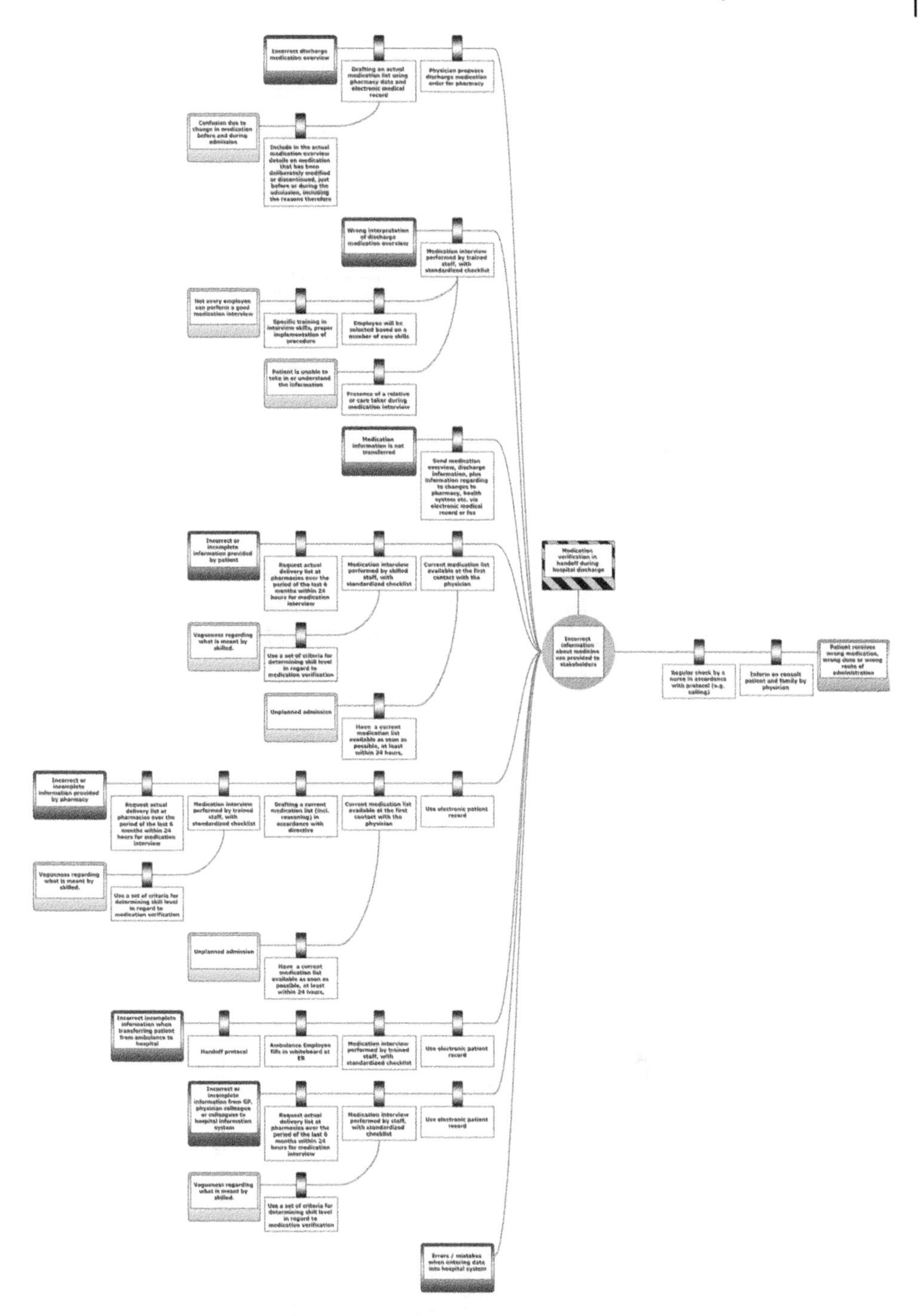

Figure 167 Medication verification in handoff during hospital discharge (2 of 2).

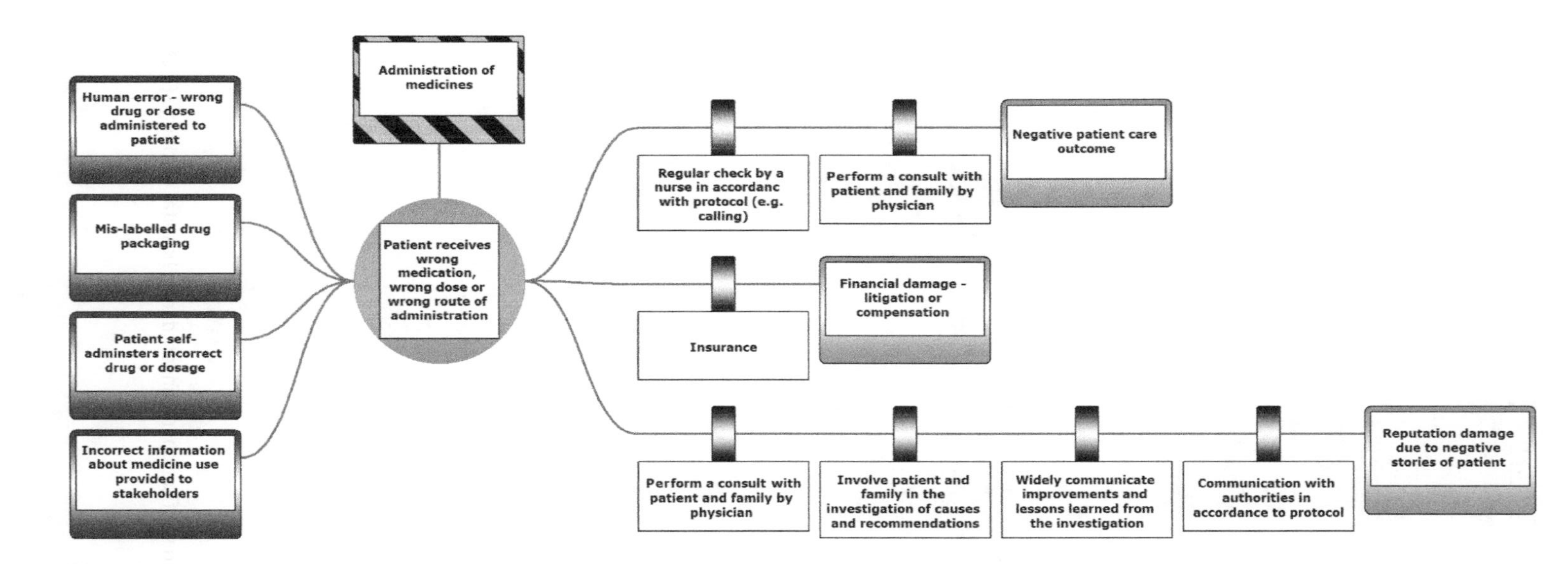

Figure 168 Administration of medicines.

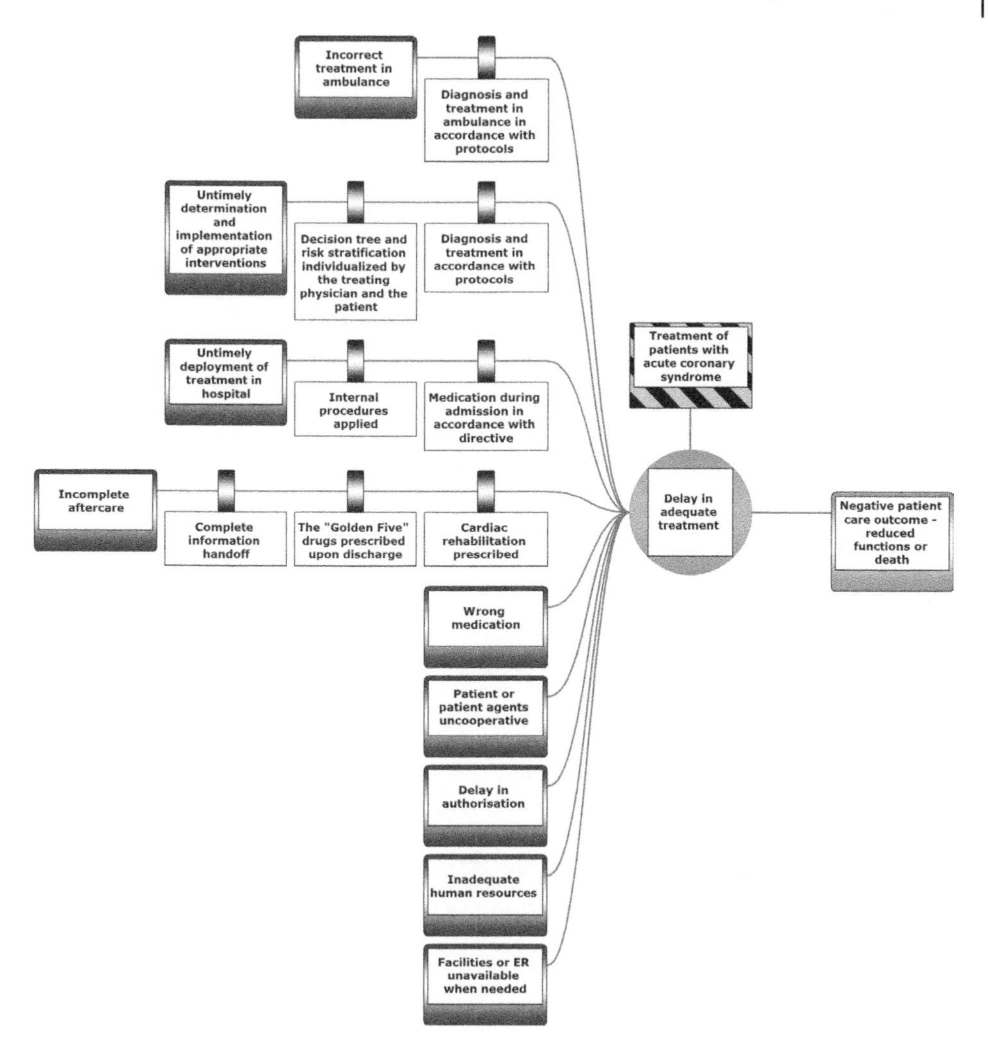

Figure 169 Treatment of patients with acute coronary syndrome.

- Time-dependent threats:
  - External or internal corrosion;
  - Cracking due to fatigue or environmentally assisted cracking (EAC).
- Time-independent threats:
  - Third party, such as excavation and ground movement activities;
  - Geo-related threats such as earthquakes or landslides;
  - Incorrect operations, such as overpressure or product deficiencies;
  - Sabotage, such as terrorist attacks.

The Bow-Tie study regarding leakage of this pipeline revealed 19 threats and 18 consequences, with a total of 286 barriers, both reducing the probability of occurrence of the failure of the pipeline and mitigating the effects of its eventual failure.

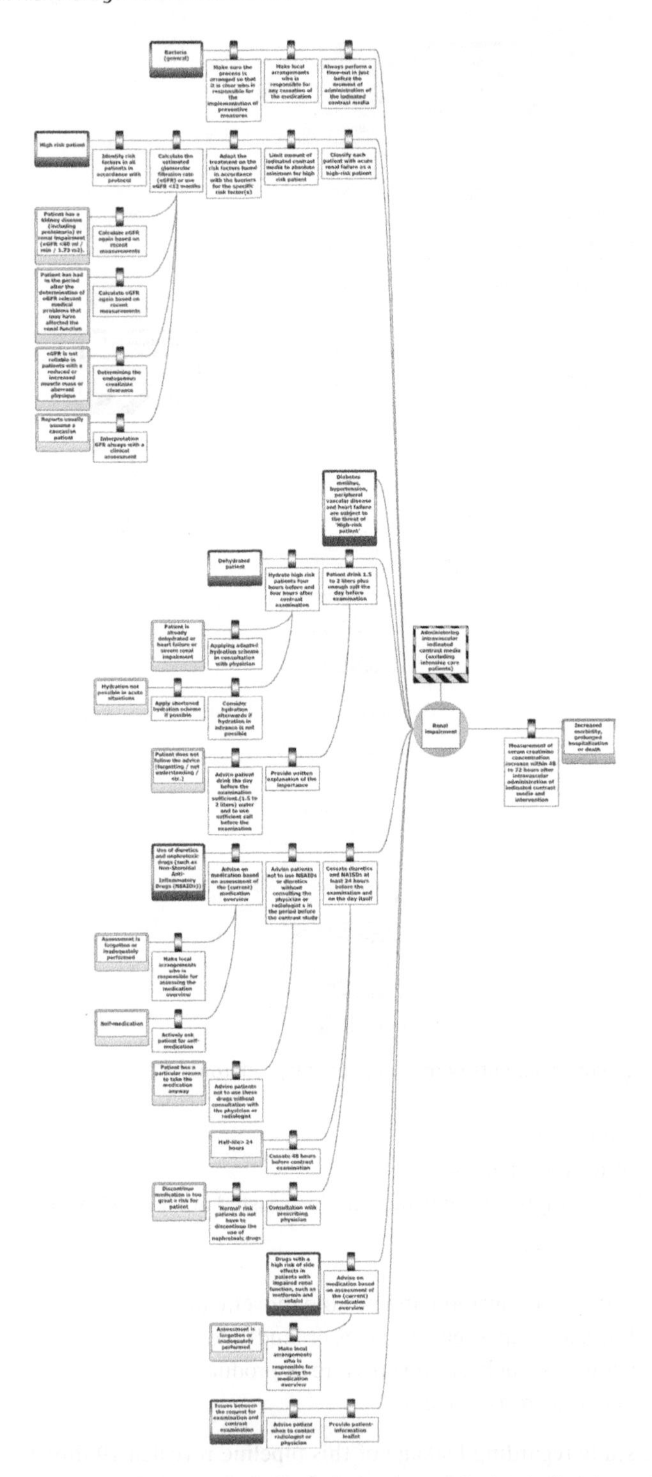

Figure 170 Administering intravascular iodinated contrast media (excluding intensive care patients).

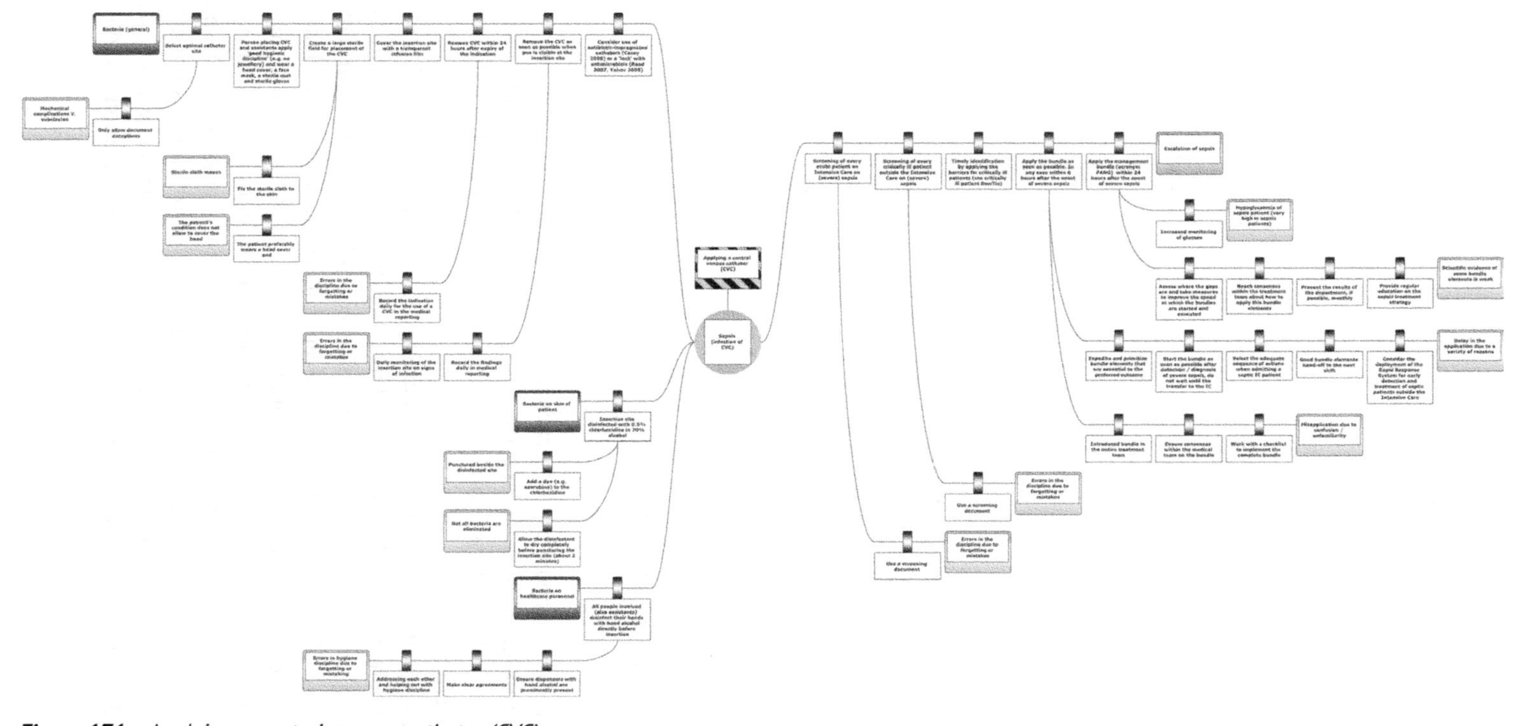

Figure 171 Applying a central venous catheter (CVC).

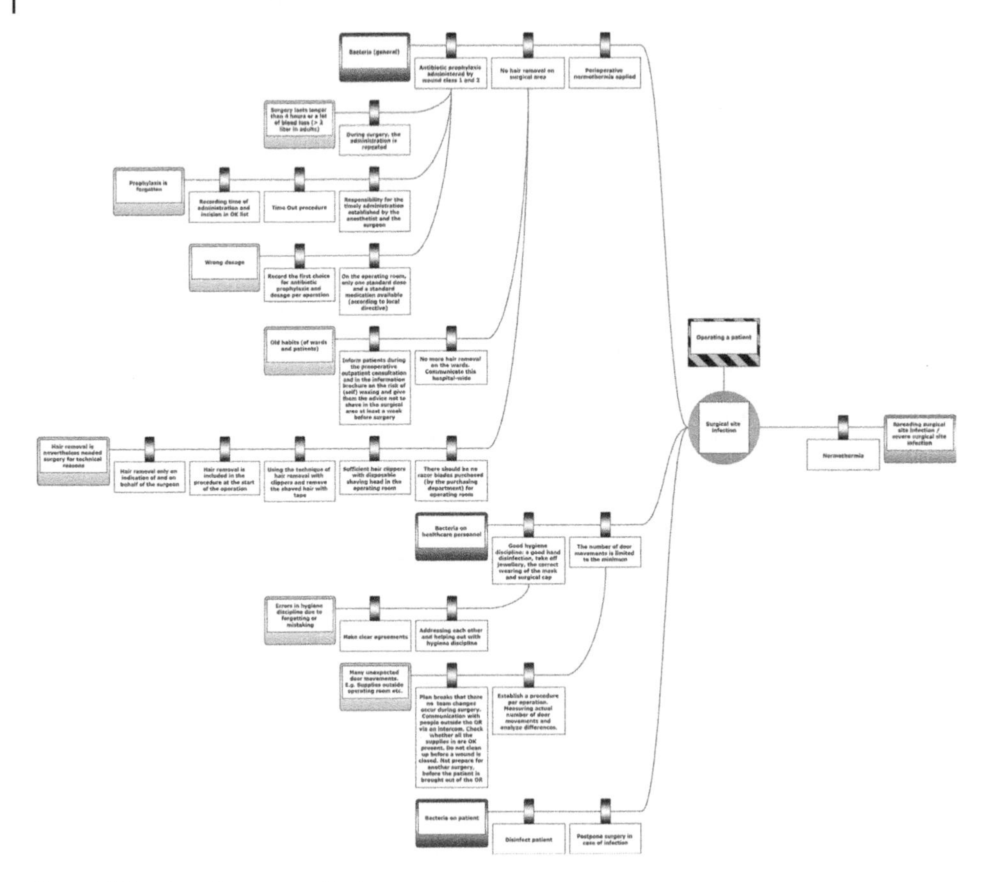

Figure 172 Operating on a patient.

In this example we elaborate one threat and its barriers and escalating factor.

The threat considered is "external corrosion"; see Figure 178. There are two main barriers to reduce the risk of external corrosion: robust coating and a condition monitoring and maintenance program.

Regarding the coating there are three escalating factors: insufficient quality of the coating application, damage to the coating, and insufficient quality of coating during operation. Regarding "coating quality," there are four controls: the coating procedure itself, visual inspection of the coating, the direct current voltage gradient (DCVG) method, and checking of the coating with special equipment. "Damage of the coating" may be controlled by cathodic protection and by the specification of a minimum thickness and type of coating material. Apart from this, for the control "cathodic protection" one more escalating factor is recognized, which may be poor quality. Controls for this are monitoring, periodic testing, periodic measurements inside areas where drinking water is subtracted, and current drainages. "Insufficient quality of coating during operation" is controlled by DCVG methods.

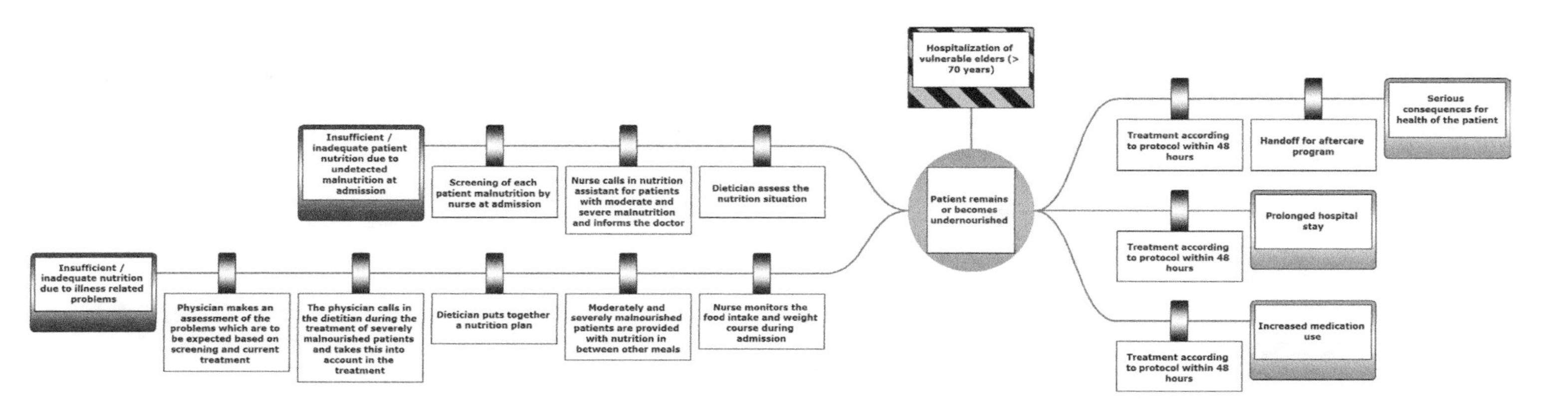

Figure 173 Hospitalization of vulnerable elders (>70 years) (1 of 4).

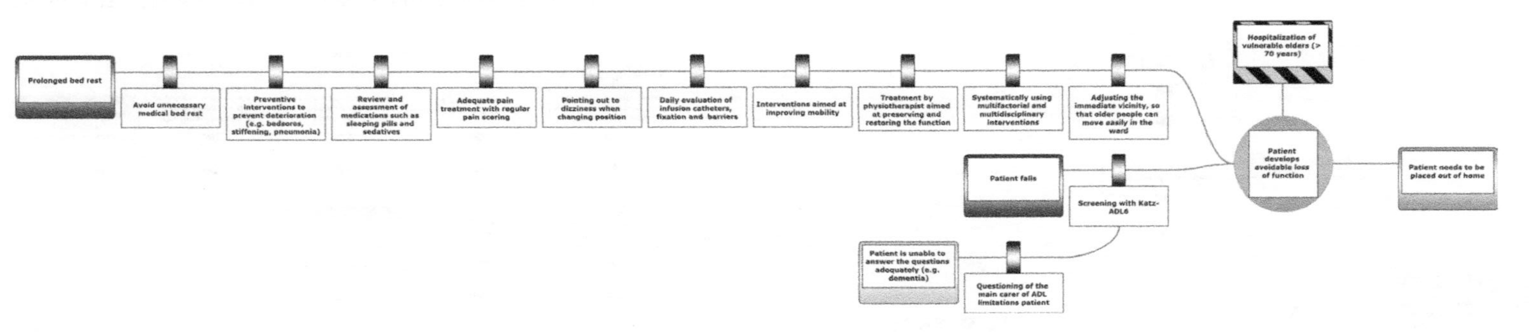

Figure 174 Hospitalization of vulnerable elders (>70 years) (2 of 4).

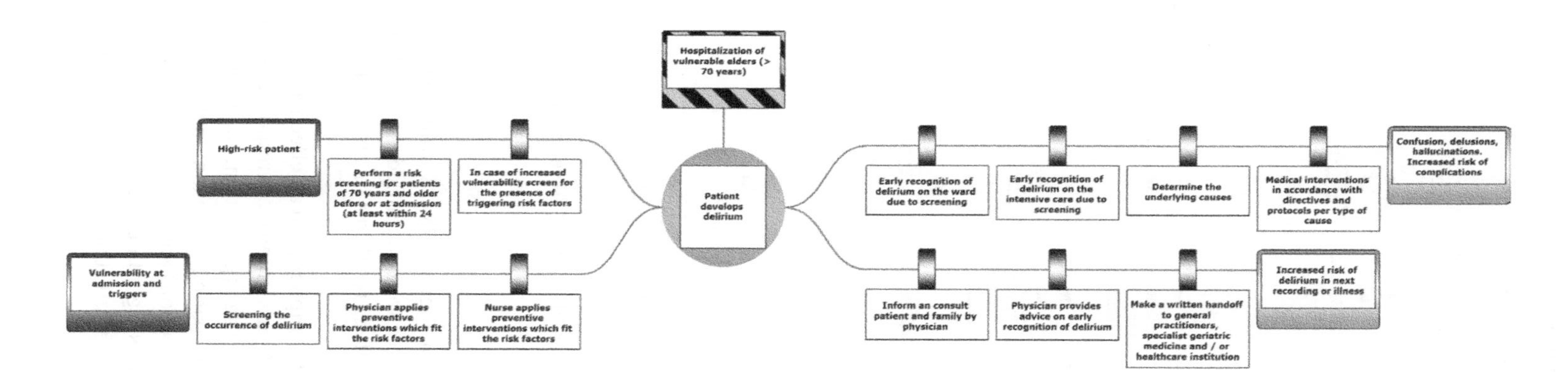

Figure 175 Hospitalization of vulnerable elders (>70 years) (3 of 4).

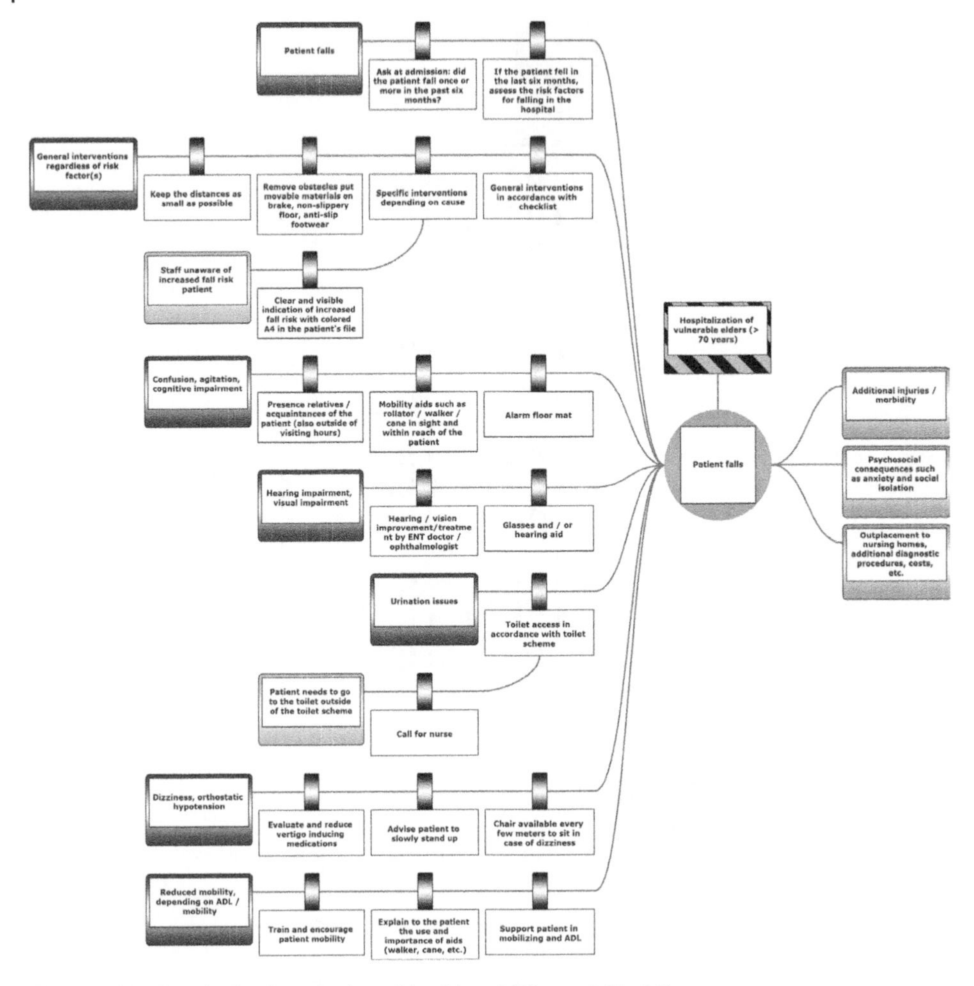

Figure 176 Hospitalization of vulnerable elders (>70 years) (4 of 4).

On the "consequence" side of Figure 178, 18 unique consequences were identified and likewise mitigating measures were identified. For each consequence a probability severity assessment was executed, resulting in an overall risk profile for the entire pipeline.

In order to manage risks and compliance in an effective way, every control and mitigation measure has to be secured by assigning clear responsibilities for monitoring its functioning and maintenance inside the organization; see Figure 179. For example, for the control "visual inspection of the coating" the maintenance department developed a clear standard (criteria for visual inspection) and an execution protocol (frequency and methods of inspection). All procedures regarding the life cycle of controls were then subjected to a general audit program to ensure that the organization will be in control of the controls.

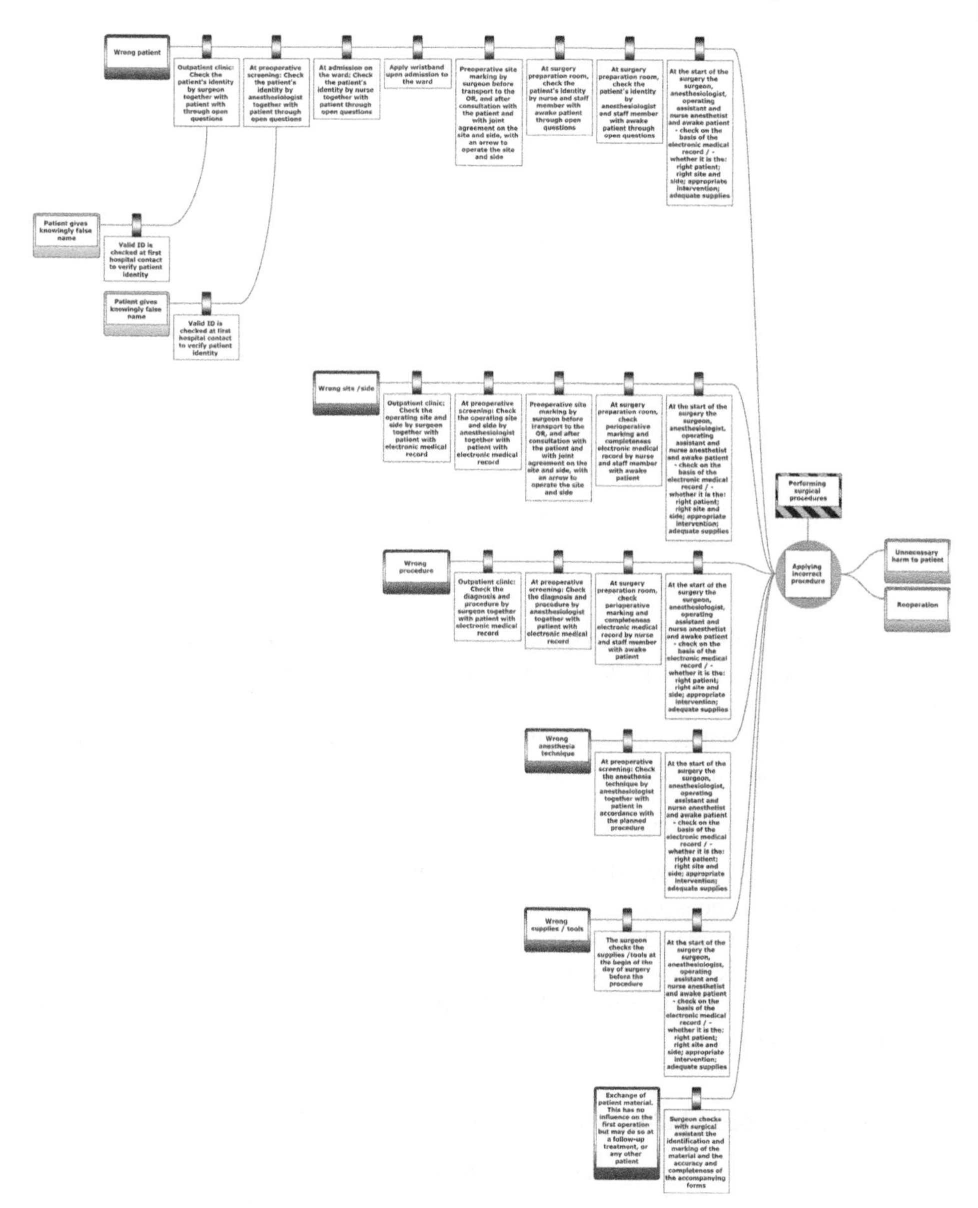

Figure 177 Performing surgical procedures.

Conclusion

The Bow-Tie analysis produced the following results:

- A risk analysis that is compliant with the following standards:
  - Dutch regulation BEVB – Besluit Externe Veiligheid Buisleidingen (ruling on the safety of pipelines);
  - Risk-based pipeline management system (NTA 8000);
  - ASME B31.8 engineering standard for pipelines.

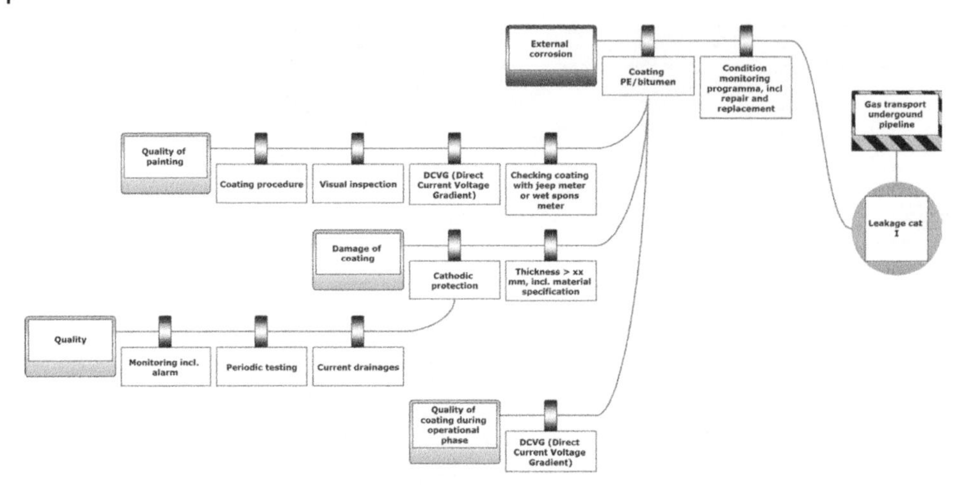

Figure 178 Elaboration of the threat "external corrosion" and main escalating factors and controls.

- A transparent means of communication allowing for demonstration of the ALARP approach, summarizing the in-depth risk analysis performed and enhancing workforce involvement;
- A link to the overall safety management system, allowing for clear barrier management and at the same time definition as well as monitoring of leading KPIs.

In summary, the Bow-Tie methodology allows for:

- Presenting at a glance an overview of complex interactions between causes, escalating factors, controls, and consequences of "loss of containment" events;
- A common language regarding such events, which allows for clear communication between company operators and staff, experts and external agencies;
- A clear link between physical and technical controls and the management procedures regarding their life cycle.

4.4.11 Famous Process Industry Incidents Analyzed with BFA

The learning from experience process is also fuelled by the dissemination of the main investigative outcomes of the most famous incidents in the industry. In the process safety context, the BFA diagrams shown in Figures 180 through 187 are proposed in order to identify events and failed barriers of some of the most important incidents in the industry.

4.4.12 Drug Administration Bow-Ties

The administration of medicines to a patient is not a risk-free activity. The Bow-Tie diagram shown in Figure 188 identifies the causes, consequences and barriers of this dangerous activity.

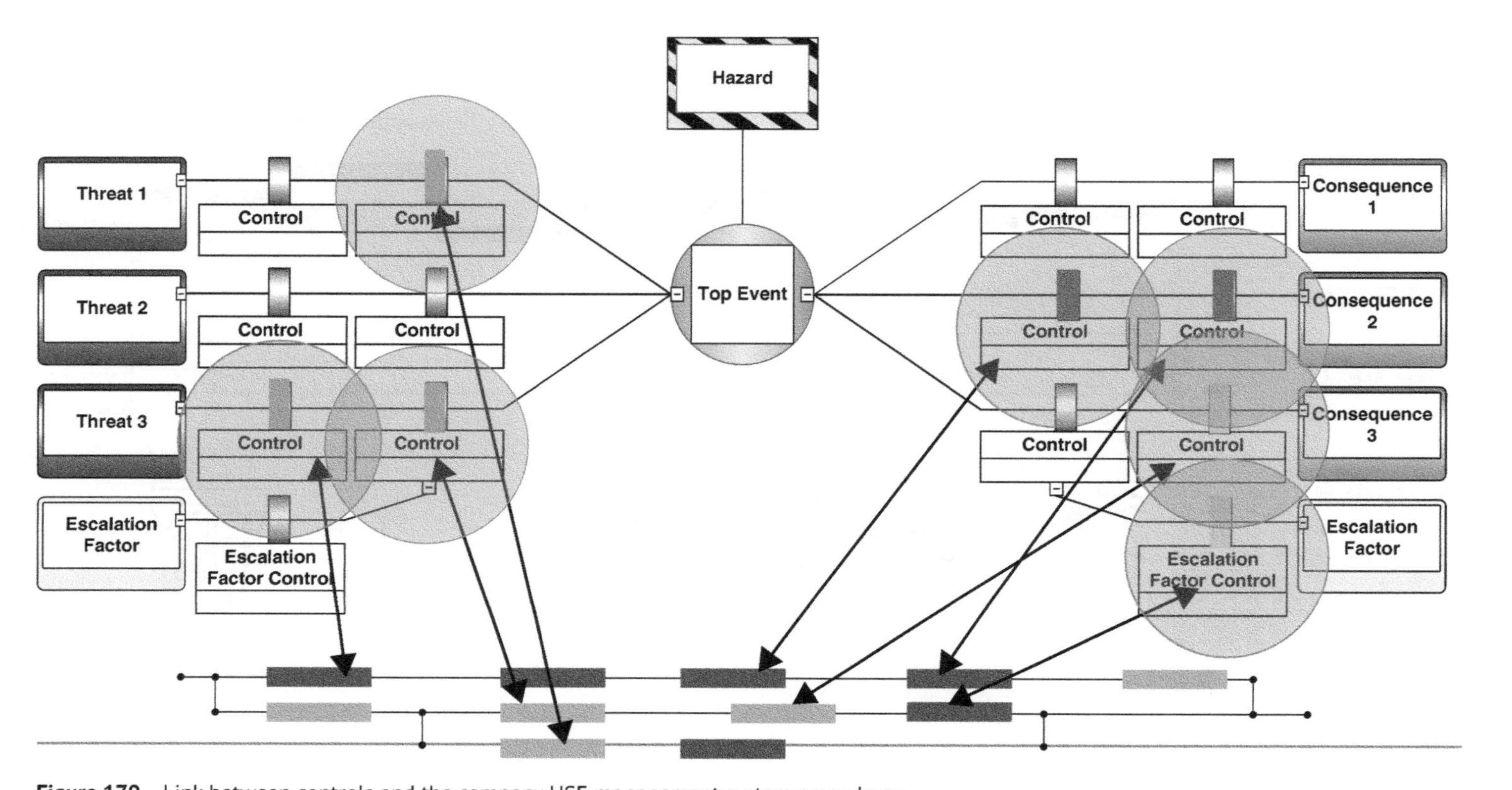

Figure 179 Link between controls and the company HSE management system procedures.

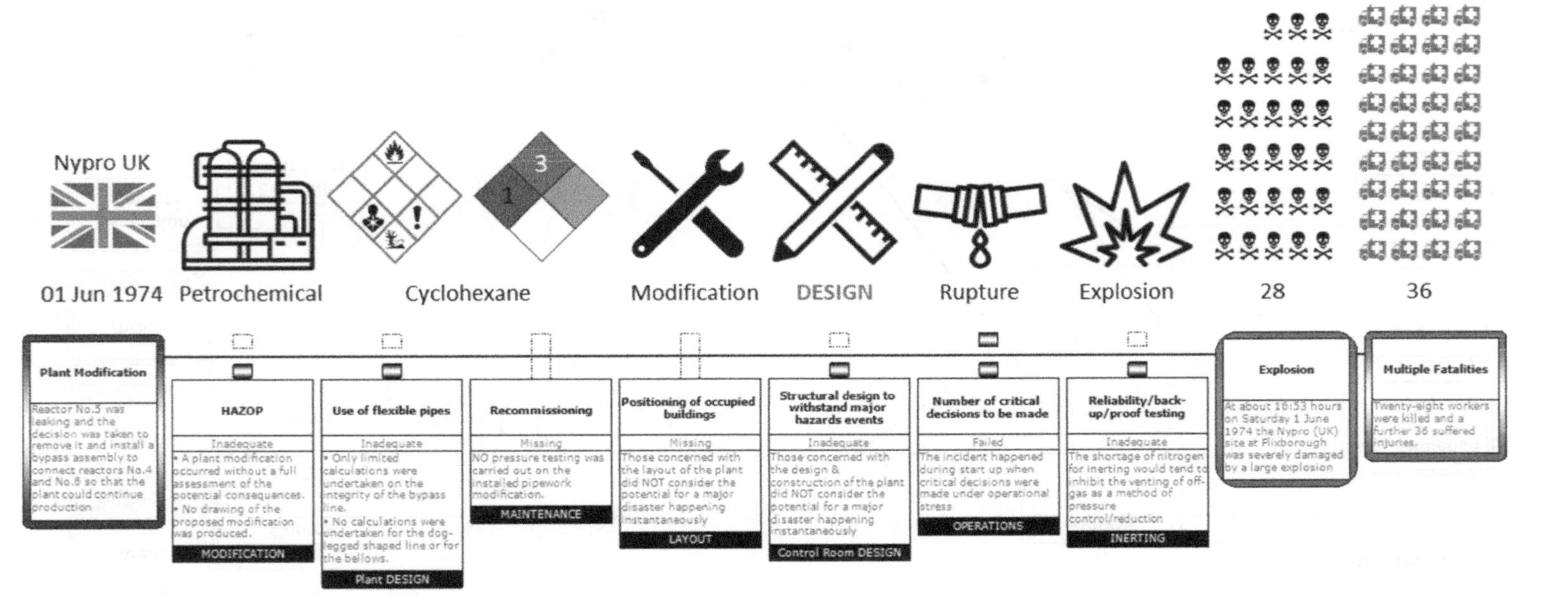

Figure 180 BFA of Flixborough (UK) incident.

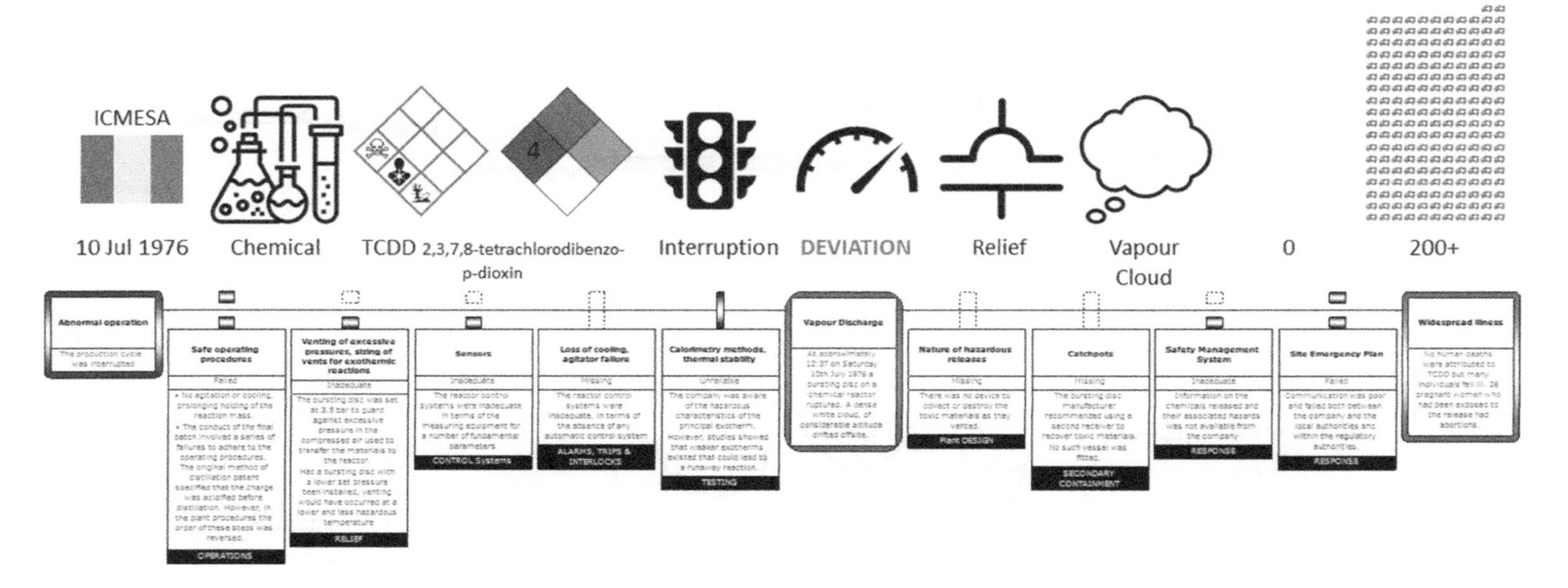

Figure 181 BFA of Seveso (Italy) incident.

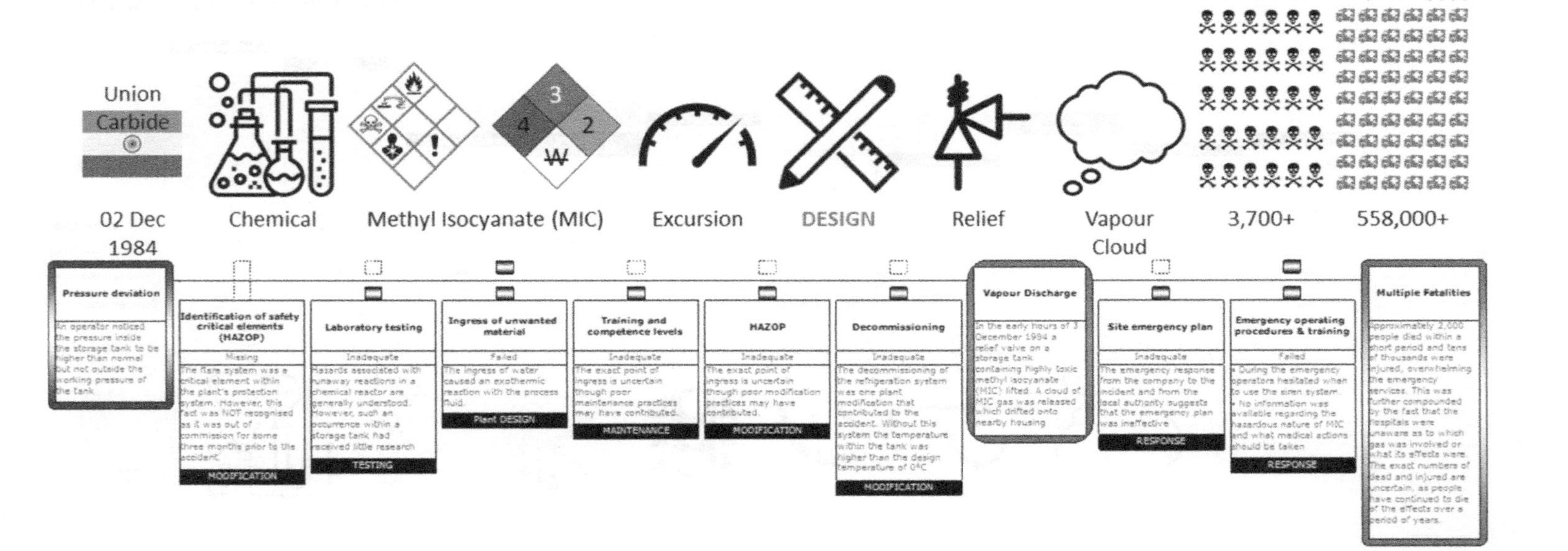

Figure 182 BFA of Bhopal (India) incident.

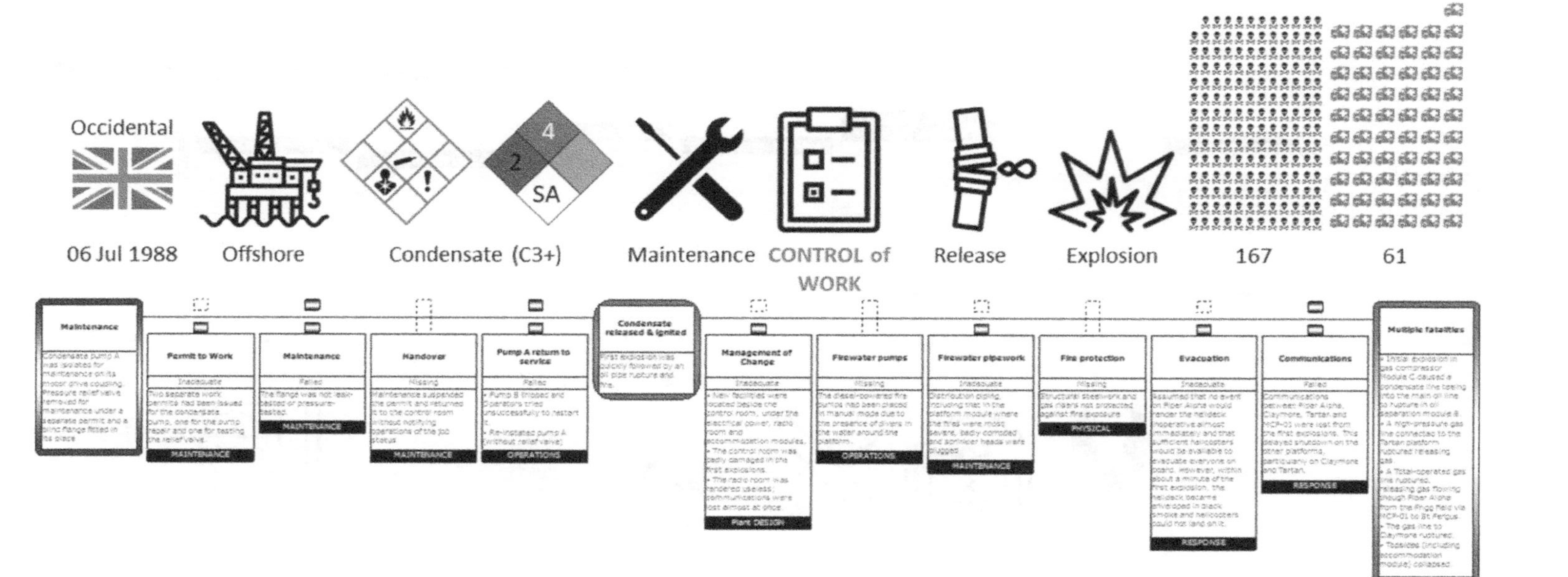

Figure 183 BFA of *Piper Alpha* (UK – offshore) incident.

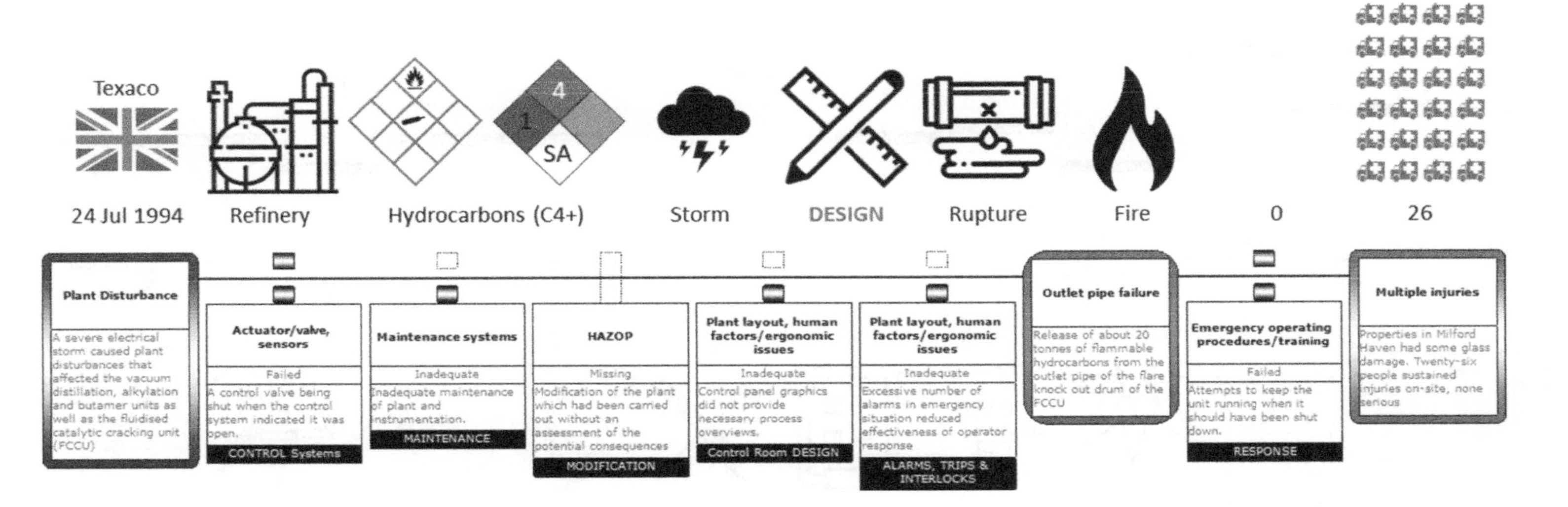

Figure 184 BFA of Pembroke Refinery (Milford Haven) (UK) incident.

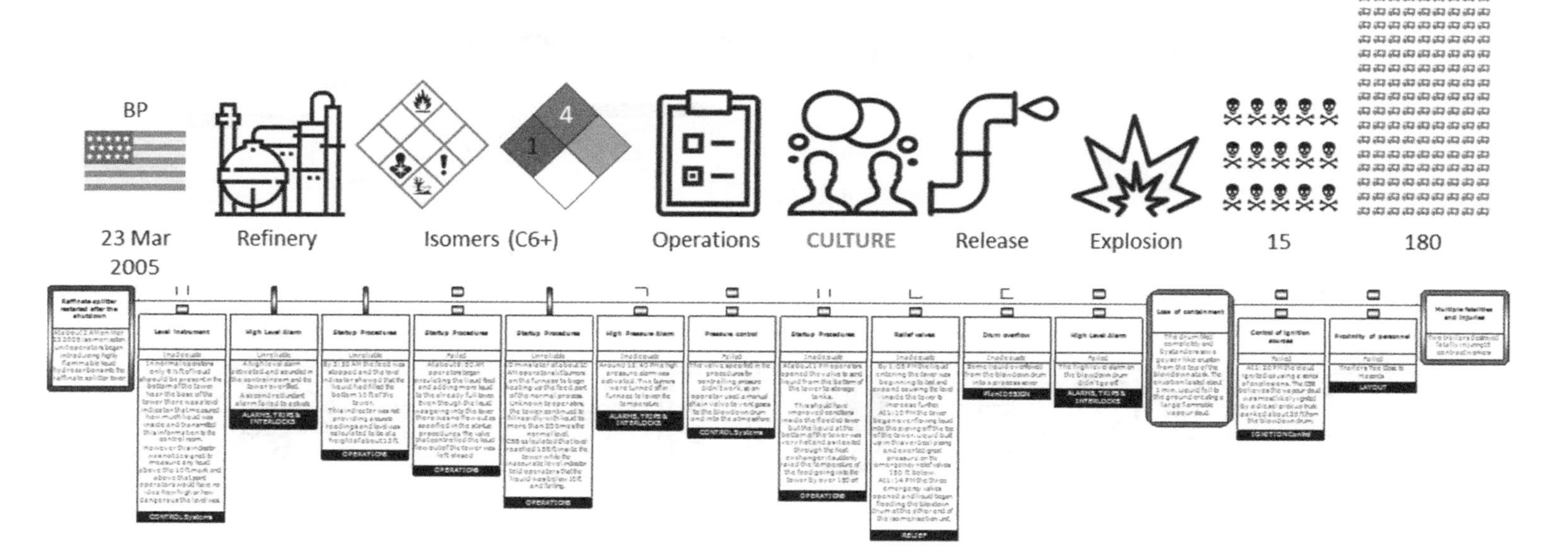

Figure 185 BFA of Texas City (US) incident.

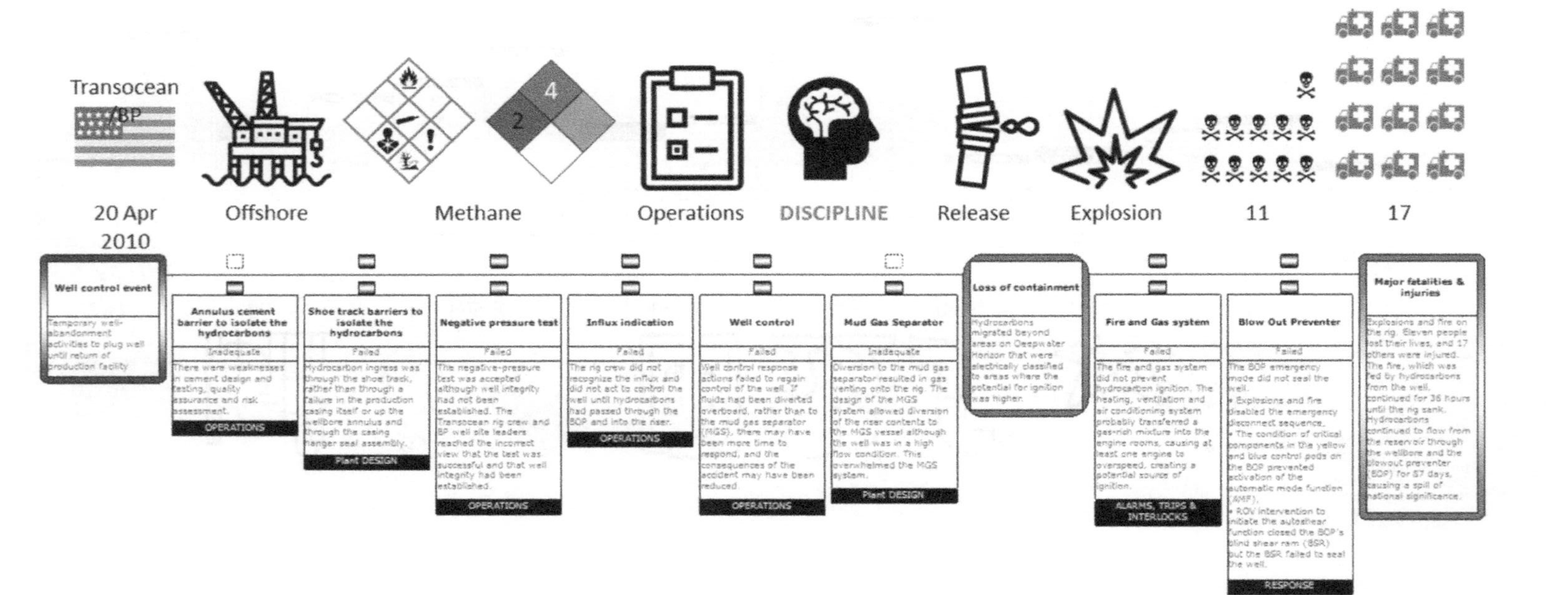

Figure 186 BFA of Macondo (*Deepwater Horizon*) (US – Offshore) incident.

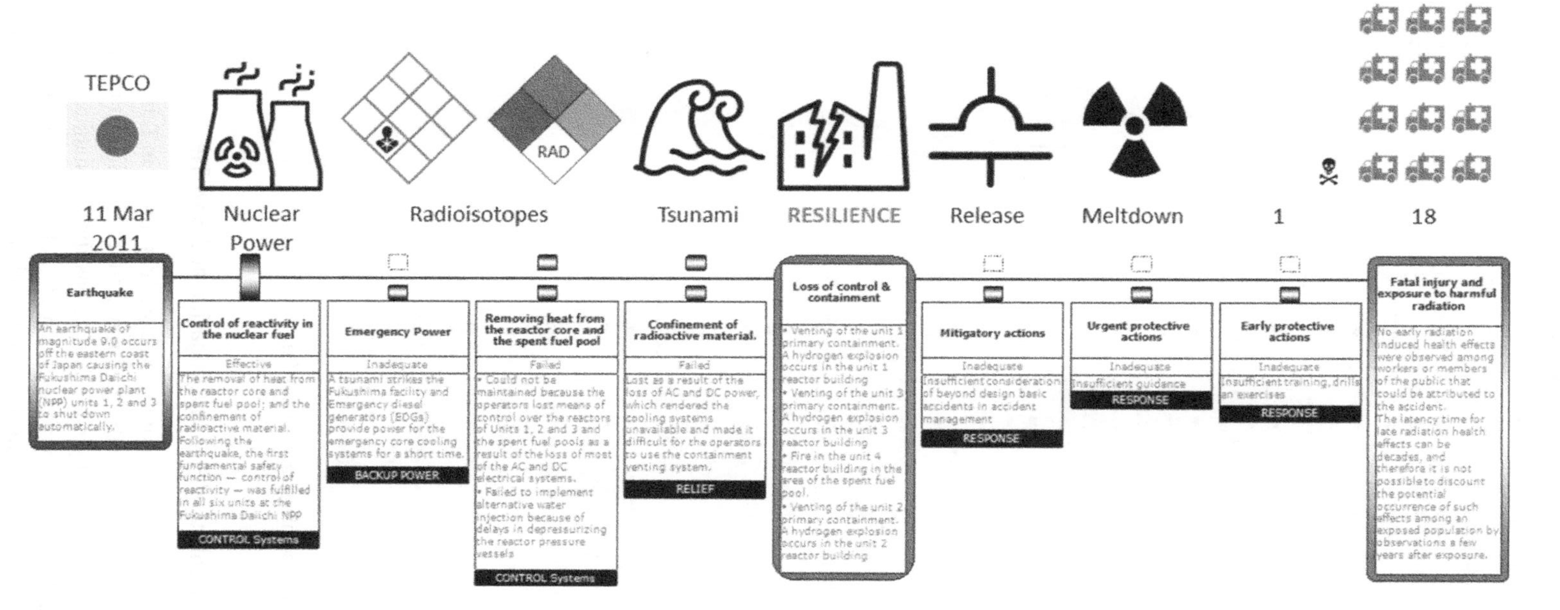

Figure 187 BFA of Fukishima (Daiichi) (Japan) incident.

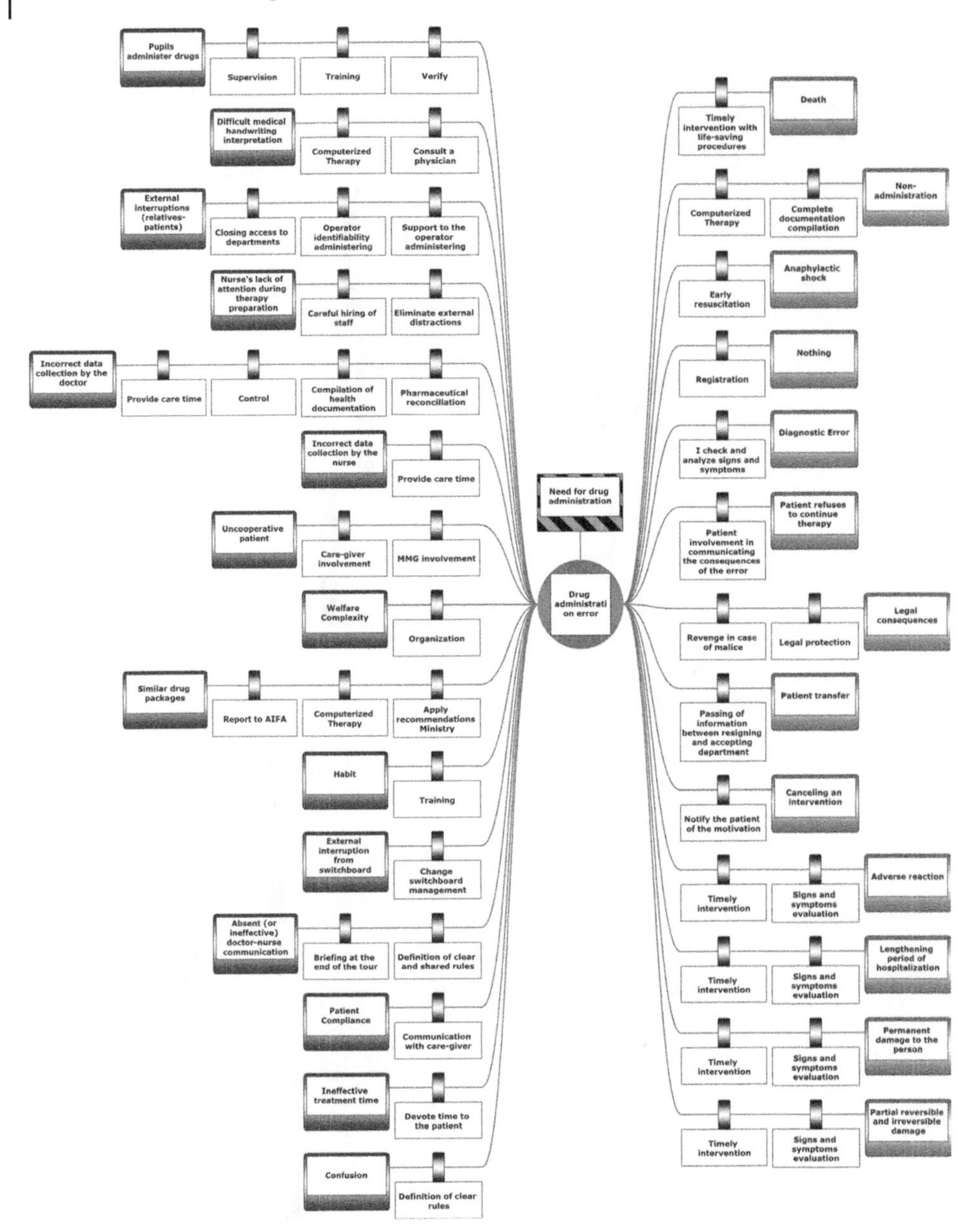

Figure 188 Drug administration Bow-Tie.

4.4.13 ThyssenKrupp Fire Investigation and Bow-Tie

The general information about the case study is shown in Table 27.

Stainless steel coils are produced throughout the world in a multitude of industrial sites. The process is conceptually simple, whereas the mechanics of the machines are rather sophisticated. The process phases, which are well known, are melting, casting, hot rolling, cold rolling, pickling and annealing. The accident discussed here occurred at a pickling and

Table 27 General information about the case study.

Who	Steel plant
What	Jet fire
When	2007
Where	Turin, Italy
Consequences	Seven fatalities
Mission statement	Determine the fire dynamics
Credits	Luca Marmo (Politecnico di Torino) Norberto Piccinini (Politecnico di Torino) Luca Fiorentini (Tecsa s.r.l.)

annealing line, two operations that are usually conducted in the same plant. Pickling and annealing (P&A) lines are conceptually very simple: steel coils have to be unrolled, then a thermal treatment is conducted in a furnace and then chemical and electrochemical pickling are performed in a series of basins. After these treatments, the coil is re-rolled. The main technical challenges of the process derive from the need to run both the thermal and the electrochemical processes continuously, even when the coils have finished. In order to comply with continuous process constraints, the subsequent coils have to be welded, and this introduces a discontinuous process. Consequently, some complications arise in the architecture of the lines. These lines should be provided with devices able to temporarily store the length of coil that must be supplied to the furnace and to the pickling section while the unrolling is suspended during welding. Further complications arise from the weight of the coils, reaching several tons, depending upon the length and width, from the need to guarantee the correct traction of the coil, and from the need to move it over several hundred meters of process line while providing adequate position control.

The coil is handled via hydraulic systems, which use mineral oil. This oil is not usually flammable, but it is of course combustible. Hydraulic circuits are fed with high-pressure oil. In this case, the pressure ranged from between 70 and 140 bar. Under these conditions, highly flammable spray/mists can originate from small leaks. Consequently, a diffused fire risk should be recognized in P&A lines. Other sources of fire hazards are the flammable/combustible materials that accompany the coils that come from the lamination process. A paper ribbon is placed between the steel coils at the end of the lamination to prevent surface damage. The paper absorbs the lamination oil residue present on the coils and sometimes sticks to it due to high temperatures and oil ageing. In this way, paper can spread along the inlet section of the P&A line, thus adding combustible material and enhancing the local fire risk.

Incident Dynamics

On December 6, 2007, five workers were on the night shift (10:00 –p.m.–6:00 a.m.) on the annealing and pickling line. Another three workers were on the line either to substitute for or train other workers. At 00:35, the line was restarted after an 84-minute stop to remove some paper lost from a previously treated coil. As there was no automatic control system on the inlet section for the axial coil position, the coil, after some time, started to rub against the line structure, which was made of iron carpentry. The location of this scraping was identified just above flattener #2, while scraping occurred to coil #1. The rubbing lasted for several minutes, and, as a consequence, produced sparks and local overheating. A local fire, which involved paper and the hydraulic oil released from previous spills, started from these circumstances. A small pool fire started in the flattener area, which is depicted in Figure 189 and Figure 190, and subsequently spread to roughly 5 m2, involving the

Figure 189 Area involved in the accident. Right, unwinding section of the line, left, the front wall impinged by flames. *Source:* Taken from Marmo, Piccinini and Fiorentini, 2013.

Figure 190 The flattener and the area involved in the accident. Details of the area struck by the jet fire, view from the front wall. *Source:* Taken from Marmo, Piccinini and Fiorentini, 2013.

flattener and its hydraulic circuits. At roughly 00:45, the workers realized there was a fire, and took some measures to fight it. First they stopped the entry section of the line, reduced the line velocity, seized some fire extinguishers, and went close to the fire to attack it from at least two directions. After some seconds, they decided to also use a fire hydrant, so one of them walked to the fire hydrant and a second one handled the fire hose. At that moment, one of the several pipes of the hydraulic circuits involved in the fire (roughly 10 mm inner diameter) collapsed and released a jet of high-pressure oil from the pipe fitting. The pipe, which is depicted in Figure 191, was fed at a pressure of 70 bar from the main pump station, which was still running. As a consequence, a spray of hydraulic oil was released into the already existing fire.

The ignition of the spray was immediate, due to the contemporary presence of the pool fire near the release point, and this resulted in a huge jet fire that struck the eight workers. Figure 192 shows a map of the site with the presumed positions of the workers (no reliable witnesses could be found concerning this topic) and the extension of the area in which the jet fire took place. The length of the jet fire has not been determined precisely, since it hit the front wall that was located at a distance of more than 10 m from the broken hosepipe. The footprint of the jet fire is clearly visible on the wall in Figure 193. The spread angle of the jet fire was roughly 30, since some scattering occurred against the various equipment. The fire also spread backwards with respect to the hose direction, and in such a manner, it involved the large area indicated in Figure 192. This spread of the fire was determined by the interaction of the fluid released at very high velocity with a number of fixed structures located in the vicinity of the release point.

A total of 13 pipes then collapsed in a few minutes. Many of these pipes were under pressure and continuously fed by the pump station, hence provoking a huge spread of oil and of the flames. The pressure in the hydraulic circuit dropped, due to the huge oil leak, thus the intensity of the jet fire diminished very quickly. At the same time, the fire reached its

Figure 191 Details of the hydraulic pipe that provoked the flash fire. *Source:* Taken from Marmo, Piccinini and Fiorentini, 2013.

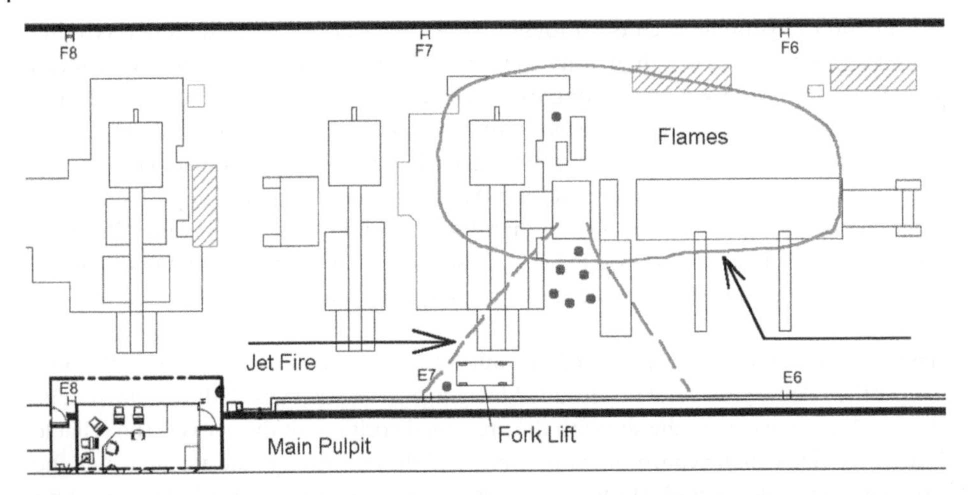

Figure 192 Map of the area struck by the jet fire and by the consequent fire. The dots represent the presumed position of the workers at the moment the jet was released. *Source:* Marmo, Piccinini and Fiorentini, 2013.

Figure 193 Footprint of the jet fire on the front wall. *Source:* Marmo, Piccinini and Fiorentini, 2013.

maximum size (see Figure 192), as it was being fed by the released hydraulic oil that burned in a pool. From an analysis of the control system records, which are summarized in Table 28, the time scale of the events resulted to be those indicated in Figure 194. The first pipe collapsed in a time interval of between 00.45'49" and 00.48'24"; the pumps were stopped by the automatic control system at 0.53'10", due to the low-level switch system on the basis of the oil level in the main reservoir. In this time interval, at least 400 L of oil escaped.

Table 28 Record of the supervisor systems (adapted from Italian).

Time	Operator/Automatic	Meaning
0.31.05	O	Set coil thickness
0.31.10–0.31.20	O	Sending data to mandrel 2
0.34.46–0.35.16	O	Start chemical section
0.35.43	O	Start inlet section
0.35.46	A	Start confirmed by field sensors
0.35.48	A	Low flow – rinsing section
0.36.06	O	Low flow acknowledged by operator
0.45.45	O	Line speed set to 18 m/min (Group 5 events)
0.45.49	O	Start pump final rinsing unit
0.48.24	A	Low oil level alarm from hydraulic station (Group 6 events)
0.48.39	A	Lubrication fault – mandrel 1
0.48.39	A	Lubrication fault – mandrel 2
0.48.39	A	Flaw fault – mandrel 2
0.48.39	A	Low pressure – mandrel 1
0.48.39	A	Low pressure – mandrel 2
0.48.39	A	Loss of control – mandrel 1
0.48.39	A	Loss of control – mandrel 2
0.48.44	A	Loss of control – mandrel 2
0.49.53	A	Fault cable inlet section
0.53.00	A	Low oil level
0.53.10	A	Emergency stop

Source: Marmo, Piccinini and Fiorentini, 2013.

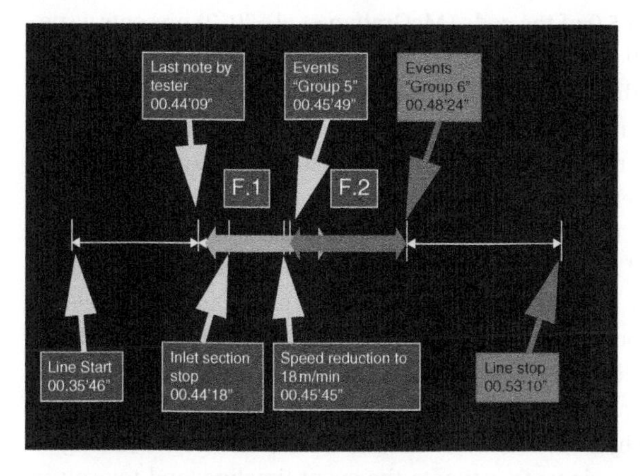

Figure 194 Timescale of the accident. F1 is the time interval in which the ignition occurred. F2 is the time interval in which it is probable that the workers noticed the fire. The group 5 and group 6 events are defined as in Table 28. *Source:* Marmo, Piccinini and Fiorentini, 2013.

The jet fire struck the eight workers who were fighting the first fire. The worker who was close to the fire hydrant (see Figure 192) was sheltered by a forklift and suffered only minor burns. Six workers were struck by the jet fire and suffered third-degree burns covering from 60% to 90% of their bodies. They died over the following months. One, who went to the back of the plant to fight the fire, was trapped and died immediately. The fire spread to the machines and lasted for approximately 2 hours before the fire brigade from the National Fire Corps could extinguish it.

Why It Happened

A specific analysis was conducted in order to understand the consequences of the accident and the level of risk for the operators. The fire scenario was modeled by means of a specific CFD calculation tool (Fire Dynamics Simulator, generally known as FDS, which was developed by the Building and Fire Research Laboratory [BFRL] of the U.S. National Institute for Standards and Technology) on the basis of the evidence and information collected during the investigation. The numerical simulation of the consequences of an accident is a useful and recognized methodology to estimate the consequences of accidental releases of hazardous chemicals in industrial premises in terms of thermal radiation, temperature rise, presence and extension of flames, smoke production, the dispersion of combustion products, and the movement of those species in the compartments under examination in order to verify what happened with a certain degree of confidence and to verify the modification of the consequences connected to the modification of the parameters that govern the accidental release. Simulation results can help technical consultants in the reconstruction of the accidental event. The well-known National Fire Protection Association (NFPA) 921 standard (NFPA, 2008) recognizes that fire behaviour numeric codes play a fundamental role for in-depth analysis in the forensic framework: both simplified routines and zone and field models are explicitly quoted. The FDS chosen by the authors is, currently, one of the most specialized and frequently used codes to assess the consequences of a fire inside a compartment, even in industrial premises, and also for forensic purposes. An extensive amount of technical literature has been published by the authors of the code (McGrattan, Baum and Rehm, 1998; McGrattan, Hamins and Stroup, 1998; McGrattan et al., 2012) and the same technical reference guide of version 5.0 of the code (McGrattan et al., 2012) presents a specific section (2.3) on the reconstruction of real fires. A number of forensic activities are listed in this section (mainly concerning the reconstruction of the consequences and dynamics of real fires). One of the fires dealt with the fire occurred in the World Trade Center on 11 September 2011 ("The collapse of the Twin Towers"). In this technical consultancy, the NIST, on behalf of the Federal Emergency Management Agency (FEMA) investigated the danger of the release of flammable liquids in the form of sprays and this danger was also assessed by conducting a number of real tests. Those tests were in fact similar to the activity that was later conducted by the U.S. Navy to test the consequences of the accidental release of hydraulic flammable oil at high pressure at a real scale and which was described in detail in Hoover, Bailey, Willauer and Williams (2008) and in a specific report (Hoover, Bailey, Willauer and Williams, 2005) that qualifies the four main objectives of the tests: to investigate the consequences of fires from hydraulic flammable fluids in submarines; to investigate the potential for hydraulic fluid explosions; to estimate the event timeline; and to acquire experimental data in order to allow a proper fire modeling to be used in engineering practice. In that report, as well as in a subsequent paper, the danger of hydraulic oil, even for releases of limited quantities of fuel, is described clearly, along with the description of the facilities used to simulate the release in the experiments. The real, full-scale experiments conducted by the U.S. Navy are comparable with the data used by the authors for the simulation of the Thyssen-Krupp accident, e.g. a pressure range from 69 bar to 103.4 bar, a released fluid with a combustion heat of 42.7 MJ/kg, and a similar viscosity. With this data, the authors found it very useful to validate the use of FDS against the results of the U.S. Navy experiments and completed the dissertation with a specific example that showed full agreement of the simulation with the results of a series of experiments conducted by the U.S. Navy, characterized by a release pressure close to 70 bar. The heat release rate and thermal conditions in the compartments can reasonably be compared. On this basis, FDS has been employed to reconstruct the accidental release and fire that actually occurred at the Thyssen-Krupp plant in Turin. The details of the conducted simulations are not the scope of this paragraph; a short description of the adopted workflow is given in the following sections in order to provide the readers with a clear picture of the procedure that has been employed by the authors as Technical Consultants of the Public Prosecutor's Office to determine the hazard associated with such an accidental event for the workers who died during the activities adopted to govern the emergency. The simulation activities helped the authors to describe the consequences of the accident in order to define the real risk for the operators and, subsequently, to verify whether the calculated risk level corresponded to the level formally declared by the owner (in the risk assessments required by law) to the authorities having jurisdiction (AHJ) and to verify whether different scenarios could have exposed the operators to similar risks (e.g. with limited releases, with retarded ignition, etc.). In particular, the authors quantified the consequences according to the legal requirements pertaining to industrial risks and identified the fire risk with respect to national law (see the threshold limits given by the national Decree dated 9 May 2001 in Table 29). Several analyses were conducted for both "jet fire" and "flash fire" cases, as defined from the extensive literature available, with the aim of comparing the results with the threshold values established by Italian law (Table 29). The simulation of the real case (i.e. a "jet fire") is described hereafter.

The simulation involved a preliminary reconstruction of the analysis domain. Several surveys were conducted to obtain a precise description of the tridimensional layout of the portion of the compartment that had to be investigated (dimensions: 12 m 10.8 m 11.2 m). The domain is presented in Figure 195 and in tridimensional form in Figure 196. The release point and the main dimensions are indicated in the plot-plan in Figure 195, while the elevation in Figure 196 shows the model and the forklift that was located in the area

Table 29 Threshold values according to Italian regulations.

Accident	High fatalities	Beginning fatalities	Irreversible injuries	Reversible injuries	Domino effect
Fire (stationary thermal radiation)	12.5 kW/m^2	7 kW/m^2	5 kW/m^2	3 kW/m^2	12.5 kW/m^2
Flash fire	LFL	½ LFL			

Source: Marmo, Piccinini and Fiorentini, 2013.

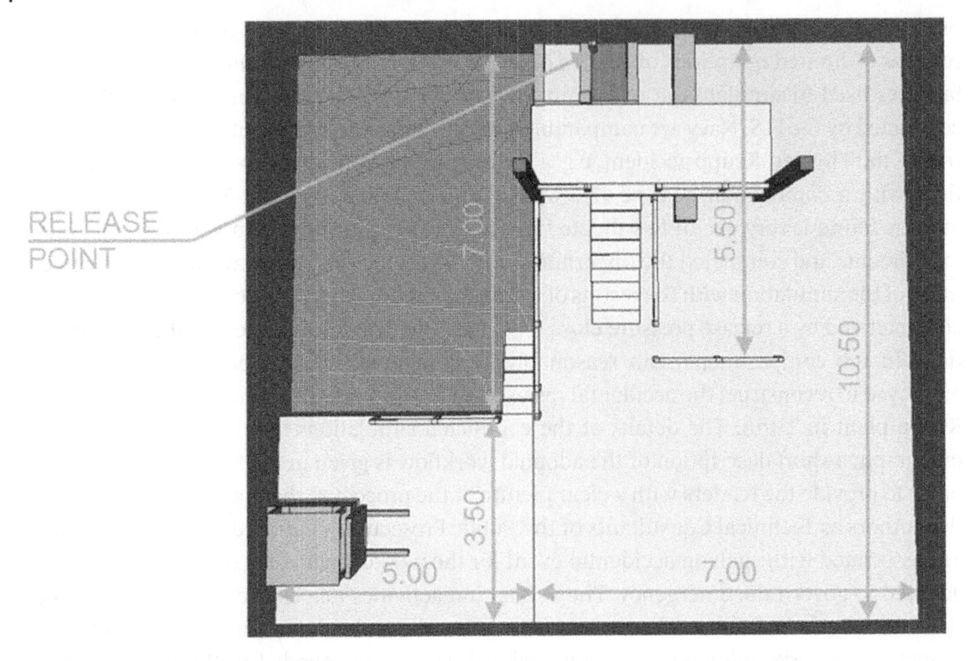

Figure 195 The domain used in the FDS fire simulations. *Source:* Marmo, Piccinini and Fiorentini, 2013.

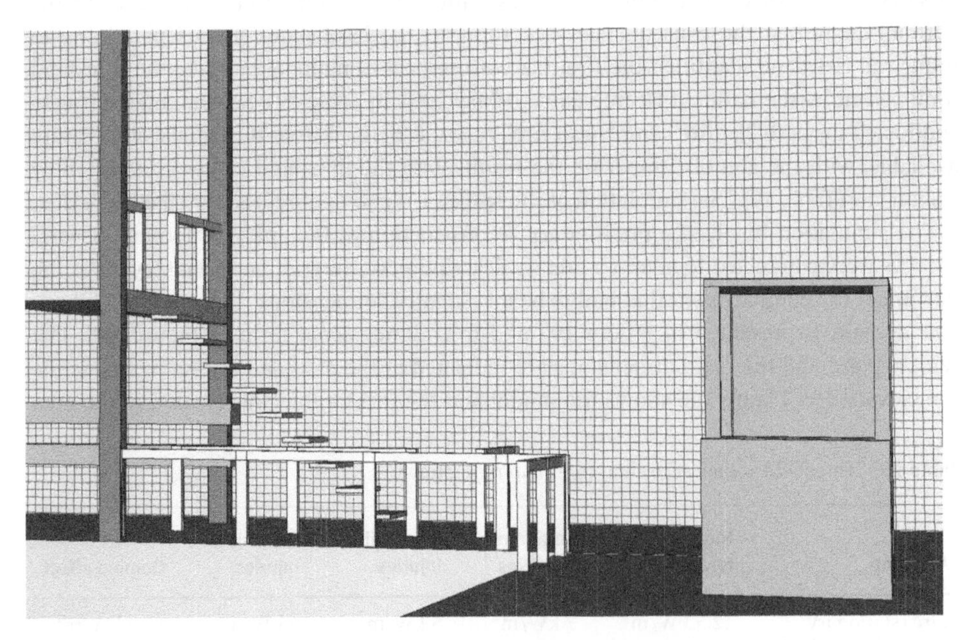

Figure 196 Simulated area, elevation. *Source:* Marmo, Piccinini and Fiorentini, 2013.

where the flames spread. The release point was located at a height of 0.5 m and identified with a circular orifice (diameter equal to 1 cm) directed toward the front wall. The analysis domain was divided into a cubic cell mesh with 1 cm sides.

The simulation of the jet fire considered an initial pressure of the hydraulic circuit of 70 bar, although a number of different simulations were run in order to verify the sensitivity of the consequences in the area where the workers were believed to be at the release time, with variations in the pressure range (up to 140 bar, which is the design pressure of the involved circuit) and a number of other parameters (e.g. direction of the release, total oil hold-up released, physical properties of the oil, etc.). This activity allowed the hazard level to be verified under different conditions and in particular to verify whether a release of a small amount of oil could have exposed the operators to danger (since a manual push button was present to limit the release via the isolation of the actuators of some hydraulic circuits).

An example of the results obtained through the use of FDS is given (the case considering a release pressure of 70 bar in the first instants from the release) in Figure 197 to Figure 202.

P&A lines at plants are commonly considered to be at high fire risk. Nevertheless, the area at major fire risk is usually considered to be the annealing furnace, where huge amounts of fuel gas (mainly natural gas) are used. Another huge fire that occurred in a P&A line in a plant located in Krefeld, Germany, in 2006, showed that the fire risk can arise from the use of annealing basins and their covers made of plastic material.

Instead, the fire risk due to hydraulic circuits, in particular in the inlet zone of the line, where hydraulic circuits are present in huge numbers, seems to have been underestimated to a great extent in the present case, but also in general in the steel industry. The inlet section of the line is a complex part of the plant, as it is composed of many devices that are activated by hydraulic circuits. Each of these circuits is generally composed of a couple of pipes, a hydraulic piston and a valve that is activated electrically. A simplified scheme of a hydraulic circuit is presented in Figure 203. The lengths of the pipes are mostly made of steel, but each single pipe is connected to a moving component (the hydraulic piston) and hence at least the last part of the pipe must be flexible. To accomplish this requirement, the terminal is usually made of a flexible, composed pipe, which is made of rubber and has a metal mesh that guarantees the mechanical performances. These terminals are connected to pistons and steel pipes via special fittings. Clearly this is, from the fire risk point of view, the weakest part of the plant. First, the hydraulic oil is combustible. Second, hydraulic circuits quite frequently suffer from leaks. There are generally two sources: the piston seals are subject to wear, which can provoke local spills, while flexible pipes are subject to fatigue which can cause cracks that can lead to sudden leaks under the form of sprays or liquid jets. Due to the frequency of these leaks, and to the huge number of hydraulic circuits, which can reach as many as several tenths in the entry section, the environment can easily become "dirty" and prone to fire if a rigorous cleaning policy is not enforced.

Other sources of combustible material can also be present in the plant. The coils can, as in the present case, come from a cold rolling line. After cold rolling, coils are re-rolled with a paper strip between the metal coils to prevent surface damage. The paper should be recovered in the entry section of the P&A line to prevent its loss along the line. Sometimes the paper sticks to the metal and its recovery is almost impossible. In this situation, the paper is spread along the line or it enters the oven. The paper is usually impregnated by the

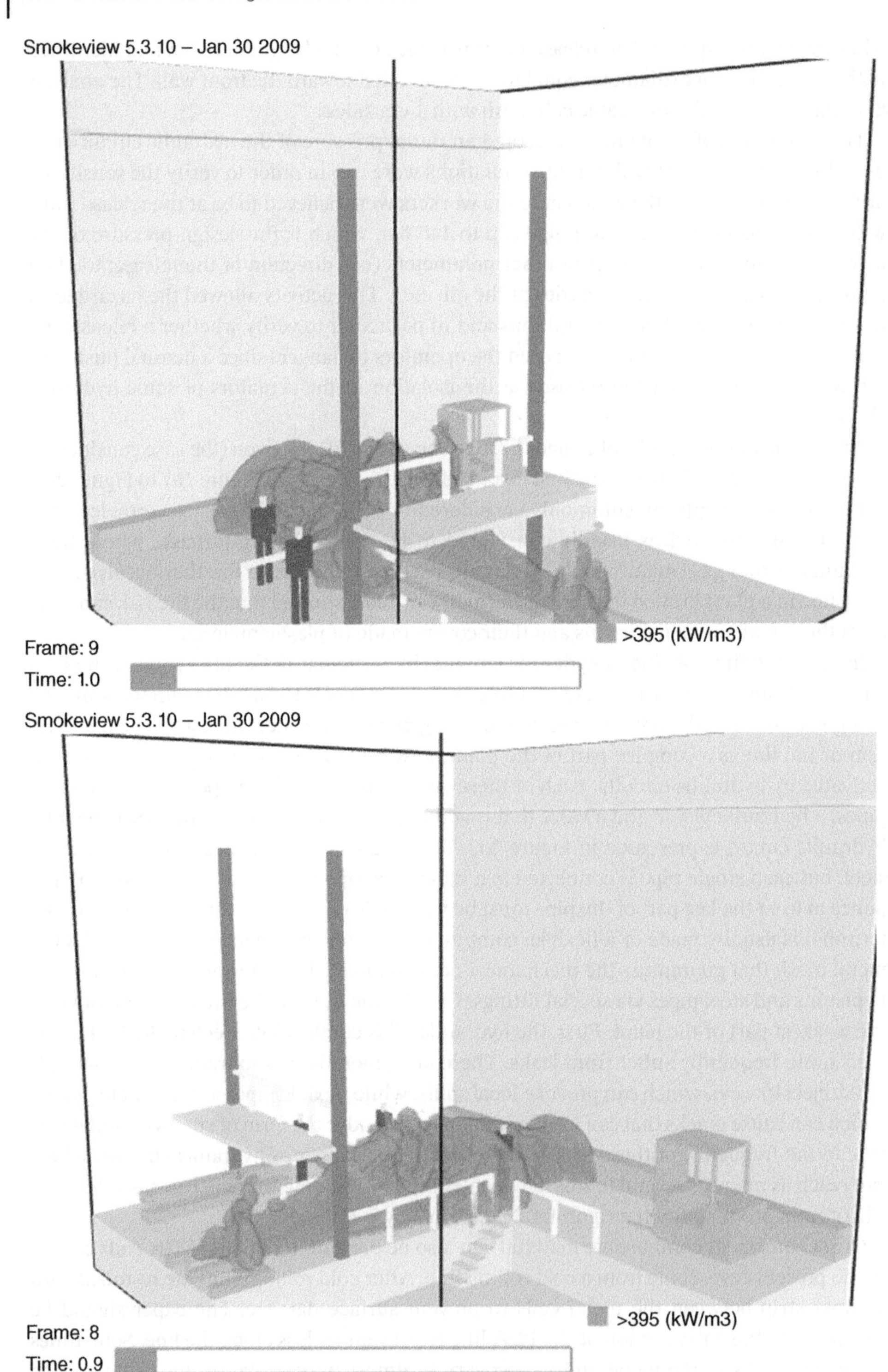

Figure 197 Jet fire simulation results: flames at 1 s from pipe collapse. *Source:* Marmo, Piccinini and Fiorentini, 2013.

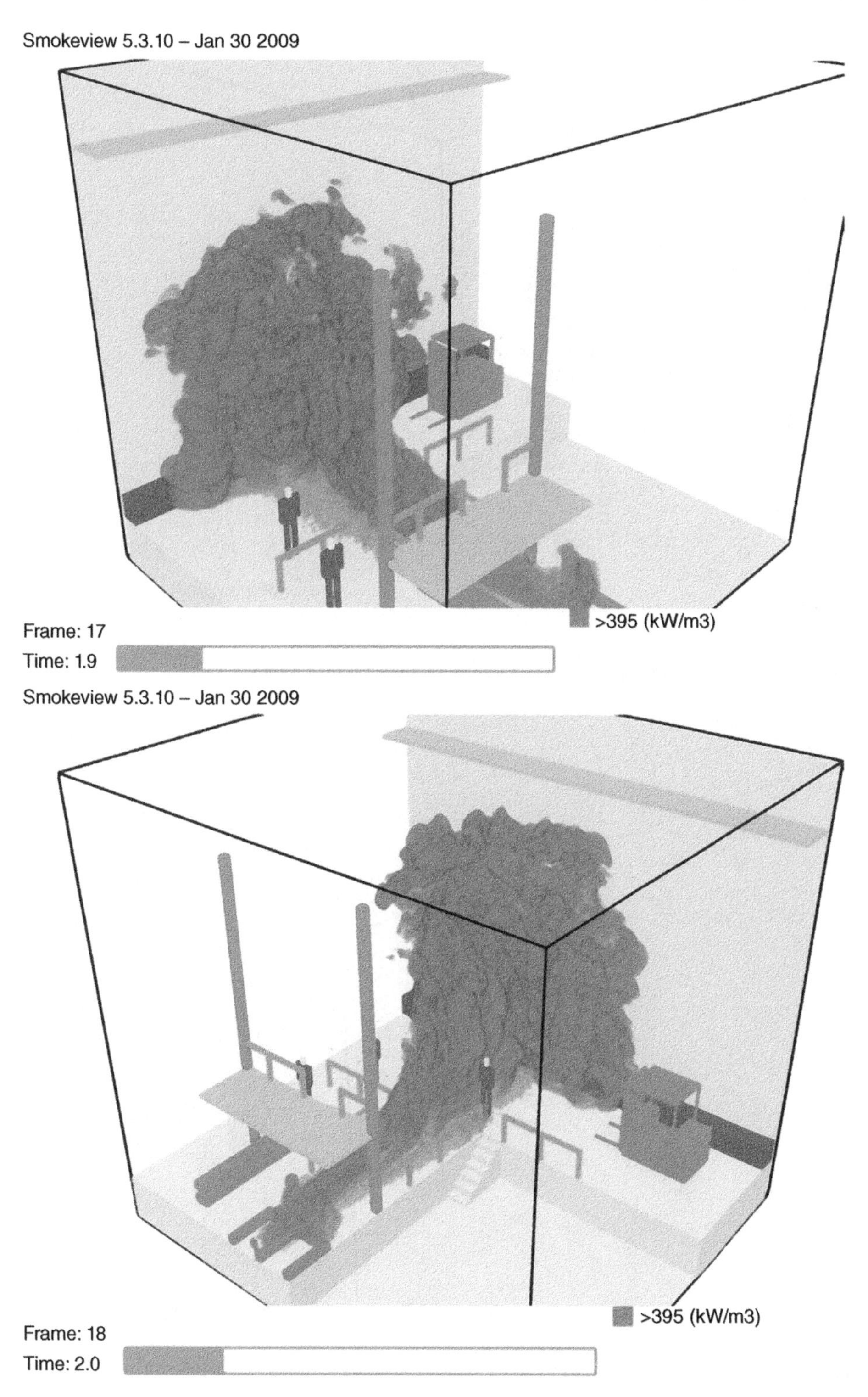

Figure 198 Jet fire simulation results: flames at 2 s from pipe collapse. *Source:* Marmo, Piccinini and Fiorentini, 2013.

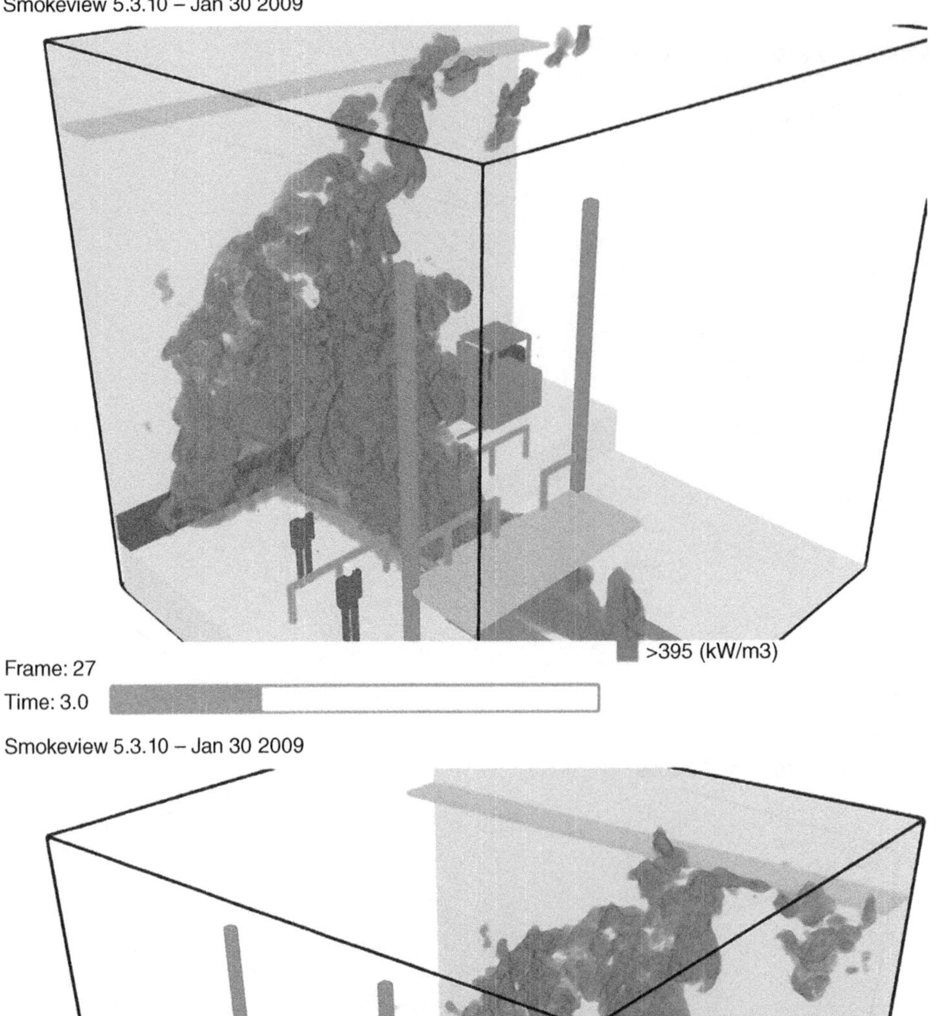

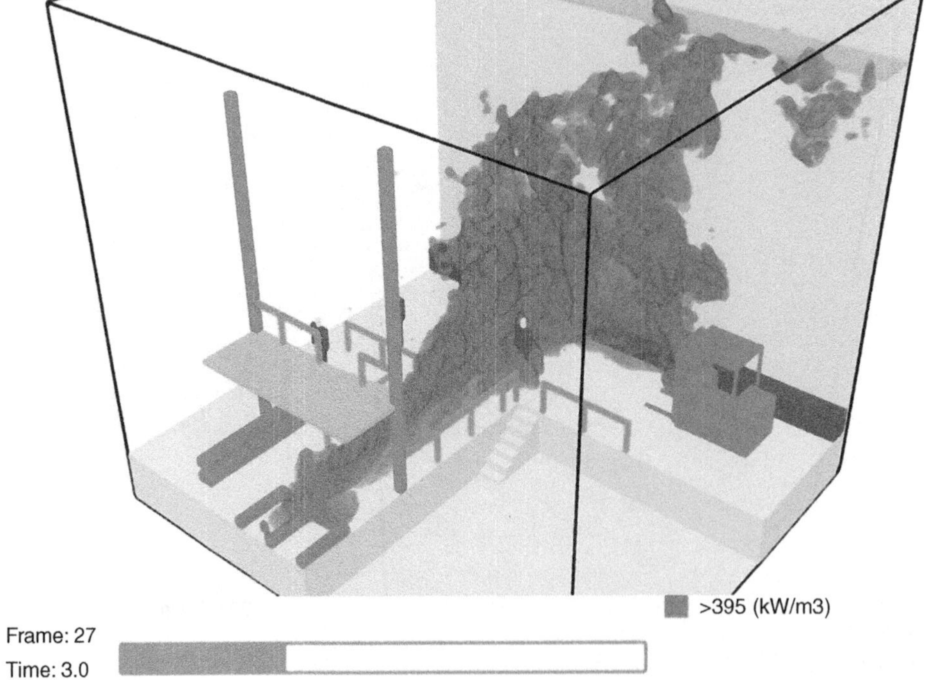

Figure 199 Jet fire simulation results: flames at 3 s from pipe collapse. *Source:* Marmo, Piccinini and Fiorentini, 2013.

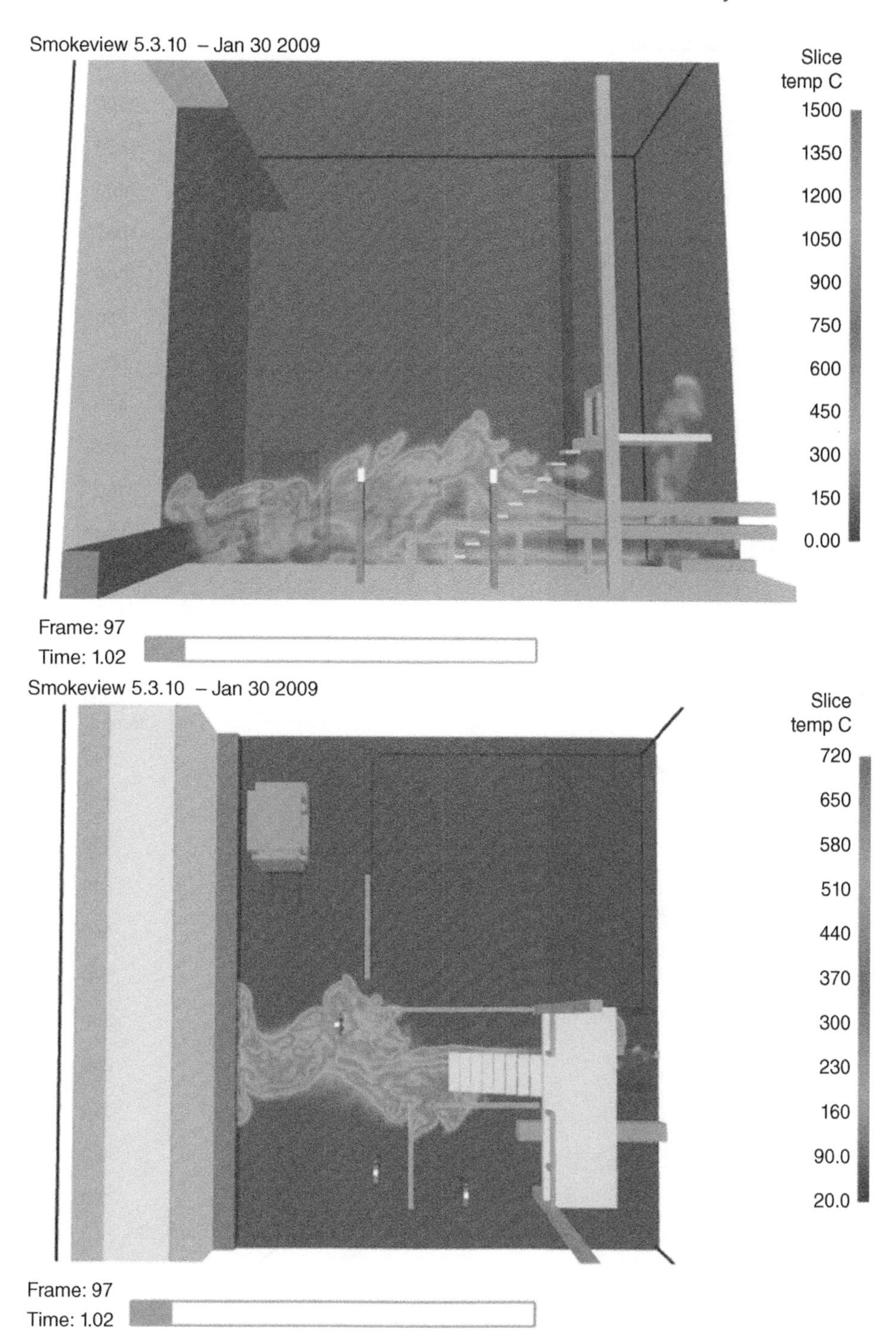

Figure 200 Jet fire simulation results: temperature at 1 s from pipe collapse. *Source:* Marmo, Piccinini and Fiorentini, 2013.

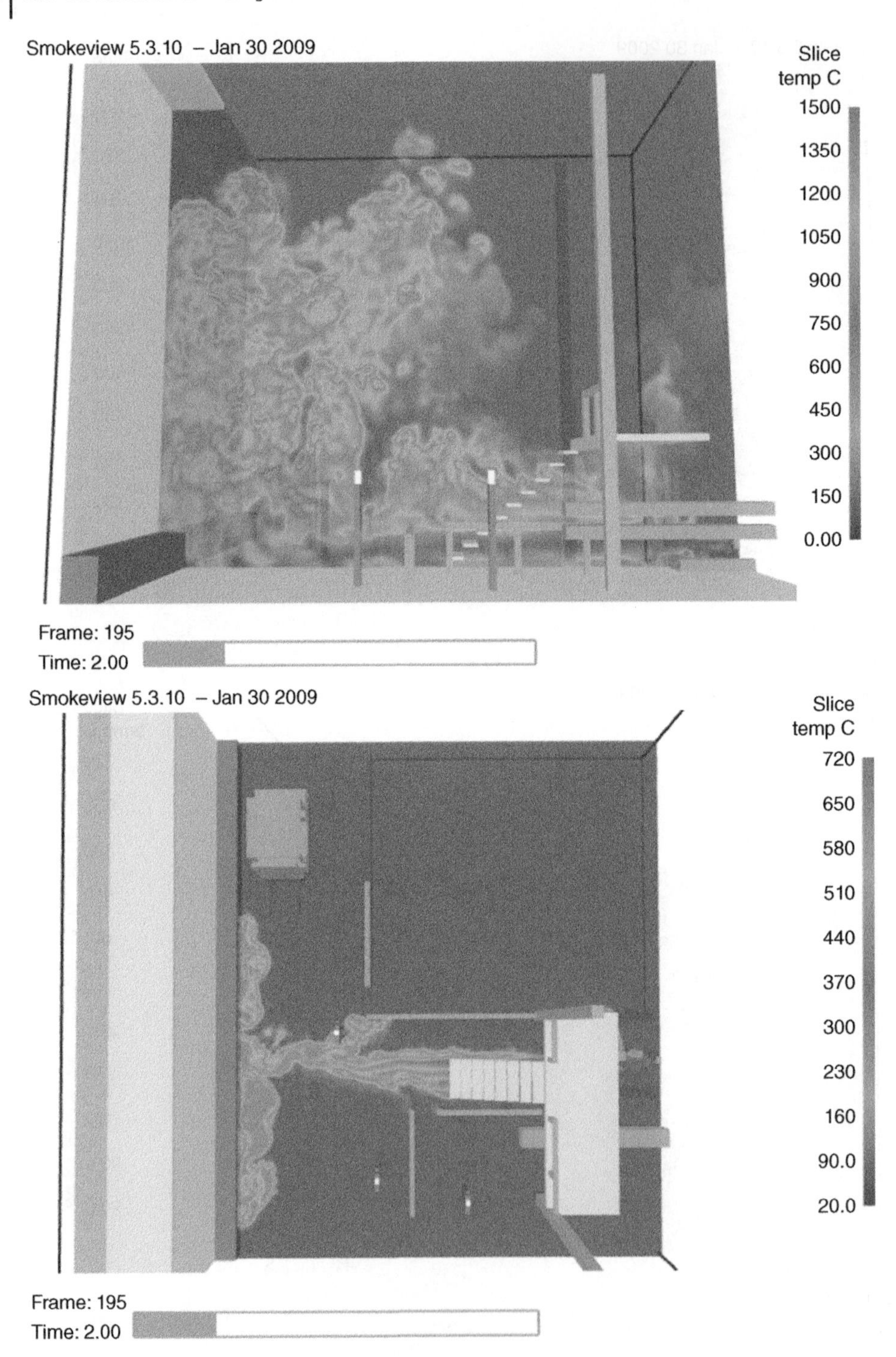

Figure 201 Jet fire simulation results: temperature at 2 s from pipe collapse. *Source:* Marmo, Piccinini and Fiorentini, 2013.

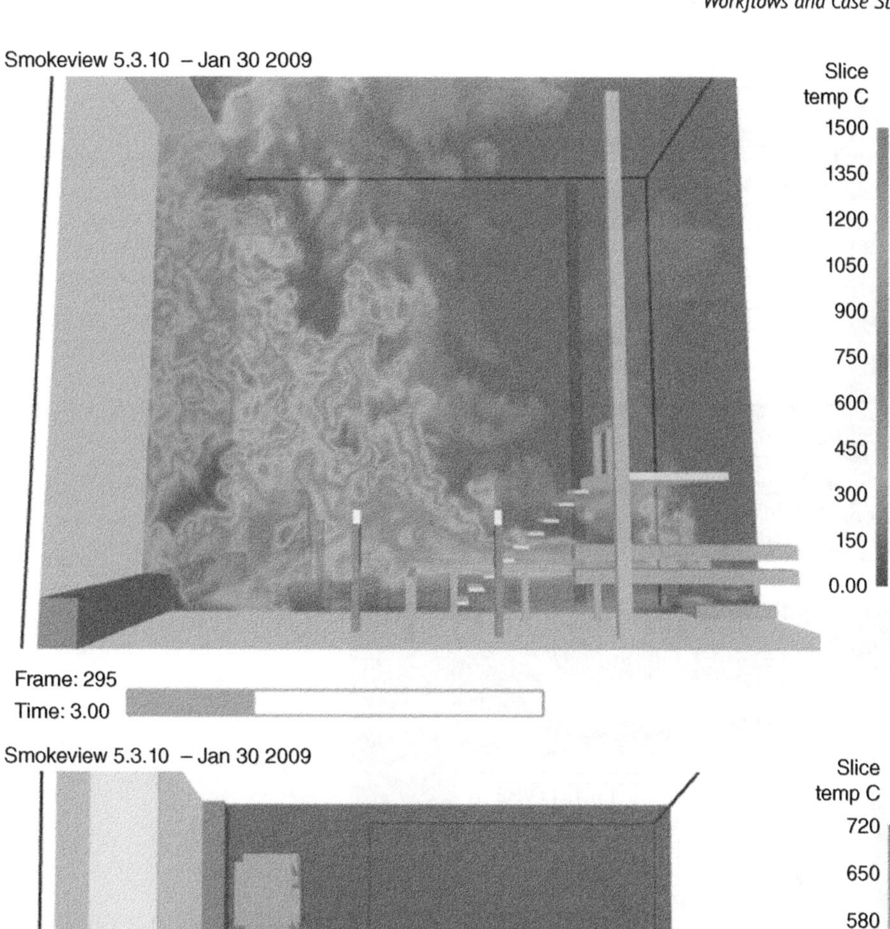

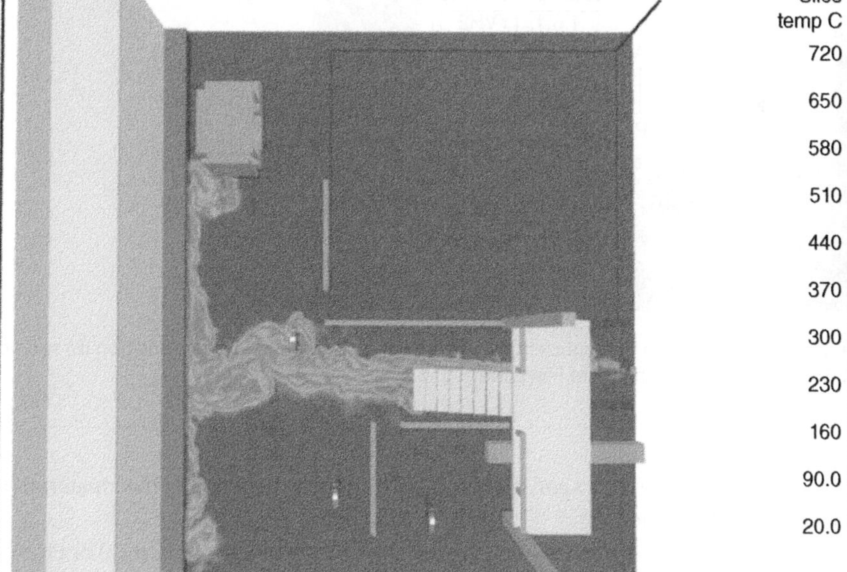

Figure 202 Jet fire simulation results: temperature at 3 s from pipe collapse. *Source:* Marmo, Piccinini and Fiorentini, 2013.

(a)

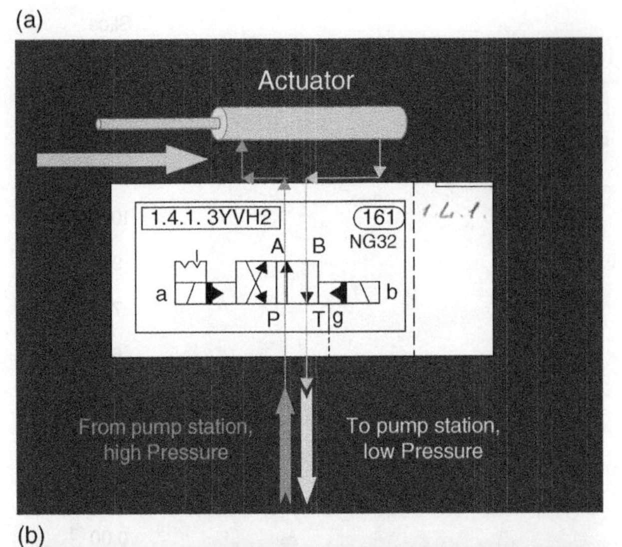

(b)

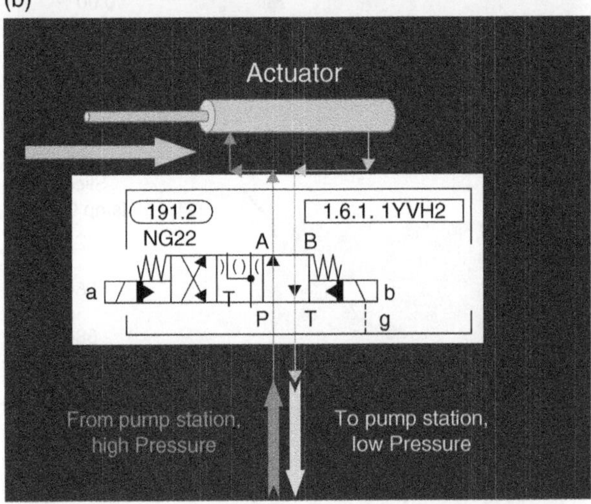

Figure 203 Scheme of the hydraulic circuits with two-position (a) and three-position (b) solenoid valves. *Source:* Marmo, Piccinini and Fiorentini, 2013.

oil used in the rolling unit. As a consequence, huge amounts of combustible materials can spread along the P&A line.

Ignition causes are also quite frequent. These are mainly due to mechanical or electrical faults or rubbing of the coil against some structural component. Scratching is far more probable in those plants, or in parts of them, where automatic coil position control devices are not present. The arc welding that is made to join each coil to the others is also a possible cause of ignition. However, this process is easy to control because welding is usually done by an operator and an eventual fire can easily be detected. Mechanical faults can occur in elements such as ball bearings, which are present in large quantities in such plants.

Findings

An accident can be the consequence of a series of undesired events, with consequences on people, objects and/or environment. The first element of the series is the primary event. There are usually many intermediate events between the primary event and the accident, which are determined by the reaction of the system and of the personnel. The dynamics of an accident generally start from a process failure, which is followed by the failure of automatic or manual protective devices. The common representation of this process, with logical trees, involves the use of logical gates (generally AND, OR).

Intermediate events are, in many cases, the condition in which the chain of events interacts with the action of protective devices. When these devices are successful, the chain of events is blocked; therefore, the intermediate events correspond to conditions that contribute to decrease the likelihood of the top event. Protective devices can of course act either automatically or manually, which implies the implementation of procedures whose success depends on the level of training of the personnel. As a consequence, the expected frequency of a top event can be reduced by first and foremost adding protection devices, and then acting on the failure rate of the involved components or improving the training of the personnel, thus reducing the probability of human failure.

Hereafter the representation of the dynamics of the ThyssenKrupp fire is proposed using an FTA (Figure 204), in which the INHIBIT gate is also used to represent the failure of the protective devices. This is substantially a variation of the AND door which is used in the case of protective means. The INHIBIT gate can distinguish between the entering events since the event entering from the bottom can propagate to the event at the top outlet if the side event, which is represented by the unavailability of a protective device, has already occurred (Demichela, Piccinini, Ciarambino and Contini, 2004; Piccinini and Ciarambino, 1997). The dynamics of the accident are represented considering the failure of the existing protective devices and also of those that the plant was not equipped with, which are indicated in the grey boxes.

These are:

- Automatic shutdown of the hydraulic circuits;
- Automatic fire extinguishing plant;
- Automatic coil control position in the inlet section;
- Fire detection systems.

Lessons Learned and Recommendations

The TK accident that occurred in Turin in December 2007 offered some very important lessons about the fire risk for A&P lines. The dispute about the risk level of these plants was ongoing, with some technical experts (the minority) affirming that prevention and protection tools, such as automatic extinguishing plants, were necessary, while the majority of technicians declared that, in most cases, they were not necessary. It seemed that, at that time, there was a general conviction that only some parts of the line needed specific fire protection equipment. The pump station (which is often located in a separate compartment together with the oil reservoir), was of course considered to be a high-risk zone. The oven zone was also considered to be at a high risk, considering the elevated temperature reached there and the presence of large amounts of natural gas. In some cases, the pickling section

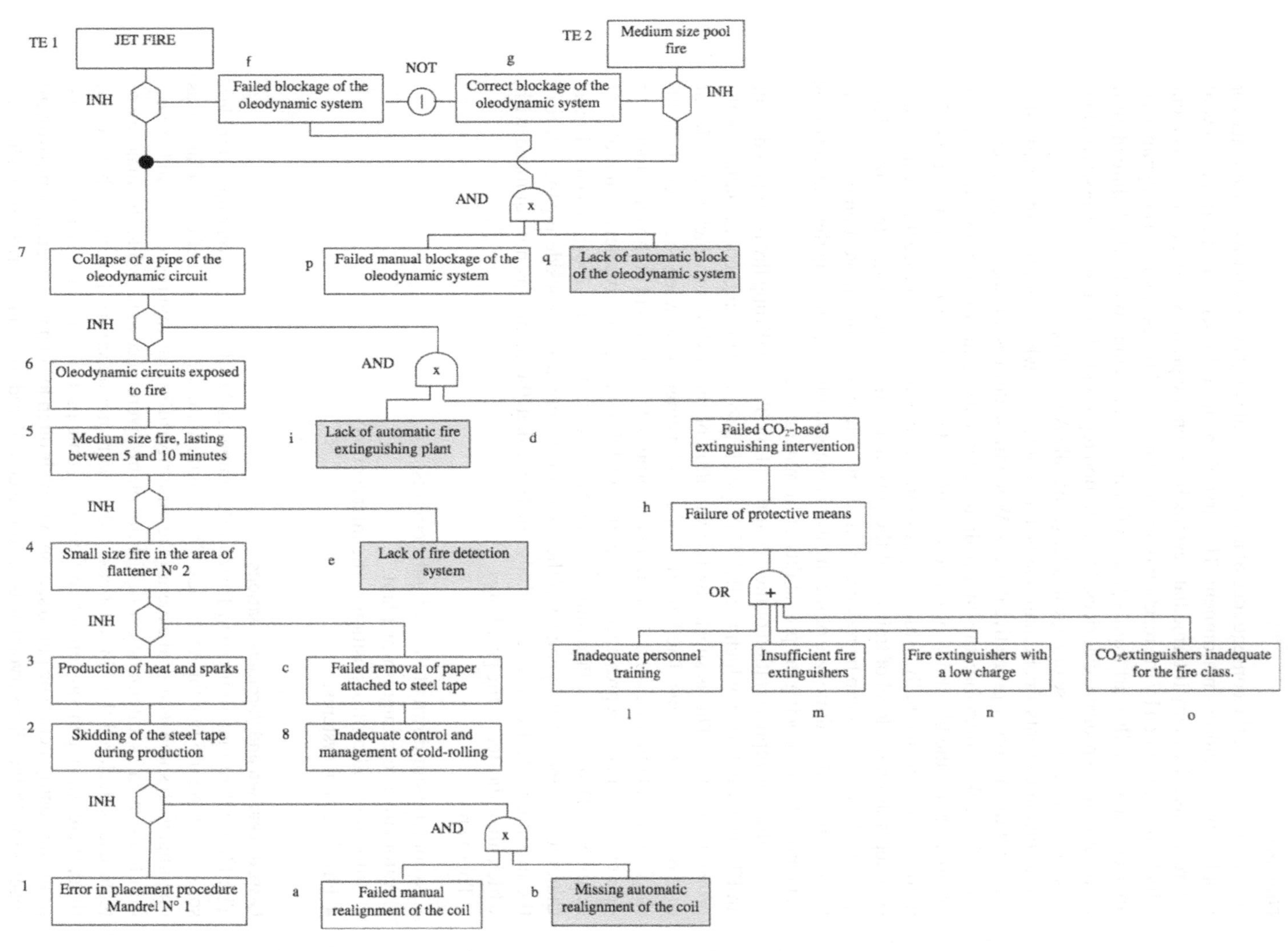

Figure 204 Event tree of the accident. The grey boxes indicate a lack of safety devices. *Source:* Marmo, Piccinini and Fiorentini, 2013.

was considered to be at a high fire risk, when the pickling pools and/or covers were made of plastic, as in the case of the fire in Krefeld.

However, the risk associated with high-pressure hydraulic circuits and with the potential release of huge amounts of oil because of pipe failure had been largely underestimated.

This case is a clear example of how simple it would have been to adopt measures to reduce risks that would have prevented the accident from occurring, also on the basis of what has already been stated in the extensive technical literature available (e.g. the NFPA considerations in the well-known *Fire Protection Handbook*, 1997).

In order to better understand the lessons that can be learned from this case, the dynamics of the accident were represented in the fault tree shown in Figure 204. It is easy to individuate the technical improvements that could have prevented the accident from the figure. Consequently, the following points can be forwarded.

- *Coil position control.* In this case, the first automatic control position was located some tens of meters downstream from the coil unrolling station. This position corresponded to deflector roll N-1. Only manual position control devices were present in the inlet area. If an automatic position control system had been located some meters downstream from the payoff mandrel, the rubbing and consequent ignition would have been much less likely.
- *Housekeeping.* This, in general, was insufficient at the moment of the accident. The spread of the fire in the early phase is a direct consequence of the presence of paper and oil on the site. As far as good housekeeping is concerned, it is generally recognized as critical to diminish the fire risk; in many fire accidents the housekeeping has been inadequate. From a more general point of view, the cleaning of such plants is not a simple task, because of the complex location of the machines. Periodic cleaning by external enterprises, as in the present case, would not have been sufficient since the spread of paper cannot be forecast as it is something occasional whose occurrence depends above all on the control of the rolling phase.
- *Fire load and fire detectors.* In this case the zone involved in the accident was not equipped with fire detectors. The reason for this is that the fire load, according to a "traditional" calculation approach, was considered negligible by the technical expert who made the analysis. This accident in fact demonstrates all the limits of such an assumption. Although the fire load was very small, the potential for a huge release of combustible and hence for a huge fire was due to the conformation of the hydraulic circuits. If a fire detector system had been present, the workers would likely have discovered the fire in its early stages, and would therefore have had a much better chance of controlling its dynamics.
- *Automatic fire extinguishing systems.* This point is very similar to the previous one. No automatic extinguishing system, which would have limited the exposure of the workers to the fire, was in place. It is evident that the fire risk depends to a great extent on the chemical-physical conditions of the substances, as well as on the plant attitude to provoke a sudden release, which in turn depends on the structure and on the control strategy of the plants. In the P&A industry, there is extensive use of hydraulic circuits in areas in which ignition can occur at a non-negligible rate. This accident has shown that the perception of the fire risk of such units had been underestimated. Traditional hydraulic units, which are not equipped with adequate emergency stop devices (see the

next point) are subject to the risk of jet fires due to sudden releases. The presence of personnel in these areas should be limited as much as possible especially when a fire breaks out.

- *Emergency stops*. The emergency stopping of complex units, such as P&A lines, is a challenging task. The emergency shutdown of hydraulic units usually leads to a loss of control of the actuators, therefore uncontrolled movements generally occur. These, which can potentially involve very heavy loads, can easily cause damages and injuries. Despite these aspects, this accident has clearly demonstrated that hydraulic units should be shut down when fires break out in order to avoid huge releases and the spread of fires. The lines are generally provided with different emergency stop levels, as in the examined case. The first level corresponds to a shutdown that is operated by the actuators and consists in the solenoid valves being placed in the central position. This kind of intervention obviously only concerns those circuits that are equipped with three position valves. This strategy cannot be considered adequate to mitigate the risk of fire since some circuits, such as the expanding mandrel, which are equipped with two position valves, are not shut down. Hence, the risk of releases and of jet fires is still considerable. Moreover, in the case of a manual shutdown strategy, such as in the present case, the personnel could be involved if an accident occurs. It seems necessary to prevent the personnel from entering the area of the accident as much as possible, and for those that are present to leave the area, until the emergency procedures are activated.
- *Emergency procedures*. The emergency procedures in the case of fire were: "In the case of fire, if the person is trained, he should immediately start to fight the fire using the available equipment. When the fire is judged to be of 'evident gravity,' the person should call the surveillance personnel and then wait for the internal emergency team to arrive. The emergency team should be activated by the surveillance personnel. The emergency team, once on site, should turn the power off the area, look for missing personnel and fight the fire with the available tools." In the present case, the procedure failed to prevent the accident because the fire was not evaluated to be of "evident gravity" by the personnel on the site. This case demonstrates, in a very evident manner, how it can be a critical task to judge the gravity of a relatively small fire in a complex industrial context such the one described here. If such a procedure fails in the most critical phase, in which the decision on how serious is a fire is must be made very quickly, personnel are exposed to serious risk. Procedures that do not involve, or at least limit, personnel judgments as much as possible should be introduced when a huge risk is detected.

Forensic Engineering Highlights

CFD analysis played an important role in this incident analysis. On the basis of the analysis of the results, which were obtained through the use of the official FDS graphical post-processor (Smokeview© by NIST) and from the record of the temperatures via virtual thermocouple devices (19 devices positioned at the height of a person), several considerations have been made:

- The jet fire involved the entire area opposite the release point in a direct manner and immediately created a serious risk for the workers who were in the area of interest.

- In fact, conditions that could have led to fatalities were recorded almost immediately.
- All the reference thresholds (Table 29) for fire thermal radiation were reached in the area (e.g. incident radiant heat, with a value of 200 kW/m^2, on the wall in front of the release point).
- A risk arose from both the direct and non-direct effects of the jet fire and it was also related to the flame extension, thermal radiation and temperature rise in the compartment area.
- The previous effects were recorded from the first instants after the release.
- The combustion of the hydraulic oil was characterized by a huge amount of smoke and soot, which made the conditions in the area worse.
- The jet was vertically and horizontally fragmented as a result of the impact with the wall (and as a result of the impact with the main plant structures in the area). This fragmentation allowed a flame wall to build up that divided the compartments into two parts (thus creating problems for the emergency procedures) and it determined more serious conditions in the area than a similar jet without the presence of fixed obstacles. The amplitude and the dispersion of the jet was in line with the status of the compartment after the real fire (see Figure 189 to Figure 193).
- The forklift could have protected a worker located behind it from the effects of the jet fire; this has been confirmed through an evaluation of the damage to the forklift itself, which presented different levels of damage on the two sides, as shown in Figure 205 (the damage is consistent with the shape and effects of the simulated flames).
- The release of hypothetically smaller quantities of hydraulic oil could have exposed the workers who were possibly located at a distance of 15 m to a serious risk (with limited effects compared to the real jet fire that occurred but which could have significant consequences on people); the same considerations were made for a hypothetical flash fire, considering that the reference thresholds were reached in the same area (a distance of 15 m for 0.5 LFL from the release point in the case of the release of 500 cc of hydraulic oil).
- The real conditions could have been judged to be worse than the simulated ones because the simulations only considered one single accidental release; it is very probable that the real evolution of the fire involved several subsequent collapses. The simulation of contemporary or slightly delayed releases from different sources would have led to a more significant impact, in terms of consequences for both accidental events (jet fire and flash fire).

Among the forensic engineering highlights of this case study, there is also the logic path leading from evidence to deductions. This effort is summed up in Table 30.

The sequence of the events has been reconstructed with a video, which was used to show the incident dynamics during the trial. In particular, the public prosecutor used the video to show the judge the dynamics of the indicent and underline the sequence of the events along with some key factors that contributed to such severe results. The video environment has been built from the photos of the real incident scenario, to highlight the site conditions. It reproduces the sounds during the work activities, and it uses 3D images to reconstruct the movement of the victims, on the base of the collected evidence (witnesses). Figure 206 collects some frames of the video.

Figure 205 Damages on the forklift.

Table 30 Summary of the investigation.

Activity	Evidence	Deductions
Site survey	Heat and fire damage. State of the coils, position. Scratching of coil against the carpentry (290 m). Paper spread along the line. Residue of carbonized paper in the area of the flattener.	Area reached by the jet fire. Area reached by the fire. Axial coil position not correct. Shift of coil No. 1 toward the side carpentry. Scratching lasted several minutes. Scratching occurred above the flattener. Combustible material in the area.
Documents on risk analysis	Fire risk evaluation. The area was considered at medium risk according to Italian regulation.	No fire detection systems were provided.
Documents examination pertaining to technical description of plants	The complete inventory of the hydraulic circuits involved in the fire. Pipe state at the moment. Circuits working pressure were identified.	The pipe that collapsed first was identified, on the basis of position, direction, and because it was under pressure at normal conditions.
Witnesses	The size and position of the initial fire. The size and shape of the jet fire. Fire growth rate (roughly).	Small fire on the flattener at the beginning. Fire grew in size after the first attempt to extinguish. Fire extinguishers unfit to control the fire. Sudden jet fire spreading "like a wave."
Electronic data	The timescale of the events.	Line start at 00.35'46". Speed reduction by workers at 00.45'45". The PLC lost the sensors located close to the flattener at 00.48'24". Line emergency stop (automatic) at 00.53'10" due to low oil level.

Bow-Tie Diagram

At the end of this complex investigation, it is possible to summarize the main relevant information in the Bow-Tie shown in Figure 207, where all the possible outcomes have been depicted, considering what would have happened in the different combinations of functioning or not of the mitigation barriers, in fact proposing a scheme that could be superimposed on an event tree.

4.4.14 Twente Stadium Roof Collapse Tripod Beta Analysis

On July 7, 2011, during work to extend the De Grolsch Veste stadium, the roof of the extension collapsed. FC Twente wanted to increase stadium capacity by further extending the L-shaped extension completed in 2008 into a U-shaped one.

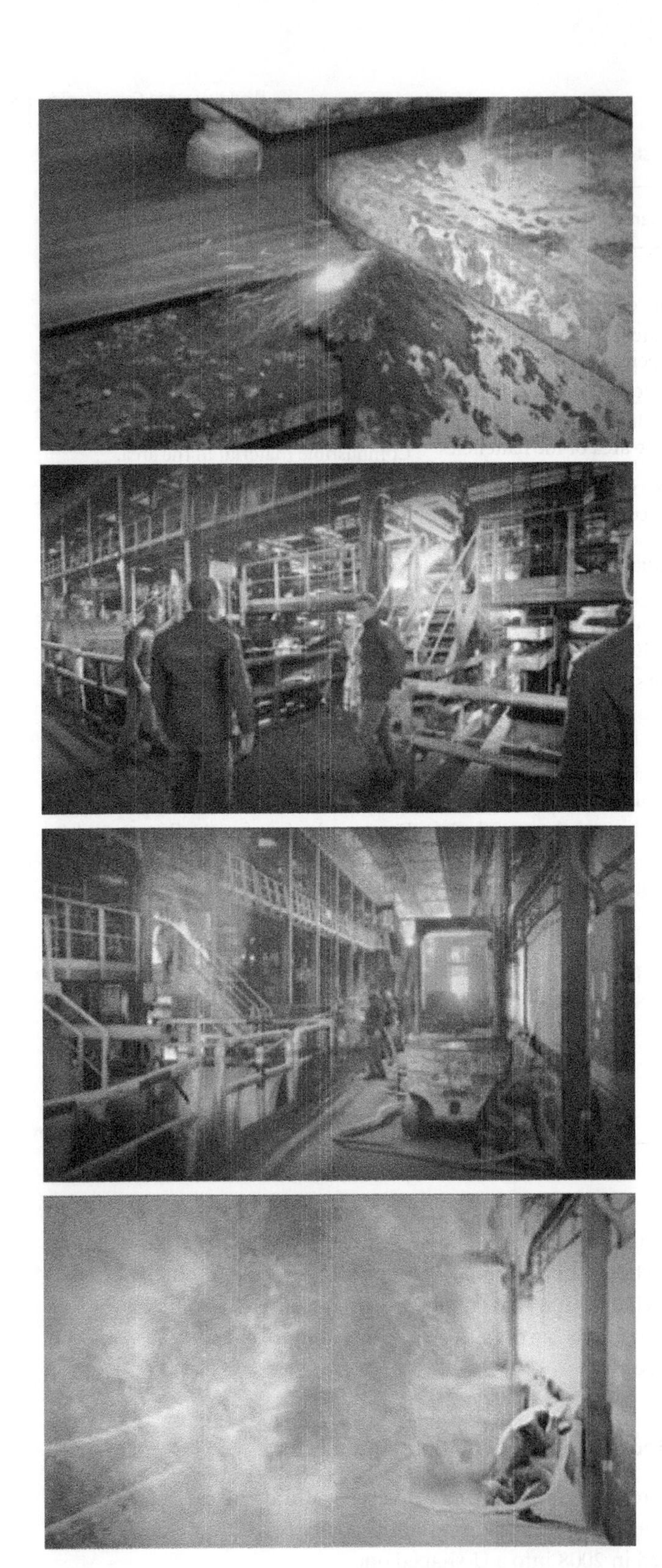

Figure 206 Frames from the 3D video, reconstructing the incident dynamics.

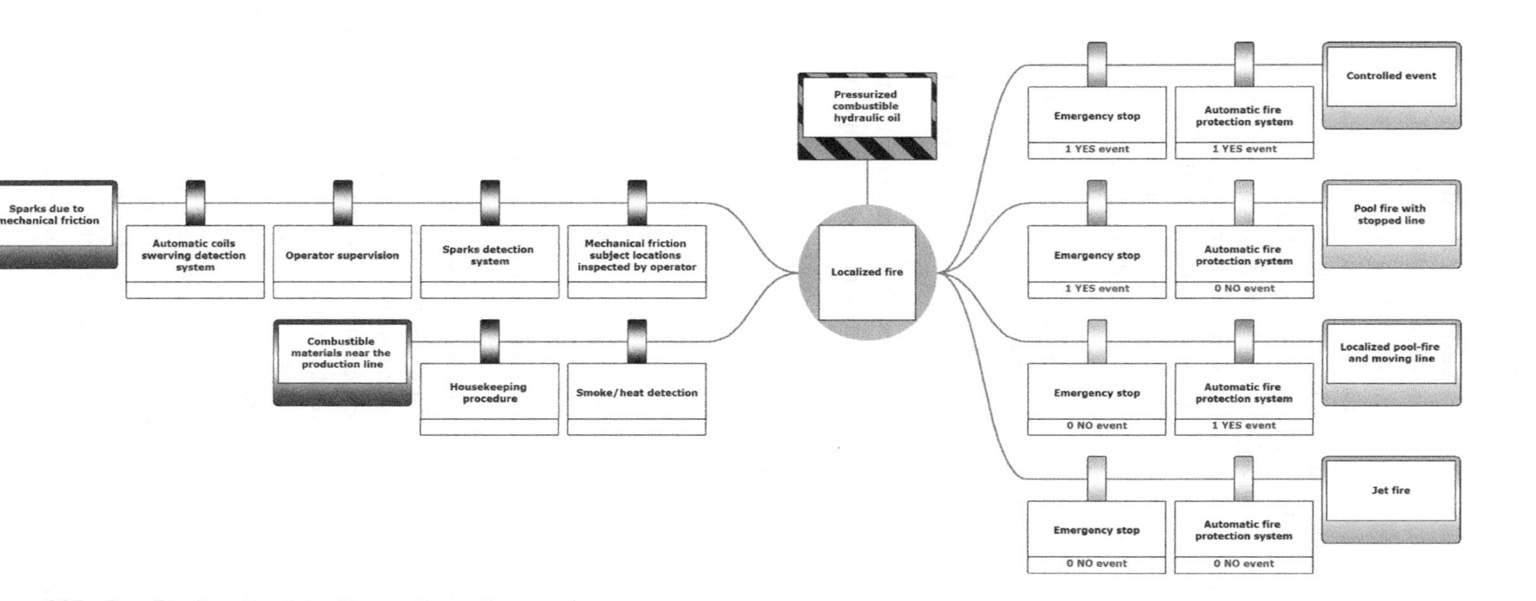

Figure 207 Bow-Tie diagram of the ThyssenKrupp fire.

The investigation conducted by the Dutch Safety Board (Figure 208) revealed that the roof structure's insufficient stability, and therefore the risk of collapse, was caused by several factors (Dutch Safety Board, 2020). The main factor was the absence of essential coupling pipes at the back ends of the roof beams and stabilizing connections in the roof structure. During assembly of the roof beams, steel cables were used as a temporary stabilizing measure. The last stabilizing cable was removed on the day of the incident. In addition, the roof structure was already being subjected to additional loading by a video wall, suspension bridges, piles of roofing sheets and the workers present. The investigation also revealed that the roof structure was being subjected to additional loading as a result of dimensional differences between the concrete beams of the stand, the foundation of the steel structure and the steel structure itself. These dimensional deviations in combination with insufficient adjustment options meant that parts of the roof structure could only be inserted by exerting deforming forces. The deformation caused additional tension that reduced the load-bearing capacity. The combination of tensions in the structure as a result of its own weight, dimensional deviations, the load already present and the absence of stabilizing measures caused one of the roof beams to fail as a result of the forces to which it was subjected, which initiated a total collapse.

As a result of this accident, twelve workers fell from a great height. Two workers were killed and nine injured, a few of them critically. One worker escaped with bodily injury.

4.4.15 Water Treatment Bow-Tie Analysis

Water is a precious good for life, including human life. The Bow-Tie shown in Figure 209 aims to analyze the risks associated with cyanobacteria blooms in a dam.

4.4.16 Operational Experience Analysis Using BFA

Deadly Explosion in a Cold Room

The example proposed in this section has been anonymized for educational purposes.

The Facts On the morning of the day 26 June 2018, some maintenance work was underway inside a room used as a refrigerator of a wholesaler, aimed at restoring the insulating layer of some walls of the room. This restoration was carried out by insufflating expanded polyurethane foam at the points where the insulation has retreated. The employer is the one to carry out the work. After a coffee break, the employer asked Operator A to replace him in the continuation of maintenance activities. At 12:06 p.m., an explosion occurred inside the premises. The burns sustained by Operator A are fatal, causing their death after three days.

Initial Survey The initial inspection of the investigators revealed that the room was not equipped with a forced ventilation system and that, on the spot, there was an electric drill used to drill holes in the walls covered with a metal sheet, in order to insufflate the polyurethane foam inside. The survey also reveals eight cans of polyurethane foam already used, and several others still not used. The room has no windows and has only one access door. No further items are detected.

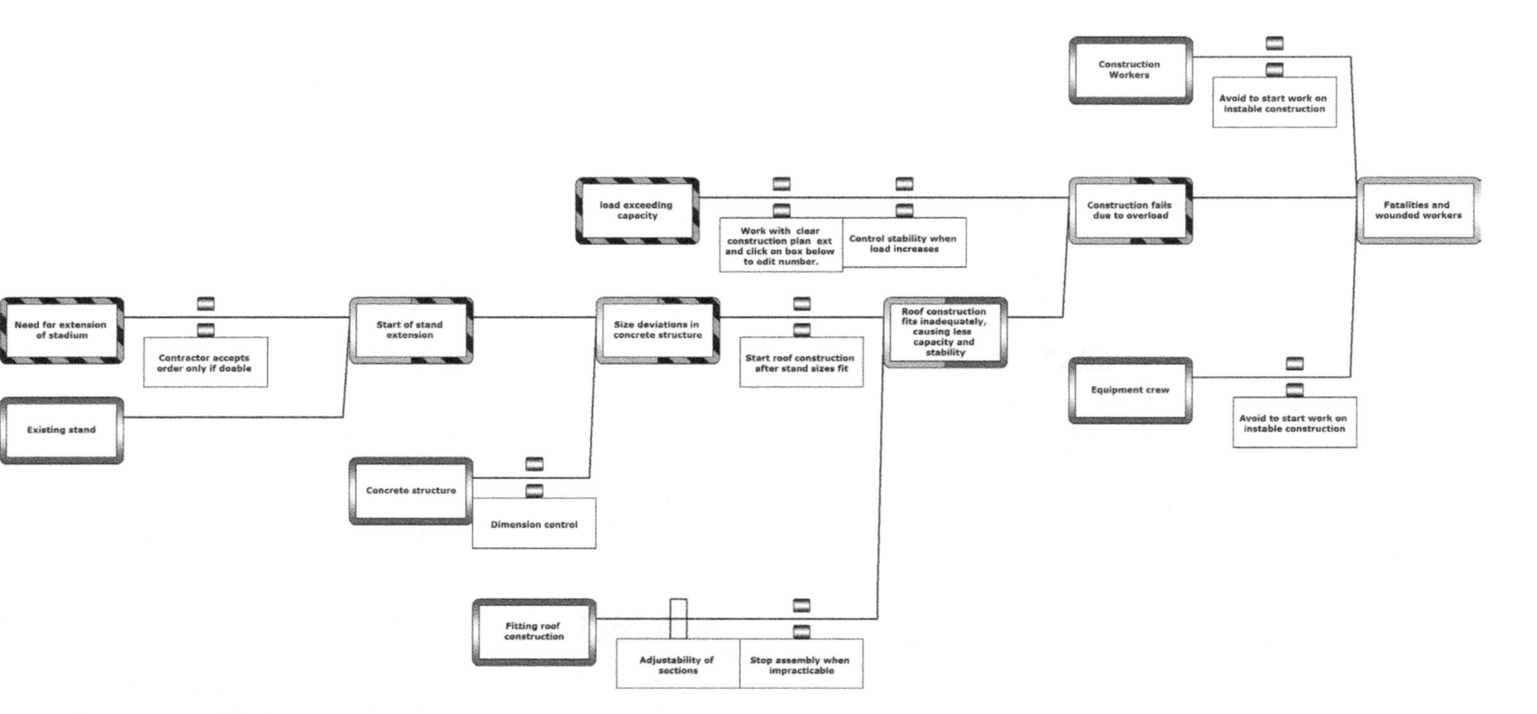

Figure 208 Twente stadium roof collapse Tripod Beta analysis.

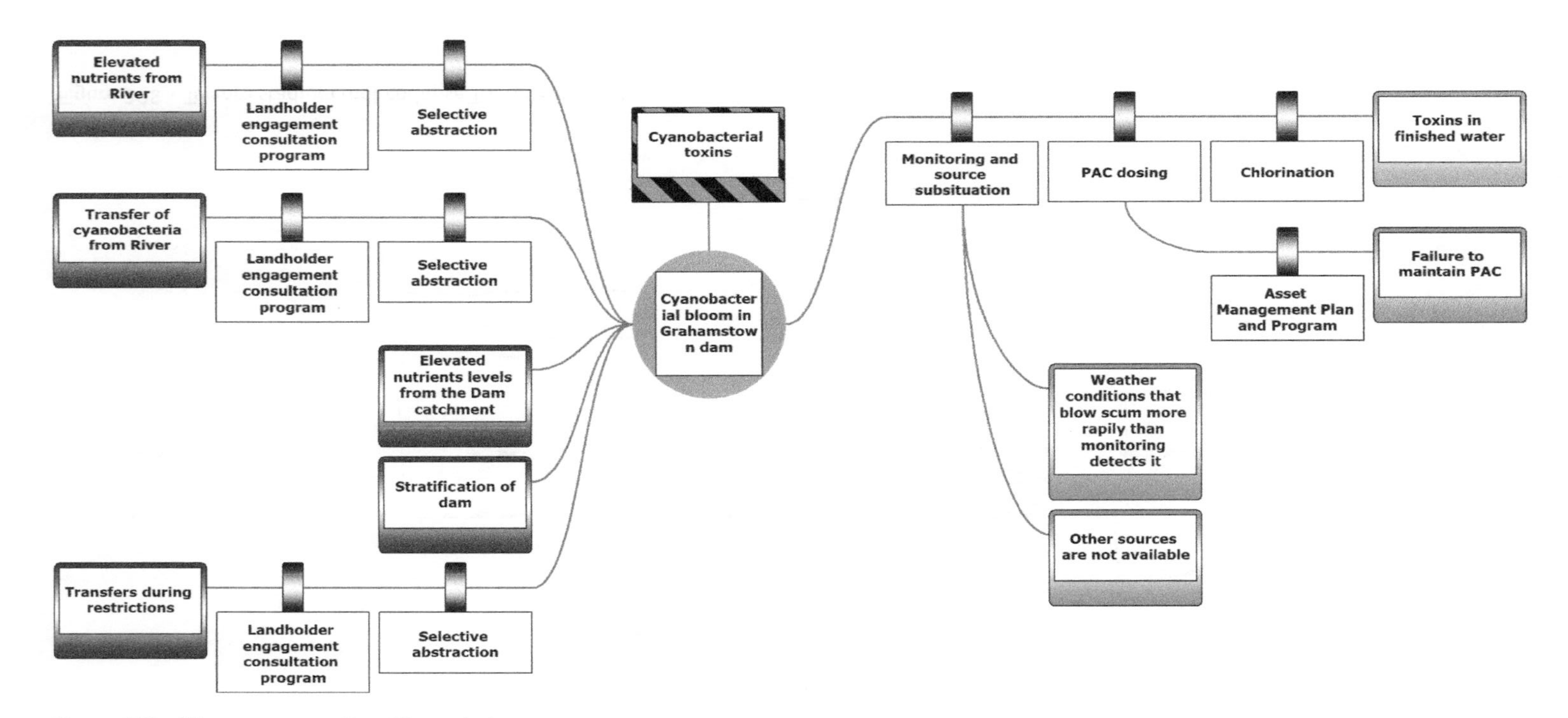

Figure 209 Water treatment Bow-Tie analysis.

Evidence Collection Investigators seized three cans of polyurethane foam, among those not yet used, in order to have the contents analyzed in the laboratory. The analysis shows that propellant gas is a mixture of highly flammable propane-butane gas and, as the product safety sheet reports, capable of generating explosive atmospheres indoors and unventilated environments. The drill, which has some partially melted plastic control levers, is also taken. In addition, the employer and other workers in the organization are questioned. The testimonies agreed in reconstructing the facts presented in the premise. One of the workers revealed that the deceased colleague was not wearing the IPR planned for processing and that, at the time of the explosion, the access door to the room was closed to facilitate the passage of personnel in the small corridor outside the room.

Incident Analysis From the acquired information, one can build the timeline of the incident. Note that one doesn't necessarily need a quantitative timeline definition. In this case, in fact, the time moments of the events reported in it are not known, but the sequence and a qualitative estimate of the durations are known. In the timeline, it is possible to add events that, at this stage, are only hypothesized in order to ensure a logical-temporal continuity to the set of events collected so far, while maintaining the need to validate or not, through the evidence, such hypotheses. In other words, it is common for the investigator to return several times to edit the timeline, based on new evidence or hypotheses. One possible timeline is the one developed in Figure 210. It is good to observe how "actors" do not necessarily correspond to individuals, but also to inanimate places or objects. Starting from the actual, incidental event (the death of Operator A from severe burns), the investigator may wonder iteratively "why" such an event took place, going so far as to build the RCA tree of Figure 211. It shows the need to investigate certain events, such as the employer's to be replaced in the work, the absence of a "work permit" procedure, and the reasons the forced ventilation system in the cold room was completely absent (such events are marked with a question mark).

Immediate causes (marked by the symbol "CC") could be identified in the use of polyurethane foam cans and the fact that the refrigerator was closed and not ventilated. In fact, from a more in-depth analysis, it was discovered that the root causes must be sought in the inadequate layout of the structure, which does not allow even an air exchange for natural ventilation, and in the organization's lack of commitment to safety (whereby both Operator A and the employer, not being trained on the risks of explosion, used the electric drill in an environment that had become at risk of explosion, without adequate protective measures such as antistatic clothing or the use of a suitable drill for use in explosive atmospheres).

Any corrective actions should therefore be developed precisely on the latest events of the ramifications of the RCA tree identified as root cause (RC), although corrective actions can also be taken at immediate cause levels (CC) in order to provide recommendations to be implemented in the short term. The need to carry out restoration work for the purpose of the investigation may not require further investigation, so the event has not been further investigated and has been designated as not a Cause (NC).

Analyzing the same incident with the Tripod Beta methodology requires not only the identification of events but also of agents and objects. The highly structured methodology requires that the barriers present also be defined, identifying their status (effective, failed, absent).

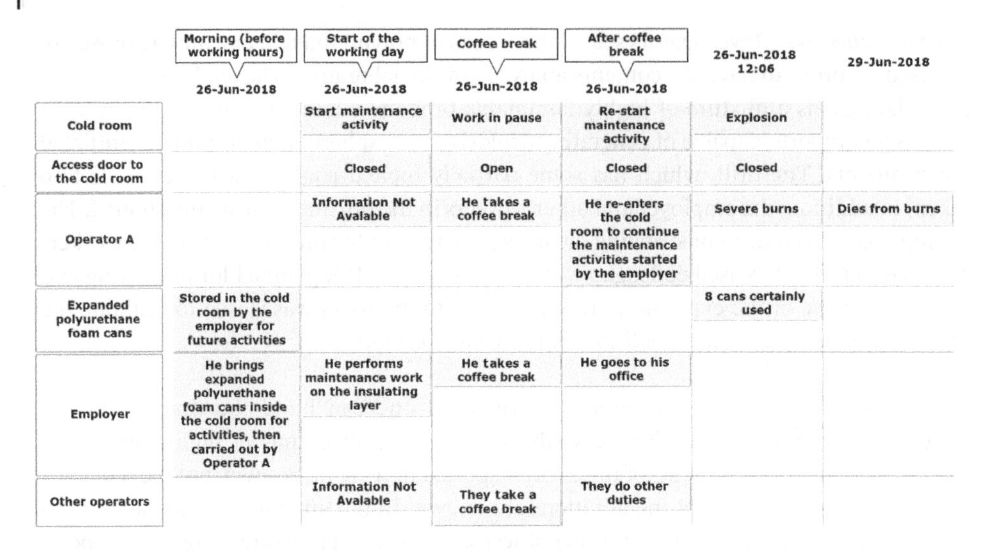

Figure 210 Timeline of the sample (developed with CGE-NL IncidentXP).

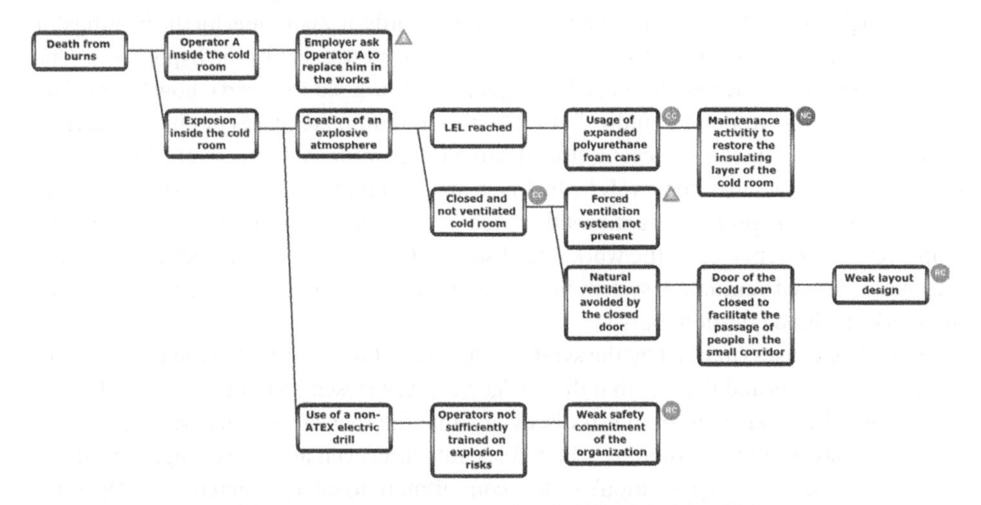

Figure 211 Possible RCA of the sample (developed with CGE-NL IncidentXP).

One possible combination of trios could be those shown Figure 212. For the convenience of representation, the immediate and latent causes of only one of the failed barriers have been identified. Finally, the same incident could also be analyzed with BFA, as shown in the deliberately simplified diagram of Figure 213.

4.4.17 Fire Risk Assessment for Companies Managing Multiple Assets

(from Fiorentini, 2018)

Railway stations are nowadays, among several other specific occupancies characterized by the presence of masses of people often in fast transit (as ports and airports), a clear example of infrastructures open to the public that are constantly subject to

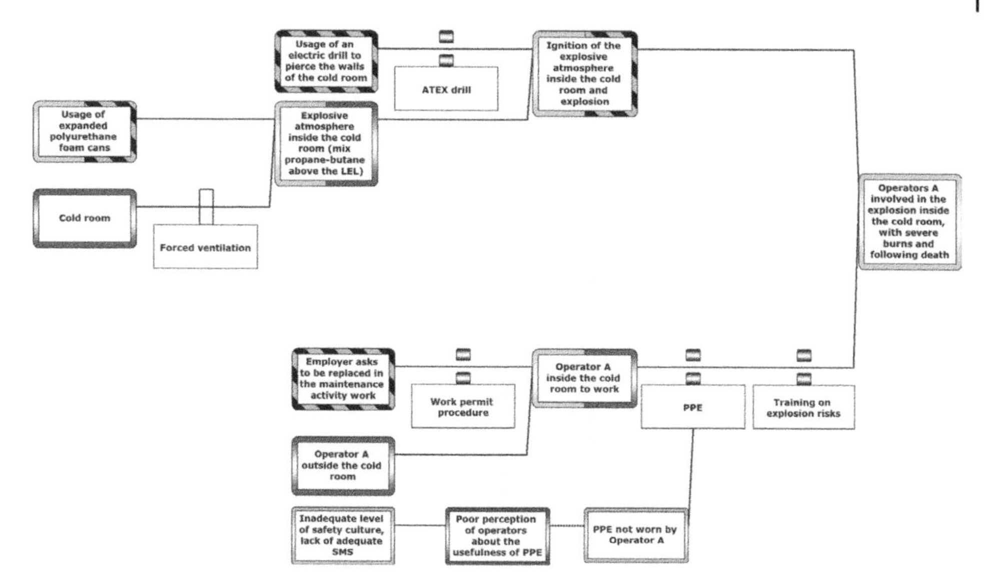

Figure 212 Possible Tripod Beta of the sample (developed with CGE-NL IncidentXP).

modifications. Changes take place to install temporary stores, new stores or facilities (even schools, healthcare facilities and hotels), advanced security measures due to recent episodes, new utility systems, and networks. Besides this, railway stations populations are particularly heterogeneous and much more different from the population in different and more controlled occupancies (ports and airports), also because railway stations are more than an infrastructure hub. They indeed are centers open to the city where people can find commercial stores, services, restaurants, and so on. This modification process, in Italy, is even dangerous because the majority of the biggest railway stations are located in heritage buildings that pose both physical and permitting related constraints. Stations are built around train tracks and composed of several buildings, with some underground floors (where connections to metro stations as well as utility occupancies, as warehouses for stores, are usually located on upper floors). In 2011, a specific decree requested railway stations to obtain their fire certificate for the entire entity and not limiting the fire certificates to single occupancies. Complexity should be considered in a uniform way across all the station with a single approach that should be properly maintained. Furthermore, it's worth mentioning the fact that up until 2011 the fire certificate application procedure in railway stations was supposed to cover only single selected occupancies on the basis of their specific activity an/or extension in surface. Since 2011 a specific fire safety regulation (national-level) enforced the requirement to extend an additional single fire certificate for the entire railway stations for those having a total public open surface greater than 5,000 m^2. This new requirement recognizes the railway station as being a single unique special occupancy and raised the need for a fire safety general design strategy to overcome the limits of specific occupancies and to find a general approach able to consider all the additional and common areas that connect and serve tracks and commercial retail stores. Focus has to be put also on station buildings and associated utilities that complement the main areas open the general public (the "station" perceived by most

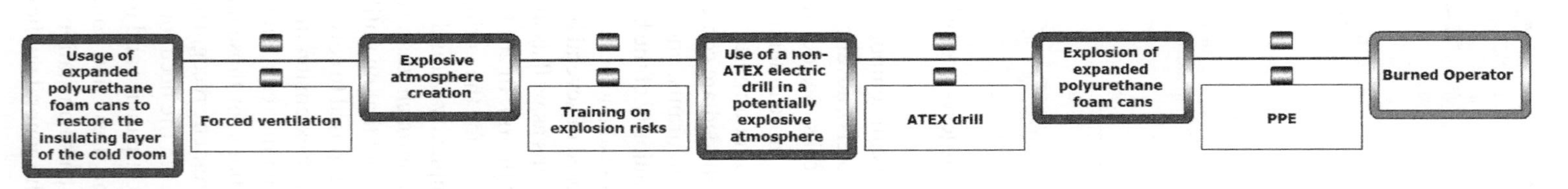

Figure 213 Possible BFA of the event (developed with CGE-NL IncidentXP).

of the traveling passengers in and out the main buildings). While these new-to-be-considered spaces are mainly offices rented to third parties, it is also true that real fire risks are associated with areas like tunnels, utilities rooms (electrical substations), and warehouses located undergrounds (several floors) as well as connections (underground tunnels with significant motor and electric vehicle traffic), car parks, and so on. Complexity becomes even worse considering that railway stations are an always-changing asset due to new commercial needs (growing number of passengers due to specific competition among fares with air traffic, especially at the national level, facilitated by the increase of high-speed train connections) and the stop operation for any revamping activities (even fire and security implementation of new measures). With this overview in mind, upper management has to face and win a specific game: in a given time (by law) and considering finite economic resources, without shutting down any part of the asset but during severe revamping projects of renovation and modernization and with general and extraordinary ongoing maintenance (and inspection) activities, gain a proper fire safety level and get the specific fire safety certificate from authorities having jurisdiction. Besides this, a parallel workflow has put in place increased security measures that collide with fire safety requirements (in particular with the emergency evacuation of large masses of people). Both the modernization processes should consider heritage buildings and specific conditions of railway stations on Italy: frequent use of historical buildings for temporary exhibitions as well as specific periods (or days) recording an extraordinary affluence of passengers due to touristic special dates, especially in summertime, in several cities that should be added to the impressive movement of tourists during the entire year in Italian cities moving from city to city by train.

On the basis of this very impressive complexity and considering a several-year renovation without any business interruption, a fire safety management plan has been defined at the central level for the main thirteen railway stations located in Italy: Bari Centrale, Bologna Centrale, Firenze S. Maria Novella, Genova Brignole, Genova Piazza Principe, Milano Centrale, Napoli Centrale, Palermo Centrale, Roma Termini, Roma Tiburtina, Torino Porta Nuova, Venezia Mestre, Venezia S. Lucia, and Verona Porta Nuova.

The fire safety plan started from the construction of a fire safety management system (FSMS) to deal both with the initial intervention phase and with the future use and modification of the railway stations. FSMS is composed of elements common to all the stations at a centralized level and customized elements specific to each single station. This plan enforces the use of a common approach, name assumptions, methods, tools with a shared dashboard with fire safety–related key performance indicators (KPIs) to track the fire safety level over time, comparing different areas of the same station as well as different stations in a global benchmarking activity for upper management. The dashboard also serves as a decision-making tool to allocate resources. FSMS encompasses several aspects: culture of safety, global policy, organization and people, training, fire risk assessment, inspection/maintenance, emergency response, audit and feedback. A key element point of the entire process is the fire risk assessment. This has been conducted with the Bow-Tie methodology (see Figure 214) to identify the top events to be considered for subsequent more in-depth assessment, also referring to the use of simulation methods (fire and evacuation) in a performance-based environment to take into account all the specific aspects of the complexities, as described in the following paragraphs.

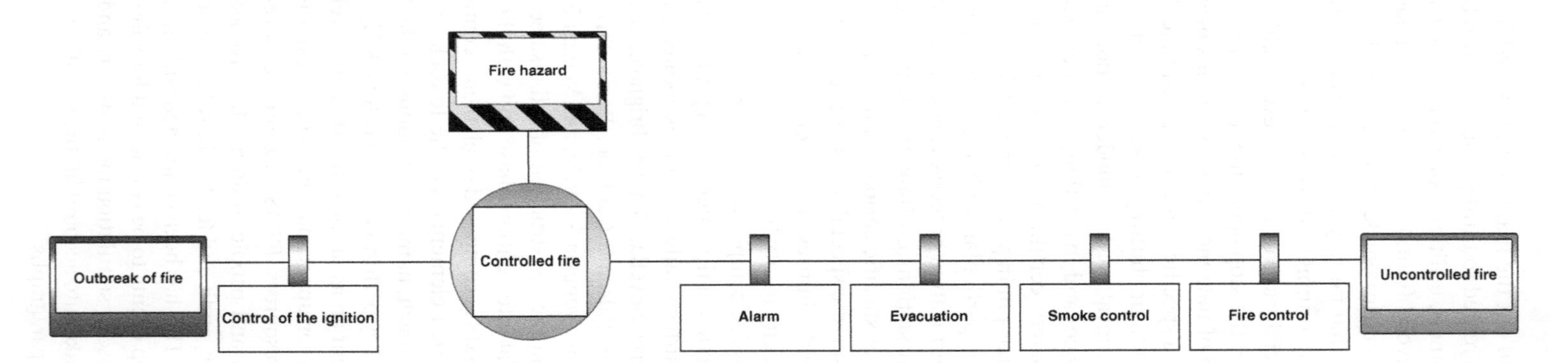

Figure 214 Bow-Ties developed to assess fire risk in multiple railway stations.

The Bow-Tie method (from the oil & gas quantitative risk assessment) is the proper tool to:

- Represent the main cause and consequences in a logic notation;
- Identify both prevention and protection barriers able to prevent the occurrence or mitigate the consequences (or reduce both) of the identified the outcomes;
- Consider both technical and management barriers.

Bow-Tie is characterized by a clear and simple methodology; it is very straightforward to use it to:

- Explain fire design considerations to both the upper management and the field operators;
- Identify the need for new/different barriers;
- Make comparisons among areas, assets, and stations at the global level;
- Describe the current and intended level of fire safety;
- Conduct technical audits using their results to visualize the current status of barriers against the minimum established level (e.g. using inspections result data on fire protection active systems to update the relevant barrier and update the related key performance indicator/s).

The top event identified can be developed with engineering tools as well as used as the basis of the emergency and evacuation plan.

Typical top events (each of them related to a single Bow-Tie diagram) have been applied to a hierarchy. Hierarchies have been defined considering for each station the single buildings and, inside each building, the various fire compartments located in floors (above and underground). Initially each fire compartment has been defined in terms of fire safety–related properties (surface, people, public presence, fire load (Figure 215), safety measures in place, etc.) and Bow-Ties have been applied to the compartments, intended as atomic units of the risk assessment, considering five standard barriers: ignition prevention, alarm, evacuation, smoke control, fire control, each of them composed of several elements to be judged:

- Ignition prevention
  - Permit-to-work
  - Electrical systems safety level
  - Fire safety quality and reliability of the building
  - Ignition sources control
  - Inspections
  - Maintenance

- Alarm
  - Fire detection
  - Call points
  - Control room
  - Emergency plan
  - Alarming system
  - EVAC system
  - Fire drills

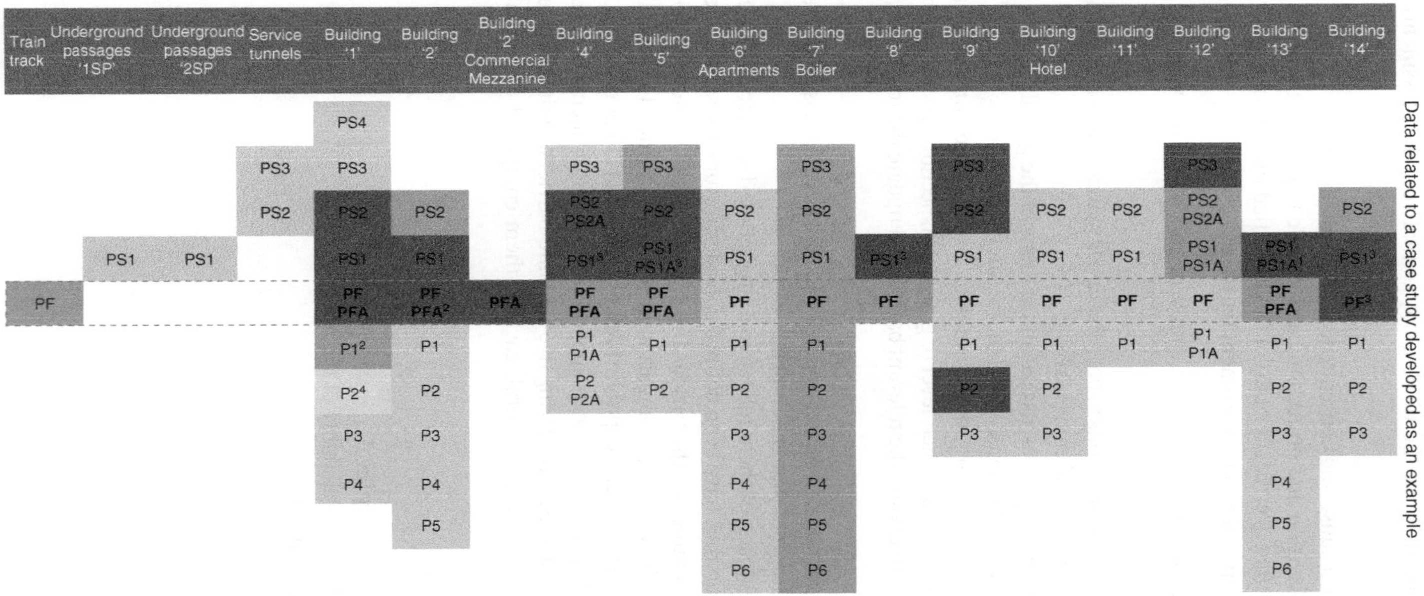

Data related to a case study developed as an example

Train track	Underground passages '1SP'	Underground passages '2SP'	Service tunnels	Building '1'	Building '2'	Building '2' Commercial Mezzanine	Building '4'	Building '5'	Building '6' Apartments	Building '7' Boiler	Building '8'	Building '9'	Building '10' Hotel	Building '11'	Building '12'	Building '13'	Building '14'
				PS4													
			PS3	PS3			PS3	PS3		PS3		PS3			PS3		
			PS2	PS2	PS2		PS2 PS2A	PS2	PS2	PS2		PS2	PS2	PS2	PS2 PS2A		PS2
	PS1	PS1		PS1	PS1		PS1[3]	PS1 PS1A[3]	PS1	PS1	PS1[3]	PS1	PS1	PS1	PS1 PS1A	PS1 PS1A[1]	PS1[3]
PF				PF PFA	PF PFA[2]	PFA	PF PFA	PF PFA	PF	PF	PF	PF	PF	PF	PF	PF PFA	PF[3]
				P1[2]	P1		P1 P1A	P1	P1	P1		P1	P1	P1	P1 P1A	P1	P1
				P2[4]	P2		P2 P2A	P2	P2	P2		P2	P2			P2	P2
				P3	P3				P3	P3		P3	P3			P3	P3
				P4	P4				P4	P4						P4	
					P5				P5	P5						P5	
									P6	P6						P6	

Figure 215 Fire load.

- Evacuation
  - Fire brigade
  - Emergency exits width
  - Emergency routes length
  - Elevators
  - People with disabilities management
  - Railway station supervisor
- Smoke control
  - Separation
  - Compartments
  - Ventilation
  - Heat/smoke extraction
  - Material control
- Fire control
  - Sprinkler
  - Separation
  - Compartments
  - Fire load requirements/HRR
  - Hydrants
  - Fire brigade connections
  - Fire brigade
  - Portable fire protection means
  - Intervention time

The specific conditions of the Bow-Tie identified barriers have been judged for each fire compartment through brainstorming sessions of experts using predefined categories about the quality of the barrier in place (or the lack of the barrier, if requested but not implemented). Due to the number of the fire compartments, classification has been made using a tabular format (Figure 216). Weights employed in expert judgement during brainstorming sessions have been defined in a qualitative way as "not necessary" (score 1), "present and fully efficient" (score 1), "compliant" (score 0.9), "to be verified" (score 0.33), "not compliant" (score 0.6), or "absent/not working" (score 0).

From collected data, structured as the underlying Bow-Tie diagram, it is possible to have an overview of the situation: specific problems of single barriers in compartments, quality problems in single buildings or even low performances across all the assets, affecting the entire station.

Taking the performance of the single barriers and representing them in the logic diagram of the station (Figure 217), it is possible to visually identify absolute weakest points and even make combinations of data to define general safety level (considering the weights of the barriers assigned in the Bow-Tie diagram) and to identify critical compartments or floors of single buildings or station areas given the relationship of the compartment's calculated performance level or the most relevant single barrier performance level and specific characteristics of the fire hazard (quality of fixed fire protection systems with significant fire load, weak alarm systems or evacuation routes where the public is located

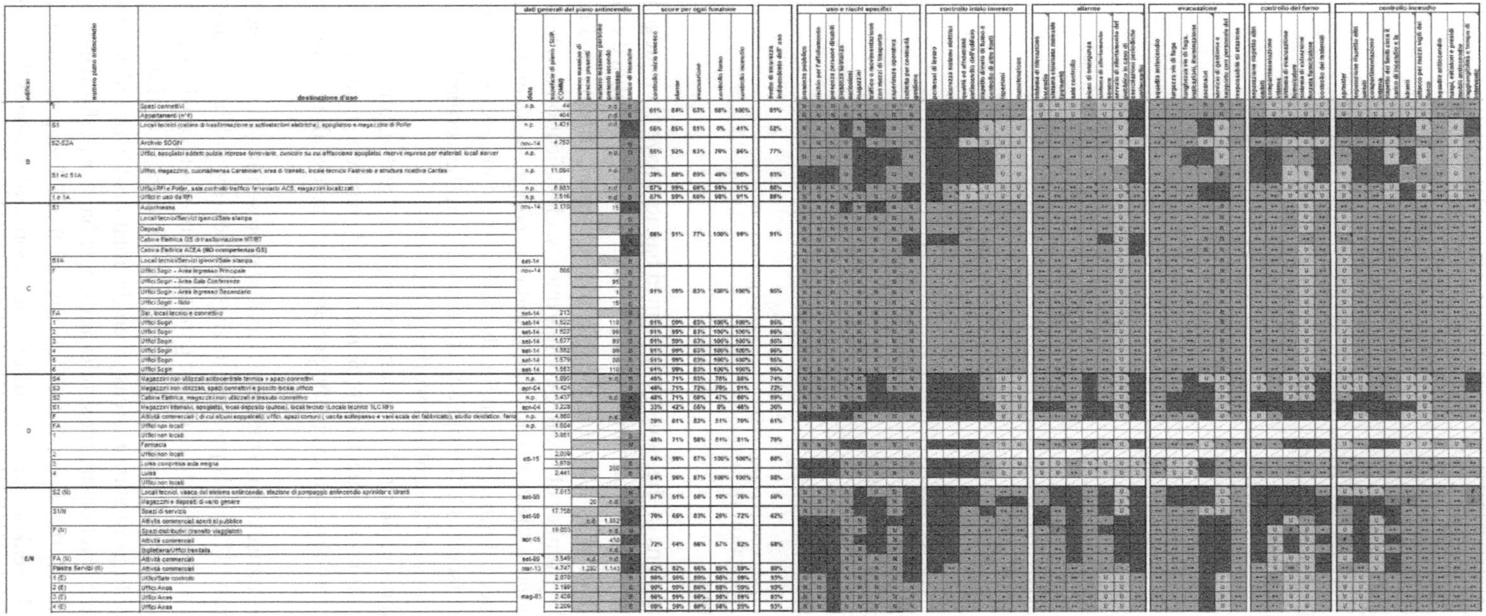

edificio	numero piano antincendio	destinazione d'uso	data	superficie di piano (SUP. COMM)	numero massimo di persone presenti	numero massimo persone presenti secondo permesso	carico di incendio	controllo inizio innesco	allarme	evacuazione	controllo fumo	controllo incendio	livello di sicurezza indipendente dall' uso
			dati generali del piano antincendio					score per ogni funzione					
	1	Spazi connettivi	n.p.	44		n.d.	B	61%	84%	63%	98%	100%	81%
		Appartamenti (n° 4)		404		n.d.	B						
B	S5	Locali tecnici (cabine di trasformazione e sottostazioni elettriche), spogliatoio e magazzino di Poller	n.p.	1.421		n.d.	[illegible]	56%	85%	81%	0%	41%	52%
	S2-S2A	Archivio SOGIN	nov-14	4.753			[illegible]	55%	92%	63%	70%	86%	77%
		Uffici, spogliatoi addetti pulizie imprese ferroviarie, cunicolo su cui affacciano spogliatoi, riserve imprese per materiali, locali server	n.p.			n.d.	[illegible]						
	S1 ed S1A	Uffici, magazzino, cucina/mensa Carabinieri, area di transito, locale tecnico Fastweb e struttura ricettiva Caritas	n.p.	11.094		n.d.	[illegible]	39%	80%	69%	48%	66%	83%
	F	Uffici RFI e Poller, sala controllo traffico ferroviario ACS, magazzini localizzati	n.p.	8.883		n.d.	B	87%	99%	66%	98%	91%	88%
	1 e 1A	Uffici in uso da RFI	n.p.	7.516		n.d.	B	87%	99%	86%	98%	91%	88%
C	S1	Autorimessa	nov-14	2.178		15	[illegible]	88%	91%	77%	100%	99%	91%
		Locali tecnici/Servizi igienici/Sale stampa					B						
		Deposito					M						
		Cabina Elettrica GS di trasformazione MT/BT					A						
		Cabina Elettrica ACEA (NO competenza GS)					A						
	S1A	Locali tecnici/Servizi igienici/Sale stampa	set-14				B						
	F	Uffici Sogin - Area Ingresso Principale	nov-14	866		3	B	91%	99%	83%	100%	100%	95%
		Uffici Sogin - Area Sala Conferenze				95	B						
		Uffici Sogin - Area Ingresso Secondario				1	B						
		Uffici Sogin - Nido				15	B						
	FA	Bar, locali tecnici e connettivo	set-14	213			B						
	1	Uffici Sogin	set-14	1.522		110	B	91%	99%	83%	100%	100%	95%
	2	Uffici Sogin	set-14	1.522		96	B	91%	99%	83%	100%	100%	95%
	3	Uffici Sogin	set-14	1.577		99	B	91%	99%	83%	100%	100%	95%
	4	Uffici Sogin	set-14	1.582		99	B	91%	99%	83%	100%	100%	96%
	5	Uffici Sogin	set-14	1.579		88	B	91%	99%	83%	100%	100%	95%
	6	Uffici Sogin	set-14	1.563		110	B	91%	99%	83%	100%	100%	96%
D	S4	Magazzini non utilizzati sottocentrale termica e spazi connettivi	n.p.	1.095		n.d.	[illegible]	48%	71%	83%	78%	88%	74%
	S3	Magazzini non utilizzati, spazi connettivi e piccolo locale ufficio	apr-04	1.424			B	48%	71%	72%	78%	91%	72%
	S2	Cabine Elettrica, magazzini non utilizzati e tessuto connettivo	n.p.	3.437		n.d.	A	48%	71%	68%	47%	60%	59%
	S1	Magazzini intensivi, spogliatoi, locali deposito (pulizie), locali tecnici (Locale tecnico TLC RFI)	apr-04	3.228			A	33%	42%	55%	0%	48%	30%
	F	Attività commerciali (di cui alcuni soppalcati), uffici, spazi comuni (uscita sottopasso e vani scala del fabbricato), studio dentistico, farm	n.p.	4.880		n.d.	A	39%	61%	83%	51%	70%	61%
	FA	Uffici non locati	n.p.	1.604									
	1	Uffici non locati		3.961			B	48%	71%	98%	81%	81%	78%
		Farmacia					M						
	2	Uffici non locati	ott-15	2.039				54%	99%	87%	100%	100%	88%
	3	Luiss compresa aula magna		3.978		200	B						
	4	Luiss		2.441			B	64%	99%	87%	100%	100%	88%
		Uffici non locati											
E/N	S2 (N)	Locali tecnici, vasca del sistema antincendio, stazione di pompaggio antincendio sprinkler e idranti	set-99	7.613			M	57%	51%	50%	10%	76%	50%
		Magazzini e depositi di vario genere			20	n.d.	M						
	S1/N	Spazi di servizio	set-99	17.758			[illegible]	70%	65%	83%	20%	72%	62%
		Attività commerciali aperti al pubblico			n.d.	1.882							
	F (N)	Spazi distributivi (transito viaggiatori)	apr-05	19.003		n.d.	M	72%	64%	66%	57%	82%	68%
		Attività commerciali				430	A						
		Biglietteria/Uffici trenitalia				n.d.	M						
	FA (N)	Attività commerciali	set-89	3.549	n.d.	n.d.	A						
	Piastra Servizi (E)	Attività commerciali	mar-13	4.747	1.292	1.143	A	82%	82%	66%	89%	59%	80%
	1 (E)	Uffici/Sale controllo	mag-03	2.078			B	90%	99%	60%	98%	99%	93%
	2 (E)	Uffici Anas		2.199			B	90%	99%	60%	98%	99%	93%
	3 (E)	Uffici Anas		2.428			B	90%	99%	60%	98%	99%	93%
	4 (E)	Uffici Anas		2.209			B	90%	99%	60%	98%	99%	93%

Figure 216 Bow-Tie worksheet developed by TECSA S.r.l. and Royal Haskoning DHV to quantify a Bow-Tie scheme with a LOPA approach. Not real scores and data presented in the image.

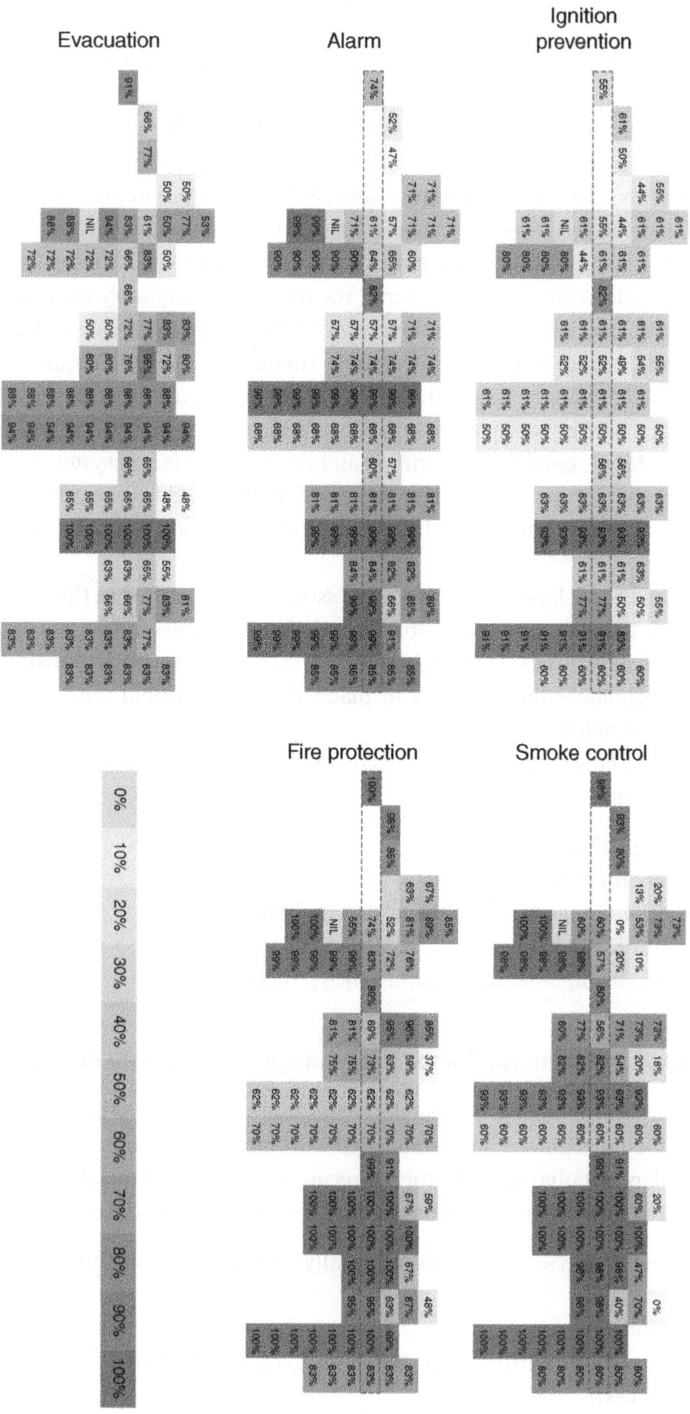

Figure 217 Barriers/protection layer scores.

(Figure 218), quality of smoke detection and smoke extraction or ventilation system underground, etc.).

This activity produces a risk-based picture of the fire safety level across the assets of the organization in order to define priorities with an improvement plan. The plan has been defined with a cost-benefit analysis on the basis of the simulation of post-mitigation new conditions.

The method (Bow-Tie, LOPA and cost-benefit analysis) has been selected as the approach to be included in a fire safety management system (FSMS) to be put in place while improving existing situations. In fact, Bow-Tie can also be used to demonstrate/verify the impact of modifications (in organizations, systems, assets, compartments), even with the support of a hazard identification (HAZID) method to verify the impact of temporary modifications on the overall fire safety level (closure of emergency exit due to temporary construction sites inside the station and impact on evacuation strategy) in order to meet the requirements of a specific management of change process. The FSM system has been defined using a standard SMS structure, with a barrier-based approach, considering responsibility, operation, inspections, maintenance, MOC, emergency planning and response, KPIs and system review.

The conducted activity showed many benefits and demonstrated how to effectively manage all the requirements coming from the application of several regulations in complex realities.

Application of the FSMS, based on barriers assessment, also resulted in the possibility to collect information, data, documents, performance indicator results to support the fire certificate request to the authorities, to make better-informed decisions, to demonstrate in real time to all the stakeholders the activities in place, the design intent with the intended results and the path to achieve those.

Bow-Tie and barrier failure analysis, applied to an organization, satisfy the requirements of an "enhanced risk management":

- Continual improvement;
- Full accountability for risks;
- Application of risk management in all decision making;
- Continual communications;
- Full integration in the organization's governance structure.

4.4.18 Fire Risk Assessment for Companies Managing Multiple PV Plants

The fire risk analysis for photovoltaic systems represents a novelty in the panorama of fire risk analysis for which recognized and established working methods are not available, both for the technical-plant peculiarities and for the recent affirmation of this industrial sector.

This methodological proposal is based on the Bow-Tie method and is based on seven steps:

1) Creation of some Bow-Tie models, conceptually applicable to all the systems under examination;
2) Extraction of critical barriers;
3) For each plant, assign the performance standards to the critical barriers and define their level of effectiveness;
4) Frequency quantification of Bow-Tie models (frequency of causes and PFD of barriers), with qualitative or semi-quantitative approach (LOPA);

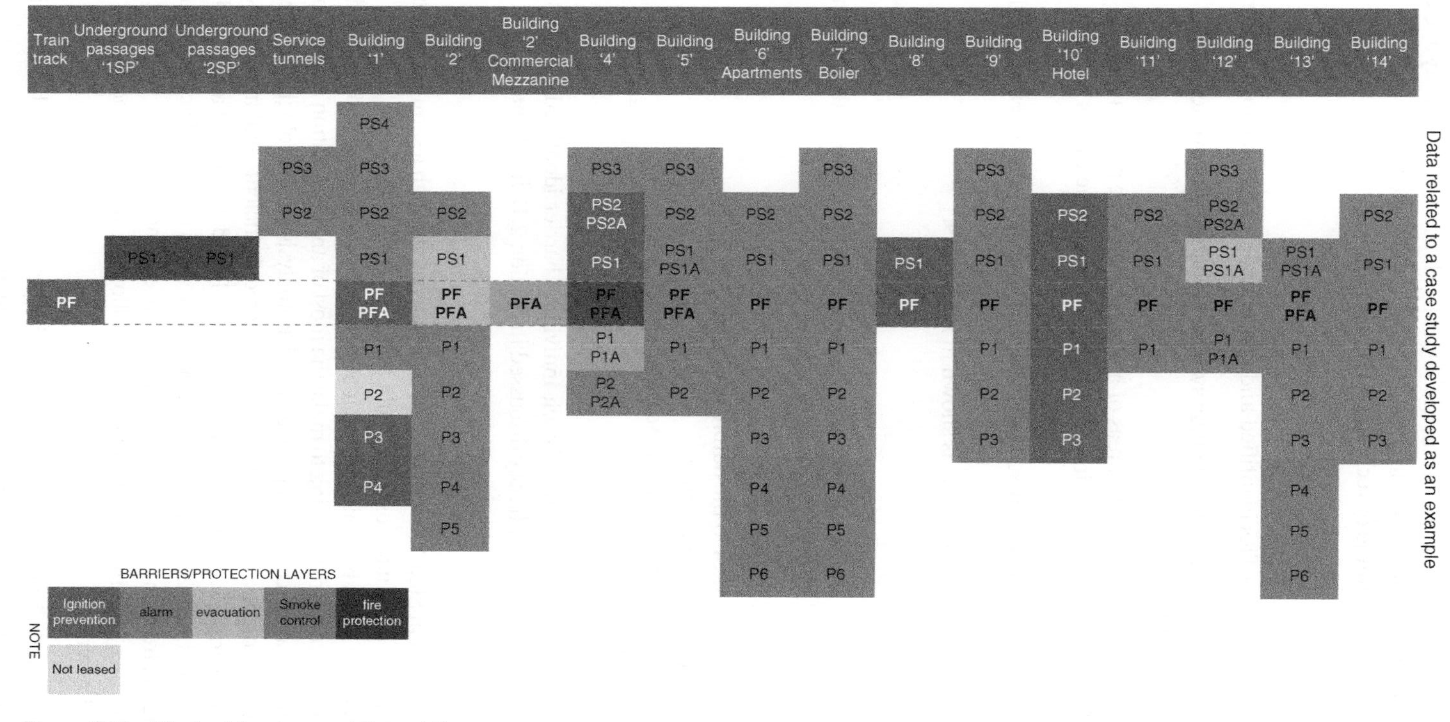

Figure 218 Weakest barriers and the public.

5) For each system, definition of parametric values like hazard, vulnerability and exposure;
6) For each system, and for each consequence, definition of the risk level;
7) Definition of a programme of recommendations.

Step 1

There are two "type" Bow-Ties identified and conceptually applicable to all systems:

- One is relative to the entire photovoltaic system, excluding the inverter cabin;
- One regards the inverter cabin only.

This distinction between the two areas of a plant has become necessary due to the characteristics of the inverter cabin, so it is necessary to diversify the causes of fire in principle with respect to outdoor plant areas.

During the Bow-Tie workshop sessions, relevant threats, consequences, and barriers have been identified for the specific top events being analyzed.

An example of the Bow-Tie model is shown in Figure 219.

Step 2

We proceed to identify which barriers, among those identified during the previous phase, are considered "critical." These barriers are defined as critical:

- Whose failure causes a fire;
- Whose failure contributes to the outbreak of fire;
- Whose purpose is to prevent a fire;
- Whose purpose is to mitigate the consequences of a fire.

Step 3

For each system, the effectiveness of each individual barrier identified in the Bow-Tie is assessed, following what was already discussed in Section 2.12 in Chapter 2.

Step 4

Bow-Ties produced in Step 1 are quantified in frequency through a semi-quantitative analysis (by orders of magnitude), using a LOPA analysis.

The frequency of cases is obtained from internationally recognized and proven databases, as well as the PFD of barriers, trying to make the most of an organization's operational experience. Given the absence of recognized fire risk analysis methodologies in the sector, a quantification by orders of magnitude is considered sufficient, making any attempt to improve the numerical accuracy of the frequency and PFD data difficult and of little benefit.

The frequency of causes and the PFD of barriers are then combined according to a LOPA to determine the frequency of occurrence of the consequences.

At the end, then, the frequency values of the following scenarios (both for BT-PV and BT-INV) are known:

- Health and safety: Uncontrolled fire within the perimeter of the photovoltaic system resulting in an accident, or fatality, or occupational disease;
- Environment: Environmental pollution (atmospheric emissions, percolation) after the fire;

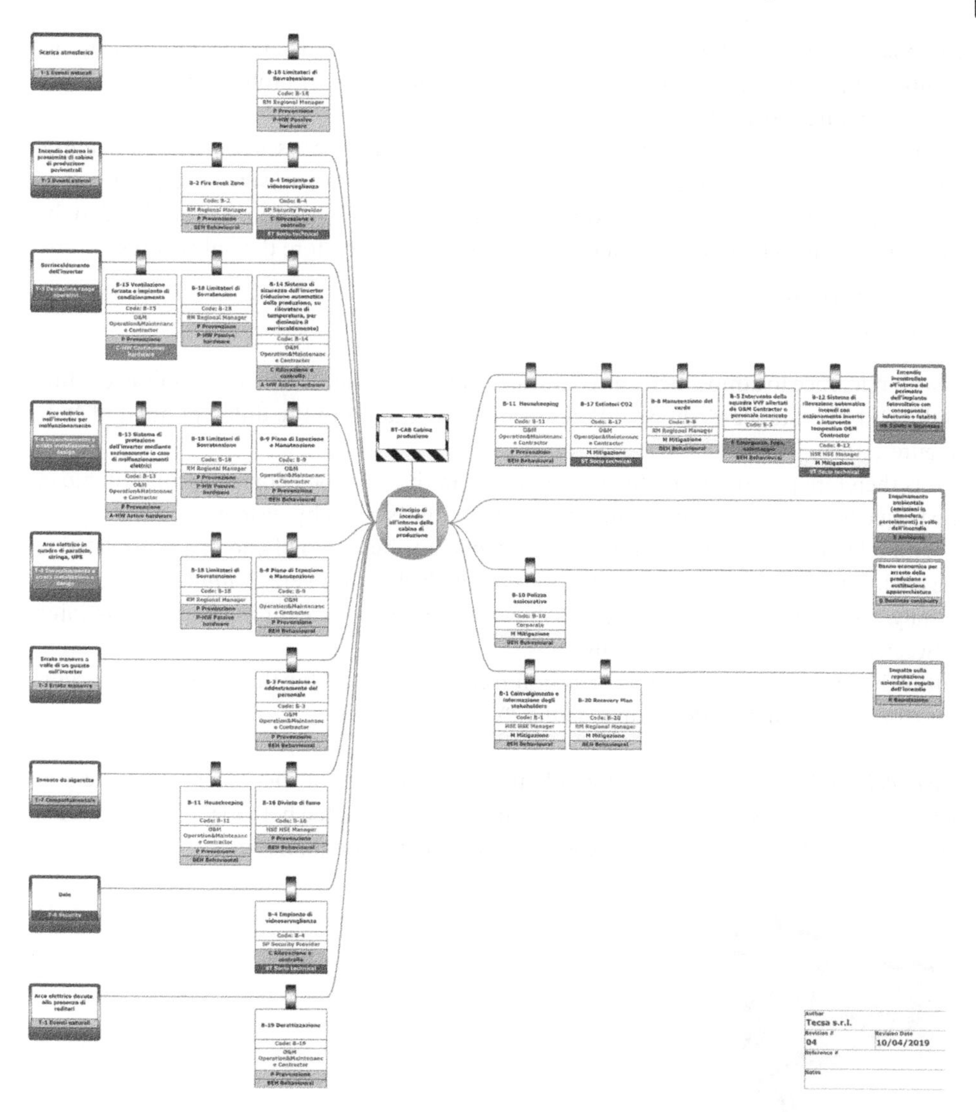

Figure 219 Bow-Tie model for fire risk assessment in PV plants.

- Business continuity: Economic damage due to production stoppage and equipment replacement;
- Reputation: Impact on the company's reputation following the fire.

In this way, Bow-Ties models have been preliminarily quantified.

Step 5

At this point it is necessary to decline the information obtained in Step 4 with the peculiarities of each system.

This option is permitted thanks to the adoption of certain "weight" factors: dangerousness (P), vulnerability (V), and exposure (E).

Dangerousness (P) Hazard is the intrinsic propensity of the site where the system is located to suffer a fire. It is therefore an expression of the greater or lesser probability of occurrence of the initiating causes of a fire.

Analyzing the causes defined in the Bow-Tie type, only the causes connected to natural external events can be weighted through a site-dependent factor, since the other causes (electrical malfunctions of various kinds and cigarette triggers) cannot be correlated to the choice of the site where the system is installed; these are therefore the only causes:

- Atmospheric download;
- External fire.

With reference to atmospheric discharges, the map of ceraunic density in Italy according to the CEI 81-3 standard shows different values in relation to the different areas of Italy (Figure 220).

It is therefore proposed that the values shown in this map be adopted as an estimate of the frequency of occurrence of atmospheric discharge, system by system.

With reference to the external fire, being able to attribute it to phenomena correlated to the environmental temperature is of interest in analyzing the average annual temperature in Italy. Figure 221 (source: Wikipedia) shows how there is a clear difference in this value between the different regions of Italy.

It is therefore proposed that a correction factor be adopted for the frequency of occurrence of an external fire, depending on the PV plant location.

Vulnerability (V) Vulnerability is the inherent characteristic of the system (regardless of where it is located) of whether it can withstand a fire. It is related both to the type of system and the way in which the control measures installed on it are managed.

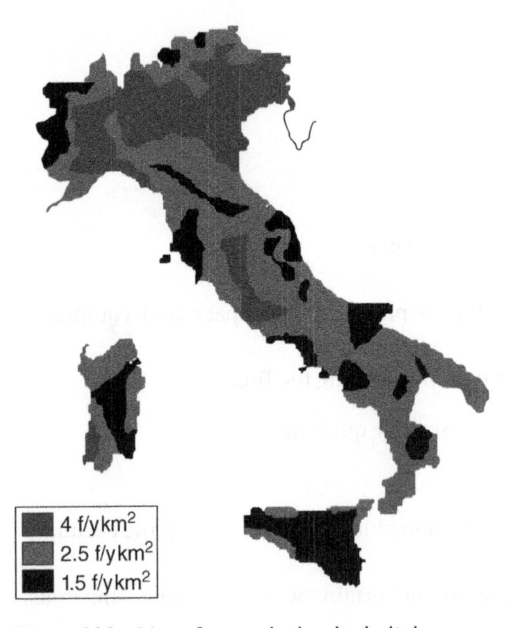

Figure 220 Map of ceraunic density in Italy.

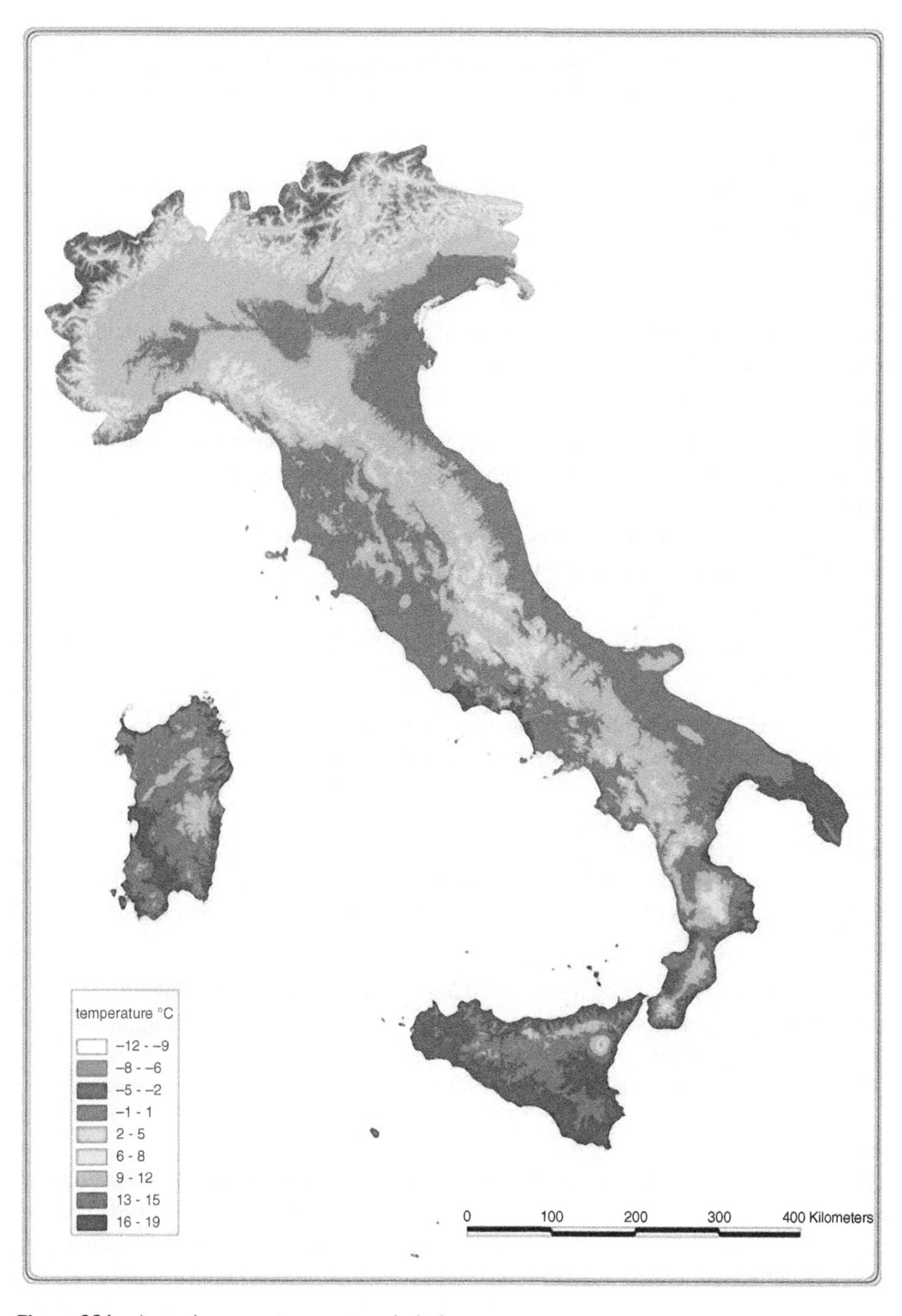

Figure 221 Annual average temperature in Italy.

First of all, the effectiveness of the barriers, evaluated in the previous phase of work, is taken into account, and then the peculiarities of the barriers plant by plant, using a "weight" factor (w) in the PFD assignment. The factor "w" is evaluated in relation to the effectiveness of the barrier.

Exposure (E) Exposure is a qualitative measure of the importance of the target in the event of a fire. It takes into account:

- Organization assets involved in a possible fire;
- Personnel involved (number of people);
- Severity of the business interruption in case of production failure from that specific plant;
- Severity of reputational risk in case of fire;
- Impact on neighbouring and neighbouring areas.

The exposure therefore contributes to increasing the level of severity defined a priori in Step 4, taking into account the peculiarities of the system, its crowding and its surroundings.

The severity of all the consequences for installations whose consequences (identified in Bow-Tie) have a significant impact on:

- Public service;
- Strategic infrastructure (e.g. airport operations, AT lines);
- Local communities.

Similarly, the severity of all the consequences for the systems is further increased for those PV plants having a big extension (greater than a defined value) or an increase in installed power.

Step 6

At this point, the level of fire risk for each system is defined, by means of the SFPE matrix and by means of the LOPA assessment previously described, suitably corrected with the weights in:

- Danger: intervenes by modifying the intial frequency of the cause "external fire";
- Vulnerability: intervenes by modifying the PFD of the barriers and changing the frequency of the top event;
- Exposure: intervenes by modifying the severity associated with the consequences.

Step 7

In conclusion, having defined the acceptability risk criteria, the plants are listed in order of decreasing risk (high, medium and low), and for each plant an action plan is defined to improve the effectiveness of the existing barriers (by first intervening on very low, then low, then good barriers).

Priority of intervention will be given to installations with high, then medium and finally low fire risk.

Conclusions

Any decision, from the most trivial to the most critical, that a human being is called to make, is based, mostly unconsciously, on the assessment of the associated risk. Organizations, of any nature, in any sector and of any size, are called to face risks that derive from internal and external factors (in practice deriving from their own operations on the market) that can influence, positively and negatively, the definition and achievement of their objectives.

Using the principles and methodologies of risk management according to consolidated paradigms, such as that provided by the international ISO 31000 standard, through the application of effective tools for the context, it is possible to improve the efficiency and effectiveness of the management system and therefore of the organization that applies it. Organizations will be able to have a clear view over time of the risks related to their activities, through the availability of useful and updated information in order to make informed decisions or react to unintended events and prevent their future occurrence. This will provide an accurate assessment of the uncertainty associated with the complexity that characterizes modern life and inevitably exposes the organization to potential imponderability in achieving the defined objectives.

A risk-based approach and risk-based thinking across the processes of an organization have, in the past decade, replaced and contrasted the traditional method that considered an organization to be composed of functions and processes, correlated with each other via input and output streams, with a specific emphasis on the hierarchy of the pieces, to guarantee focused and specialized processes and people, to assure control over people, processes and assets. This approach showed some limits: decomposed organizations with competing processes (and process owners), inability to manage changes (even sudden and unexpected ones), limited reaction to anomalies, partial vision based on specific and immediate objectives, failure in assessing common causes of disruptions, unbalanced focus by internal and even external stakeholders on single disciplines, strategic structure and documentation not related with the probability of the events and with the organization's experience – in a few words, the inability of organizations to evolve, to adapt themselves against performance criteria; management opposed to real organization governance and organization assessment opposed to audit.

Bow-Tie Industrial Risk Management Across Sectors: A Barrier-Based Approach,
First Edition. Luca Fiorentini.

Governance is the system through which an organization makes and implements decisions to pursue its objectives. This definition, derived from the ISO 26000 standard, highlights the centrality of the decision-making aspect, leading it back to the organization's objectives. The achievement of these objectives in today's increasing complexity is modified by uncertainty, which poses risks for the organization in one or more processes simultaneously in different areas.

Governance of an organization is the next upcoming challenge for organizations.

Governance of an organization, as will be discussed in the future ISO 37000 standards series, helps to:

- Reflect the identity, mission or purpose of the organization with respect to society and its stakeholders;
- Increase organizational effectiveness, sustainability, accountability and fairness;
- Fulfil the organization's purpose;
- Avoid major incidents.

From "Vision 2000," the first systemic approach to the Plan-Do-Check-Act (PDCA) model (Figure 222) given by Edward Deming ("Without data you are just another person with an opinion") to obtain the "perfect organization" by the composition of "perfect internal processes" dealing with the external context, we are now in the process of reaching modern high-level systems (HLS), introduced in 2012 with ISO 22301 on business continuity, that should face every day three specific issues: complexity, uncertainty and risk. This is achieved using a system valuing three pillars:

1) Systemic and adaptive vision;
2) Focus on efficiency strategic objectives;
3) A performance-based approach to requirements definition.

In a systemic and adaptive vision, attention is placed on the organization of its entirety: assets, processes, internal relations and relations with the external context. The strategic organization invests in the knowledge of the mechanisms to achieve the expected results

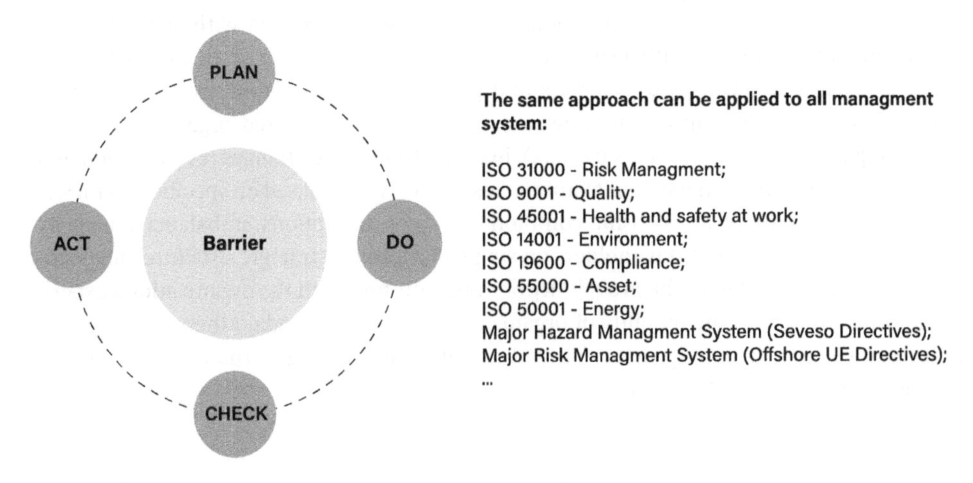

PLAN
DO
CHECK
ACT
Barrier
The same approach can be applied to all managment system:
ISO 31000 - Risk Managment;
ISO 9001 - Quality;
ISO 45001 - Health and safety at work;
ISO 14001 - Environment;
ISO 19600 - Compliance;
ISO 55000 - Asset;
ISO 50001 - Energy;
Major Hazard Managment System (Seveso Directives);
Major Risk Managment System (Offshore UE Directives);
...

Figure 222 Deming Cycle from a barrier-based perspective.

through the overall involvement of all resources, from leadership to operational resources, always with specific attention to periodic verification and measurable improvement. The organization defines its objectives according to a performance-based logic with better flexibility and responsibility in the ways of achieving the objectives according to the business logic. These three pillars guarantee the topicality of the organization itself with respect to the external context in which it operates, allowing it to evolve in all its components, including the modification of strategic objectives in the face of new needs. It is the organization that adapts to its needs, unlike the "Vision 2000" approach, which, although modern for the definition of processes, often led to the definition of rigid structures that were difficult to manage, especially in relation to the new and daily needs or experiences of the organization itself.

New HLS is characterized by transversal items rather them individual components or components of systems having input and output relations with specific internal mechanisms. These items are recognizable in a risk-based approach: more general risk-based thinking, advanced process segmentation with focus on relationships about processes, context understanding, responsible leadership (including policies, roles and responsibilities in the organization) and documented information (where evidence and data can replace documents, procedures, and records that are not useful).

These principles become the basis of a new understanding of organizations, regardless of their dimension, application domain, context, and so on: in order to work in an effective way (quality as per ISO 9001) they should create and protect value (as per ISO 31000), define value drivers (as per EN 12973) and create new value (as per ISO 56000 related to innovation management). This basis is a common framework to work within to achieve goals with a modern approach.

Risk understanding plays, in this framework, a vital role.

Every day, organizations should answer three main questions:

- "Do we understand what can go wrong?
- "Do we know what our systems are to prevent this happening?"
- "Do we have information to assure us they are working effectively?"

Forty years ago a first attempt to answer these questions was made by Haddon (1980) that developed, in his "energy model," the concept of the barrier among an energy source (hazard) and the vulnerable target (victim). He, indeed, moves from Gibson's simplified earlier model, defined in 1961, adding to the initial barrier model 10 very powerful rules composed of accident prevention, transition and control strategies around the three main components of the model: energy source, barrier and vulnerable target. From these initial approaches, Johnson, in his famous Management Oversight and Risk Tree (MORT) method (1980), clearly recognized the importance of non-physical barriers. Barriers are "the physical and procedural measures to direct energy in wanted channels and control unwanted releases," where releases are unwanted events of modern management systems.

Hollnagel, in 2004, states that "whereas the barriers used to defend a medieval castle mostly were of a physical nature, the modern principle of defense in depth combines different types of barriers – from protection against the release of radioactive materials to event reporting and safety policies."

But, in between Hollnagel and Johnson, Reason (1997) developed the theory most referenced in this book, the Swiss Cheese Model. There are multiple control barriers in systems and every layer has its own faults and failure. If the hazard (energy) transits through all the holes of the barriers, it may result in being converted into a failure of the system (an accident, a safety issue, etc. depending on the considered domain and management system). Barriers can be seen as systems themselves (Sklet, 2006). The barrier system has been designed and implemented to perform one or more barrier functions, that is the function planned to prevent, control or mitigate undesired events. Barrier elements are those components that, by themselves, are not sufficient to perform a barrier function: each barrier function requires 1 to *n* barrier elements to function on demand.

If the undesired event raises the barriers have to perform as they have been planned, on demand. They can be described in terms of their "probability of failure on demand" (PFD) and in terms of their "risk reducing factor" (RRF). In recent years, after the *Deepwater Horizon* accident in 2010 caused by the failure of a blow-out preventer, the barrier-based method gained an interest, even in application domains different from industrial risk and process safety: barriers include physical and non-physical means in different industries for preventing the occurrences of unwanted events and mitigating the consequences in case they occur.

Unwanted or undesirable events are those preventing an organization from reaching its strategic objectives. Those events are characterized by uncertainty. Uncertainty on objectives is risk, as defined by ISO 31000. Risk increases with uncertainty and uncertainty increases with complexity; that is part of our everyday life and different each day.

Simplicity behind the barrier-based approach can be a good starting point to implement risk management intended, as per ISO 31000, as all the coordinated activities to direct and control an organization with regard to the risk factor that affects all the processes, even with common cause failures due to related risk sources ("element that alone or in combination has the potential to give rise to risk").

These methods have their limitations and for very complex situations they could be challenging to implement (they are based on linear progression of failures, there are interdependencies among barriers from different barrier systems, and they do not well deal with over conservative barriers from the design process).

In any case they have advantages that are obvious in the majority of the domains and situations (in particular through sound tools), they can be used in both the active and reactive phases, they form a basis for analytical risk control, they can be easily explained, and they can take into account the human factor that is a precise part of the complexity.

As stated by ISO 31000, "organizations of all types and sizes face internal and external factors and influences that make it uncertain whether and when they will achieve their objectives. The effect this uncertainty has on organization's objectives is a risk. All activities of an organization involve risk." The International Organization for Standardization (ISO), together with the International Trade Centre and the United Nations Industrial Development Organization, published a specific guide about the importance of the implementation of sound risk management practices in small and mid-sized enterprises. Risk management is an integral part of all organizational processes and of decision making. It should be systematic, structured and timely. It also should be based on the best available information and tailored. It should consider human and cultural factors ("soft" factors) together with technical and organizational factors ("hard" factors) (risk management takes human and cultural factors into account). Reporting mechanisms and internal and external communication play a fundamental role in considering all the involved stakeholders. A detailed implementation plan for risk management is needed to ensure that the necessary changes occur in a coherent order and that the necessary resources can be provided and applied. Progress against the plan should be tracked, analyzed and reported to top management on a timely basis (monitoring can be achieved with specific key performance indicators).

Risk management should address various risks as well identified in financial, organizational, security, safety, environment, reputation, contractual, service delivery, commercial, business continuity, customer relationship, technology, and so on.

The management of risk enables an organization to increase the likelihood of achieving objectives, encourage proactive management, be aware of the need to identify and treat risk throughout the organization, improve the identification of opportunities and threats, comply with relevant legal and regulatory requirements and international norms, improve mandatory and voluntary reporting, improve governance, improve stakeholders' confidence and trust, improve operational effectiveness and efficiency, improve loss prevention and incident management, and improve organizational learning and resilience.

In simple words: risk-based thinking.

Risk-based thinking, requested by modern management systems as a fundamental principle to govern organizations, can find an initial approach with tools that demonstrated a good compliance with risk assessment and risk management over time phases as the Bow-Tie method explained with examples from various industries and domains in this book.

Barrier-based approach is the founding principle of two specific assessment methods:

1) Bow-Tie, used for risk assessment and management;
2) Barrier failure analysis (BFA), used for the investigation of near-accidents, near-misses and accidents or unwanted events.

The Bow-Tie diagram is the core of the method and one of the most used diagrams within barrier-based management. With the Bow-Tie diagram one can visualize a risk scenario that would be very difficult to explain otherwise (risk communication is a specific requirement of and it is defined as a: continual and iterative process that an organization conducts to provide, share and obtain information and to engage in dialogue with stakeholders regarding the management of risk). A Bow-Tie is a diagram that represents the risk you are dealing with in just one, easy-to-understand the picture. The diagram is shaped like a bow tie, creating a clear differentiation between proactive and reactive risk management.

The power of a Bow-Tie diagram is that it gives you an overview of multiple plausible scenarios, in a single picture. In short, it provides a simple, visual explanation of a risk that would be much more difficult to explain otherwise. Bow-Tie is defined as a simple diagrammatic way of describing and analyzing the pathways of a risk from hazards to outcomes and reviewing controls. It can be considered to be a combination of the logic of a fault tree analyzing the cause of an event and an event tree analyzing the consequences. Actually the fault-tree part can be seen also as a structured cause-and-effect analysis, traditionally organized in either a Fishbone (also called Ishikawa) or sometimes a tree diagram, that alone cannot be considered complete as a Bow-Tie due to the fact that it is a display technique for brainstorming rather than a separate analysis technique and the separation of causal factors into major categories at the start of the analysis means that interaction between the categories may not be considered adequately, e.g. where equipment failure is caused by human error, or human problems are caused by poor design.

The Bow-Tie method overcomes both these limits. The process of going from having identified the controls to assessing the performance based on actual data is an organization objective that can be identified as advanced barrier management.

It is advanced in two ways: (1) Companies need to have an excellent understanding of what their controls are and identify them with sufficient confidence in a risk assessment such as Bow-Tie. (2) Performance is tested using a range of different data sources such as incidents, audits and maintenance systems. Those data need to be aggregated. So, the process is advanced in the sense of both data collection and analysis.

The payoff is tremendous:

- Risk assessments come to life. Instead of being forgotten and archived, risk assessments are actually used because they are relevant in day-to-day operations.
- The aggregation of various data sources allows a level of understanding and insight into risks, which is unprecedented in risk management until now.

BFA is a pragmatic, un-opinionated, general-purpose incident analysis method. It has no affiliation with any particular organization. BFA is a way to structure an incident and to categorize parts of incident taxonomy. The structure has events, barriers and causation paths. Events are used to describe an unwanted causal sequence of events. This means that each event causes the next event. There can also be parallel events that together cause the next event.

Bow-Tie and BFA methods have a relationship and, used together, can improve the risk management system in place:

- Bow-Tie risk assessment provides input to incident analysis.
- BFA incident analysis outputs lessons learned to Bow-Tie risk assessment.

Bow-Tie and BFA, applied to an organization, satisfy the requirements of an enhanced risk management toward a real continuous improvement:

- Full accountability for risks;
- Application of risk management in all decision making;
- Continual communications;
- Full integration in the organization's governance structure.

In conclusion, Bow-Tie and BFA could be an initial approach to risk-based thinking for an evolving organization to meet its goals, finding a way to comply with the highly demanding PDCA approach, very well described by a number of standards, in the challenging framework of an organization's internal and external context complexity that modifies and constantly drives away the possibility of achieving goals with sudden, unexpected, underevaluated undesired events.

Keep it simple!

Appendix 1

Bow-Tie Easy Guide

In Figure 223 the core elements of a Bow-Tie diagram have been identified along with their position in the assessment schema; in Figure 224 some characteristics of each core element have been pointed out with some easy examples. The two figures may serve as a readily available handbook during Bow-Tie assessment sessions.

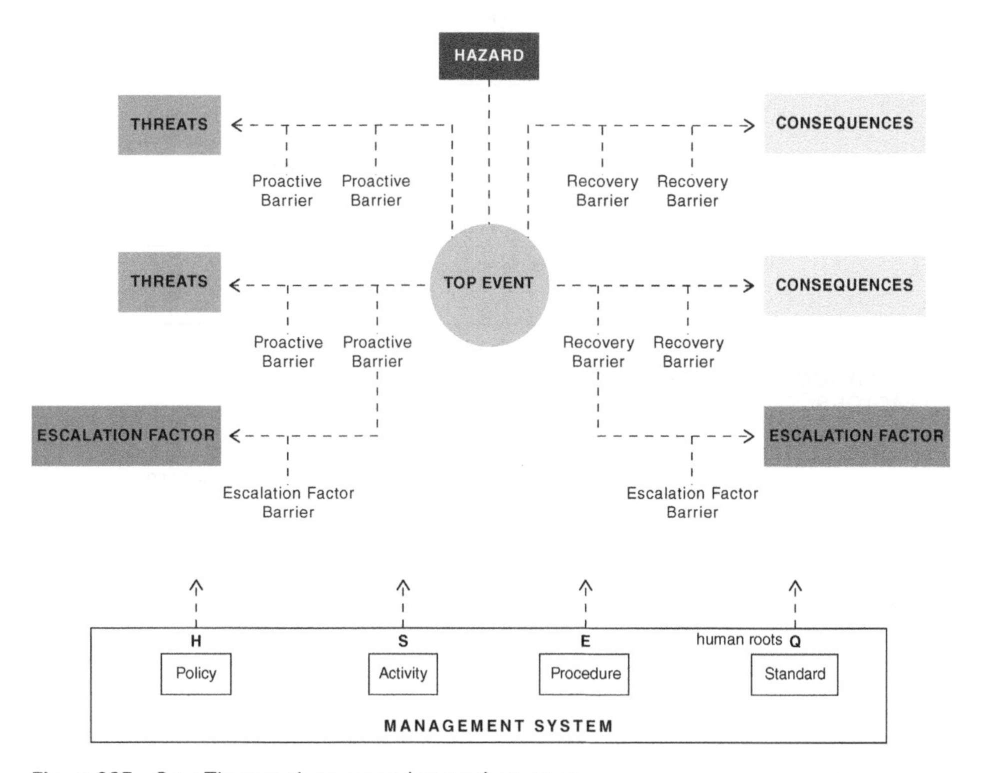

Figure 223 Bow-Tie core elements and general structure.

Bow-Tie Industrial Risk Management Across Sectors: A Barrier-Based Approach,
First Edition. Luca Fiorentini.

HAZARD

- Describes the desired state or activity
- Is part of normal business
- Has the potential to cause harm if control's lost
- Defines the context and scope of the Bow-Tie diagram

e.g.: Driving a car, hydrocarbons in containment, landing an aircraft

TOP EVENT

- Is a deviation from the desired state or activity
- Happens before major damage has occurred
- It's still possible to recover
- Hazards can have multiple top events

e.g.: Losing control over the car, loss of (hydrocarbons) containment, deviation from intented flight part

THREATS

- Are credible causes for the top event
- Are not barrier failures
- Should lead directly to the top event
- Should be able to lead independently to the top event

e.g.: Driving on a slippery road, pipeline corrosion, loss of positional awareness

CONSEQUENCES

- Are the hazardous outcomes arising from the top event
- Describe the direct cause for loss or damage
- Describe how the damage occurs

e.g.: Car rollover, ignition of vapor cloud, mid-air collision

ESCALATION FACTORS

- Are factors that reduce the effectiveness of a barrier
- Should be used sparingly to highlight real issues
- Tip: focus on critical barriers
- Tip: avoid repetition and duplication

e.g.: Forgetting to wear the seatbelt, no maintenance done, person not trained

BARRIERS

- Prevent, control or mitigate undesired events or accidents
- Can be (a combination of) behaviour and hardware
- A barrier system contains a detect, decide & act component

e.g.: Wearing a seatbelt, Blow-Out Preventer, Ground Proximity Warning System

---

Preventive Barrier – Eliminates the Threat or prevents the Top Event

Recovery Barrier – Avoids or mitigates the Consequence

Escalation Factor – Reduces the effect of the Escalation Factor

Figure 224 Bow-Tie guiding principles.

Appendix 2

BFA Easy Guide

In Figure 225 some characteristics of each core element of a barrier failure analysis have been pointed out with easy examples, while in Figure 226 the possible states of a barrier have been listed together with the relationship among a barrier state and the entire barrier life cycle. Figure 227 provides the reader with a simple decision support tree to evaluate barrier effectiveness following an incident and Figure 228 provides for some guiding principles in the general schema of real event. Figures may serve as a readily available handbook during BFA assessment sessions.

Bow-Tie Industrial Risk Management Across Sectors: A Barrier-Based Approach,
First Edition. Luca Fiorentini.

EVENT

- A happening or a "change of state" in which the incident sequence changes
- Place events in chronological order
- Usually the first event is "normal" business activity or process
- The amount of events depends on the scope of the analisys and complexity of the incident.
- There can be parallel sequences for events

POTENTIAL EVENT

- Events can have various appearances
- Potential event appearance can be especially powerful to communicate about effective barriers and near misses

e.g.: Event appearances: event, threat event, top event, consequence event or potential event

BARRIER

- A barrier is a measure which should prevent one event leading to another event
- A barrier should be defined in the normal or wanted state
- Place barriers in the order of their effect
- Apply a barrier state to the barrier

e.g.: Defining a barrier in normal state: "Wearing a seatbelt" instead of seatbelt not worn
e.g.: Barrier state: effevctive, unreliable, inadequate, failed or missing

BARRIER
PRIMARY CAUSE
SECONDARY CAUSE
TERTIARY CAUSE

The causation assessment helps you to analyze why the barrier did not function as desired.

- Primary cause is a direct act or omission from an actor.
- Secondary cause shows the context of the work environment which influenced the primary cause
- Tertiary cause shows the underlying organizational influences, this is where the improvement actions should be made
- Link your incident analisys causation categories to the causation assesment to be able to perform trand analisys

e.g.: Actor: human or equipment
e.g.: Primary cause: Procedure Y not performed.
e.g.: Secondary cause: Contractor was not aware of need to perform procedure Y.
e.g.: Tertiary cause: Contractors are not in the communication system of the organization

---

ACTION

- Recommandation can be placed anywhere in the diagram to improve the SMS.
- If it was never consiered to be an industry standard or described in the SMS, it is not a missing barrier but recommendation

Figure 225 BFA core elements.

EFFECTIVE
The barrier functioned as planned and topped the next event in the incident scenario
e.g.: Seatbelt which prevented fatality
UNRELIABLE
The barrier stopped the next event in the incident sequence, but organisation is uncertain if it will do so in the future
e.g.: Seatbelt prevented fatality, but the seatbelt is not always worn in the organization
INADEQUATE
The barrier functioned as intended by its design (envelope), but was unable to stop the sequence of events
e.g.: seatbelt was worn but it broke because it was not designed to witstand forces of impact which it encountered during the incident
FAILED
The barrier was implemented, but did not function ccording to its intended design
e.g.: Seatbelt did not prevent a fatality, because is not worn.
MISSING
The barrier was described in the organization's SMS or was considered an industry standard, but it was not succesfully implemented
e.g.: Seatbelt is described in policy and acquired, but was not yet placed in the vehicle
RELATION BETWEEN BARRIER STATE AND BARRIER LIFECYCLE
concept
design
implementation
operations/upkeep
ACTION
INADEQUATE
MISSING
FAILED
UNRELIABLE
EFFECTIVE

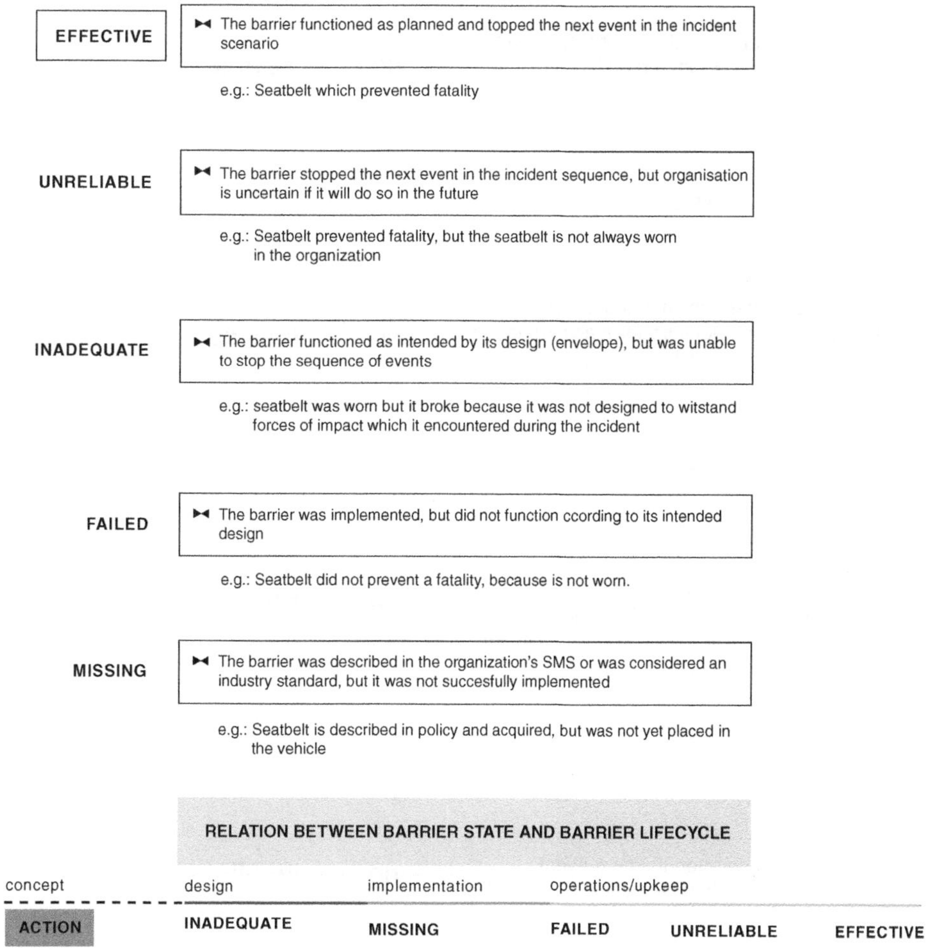

Figure 226 Incident barrier state.

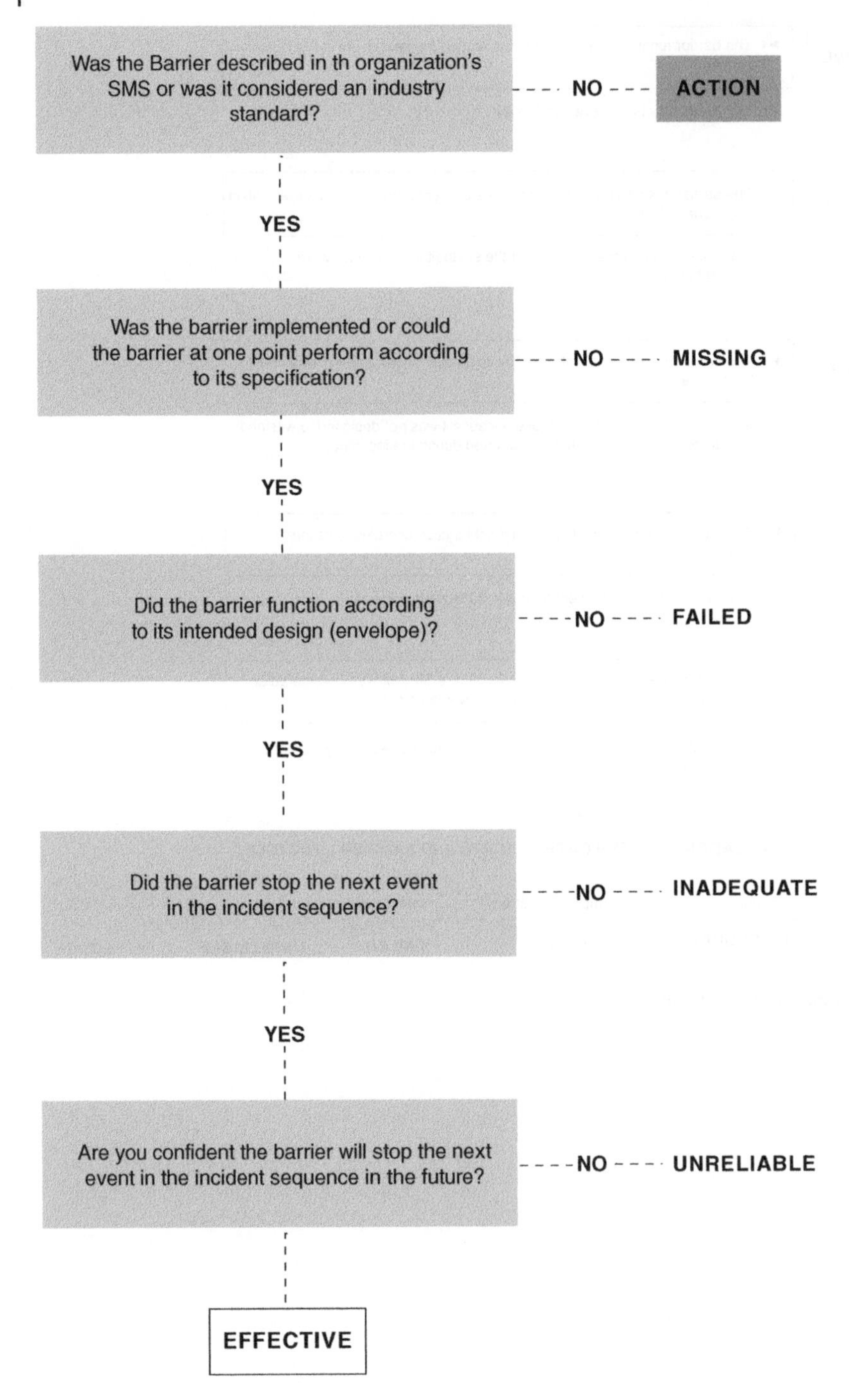

Figure 227 Incident barrier state decision support tree.

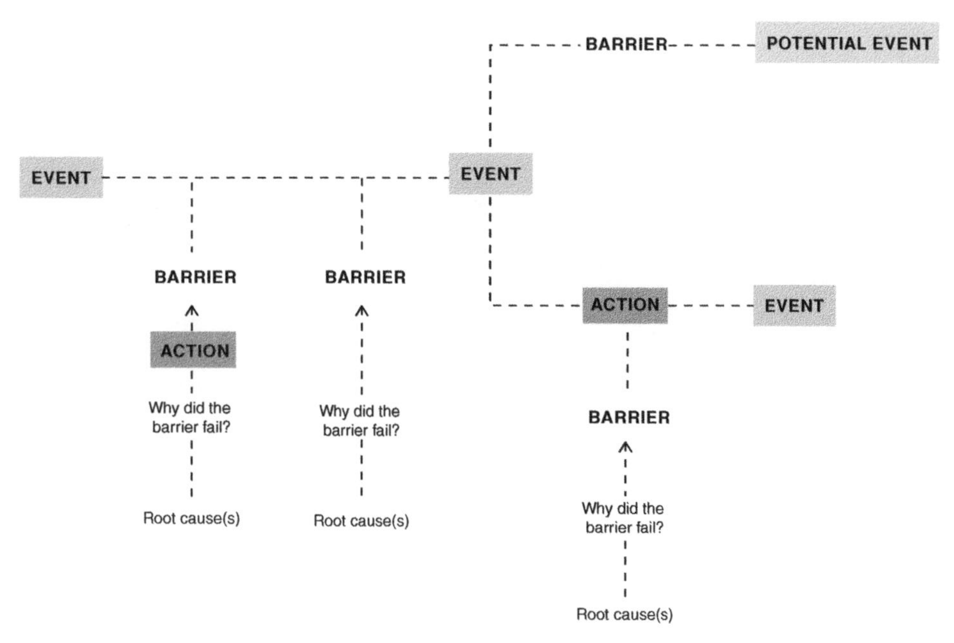

Figure 228 BFA guiding principles.

Appendix 3

Human Error and Reliability Assessment (HRA)

Human Errors and Violations

It has already been anticipated that talking about human error as the cause of a negative event can only be considered a first approach to the analysis of the event.

By "human error" we mean an incorrect action or, more generally, a failure to act. This view is obviously overly simplistic and requires a deeper analysis to understand the factors that led to such a failure. According to an obsolete way of reasoning, human failures can cause accidents, systems – even complex ones – are intrinsically safe, and human errors are attributable to the personality of the individual who commits them. Obviously, this view has now been radically abandoned, as it is known that systems cannot be inherently safe and that people play a central role in creating a safe environment, so human failures are indicators of irreconcilable pressures or negative influences on human performance. Therefore, the most modern theories pay attention to human limitations, the way humans interact with their tasks, machines and the environment.

People do not make mistakes deliberately (except for specific cases, of course) and their actions are influenced by many factors, among them:

- The way the brain processes information;
- The training;
- The machines and procedures or work instructions used;
- The culture of the organization they work for;
- Their values and their opinions or beliefs.

Based on this more modern perspective, it is possible to state that human failures are the actions and omissions of individuals that in turn cause partial or total failure of barriers (risk control measures). They can take place in various forms:

- Not performing a routine action because of distraction or a momentary memory failure.
- Applying an inappropriate procedure, rule, standard or practice for a specific situation that would have required a different one. Or similarly, applying the correct rule but doing so incorrectly or inappropriately or even untimely.

Bow-Tie Industrial Risk Management Across Sectors: A Barrier-Based Approach,
First Edition. Luca Fiorentini.

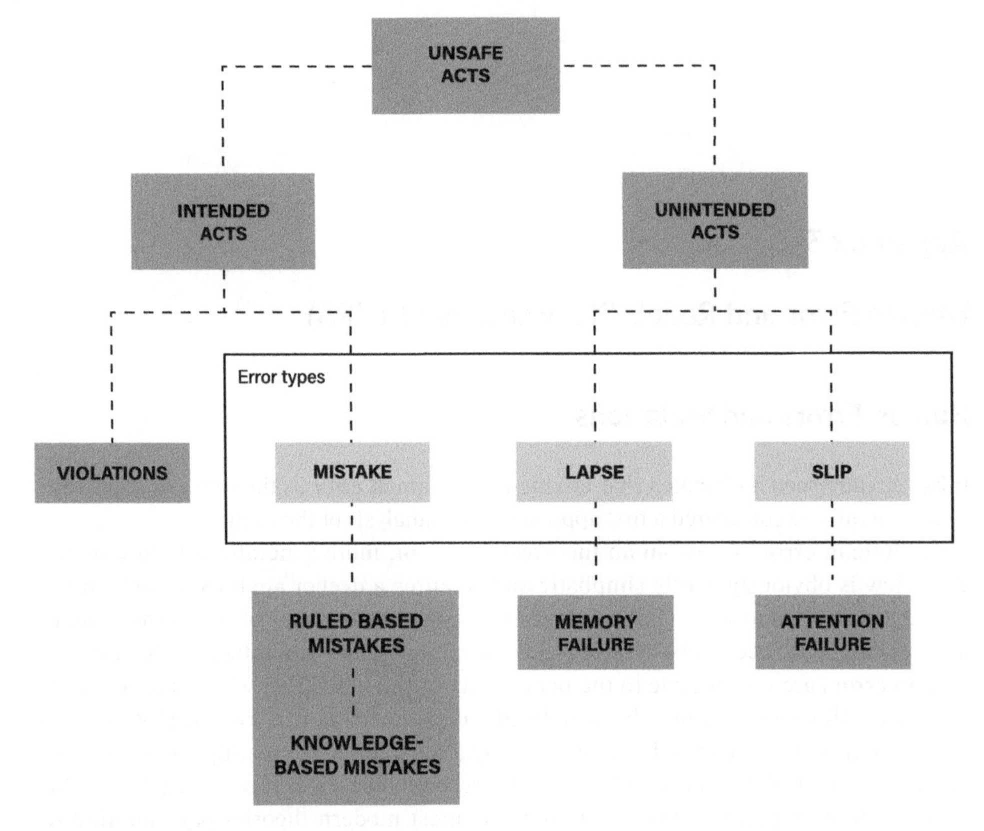

Figure 229 Classification of human failure.

- Incorrectly assessing a situation for which there is no procedure or rule but specific knowledge (understood as a combination of skills and experience) is required to achieve the correct action.
- Ignoring or deliberately deviating from a known prescribed rule.

These human failures can be categorized into two sets (Figure 229): human errors and violations.

A human error is an unintentional wrong action or decision by a person that leads to an unintentional and unwanted consequence. Human errors are grouped, as clarified in the following section, into three categories: skills-based errors (forgetting and inattention), rules-based errors, knowledge-based errors.

A violation is an intentional and deliberate deviation from a rule or procedure, i.e. when a person decides to act without respecting a known rule or procedure. Violations arise from motivational problems and are classified into five categories: unintentional, routine, situational, optimization and exceptional.

The Rasmussen Skills-Rules-Knowledge Model of Human Error

The Generic Error Modelling System (GEMS) is a conceptual model of actions or omissions used to classify human errors. It was developed by Professor James Reason and is based on Rasmussen's three-level performance model:

1) Skill-based: requires little or no mental effort, as these are "automatic" actions, such as writing or cycling;
2) Rule-based: concerns those actions that explicitly follow a procedure;
3) Knowledge-based: concerns those actions that require mental effort to solve a problem.

At the skill-based level, human performance is performed in the absence of conscious control (which intervenes only to shape the intention, not its practical implementation) and is generally used to cope with routine, non-problematic activities in familiar contexts. At this level, inattention and forgetfulness may occur.

Rule-based and knowledge-based performance come into play only after an individual becomes aware of the problem.

The GEMS model therefore shows how humans process information for a particular task and how they can move from one level to another in performance. For example:

- *Skill-based performance.* A worker is monitoring the control panel of a process plant. He performs a series of routine operations such as opening and closing valves and starting agitators and heat exchangers. As the worker has a lot of experience, he performs these operations in an automatic mode and only occasionally checks the situation to confirm that everything is okay.
- *Rule-based performance.* If during one of these occasional checks a problem emerges, for example indicated by an alarm, the worker enters the rule-based level to determine the nature of the problem. This may require collecting information and data from different sources that will then be used as input to the "diagnostic rule," which will take the following form: "If the symptoms are X, then the cause of the problem is Y." Having established the possible causes of the problem on the basis of these indications, a "rule of action" of the type "If the cause of the problem is Y, then do Z" can also be applied.
- *Knowledge-based performance.* If, as a result of applying the action rule, the problem is resolved, then the worker will return to the original skill-based sequence. If the problem is not solved, then more information will need to be gathered to try to identify symptoms corresponding to a known cause. If the cause of the problem cannot be established by applying every available diagnostic rule, the worker must go as far as the knowledge-base level, using their own or others' knowledge (in the example, chemistry or engineering) to manage the situation.

Some examples of tasks at the three different levels of performance are:

- Skill-based
  - Use a manual tool such as a hammer;
  - Manually check routine practices and processes;
  - Use familiar equipment to perform measurements and tests;
  - Open or close a valve;
  - Routinely replace a specific part during maintenance.
- Rule-based
  - Decide whether to replace a gasket inspected during preventive maintenance;
  - Respond to a control panel alarm;
  - Estimate the level change in a tank due to a temperature change;
  - Use emergency operating procedures.
- Knowledge-based
  - Review a procedure subject to change;
  - Resolve conflicting indications;
  - Plan strategies, objectives, business policies;
  - Conduct emergency response activities.

Human errors can occur at each of these three levels. At the skill-based level they are called "slips" and "lapses", i.e. carelessness and forgetfulness; at the rule and knowledge-based levels they are expressly referred to as "mistakes."

Slips and Lapses

Inattentions (slips) and forgetfulness (lapses) are involuntary actions, or inactions, that lead to unintended consequences and occur during the performance of highly familiar and routine tasks. Slips occur when a person incorrectly performs a task, e.g. presses the wrong button or reads data from the wrong indicator. Slips are caused by "short circuits" in our memory, e.g. forgetting to perform a step in a procedure. For example, slips and lapses can occur:

- When two similar tasks are confused;
- When the tasks are too complicated;
- When the main part of a task is performed, but not its finer details;
- When the steps of a procedure are not arranged in a "natural" sequence;
- When there are distractions and interruptions.

Slips are involuntary deviations of a person from the correct plan of action and are caused by attention failures. They occur when a person intends to perform a certain action but performs another. They usually arise when a simple and frequently performed action goes wrong. Examples of slips are:

- Operating the car's flashing lights instead of the windscreen wash buttons;
- Operating a switch in the wrong direction, e.g. upwards instead of downwards (wrong action on the correct object);
- Taking a measurement from the wrong instrument (correct action on the wrong object);
- Operating the wrong switch;

- Reversing two digits when copying numbers;
- Taking the wrong component from a box of mixed objects;
- Anticipating, postponing or omitting certain steps of a procedure from the expected sequence.

Lapses are memory failures that lead a person to forget to perform a certain action. Examples of lapses are:

- Forgetting to use the directional signals at a road junction;
- Leaving a medical tool inside a patient after surgery;
- Not performing a crucial stage of a safety-critical power of attorney;
- Forgetting to nail a joist;
- Removing an airway mask to talk to a colleague and then forgetting to put it back on;
- Not securing a scaffold because of a work stoppage;
- Driving a tanker truck with the transfer hose still connected.

In general, slips and lapses can be minimized with a few precautions:

- Design of the man–machine interface: consistency (e.g. upward switches are always "off"); intuitive layout of controls and instrumentation; level of automation, etc.;
- Control lists and reminders; procedures with the possibility of marking the execution of every single step with a tick;
- Independent cross-checks on critical activities, e.g. work permits;
- Reduce distractions and interruptions from assigned tasks;
- Ensure that sufficient time is available for the completion of the assigned task;
- Alarms, warnings and alerts to help identify possible errors.

Although lapses and slips can affect every person, their likelihood of occurrence increases due to certain conditions that have a negative impact on people, e.g. fatigue, lighting and noise levels, ergonomic conditions, sudden changes from routine, illogical designs, inconsistent human-machine interfaces and so on. These conditions are often the result of an ill-conceived plan by someone else, e.g. the person who designed the equipment or organized the work shifts. These human failures are often committed by experienced, highly trained and well motivated personnel, so supplementary training in such cases is ineffective in minimizing these types of errors.

Mistakes

Errors in the strictest sense (aka "mistakes") are errors of assessment or failures in the decision-making process that arise when the wrong thing is done while believing that it is right when doing it. They are therefore intentional actions that occur when a person does what he or she is thinking about doing (believing it to be the right thing) but should have performed another action, thus leading to an unintended consequence. Typically such errors involve misinterpretation or lack of knowledge and involve a person's mental process, controlling their planning, evaluating the information available, making decisions and assessing the expected consequences. They tend to occur when the person does not know the correct way to perform a task assigned to them, either because it is a new and unexpected task or because the person has not been properly trained for it (or both). In such circumstances, people often tend to behave using rules extracted from similar situations but which are not the correct ones. Mistakes in the strict sense can be caused by:

- Pressures to finish a job in a limited time;
- An unpleasant work environment, e.g. too hot, too cold, poorly lit, narrow, noisy;
- Confusing and inaccurate instructions and procedures;
- Excessive demand for work, e.g. high workloads, boring and repetitive work, work requiring a lot of concentration, too many distractions present;
- Social problems, e.g. pressure from a colleague, attitudes in conflict with safety and health, attitudes of workers in conflict about how to complete a job;
- Individual stressors, e.g. drugs and alcohol, lack of sleep, family problems, health problems;
- Equipment problems, e.g. poor design of human–machine interfaces or inconsistent interface;
- Organizational problems, e.g. an inability to understand where errors can occur and neglecting to implement related controls such as training and monitoring.

Errors can typically be minimized by improving the information provided to a person, e.g. through training or instructions in the workplace. There are two types of mistakes: rule-based mistakes and knowledge-based mistakes. Both are intentional actions with unintentional consequences.

Rule-based mistakes are intentional actions by a person who selects and follows a wrong procedure (or rule, standard, etc.) for a situation in which it was thought that rule should have been followed. The error therefore manifests itself as either a failure to apply the correct rule, or the application of an incorrect rule. These errors arise when a person's behaviour is based on rules they remember, procedures and familiar solutions, and when there is a strong tendency to use them even when they are not correct. Rule-based errors arise from situations that are guided by general, simple rules that can be applied several times in specific situations. The person then makes a choice, as the rules are not activated automatically but are specifically selected. This type of error occurs when:

- A procedure (or rule, standard, etc.) not appropriate to the situation is chosen and applied;
- The correct procedure is selected but it is not followed correctly, e.g. because it is written ambiguously or too complexly;
- The defects are not identified within a properly selected procedure;
- A procedure is applied for a situation that is not appropriate for a procedure and the situation requires knowledge, analysis and evaluation skills typical of unusual circumstances.

Examples of rule-based mistakes are:

- Selecting the wrong procedures for the insulation of a system line in order to temporarily exclude it from the circuit normally in use;
- Ignoring an alarm in a real emergency, letting oneself be guided by the history of previous spurious alarms;
- Incorrectly evaluating an overtaking manoeuvre with an unfamiliar and less powerful car than the one a person normally drives.

Knowledge-based mistakes occur when an individual does not have rules or routine behaviours to follow in order to manage an abnormal situation and must rely on theoretical principles, creative and analytical thinking, and experience to solve the problem. But bad diagnoses and incorrect calculations can lead to a wrong assessment. Examples of knowledge-based errors are:

- Misdiagnosing a problem in a business process and taking inappropriate corrective action (e.g. due to lack of experience or insufficient and incorrect information);
- Relying on an outdated map to plan an unusual route.

Mistakes can be minimized:

- By planning the actions to perform for all possible "what-if" scenarios (procedures for abnormal scenarios, deviations from normal operating conditions, emergency scenarios);
- With competence (knowledge and understanding of the system, training in decision-making techniques);
- By performing regular exercises/simulations of abnormal and emergency scenarios;
- With effective feedback systems and shift change procedures;
- By providing up-to-date tools and information for problem diagnosis and decision-making support (flowcharts, diagrams, etc.);
- By ensuring the learning of the whole organization (capturing and sharing experiences on the management of abnormal events).

Violations

Violations are voluntary deviations from known rules, procedures or policies by a person or group of people. They are deviations from expected behaviour deliberately carried out. Unlike human errors, which are involuntary deviations, violations are the expression of the voluntary and intentional decision to deviate. They occur when a person knowingly performs a short-cut or does not follow a procedure in order to save time or effort (whether physical or mental), believing that the rules are too restrictive and non-binding. These actions are usually well-thought-out but reckless and often exacerbated by involuntary encouragement from management, who cares about the work being done and puts in the background how it should be done.

Many violations occur out of a healthy and honest desire to do the work according to how management's wishes are perceived. Such voluntary actions may involve both an omission (not doing something that should have been done) and misconduct. Any negative consequence of this is obviously unintended, unless it is sabotage. The latent cause of a violation is often cultural or motivational and is generally a function of the practices and values accepted and shared by the team, its leadership and the character of the individual. All groups can commit violations, including operational staff, supervisors, managers, engineers and even company management.

Different types of violations can be identified:

- Understanding (unintentional): One is aware of the existence of a procedure, but does not know how to apply it (difficulty in language, excessive use of cross-references, complexity in writing);

- Understanding (unintentional): One acts as if there were no procedure to follow (because one does not know that it exists or is not familiar with it, or because it is not available on the site where it is requested);
- Routine: One is used to performing a task in a given way because one has always done so;
- Situational: It is impossible to perform the work following the procedure. The only way to get the job done is to violate the procedure.
- Optimization for the person: By committing the violation, the person performing the violation gains an advantage (e.g. finishing before work and going home earlier);
- Optimization for the organization: By committing the violation, the company gains an advantage (e.g. productivity is increased or profits are increased or costs are reduced);
- Exceptional: A person solves a problem for the first time, but fails to follow a procedure (e.g. when responding to an emergency, one may be acting inconsistently with a procedure).

Violations can also be minimized by:

- Eliminating the reasons that cause them (bad work scheduling; uncomfortable and annoying demands; unrealistic objectives and workloads, impractical or unrealistic procedures, adverse environmental factors);
- Improving attitudes, organizational culture, values, morals and teamwork rules (active participation of the workforce in the drafting of procedures, encouraging reporting of violations, making non-compliance socially unacceptable);
- Improving the perception of risk, promoting its understanding and increasing awareness (explaining the rationality behind the rules and procedures, providing specific training);
- Ensuring efficient supervision and performance monitoring;
- Increasing the probability of detection of violations.

Reducing the Risk of Human Error

Reducing the probability of human error is possible by taking appropriate action to improve previously identified form factors. Such interventions should find their place within a modern safety management system. Examples of such measures include:

- Minimization of an operator error:
  - Make the procedures and instructions logical and explain why the individual steps identified in the sequence are necessary as well as correct;
  - Make the sequences of equipment use logical;
  - Standardize operational approaches within the organization (thus also facilitating the replacement of operators in different areas of action);
  - Design simple and intrinsically safe operations as far as possible;
- Provide (and verify) training plans for reliable and safe operations while constantly (periodically) monitoring the level of skills;
- Provide good displays, in step with technology and in general correct, consistent and modern human–machine interfaces from an ergonomic point of view;
- Equipping itself with a system for the management, rationalization and prioritization of alarms (especially for process and manufacturing companies), in order to:

  - Eliminate alarms that have no function;
  - Use some alarms only when actually useful (e.g. by suppressing low-temperature alarms during the start-up phases of a process plant, before heat development takes place);
  - Filter repetitive alarms;
  - Group multiple alarms under one alarm signal.
- Use, where applicable, safety instrumented systems (Safety Instrumented System) in accordance with the requirements of IEC 61508 and IEC 61511 standards which, given the degree of risk reduction (RRF) required, make it possible to ensure a suitable PFD of the system indicating the requirements for maintaining performance over time;
- Provide a system of work permit management and safety review before start-up (–pre start-up safety review [PSSR]).

Probability of Error and Evaluation Techniques

While technological evolution and suitable prevention systems have led to a reduction in accidents and incidents due to technical failures, it is impossible to speak of the reliability of a system without taking into account the possibility of failure of all its components and, therefore, the human component as well, also in relation to what has been stated in the previous paragraphs.

There are various techniques for the analysis of human reliability (human reliability assessment [HRA]), i.e. techniques developed to provide estimates of human error probability values related to operators' tasks to be included in the wider context of system risk assessment.

The study of human reliability consists of the study of all factors, internal and external, that influence the efficiency and reliability of workers' performance, similar to what can be done with analytical methods for technical systems.

The first are all the random technical or systemic events (due to the environment: work equipment, materials used, workplace, work organization) that influence and alter working conditions, inducing operators to commit incorrect behaviour; the second ones, more difficult to predict because they are linked to individual characteristics, are related to psycho-physical conditions that, by their nature, do not lend themselves to be structured in systemic behaviour models and are therefore difficult to model.

There are several HRA techniques that have been developed by human reliability specialists. Among those most widely used is possible to mention:

- **Technique for Human Error Rate Prediction (THERP)**, which makes it possible to predict the probability of human error and to evaluate the possibility of degradation of a man–machine system as a result of human errors considered on their own or in association with the operation of equipment, operating procedures and practices, or with other characteristics of the system or people, which affect system behaviour. In fact, each task of the operator is analyzed in the same way as the reliability of these components is evaluated, with additional adjustments to take into account the peculiarity of human performance. Operators' wrong actions are divided into omission errors and commission errors: the former are related to an assigned action that is not performed at all; the latter, on the other hand, are those related to the wrong performance of an assigned action.

- **Human Error Assessment and Reduction Technique (HEART)** is a method based on certain assumptions. In particular, according to this method, basic human reliability depends on the generic nature of the task to be performed and under "perfect" conditions. This level of reliability will tend to be achieved consistently with a given nominal probability within probabilistic limits; since these perfect conditions do not exist in all circumstances, the expected human reliability may worsen depending on the extent to which the identified error conditions (EPC) could be applied. There are nine types of generic activities (GTT) described in HEART, each with an associated nominal human error potential (HEP) and 38 error production conditions (EPC) that can affect the reliability of the activities, each with a maximum value by which the nominal HEP can be multiplied.
- **Accident Sequence Evaluation Programme (ASEP)** is a simplified and slightly modified version of THERP.
- **Human Cognitive Reliability (HCR/ORE)** The HCR/ORE method uses a normalized time-to-reliability correlation (T/RC) to estimate the probability of failure of the operator to understand the magnitude of the event based on the relationship between the time available to decide on and the time taken to start the necessary actions. The model is a set of decision trees through which the analyst evaluates a set of combinations to provide estimates of cognitive failure. Failures in the execution of the operator's actions are evaluated using the THERP method.

Among the various methods for assessing reliability and estimating the probability of human error is the **Standardized Plant Risk Human Reliability Analysis (SPAR-H)**, developed in the United States to support this type of analysis in the nuclear industry, but also applicable to different contexts. This method has found a strong application in Italy in the oil & gas sector, driven by the guidelines of Legislative Decree 145/2015 implementing Directive 2013/30/EU, the so-called "Offshore Directive" on the safety of operations at sea in the hydrocarbons sector, and is therefore described in detail in the following section.

Another applicable methodology that is described in detail in a specific paragraph is the **Success Likelihood Index Method (SLIM)**. This technique consists in correlating the probability of operational error with certain factors, called performance-influencing factors (PIFs), that influence human behaviour. In particular, it will show its application to a real case study, appropriately anonymized.

The SPAR-H Method The application of the SPAR-H methodology involves the preliminary identification of the barriers to which the human factor is attributable. To facilitate this identification, all barriers identified in the Bow-Tie analysis can be grouped into different categories. Discriminating in one or the other category, as already illustrated in this volume, is the person who performs the detection, decision and action steps. The following categories are identified:

- *Behavioural:* These are those barriers that rely 100% on the human factor (e.g. human actions foreseen by operational procedures). Detection, decision and action are entirely entrusted to people.
- *Socio-technical:* These are barriers in which technical-instrumental and human factors are simultaneously present (e.g. operational intervention on alarm) belong to this

category. Typically, detection and action are entrusted to technology, while the decision (sometimes also action) is up to people. In any case, any barrier where technology and people are combined falls into this category;
- *Active hardware:* These are those barriers that do not foresee the intervention of people, but entrust exclusively to technology the phases of detection, decision and action (e.g. an automatic block);
- *Passive hardware:* These are those barriers that do not detect, decide or act. They are simply present and serve to absorb any release of energy or products (e.g. containment basins);
- *Continuous hardware:* These are those barriers that do not detect or decide, but are always in operation (e.g. inertisation or ventilation systems).

It is therefore clear that the human factor is to be sought in behavioural and socio-technical barriers. From the totality of the barriers identified in a Bow-Tie we extract those belonging to the above-mentioned categories and for each barrier we apply the SPAR-H methodology for the quantification of the probability of failure on demand (PFD) of the human component.

The methodology foresees the possibility to identify two phases of human intervention for each human failure error (HFE): the diagnosis phase and the action phase. The first includes the whole spectrum of the cognitive process, from the interpretation of information and understanding of the situation to the formulation of the decision to act. The action phase does not include any cognitive process, but only the executive process. Depending on the specific barrier, it is possible that the human factor concerns only the diagnosis phase, the action phase or both.

Once the human failure events have been categorized as diagnosis and/or action, the scores to be assigned to eight performance-shaping factors (PSFs) are assessed. Before proceeding with the assignment of scores, one must ask oneself:

- Is sufficient information available to assess the performance factor?
- Does the performance factor under consideration really influence, as a performance driver, the probability of failure for the human operator?

Only those PSFs for which sufficient information is available to make a judgement and only those that have been identified as performance drivers need to be assessed and quantified.

The PSFs are as follows:

- *Available time* is an expression of the relationship between the time available and the time required to complete the diagnosis/action. That is, it intends to quantify the time margin, to understand whether or not there is sufficient time to diagnose the problem or take action.
- *Stress/Stressors* are the performance factor refers to the level of undesirable conditions and circumstances that prevent the completion of a task. It may include mental stress, excessive workloads, or physical stress such as that imposed by environmental factors (excessive heat, noise, poor ventilation, etc.). This performance factor is not independent from the others, so care should be taken to avoid counting the same influence factor twice.
- *Complexity* refers to how difficult it is to perform the task (diagnosis or action) in a given context.

- *Experience/Training* takes into account the years of experience of the individual or team, the training on the type of accident assumed, the time elapsed since the last training, and the relative frequency of updating.
- *Procedures* refers to the existence and use of formal operational procedures for the tasks under examination, diagnosis or action;
- *Ergonomics/Human-Machine Interface (HMI)* refers to the instrumentation, displays and controls, arrangements, quality and quantity of information available from the instruments, and the interaction of the operator or team with the instrumentation, in order to perform and complete the specific task. Examples of poor ergonomics may include mislabelling alarms or valves, incorrect layout of displays in the control room, remote location of buttons and so on.
- *Fitness for duty* refers to the mental and physical suitability of the individual who has to perform the assigned task (diagnosis or action). Fatigue, tiredness, distractions, overestimation of one's abilities are examples of factors that can degrade this performance factor.
- *Work processes* refers to organizational aspects, safety culture, work planning, communication, support and management policies.

Each PSF is assigned a level and the relative numerical value, choosing among those provided by the methodology (for further details on quantification see Idaho National Laboratory, 2005).

Once PSF levels have been assigned, the human error probability (HEP) is simply the product of the nominal human error probability (nominal HEP [NHEP], assumed to be 0.001 for action and 0.01 for diagnosis) for the product of the scores assigned to PSF ($PSF_{composite}$).

$$HEP = NHEP \cdot PSF_{composite}$$

This value should be corrected if three or more PSFs adversely affect the $PSF_{composite}$ value, making it likely that the final HEP value may be greater than one. In this case a correction factor is adopted:

$$HEP = \frac{NHEP \cdot PSF_{composite}}{NHEP \cdot \left(PSF_{composite} - 1\right) + 1}$$

If diagnosis and action are combined into a single HFE, then the two HEPs are calculated separately and then combined to produce the single probability value, as calculated below:

$$HEP_{diagnosis\ and\ action} = HEP_{diagnosis} + HEP_{action} - \left(HEP_{diagnosis} \cdot HEP_{action}\right)$$

When assessing the final value of the HEP, the dependency is taken into account. It exists when the occurrence of an event produces a change in the probability of a subsequent HFE. For example, if the training is poor, the stress is high, or the time available is short, more events may be affected and this dependency must be taken into account. The simple fact that there are two HFEs in sequence does not make them dependent: HFEs must be psychologically connected. In practice, the following factors are taken into account:

- *Time:* Any temporal dependencies between two psychologically connected HFEs are assessed, such as more or less reduced, or, at the limit, overlapping time intervals.
- *Location:* Any spatial dependencies between two psychologically connected HFEs are assessed (are the tasks carried out within the same spaces or in different spaces?).
- *Crew:* Any dependencies attributable to the person performing the tasks (same person? same team?) are assessed.
- *Cues:* The presence or absence of additional ideas that stimulate people to think differently is assessed.

In relation to the combination of these parameters, a variable level of dependence is obtained between complete, high, moderate and low. In relation to the level of dependence, different formulas of corrective factors for the probability of human error are applied, for which reference is made to the technical literature mentioned above.

In Table 31, the calculation tabulate is shown with an applicative example of the SPAR-H methodology.

The SLIM Method When developing corrective action following an accident, it is stressed that the proposed recommendations should be effective. The following is a brief case study to define the process for selecting improvements following an accident using risk analysis techniques. In particular, the choice between possible alternatives for improvement is given by the solution that ensures a greater decrease in the probability of occurrence of the accident event.

Following an accident event on the gasifier of an Italian refinery, the possibility of introducing modifications is analyzed for the Start-up procedures, training of the personnel entrusted with the start-up of the plant, and verification procedures for third-party companies involved in maintenance, in order to reduce the probability of human error in the performance of operations related to the start-up and maintenance of the unit.

In particular, those causes (which can be traced both to operating errors and instrumental failures) that can lead to oxygen being sent to the plant's blow-down system and torches are analyzed.

Plant modifications are also proposed, which can be summarized in the following points:

- Installation of a second valve (XV) on the fuel oil recirculation line in series with the oil recycling valve (V1) if this second valve is not already present;
- Installation of a fuel oil flow meter on the inlet line to the gasifiers with associated low flow alarms in the control room.

It should be noted that, with reference to the technological and safety adjustments envisaged by the licensee of the process, the plant configuration:

- Includes the presence of a second valve installed on the recycling line (manual valve) for which an evaluation was made of the expected benefits following the use of this valve in the start-up sequence both in the case of use by the operator and considering its automatic closure in the start-up sequence;
- Does not allow the installation (for system layout reasons) of instruments to measure the fuel oil flow rate entering the gasifiers for which an evaluation of the expected benefits has been carried out following the installation of instruments to measure the fuel oil flow rate on the recycling line, equipped with high oil flow alarms reported in the control room.

Table 31 Example of calculating HEP with the SPAR-H Method.

ACTION PSF Levels		
PSFs	**PSF Levels**	**Multiplier for Action**
Available Time	Inadequate time	P(failure)=1
	Time available is equal to the time required	10
	Nominal time	1
		0.1
		0.01
	Insufficient information	1
Stress/Stressors	Extreme	5
	High	2
	Nominal	1
	Insufficient information	1
Complexity	High complex	5
	Moderately complex	2
	Nominal	1
	Insufficient information	1
Experience/ Training	Low	3
	Nominal	1
	High	0.5
	Insufficient information	1
Procedures	Not available	50
	Incomplete	20
	Available, but poor	5
	Nominal	1
	Insufficient information	1
Ergonomics/HMI	Missing/Misleading	50
	Poor	10
	Nominal	1
	Good	0.5
	Insufficient information	1
Fitness for duty	Unfit	P(failure)=1
	Degraded fitness	5
	Nominal	1
	Insufficient information	1

Table 31 (Continued)

ACTION PSF Levels		
PSFs	**PSF Levels**	**Multiplier for Action**
Work Processes	Poor	5
	Nominal	1
	Good	0.5
	Insufficient information	1
Action HEP		
Final Action HEP	2.50E-3	

Following is a brief description of the unit 300 (gasification and washing) and the start-up procedure.

In Unit 300 gasification there is the production of synthesis gas, based on licensed gasification technology.

The charging fuel oil coming from Unit 200 goes to the T1 oil/steam mixer for mixing with high-pressure steam.

The mixture is sent to the process burners positioned at the head of the GAS1 gasifier.

The oxygen coming from the air separation unit through piping is filtered and divided into two currents:

- to Unit 300;
- to Unit 510.

The reagents (steam, oxygen and charge oil) are fed into the reactor chamber via the process burner. Oxygen is supplied with a flow rate below the amount required for complete charge combustion. The moderation steam, pre-mixed with the filling oil, mitigates the temperature in the reactor chamber and reacts partially.

The gasification reaction is not catalytic but exothermic and the temperature of the gas at the exit of the gasifier chamber varies from 1200 to 1450°C. The main products of the reaction are carbon monoxide, hydrogen, carbon dioxide, water vapour, methane and carbon black.

The gas mixture obtained from the gasification chamber passes into the cooling chamber through a tube immersed in water. The mixture, coming into intimate contact with the cooling water, exits the GAS1 gasifier at a temperature of about 210°C.

The synthesis gas is sent to the scrubber connected to the gasifier, after mixing with the recirculating water coming from the scrubber itself. The first gas flushing is carried out in the scrubber.

The process burners, given the high temperature resulting from the reactions, are cooled with a coil fed with cold water.

The start-up operations of the gasifiers require the heating of the refractory. For this purpose each gasifier is equipped with a preheating burner.

The gasifier start-up procedure following an ordinary or extraordinary stoppage provides for a sequence of automatic operations whose correct execution is verified by the operator.

The pre-start operations (instrumentation tests on valves and sequences) make it possible to verify that all the appropriate permissives are met and, consequently, to authorize the panel builder to launch the initialization sequence from DCS.

The subsequent sequence of operations is as follows:

1) Reset of the safety system with a gasifier pressure of less than 4 barg;
2) Adjustment of the start-up flows of moderator steam, oxygen and charging oil through the respective vent and recycling valves by the operator;
3) Replacement of the fuel gas burner with the TAR burner;
4) Purging with low-pressure nitrogen.

At the end of this part of the start-up procedure, after checking that all the permissives are satisfied, the panel builder is authorized by DCS to launch the start-up sequence that it envisages:

1) Closing of the steam vent valve (V2) and subsequent timed opening of the main steam shut-off valve (V3);
2) Opening of the fuel oil block valve (V4) and closing of the recycling oil valve (V1) when the first one is open at 5%;
3) Opening of the downstream oxygen valve (V5), subsequent closing of the oxygen vent valve (V6) and finally opening of the upstream oxygen valve (V7).

In order to estimate the frequency of occurrence of the accidental event, a risk analysis was carried out and elaborated, considering the possible procedural and plant modifications mentioned in the introduction and evaluating the expected benefits.

In particular, the analysis was carried out considering:

- The procedures in place and the plant layout at the date of the event;
- The implementation of some procedural and plant engineering changes, as detailed below:

B1. Updating of the start-up procedure and related staff training;

B2. Identification of critical elements (and critical operations) of the plant. Critical elements are those valves, instrumentation, and so on whose malfunction (failure, incorrect assembly, incorrect maintenance) can directly lead to a significant accidental event. Revision of the Operating Instruction I.O.001 relating to maintenance management entrusted to third-party companies. A specific control plan for safety-critical components must be codified by the third-party company that made the modification. The verification of correct maintenance on critical components must be carried out in the presence of refinery personnel and, in any case, the personnel of the third party company that activates the quality control plan must be different from those who have carried out the maintenance.

B3. Implementation of the plant modifications that foresee, in brief, in the gasifier start-up sequence, the automatic closure of two valves on the fuel oil recirculation line, the existing V1 and a second valve of the same type, in series. A second manual valve (V8) is currently installed on the recirculation line of the fuel oil. This valve, after feasibility check, could be automatically managed by the start-up sequence in order to achieve the recommendations. Further recommendations are related to the installation of a flow rate transmitter on the fuel oil inlet line to the gasifiers alarmed for low oil flow: in consideration of the system layout, an equivalent modification is implemented, which consists of measuring the flow rate of the fuel oil on the recycling line by means of an alarmed transmitter in the control room for high flow rate.

B4. modification of the system start-up procedure in order to foresee the closure by the operator of the V8 instead of the proposed automatic closure.

The updating and adoption of the new start-up procedure and the related specific training activities for operating personnel identified during the safety analysis carried out following the accident led to the updating and adoption of the new procedure:

- The reduction of the probability associated with a lack of operational intervention in the event of operating anomalies during the start-up phase of the GAS1 gasifier;
- The reduction of the frequency of occurrence of the accidental hypothesis by one order of magnitude;
- An increase of a further two orders of magnitude in the plant's safety level can be obtained by modifying the maintenance management procedure carried out by third-party companies on components classified as safety critical, in the presence of internal refinery personnel and in accordance with I.O.001 to be formalized as a safety management system procedure.

In particular, following the implementation of the expected safety adjustments, the frequency of occurrence of the accidental hypothesis relating to "Blow down oxygen delivery during start-up of the GAS1 gasification reactor" would be of the order of magnitude of 10^{-5} occ/year, falling within a probability class defined as "Unlikely" against a frequency of occurrence in the configuration following the first level of changes that places the accidental hypothesis in the probability class defined as "Rather unlikely."

Finally, the adoption of the plant engineering recommendations, i.e. the use, with automatic closure, of a second valve on the fuel oil recirculation line, a fuel oil flow meter with alarm in the control room and the consequent further updating of the unit start-up procedure, entail:

- The optimization in ergonomic terms of operational interventions thanks to the alarm signal;
- The reduction of intervention times by operators in case of deviation of critical operating parameters from normal start-up conditions;
- The reduction in the frequency of occurrence of the hypothesis of oxygen blow down during start-up of the GAS1 gasification reactor.

The use, according to specific operating instructions, of the existing manual valve V8 in the start-up phase by the operator in the field instead of installing a second block valve, also by implementing the remaining plant recommendations, would result (with respect to full compliance with the licensee's requirements) in a higher frequency of occurrence for about three orders of magnitude. In view of this, it has been suggested that the licensee's definition of the process should be fully adopted and that the latter should be involved in the implementation details of each of the recommendations identified.

The adoption of new blocks and alerts would make it possible to reduce the frequency of occurrence of the accident hypothesis identified by a further three orders of magnitude, falling within a probability class defined as "Extremely unlikely."

SLIM was applied in order to assess the probability of operational error during the plant start-up sequence currently foreseen in the refinery manuals. This technique consists in correlating the probability of operational error with PIFs that influence human behaviour.

In this case the significant factors can be summarized in the following points:

- Unusual and/or complex operation;
- Noise or sources of distraction in general;
- Control panel design;
- Work supports and procedures;
- Training;
- Operator experience;
- Group work.

The operations carried out by operators to complete the start-up procedure can be summarized as follows:

1) Follow the sequence of steps before starting the gasifier;
2) Once all the appropriate permissives (instrumentation tests on valves and sequences) have been satisfied, start the system;
3) Follow the sequence of valve opening and closing to verify the correct operation;
4) Verify a rapid increase in pressure and temperature in the gasifier and scrubber;
5) Verify an increase in the torch flame (assistance to the panel builder for this operation);
6) Check the gasifier level by acting on the quench ring flow rate, blow-down flow rate and pressure increase in the system.

For carrying out the risk analysis, the SLIM methodology was used to evaluate the probability of error in the execution of operations 3, 4, and 5 of the sequence.

Using operations 2 and 6 of the above sequence, the parameters A and B that characterize the equation of the SLIM method below have been estimated:

$$\log(\text{HEP}) = A \cdot \text{SLI} + B$$

where HEP is the probability of human error and SLI is the success probability index obtained from the combination of the values assumed by BIPs.

On the basis of the PIF parameters defined for operations 2 and 6, the following equation has been derived that links the probability of operational error and the PIF, through the SLI:

$$\text{Log}(\text{HEP}) = -31{,}5\,\text{SLI} + 14{,}4$$

The equation obtained was used to estimate the probability of error in operations 4 and 5 of the sequence in the current configuration.

Based on the available documentation, operations 4 and 5 of interest are characterized by the PIFs in the current configuration (hereinafter referred to as "ANTE"), shown in Table 32.

The values in Table 32 have been assigned on a relative scale between 1 and 9 where the upper end represents the optimum. For example, in the case of the first operation the value 9 represents the case of a habitual and not complex operation.

Table 32 PIF (current configuration).

PIF	Value Operation 4 (ANTE)
a) Unusual and/or complex operation	2
b) Noise or distractions in general	7
c) Designing the control panel	5
d) Support at work and procedures	5
e) Training	5
f) Operator experience	5
g) Group work	5

This results in a probability of operational error in the execution of the operation itself equal to:

$$HEP_{ANTE} = P_{ANTE} = 1{,}46 \cdot 10^{-1}$$

In the first instance, it is proposed to update the start-up procedure and the training of plant personnel, with:

- A written and detailed procedure that supports the interpretation of the temperature and pressure curves and the increase in torch flame in order to identify anomalous conditions and that describes the operations to be carried out to make the plant safe in the event of the identification of critical points;
- Training on the start-up procedure of the plant as a whole and on the procedure referred to in the previous point, particularly aimed at all personnel involved in the operation before each start-up of the GAS1 gasifier.

Taking into account the changes listed, in the new configuration (A), operations 4 and 5 of interest are characterized by the PIFs shown in Table 33.

The probability of human error consisting in the failure to interrupt the start-up sequence in the event of anomalies has therefore been assessed at:

$$P_A = 1{,}1 \cdot 10^{-2}$$

In addition to the protections currently provided, during the start-up phases, the activation in the control room of a high flow rate fuel oil alarm in the recirculation line controlled by a new flow meter located downstream of the recirculation oil sectioning and relative updating of the start-up procedure and the training of personnel in the plant is evaluated:

- A written and detailed procedure describing the operations to be carried out to make the system safe following the high flow alarm signal on the fuel oil recirculation line.

The adoption of the above-mentioned alarm allows the panel builder a better support in the evaluation of the correct execution of the start-up procedure.

This adjustment is equivalent to the licensee's recommendation to activate an alarm signal for low oil flow in charge to the gasifiers, which cannot be reached because the system layout does not allow the positioning of measuring instruments on the charge line.

Table 33 PIF (Configuration A).

PIF	Value Operation 4 (A)
a) Unusual and/or complex operation	2
b) Noise or distractions in general	7
c) Designing the control panel	5
d) Supports at work and procedures	**6**
e) Training	**6**
f) Operator experience	5
g) Group work	5

Table 34 PIF (POST configuration).

PIF	Value Operation 4 (POST)
a) Unusual and/or complex operation	2
b) Noise or distractions in general	7
c) Designing the control panel	**6**
d) Supports at work and procedures	6
e) Training	6
f) Operator experience	5
g) Group work	5

Taking into account the changes listed, in the new configuration operations 4 and 5 of the procedure are characterized by the PIFs shown in Table 34.

The probability of human error of failure to interrupt the start-up sequence due to anomalies has therefore been assessed at:

$$P_B = 3 \cdot 10^{-3}$$

Expected benefits are analyzed in terms of a decrease in the frequency of occurrence of the accident hypothesis identified as a result of the event, by combining the above human error probabilities with the component failure rates in Figure 230, Figure 231, and Figure 232.

EVENT NAME	DATA REF	DESCRIPTION	FAILURE RATE	Trep/ Ttest	PROB	R/U
XV006	EXIDA	Guasto valvola in chiusura	5,20E-02	9,13E-04	4,75E-05	R
LOGICAB		Guasto logica di blocco	2,77E-03	5,00E-01	6,93E-04	U
ERR_RIM	LEES	Errato rimontaggio post-manutenzione			3,00E-02	-
MIO		Mancato intervento operativo			1,46E-01	-

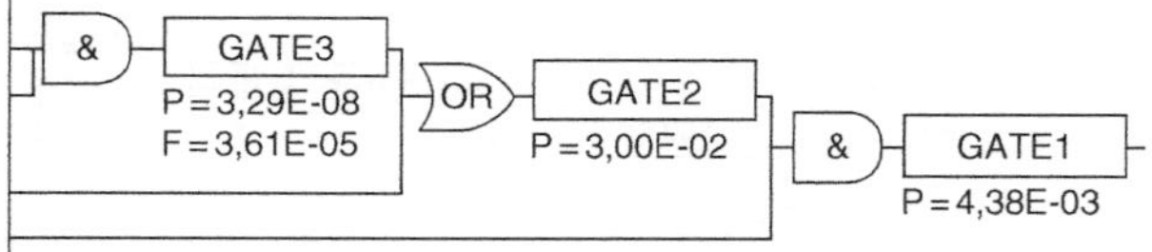

Figure 230 Fault tree Analysis, current configuration (ANTE).

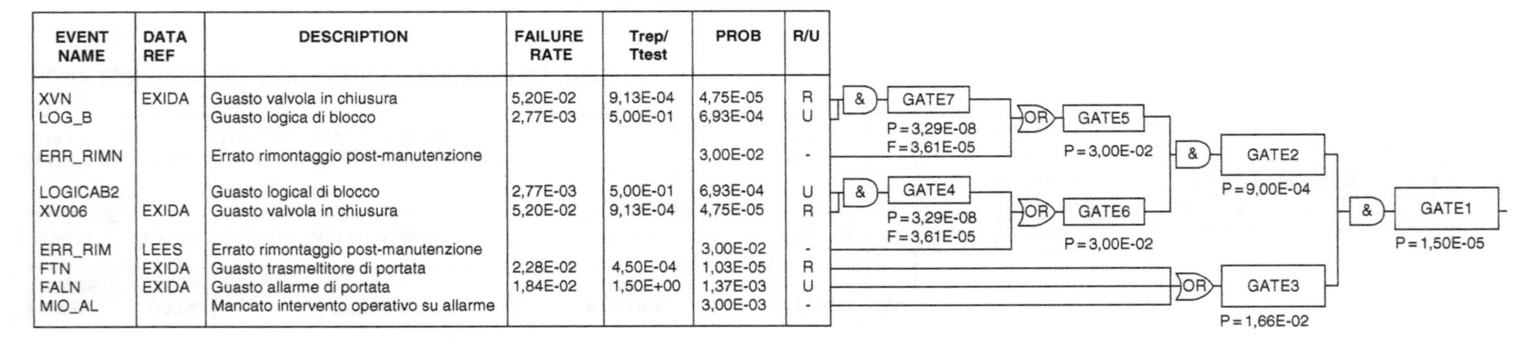

EVENT NAME	DATA REF	DESCRIPTION	FAILURE RATE	Trep/ Ttest	PROB	R/U
XVN	EXIDA	Guasto valvola in chiusura	5,20E-02	9,13E-04	4,75E-05	R
LOG_B		Guasto logica di blocco	2,77E-03	5,00E-01	6,93E-04	U
ERR_RIMN		Errato rimontaggio post-manutenzione			3,00E-02	-
LOGICAB2		Guasto logical di blocco	2,77E-03	5,00E-01	6,93E-04	U
XV006	EXIDA	Guasto valvola in chiusura	5,20E-02	9,13E-04	4,75E-05	R
ERR_RIM	LEES	Errato rimontaggio post-manutenzione			3,00E-02	-
FTN	EXIDA	Guasto trasmeltitore di portata	2,28E-02	4,50E-04	1,03E-05	R
FALN	EXIDA	Guasto allarme di portata	1,84E-02	1,50E+00	1,37E-03	U
MIO_AL		Mancato intervento operativo su allarme			3,00E-03	-

Figure 231 Fault tree analysis, better configuration (configuration A).

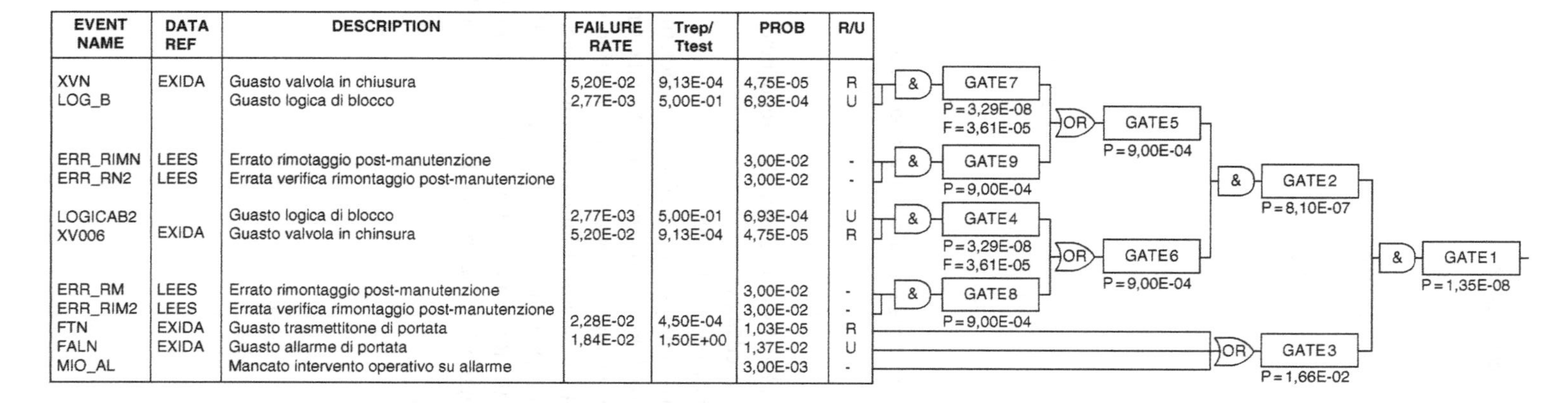

EVENT NAME	DATA REF	DESCRIPTION	FAILURE RATE	Trep/ Ttest	PROB	R/U
XVN	EXIDA	Guasto valvola in chiusura	5,20E-02	9,13E-04	4,75E-05	R
LOG_B		Guasto logica di blocco	2,77E-03	5,00E-01	6,93E-04	U
ERR_RIMN	LEES	Errato rimotaggio post-manutenzione			3,00E-02	-
ERR_RN2	LEES	Errata verifica rimontaggio post-manutenzione			3,00E-02	-
LOGICAB2		Guasto logica di blocco	2,77E-03	5,00E-01	6,93E-04	U
XV006	EXIDA	Guasto valvola in chinsura	5,20E-02	9,13E-04	4,75E-05	R
ERR_RM	LEES	Errato rimontaggio post-manutenzione			3,00E-02	-
ERR_RIM2	LEES	Errata verifica rimontaggio post-manutenzione			3,00E-02	-
FTN	EXIDA	Guasto trasmettitone di portata	2,28E-02	4,50E-04	1,03E-05	R
FALN	EXIDA	Guasto allarme di portata	1,84E-02	1,50E+00	1,37E-02	U
MIO_AL		Mancato intervento operativo su allarme			3,00E-03	-

Figure 232 Fault tree analysis, the best configuration (POST configuration).

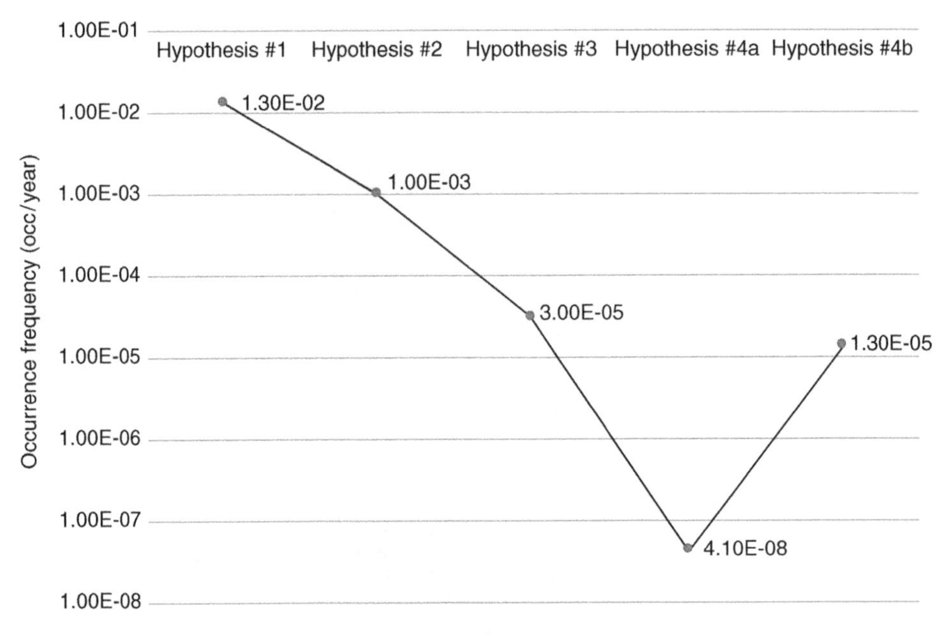

Figure 233 Frequency estimation of the scenario "Oxygen sent to blow down, during start up of reactor of GAS1".

Figure 233 shows the graphical summary of the frequencies of occurrence obtained downstream of each change listed above.

The results are also summarized in Table 35.

It therefore emerges that risk analysis is a useful tool in determining which recommendation to adopt in order to avoid the recurrence of accidental events.

In conclusion, the human factors analysis and classification scheme (HFACS) is also mentioned as a method for the analysis of human performance, as mentioned in the IEC 62740 standard. This method (Shappel and Wieggmann, 2001) analyzes the causes of human error on the basis of the Reason model (Reason, 1990) (Figure 234), distinguishing four levels, represented in Figure 235, Figure 236, Figure 237, and Figure 238:

- Organizational influences;
- Supervision;
- Preconditions for unsafe acts;
- Unsafe acts.

Table 35 Frequency of incidental assumptions considered.

Ref. Hypothesis	Description	Frequency of occurrence (occ/year)	Probability class	Changes
1	Sending oxygen to blow down during gasification reactor start-up GAS1 (current configuration)	$1{,}3 \cdot 10^{-2}$	Quite likely	
2	Sending oxygen to blow down during gasification reactor start-up GAS1 (following procedural changes)	$1 \cdot 10^{-3}$	Quite unlikely	Operator-specific start-up and training process update
3	Sending oxygen to blow down during gasification reactor start-up GAS1 (following additional procedural changes for third-party maintenance checks)	$3 \cdot 10^{-5}$	Unlikely	Maintenance of components classified as critical for the safety carried out by third-party companies in the presence of personnel inside the refinery
4a	Sending oxygen to blow down during gasification reactor start-up GAS1 (following plant recommendations)	$4{,}1 \cdot 10^{-8}$	Extremely unlikely	• Installing a second valve on the fuel oil recirculation line • Installing a fuel oil flow meter on the recirculation line with alarm in the control room • Updating start-up procedure and specific training of operators
4b	Sending oxygen to blow down during gasification reactor start-up GAS1 (no new XV installation)	$1{,}3 \cdot 10^{-5}$	Unlikely	As changes from Hypothesis 4a but with use according to the start-up procedure of the existing V8 manual valve instead of the installation of a second valve XV.

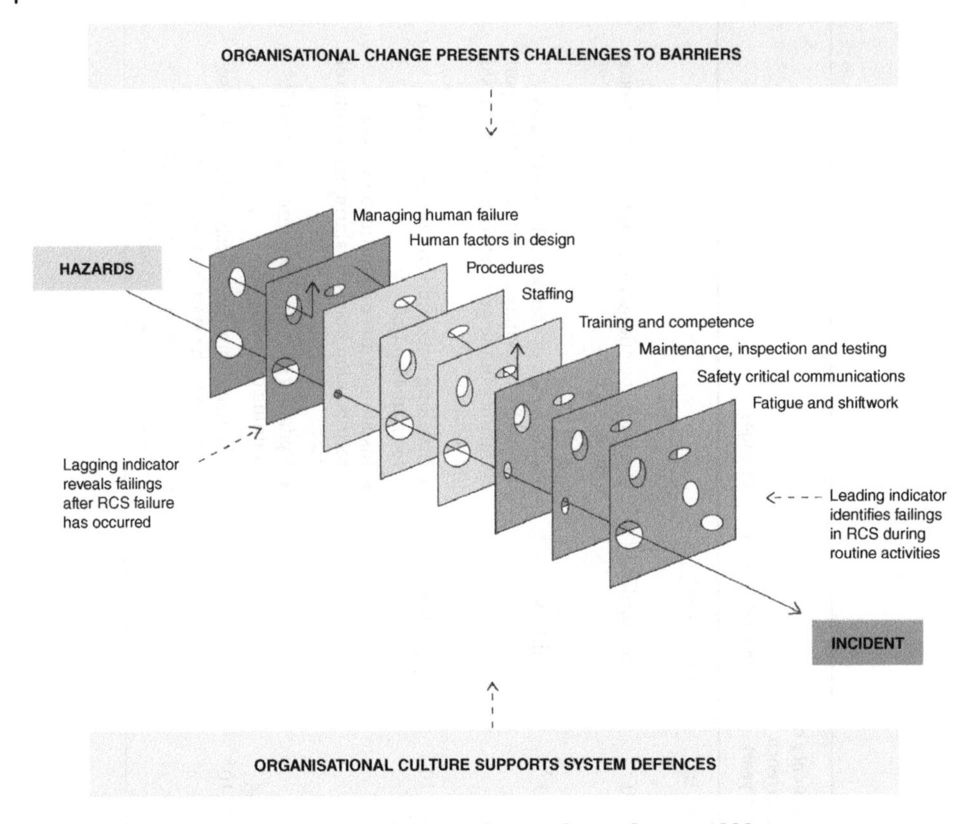

Figure 234 The Swiss Cheese Model by James Reason. *Source:* Reason, 1990.

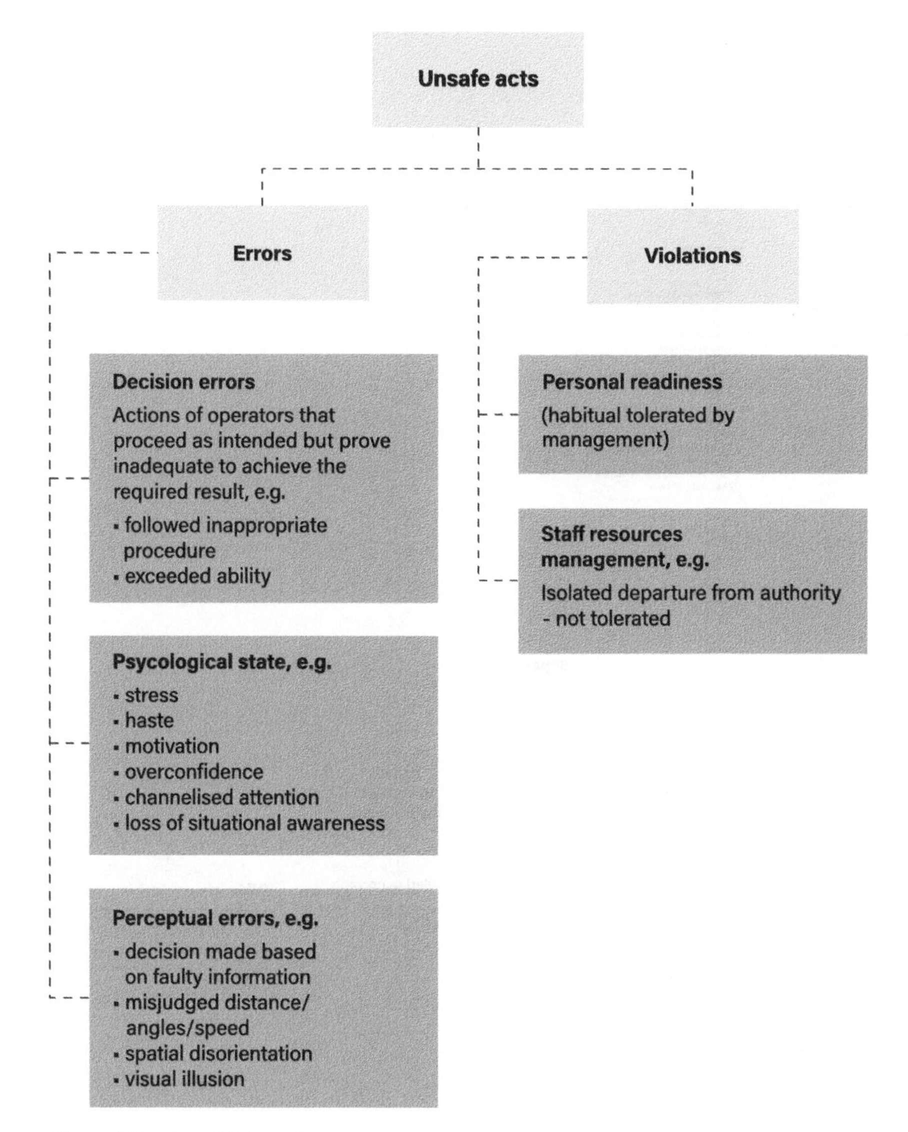

Figure 235 Level 1: Unsafe acts.

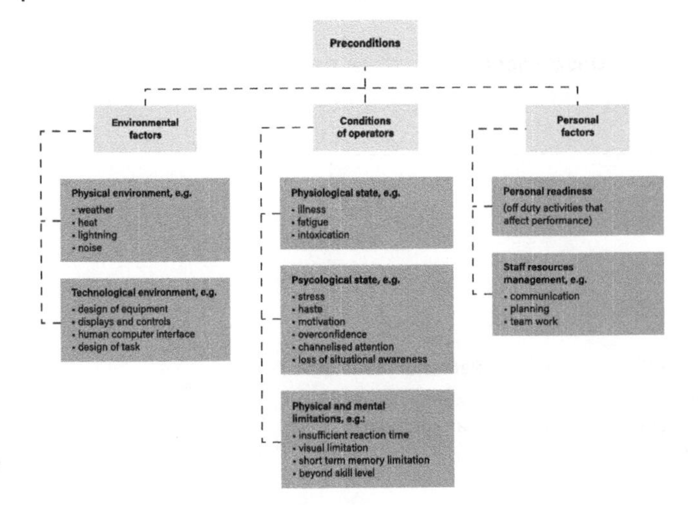

Figure 236 Level 2: Preconditions.

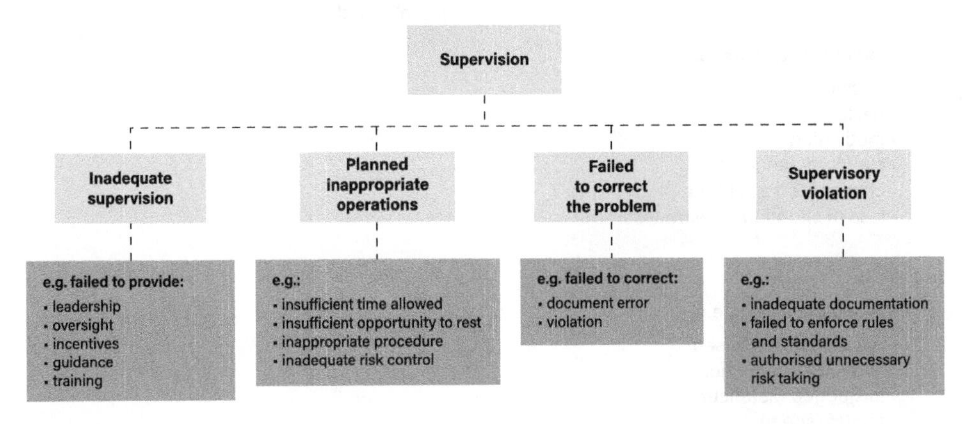

Figure 237 Level 3: Supervision Issues.

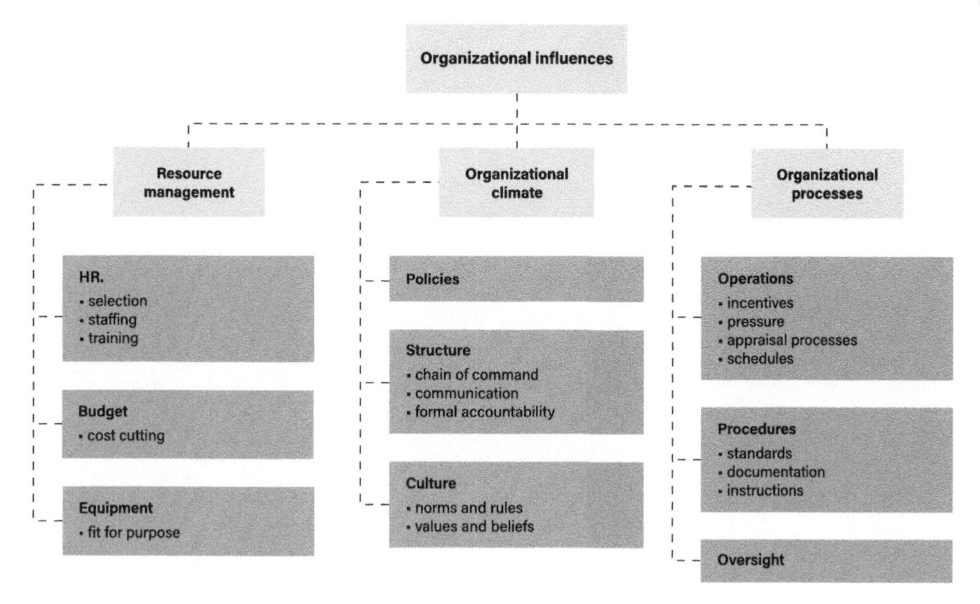

Figure 238 Level 4: Organizational Issues.

References and Further Reading

References

AIChE-CCPS. 2001. *Layer of Protection Analysis*. New York: Center for Chemical Process Safety of the American Institute of Chemical Engineers.

AIChE-CCPS. 2003. *Guidelines for Investigating Chemical Process Incidents*. 2nd edition. New York: American Institute of Chemical Engineers.

AIChE-CCPS. 2009. *Guidelines for Developing Quantitative Safety Risk Criteria*. New York: Center for Chemical Process Safety (CCPS).

AIChE-CCPS. 2013. *Guidelines for Enabling Conditions and Conditional Modifiers in Layers of Protection Analysis*. Wiley.

AIChE-CCPS. 2016. *Introduction to Process Safety for Undergraduates and Engineers*. Hoboken, NJ: Wiley.

AIChE-CCPS. 2018. *Bow Ties in Risk Management: A Concept Book for Process Safety*. Hoboken, NJ: Wiley.

Assael, M. and Kakosimos, K. 2010. *Fires, Explosions, and Toxic Gas Dispersions: Effects Calculation and Risk Analysis*. Boca Raton, FL: CRC Press/Taylor & Francis.

Bansal, A., Kauffmann, R., Mark, R. and Peters, E., 1991. *Financial Risk and Financial Risk Management Technology (RTM): Issues and Advances*. New York: Leonard N. Stern School of Business, New York University.

Bento, J. 2003. "Review from an MTO-Perspective of Five Investigation Reports from BP" (Draft). Slavangerm, Norway.

Committee on Sponsoring Organizations of the Treadway Commission (COSO). 2017. *Enterprise Risk Management—Integrating with Strategy and Performance*. New York: AICPA.

Dekker, S., Cilliers, P. and Hofmeyr, J. 2011. "The Complexity of Failure: Implications of Complexity Theory for Safety Investigations." *Safety Science* 49 (6): 939–945.

Demichela, M., Piccinini, N., Ciarambino, I. and Contini, S., 2004. "How to Avoid the Generation of Logic Loops in the Construction of Fault Trees." *Reliability Engineering & System Safety* 84 (2): 197–207.

DNV GL. 2014. "Good Practices: Barrier Management in Operation for the Rig Industry." Norwegian Shipowners Association.

Bow-Tie Industrial Risk Management Across Sectors: A Barrier-Based Approach,
First Edition. Luca Fiorentini.

Dutch Safety Board. 2020. "Roof Collapse During Extension Work at the Stadium of FC Twente in Enschede, The Netherlands, July 7, 2011."

ESReDA Working Group on Accident Investigation. 2009. "Guidelines for Safety Investigations of Accidents." European Safety Reliability and Data Association.

Fleming, K.N. and Silady, F.A. 2002. A risk informed defense-in-depth framework for existing and advanced reactors. *Reliability Engineering & System Safety* 78(3), 205–225.

Fiorentini, L., 2018. "Fire Risk Assessment as the Key Element of a Bow-Tie Based Fire Safety Management System (FSMS) to Design Fire Safety Strategy for Railway Stations. Case Studies in Italy." SFPE Europe Conference: Fire Safety Engineering, Rotterdam, The Netherlands.

Fiorentini, L. and Marmo, L. 2018. *Principles of Forensic Engineering Applied to Industrial Accidents*. Wiley.

Fiorentini, L., Pinetti, G., Sicari, R., Farinella, M. and Marmo, L. 2018. "'Offshore Directive' on Major Accidents: A Barrier-Based Safety Management System Built on Shared Ontologies and Taxonomies. Real Applications in Italy." *Chemical Engineering Transactions* 67: 355–360.

Forck, F. and Noakes Fry, K. 2016. *Cause Analysis Manual*. Brookfield, CT: Rothstein Publishing.

Forck, F. and Noakes Fry, K. 2016. *Cause Analysis Manual*. Brookfield, CT: Rothstein Publishing.

Franks, A. 2003. "*Lines of Defence/Layers of Protection Analysis in the COMAH Context*." Warrington, UK: Amey VECTRA.

Gertman, D. and Blackman, H. 1994. *Human Reliability and Safety Analysis Data Handbook*. New York: Wiley.

Haddon, W. 1980. "Advances in the Epidemiology of Injuries as a Basis for Public Policy." *Public Health Reports* 95 (5): 411–421.

Hale, A. 2003. "Note on Barriers and Delivery Systems." PRISM Conference, Athens.

Health and Safety Executive. 2004. "Investigating Accidents and Incidents: A Workbook for Employers, Unions, Safety Representatives and Safety Professionals." HSG245. HSE Books.

Hollnagel, E. 1995. *The Art of Efficient Man-Machine Interaction: Improving Coupling Between Man and Machine*. Hillsdale, NJ: Lawrence Erlbaum Associates.

Hollnagel, E. 2004. *Barrier and Accident Prevention*. Hampshire, UK: Ashgate.

Hoover, J., Bailey, J., Willauer, H. and Williams, F. 2005. "Evaluation of Submarine Hydraulic System Explosion and Fire Hazards." Ft. Belvoir, VA: Defense Technical Information Center.

Hoover, J., Bailey, J., Willauer, H. and Williams, F. 2008. "Preliminary Investigations into Methods of Mitigating Hydraulic Fluid Mist Explosions." *Fire Safety Journal* 43 (3): 237–240.

Idaho National Laboratory. 2005. "The SPAR-H Human Reliability Analysis Method." Washington, DC: US Nuclear Regulatory Commission.

IEC. 2019. IEC 31010, *Risk Management – Risk Assessment Techniques*. International Electrotechnical Commission.

INCONTROL Simulation Solutions, 2019. Pedestrian Dynamics Technical Background.

ISO. 2008. ISO 31000, *Risk Management – Guidelines*. International Organization for Standardization.

ISO. 2013. ISO/TR 31004, *Risk Management – Guidance for the Implementation of ISO 31000*. International Organization for Standardization.

ISO. 2018. ISO 31000, *Risk Management – Guidelines*. International Organization for Standardization.

ISO/IEC. 2019. ISO/IEC. *Guide 73, Risk Management – Vocabulary*. International Organization for Standardization.

Johansen, I. and Rausand, M. 2015. "Barrier Management in the Offshore Oil and Gas Industry." *Journal of Loss Prevention in the Process Industries*, 34, 49–55.

Johnson, W. 1980. *MORT Safety Assurance Systems*. New York: Dekker.

Katsakiori, P., Sakellaropoulos, G. and Manatakis, E. 2009. "Towards an Evaluation of Accident Investigation Methods in Terms of Their Alignment with Accident Causation Models." *Safety Science* 47 (7): 1007–1015.

Kecklund, L., Edland, A., Wedin, P. and Svenson, O. 1996. "Safety Barrier Function Analysis in a Process Industry: A Nuclear Power Application." *International Journal of Industrial Ergonomics* 17 (3): 275–284.

Kjellén, U. 2007. "Safety in the Design of Offshore Platforms: Integrated Safety versus Safety as an Add-on Characteristic." *Safety Science* 45 (1–2): 107–127.

Klein. 2016. "The ChE as Sherlock Holmes: Investigating Process Incidents." *Chemical Engineering Progress* 112 (10): 28–34.

Kletz, T. 2001. *Learning from Accidents*. Oxford, UK: Gulf Professional.

Latino, R., Latino, K. and Latino, M. 2011. *Root Cause Analysis*. 4th edition. Boca Raton, FL: CRC Press/Taylor & Francis.

Li, Y. and Guldenmund, F. 2018. "Safety Management Systems: A Broad Overview of the Literature." *Safety Science* 103: 94–123.

Liu, Y. 2020. "Safety Barriers: Research Advances and New Thoughts on Theory, Engineering and Management." *Journal of Loss Prevention in the Process Industries* 67: 104260.

Louisot, J. and Ketcham, C., 2014. *ERM - Enterprise Risk Management*. Hoboken: Wiley.

Marmo, L., Piccinini, N. and Fiorentini, L. 2013. "Missing Safety Measures Led to the Jet Fire and Seven Deaths at a Steel Plant in Turin: Dynamics and Lessons Learned." *Journal of Loss Prevention in the Process Industries* 26 (1): 215–224.

McGrattan, K., Baum, H. and Rehm, R. 1998. "Large Eddy Simulations of Smoke Movement." *Fire Safety Journal* 30 (2): 161–178.

McGrattan, K., Hamins, A. and Stroup, D. 1998. "Sprinkler, Smoke & Heat Vent, Draft Curtain Interaction. Large Scale Experiments and Model Development." *NISTIR 6196-1*. Gaithersburg, MD: National Institute of Standards and Technology.

McGrattan, K., Hostikka, S., Floyd, J., Baum, H., Rehm, R., Mell, W. and McDermott, R. 2012. *Fire Dynamics Simulator (Version 5)*. Gaithersburg, MD: U.S. Department of Commerce, NIST.

National Fire Protection Association (NFPA). 2017. *NFPA 921: Guide for Fire and Explosion Investigations*. NFPA.

Noon, R. 2009. *Scientific Method*. Boca Raton, FL: CRC Press.

National Research Council (NRC). 2007. *Assessment of the Performance of Engineered Waste Containment Barriers*. Washington, DC: The National Academies Press.

Petroleum Safety Authority. 2011. "Regulations Relating to Management and the Duty to Provide Information I: The Petroleum Activities and at Certain Onshore Facilities." Petroleum Safety Authority Norway.

Piccinini, N. and Ciarambino, I. 1997. "Operability Analysis Devoted to the Development of Logic Trees." *Reliability Engineering & System Safety* 55 (3): 227–241.

Pitblado, R., Fisher, M., Nelson, B., Fløtaker, H., Molazemi, K. and Stokke, A. 2016. "Concepts for Dynamic Barrier Management." *Journal of Loss Prevention in the Process Industries* 43: 741–746.

Prashanth, I., Fernandez, G., Sunder, R. and Boardman, B. 2017. "Factors Influencing Safety Barrier Performance for Onshore Gas Drilling Operations." *Journal of Loss Prevention in the Process Industries* 49: 291–298.

Rasmussen, J. 1983. "Skills, Rules, and Knowledge: Signals, Signs, and Symbols, and Other Distinctions in Human Performance Models." *IEEE Transactions on Systems, Man, and Cybernetics*, SMC-13 (3): 257–266.

Rathnayaka, S., Khan, F. and Amyotte, P. 2011. "SHIPP Methodology: Predictive Accident Modeling Approach. Part II. Validation with Case Study." *Process Safety and Environmental Protection* 89 (2): 75–88.

Reason, J. 1990. *Human Error*. Cambridge, UK: Cambridge University Press.

Reason, J. 1997. *Managing the Risks of Organizational Accidents*. Ashgate, UK: Aldershot.

Sanders, R. 2015. *Chemical Process Safety: Learning from Case Histories*. 4th edition. Oxford, UK: Elsevier Science.

Shappel, S. and Wieggmann, D. 2001. "Applying Reason: The Human Factors Analysis and Classification System (HFACS)." *Human Factors and Aerospace Safety* 1 (1): 59–86.

Sklet, S. 2002. *Methods for Accident Investigation*. Trondheim: Norwegian University of Science and Technology.

Sklet, S. 2004. "Comparison of Some Selected Methods for Accident Investigation." *Journal of Hazardous Materials* 111 (1–3): 29–37.

Sklet, S. 2006. "Safety Barriers: Definition, Classification, and Performance." *Journal of Loss Prevention in the Process Industries* 19: 494–506.

Strobhar, D. 2013. *Human Factors in Process Plant Operation*. New York: Momentum Press.

Sutton, I. 2010. *Process Risk and Reliability Management: Operational Integrity Management*. Burlington, MA: William Andrew, Inc.

Svenson, O. 1991. "The Accident Evolution and Barrier Function (AEB) Model Applied to Incident Analysis in the Processing Industries." *Risk Analysis* 11 (3): 499–507.

Swain, A. and Guttmann, H. 1983. *Handbook of Human Reliability Analysis with Emphasis on Nuclear Power Plant Applications*. Washington, DC: US Department of Energy.

Taylor, J. 2016. *Human Error in Process Plant Design and Operations*. Boca Raton, FL: Taylor & Francis.

Taylor, R. 1988. *Methods for Assessment of Weapon Safety*. Glumsø, Denmark: Institute for Technical Systems Analysis.

US Department of Energy. 1996. "Hazard and Barrier Analysis." Guidance Document. Washington, DC: DOE.

US Department of Energy. 1997. "Implementation Guide for Use with DOE Order 225.JA, Accident Investigation, DOE G 225.JA-L, Rev. J." Washington, DC: DOE.

US Department of Energy. 1999. *Conducting Accident Investigations*. Washington, DC: DOE.

Vanden Heuvel, L., Lorenzo, D., Jackson, L., Hanson, W., Rooney, J. and Walker, D. 2008. *Root Cause Analysis Handbook: A Guide to Efficient And Effective Incident Investigation*. 3rd edition. Brookfield, CT: Rothstein Publishing.

Wahlström, B. and Gunsell, L. 1998. "Reactor Safety: A Description and Assessment of Nordic Safety Work." Nordic Nuclear Safety Research.

Woods, D., Dekker, S., Cook, R., Johannesen, L., and Sarter, N., 2010. *Behind Human Error*, 2nd edition. Boca Raton, FL: Taylor & Francis.

World Economic Forum. 2020. *The Global Risks Report 2020*. World Economic Forum.

Xie, L., Lundteigen, M. and Liu, Y. 2018. "Common Cause Failures and Cascading Failures in Technical Systems: Similarities, Differences and Barriers." European Safety and Reliability Conference. Trondheim, Norway.

Xie, L., Lundteigen, M. and Liu, Y. 2018. "Safety Barriers against Common Cause Failure and Cascading Failures: A Review and Pilot Analysis." IEEE International Conference on Industrial Engineering and Engineering Management. Bangkok, Thailand.

Further Reading

AIChE-CCPS. 2018. *Bow Ties in Risk Management: A Concept Book for Process Safety*. Hoboken, NJ: Wiley.

AIChE–CCPS. 2001. *Layer of Protection Analysis*. New York: Center for Chemical Process Safety of the American Institute of Chemical Engineers.

AIChE–CCPS. 2018. *Bow Ties in Risk Management: A Concept Book for Process Safety*. Hoboken, NJ: Wiley.

Assael, M. and Kakosimos, K. 2010. *Fires, Explosions, and Toxic Gas Dispersions: Effects Calculation and Risk Analysis*. Boca Raton, FL: CRC Press/Taylor & Francis.

Aven, T. 2012. *Foundations of Risk Analysis*. 2nd edition. Chichester, UK: Wiley.

Aven, T. and Zio, E. 2018. *Knowledge in Risk Assessment and Management*. Wiley.

Bragg, S. 2015. *Enterprise Risk Management: Practical Applications*. Centennial, CO: Accounting Tools.

BSI. 2014. ISO 55001:2014, *Asset Management Management Systems – Requirements*. British Standard Institution.

Castaldo, D. 2009. *Project Risk Management*. Franco Angeli.

Castaldo, D. 2011. *Project Sustainability Management*. Franco Angeli.

Center for Chemical Process Safety. 2013. *Guidelines for Enabling Conditions and Conditional Modifiers in Layers of Protection Analysis*. Wiley.

CGE Risk, 2020. BowtieXP Software Manual.

Cote, A. 2008. *Fire Protection Handbook*. Quincy, MA: National Fire Protection Association.

Dattilo, F., Tafaro, S. and Fiorentini, L. 2019. Matera, capitale europea della cultura 2019: integrare l'arte e la sicurezza nella gestione dei grandi eventi. Antincendio - EPC Editore, (May and June).

de Ruijter, A. and Guldenmund, F. 2016. "The Bowtie Method: A Review." *Safety Science* 88: 211–218.

Fiorentini, L. and Marmo, L. 2018. "Sound Barriers Management in Process Safety: Bow-Tie Approach According to the First Official AIChE–CCPS Guidelines." *Chemical Engineering Transactions*, 67.

Fiorentini, L. and Marmo, L. 2018. *Principles of Forensic Engineering Applied to Industrial Accidents*. Wiley.

Fiorentini, L. and Sicari, R. 2020. *Analisi, Valutazione E Gestione Operativa Del Rischio*. Rome: EPC editore.

Fraser, J. and Simkins, B. 2010. *Enterprise Risk Management: Today's Leading Research and Best Practices for Tomorrow's Executives*. Wiley.

Gonzalez, F. 2020. "2011: Enschede Stadium Roof Collapse." Learning from Building Failures. Available at: https://buildingfailures.wordpress.com/2011/07/07/2011-enshede-stadium-roof-collapse (accessed 13 October 2020).

Health and Safety Executive (HSE). 2020. "Guidance on ALARP Decisions in COMAH – SPC/Permissioning/37." Available at: https://www.hse.gov.uk/foi/internalops/hid_circs/permissioning/spc_perm_37/ (accessed 5 October 2020).

Hopkin, P. 2018. *Fundamentals of Risk Management*. 5th edition. London/New York: Kogan Page.

Hunziker, S., 2019. *Enterprise Risk Management – Modern Approaches to Balancing Risk and Reward*. Rotkreuz: Springer Gabler.

IEC. 2019. IEC 31010:2019, *Risk Management – Risk Assessment Techniques*. International Electrotechnical Commission.

IEC. 2020. IEC 61511, *Functional Safety – Safety Instrumented Systems for the Process Industry Sector*. 2nd edition. International Electrotechnical Commission.

ISO 2011. ISO 19011, *Guidelines for Auditing Management Systems*. International Organization for Standardization.

ISO. 2009. ISO Guide 73:2009, *Risk Management – Vocabulary*. International Organization for Standardization.

ISO. 2012. ISO 22301, *Societal Security – Business Continuity Management Systems – Requirements*. International Organization for Standardization.

ISO. 2012. ISO 39001, *Road Traffic Safety Management Systems*. International Organization for Standardization.

ISO. 2014. ISO 19600, *Compliance Management Systems*. International Organization for Standardization.

ISO. 2014. ISO 55000, *Asset Management: Overview, Principles, and Terminology*. International Organization for Standardization.

ISO. 2015. ISO 14001, *Environmental Management Systems – Requirements with Guidance for Use*. International Standard Organization.

ISO. 2015. ISO 31000, *Risk Management – A Practical Guide for SMEs*. International Organization for Standardization.

ISO. 2015. ISO 9000, *Quality Management Systems – Fundamentals and Vocabulary*. International Organization for Standardization.

ISO. 2016. ISO 17776, *Petroleum and Natural Gas Industries – Offshore Production Installations – Major Accident Hazard Management During the Design of New Installations*. International Organization for Standardization.

ISO. 2018. ISO 31000:2018, *Risk Management – Guidelines*. International Organization for Standardization.

ISO. 2018. ISO 45000, *Occupational Health and Safety*. International Organization for Standardization.

ISO. 2020. ISO 31022, *Risk Management – Guidelines for the Management of Legal Risk*. International Organization for Standardization.

ISO/IEC. 2012. ISO/IEC 17024, *Conformity Assessment – General Requirements for Bodies Operating Certification of Persons*. International Organization for Standardization/ International Electrotechnical Commission.

ISO/IEC. 2013. ISO/IEC 27001, *Information Technology – Security Techniques – Information Security Management Systems – Requirements*. International Organization for Standardization/International Electrotechnical Commission.

ISO/IEC. 2013. ISO/IEC 27002, *Information Technology – Security Techniques – Code of Practice for Information Security Controls*. International Organization for Standardization/ International Electrotechnical Commission.

ISO/IEC. 2018. ISO/IEC 27000, *Information Technology – Security Techniques – Information Security Management Systems – Overview and Vocabulary*. International Organization for Standardization/International Electrotechnical Commission.

ISO/IEC. 2018. ISO/IEC 27005, *Information Technology. Security Techniques. Information Security Risk Management*. International Organization for Standardization/International Electrotechnical Commission.

IWA. 2020. IWA 31, *Risk Management – Guidelines on Using ISO 31000 in Management Systems*. Geneva: International Workshop Agreement.

Lumbe Aas, A. 2008. "The Human Factor Assessment and Classification System (HFACS) for the oil & gas industry." International Petroleum Technology Conference, 3–5 December, Kuala Lumpur, Malaysia.

Maritan, F., Masiero, C. and Rossato, M. 2020. Un metodo di valutazione dei rischi basato su riferimenti normativi: dalla BS 18004 alla ISO/TR 14121-2. Igiene & Sicurezza del Lavoro, 8–9, pp. 455–460.

McLeod, R. 2016. "Human Factors in Barrier Management." A CIEHF White Paper. Chartered Institute of Ergonomics and Human Factors.

Olson, D. and Wu, D. 2017. *Enterprise Risk Management Models*. Springer.

Perry, M. 2019. "Why Reputation Could Be Your Biggest Future Risk." *Racounter* 0583, 1–4.

Philley, J. 2020. "Collar Hazards with a Bow-Tie." *Chemical Processing*. Available at: www.chemicalprocessing.com/articles/2005/612.html?page=1 (accessed 28 September 2020).

Popov, G., Lyon, B. and Hollcroft, B. 2016. *Risk Assessment: A Practical Guide to Assessing Operational Risks*. Wiley.

Rausand, M. and Haugen, S. 2013. *Risk Assessment*. Wiley.

Rees, M. 2016. *Business Risk and Simulation Modelling in Practice*. Wiley.

Shoman, H. and Ellahham, S. 2018. "An Innovative Risk Assessment Tool for Prospective Risk Analysis to Improve the Quality in Health Care: The Bow-Tie." *International Journal of Pharmacovigilance* 3 (1): 1–3.

Sklet, S., 2002. "Methods For Accident Investigation." Trondheim: Norwegian University of Science and Technology. © 2002, Norwegian University of Science and Technology CC by 2.0.

Sklet, S. 2006. "Safety Barriers: Definition, Classification, and Performance." *Journal of Loss Prevention in the Process Industries* 19 (5): 494–506.

Tranchard, S. 2018. "The New ISO 31000 Keeps Risk Management Simple." ISO.org. Available at: www.iso.org/news/ref2263.html (Accessed 21 September 2020).

Index

a

absolute risk, BT, LOPA 181
acceptability criteria
  BT workflows 221
  LOPA IPLs 247
acceptable risk
  ALARP 41, 89
  BT 102
  BT workflows, LOPA 248–249
  ISO 31000 38–39
  risk indices 130
  risk matrix 222
access control, BT workflows, web-based software development 278
Accident Sequence Evaluation Programme (ASEP) 388
accidents 185–186
  BFA 369
  BT workflows, LOPA 244
  domino effects 187
  human error 53
  investigation 189–194
  major 87, 220
  root causes 53
  Swiss Cheese Model 368
active hardware category, SPAR-H 389
activities
  BT 371
    LOPA barriers 175
  BT workflows 228
  HEART 388
  organizational 16
AHJ. *See* authorities having jurisdiction
AIChE-CCPS. *See* American Institute of Chemical Engineers - Center for Chemical Process Safety
alarm management 90
ALARP. *See* as slow as reasonably practical
American Institute of Chemical Engineers - Center for Chemical Process Safety (AIChE-CCPS) 5
  BT 101, 136
    barriers 160, 162, 163
    LOPA 167, 169–175, 176, 177
  BT workflows 221
  RM systems approach, industrial safety 87
AMHC. *See* annual mitigated hazard cost
Amyotte, P. 160
annual cost of the unmitigated hazard (AUHC) 133, 134
Annual Global Risk Report, by World Economic Forum 7
annual mitigated hazard cost (AMHC) 134
apparent cause analysis 210, 211
as slow as reasonably practical (ALARP)
  acceptable risk 41, 89
  best practices 43
  BT
    consequences 156
    LOPA 167
    RM risk evaluation 128
  BT workflows
    BFA 262
    LOPA 248, 249

Bow-Tie Industrial Risk Management Across Sectors: A Barrier-Based Approach, First Edition. Luca Fiorentini.

as slow as reasonably practical (ALARP) (*cont'd*)
LOPD 248
risk matrix 222
ISO 31000 37, 41–44
ISO IE 27001 95
RM systems approach, RAGAGEP 89
tolerable risk 222, 248
ASEP. *See* Accident Sequence Evaluation Programme
Assael, M. 108, 124
asset management and integrity. *See* ISO 55000
audits
BT, LOPA 173
BT workflows 236–240
ISO 19011 95–97
ISO 31000 48
RM systems approach 79, 81
AUHC. *See* annual cost of the unmitigated hazard
authentication management, web-based software development 279
authorities having jurisdiction (AHJ)
BT workflows 325
RM systems approach, RAGAGEP 91
automatic deluge fire-fighting system 157
available time, SPAR-H 389, 392

b

BAC. *See* barrier annual cost
Bailey, J. 324
bar coding, BT workflows, web-based software development 275
barrier annual cost (BAC) 134
barrier decay mechanism 214, 225
barrier failure analysis (BFA) 4, 5, 194–208
accidents 369
barriers 374, 377
identification 196, 201
BT 195–196, 202, 213–216
BT workflows 249–265, 308, 309–317
barrier identification 255
barrier state assessment 255–258
causation analysis 258
cause-and-effect analysis 257
communication 251–252, 260–261
control barriers analysis 257
decision-making 262
event chaining 254–255
fact-finding 253–254
leadership 251–252
operational experience 344, 347–350
RCA 257, 347–348
reporting 262–263
risk treatment 261
root causes 262
SMART 262
Tripod Beta 257, 258
causation path 195, 207–208
cause-and-effect analysis 202–206
easy guide 373–377
event chaining 195, 196
events 199, 202, 374, 377
governance 370
guiding principles 377
incident analysis 347–348, 370
incident barrier 375
ISO 22301 94
near-misses 369
PDCA 370
potential events 374, 377
primary cause 196, 199, 208, 374
RCA 208–213
recommendations 196, 199, 202, 374
BT workflows 258–263
secondary cause 196, 199, 208, 374
taxonomies 370
tertiary cause 196, 199, 208, 374
timeline 202–206
barrier identification
BFA 196, 201
brainstorming 355
BT workflows, BFA 255
barrier life cycle
incident barrier 258
RM systems approach, RAGAGEP 91
barrier state assessment
BFA 196
BT workflows, BFA 255–258
barriers. *See also specific types*

BFA 374, 377
BT 125–127, 156–163, 372
LOPA 171–175, 179–183
BT workflows 221, 223–224, 225
IT operations 281–282
web-based software development 275–279
control 368
Deepwater Horizon 368
defense in depth 367
escalation factors, BT 150, 163–165
hazard 368
basic control process system (BPCS), BT
calibrated risk graph 132
LOPA 165
basic-risk factors (BRF), Tripod Beta 194, 195
batch operations, BT workflows, LOPA 230, 246
BCM. *See* business continuity management
behavioural category, SPAR-H 388
best practices
ALARP 43
BT workflows, web-based software development 276
culture 90
ISO 19600 98
ISO 55000 90
preventive barriers 275, 276, 277
BFA. *See* barrier failure analysis
BIA. *See* business impact assessment
BIPs, SLIM 396
Blackman 177
bottom-up approach, BT workflows 241–242
Bow-Tie (BT) 4, 5
activities 371
LOPA barriers 175
advanced 183–184
application 140–146
barrier-based methods 125–127
barriers 125–127, 156–163, 372
escalation factors 150, 163–165
LOPA 171–175, 179–183
BFA 195–196, 202, 213–216
brainstorming 369–370
building 150–152
chaining 183–184
checklists, risk identification 106
combination 183–184
consequences 150, 156, 371, 372
BFA 202
LOPA 170–171, 244
controls 370
danger 152–153
DC 140–145
easy guide 371–372
escalation factors 371, 372
barriers 150, 163–165
LOPA, secondary barriers 174–175
governance 370
HAZAN, 135n1
hazard 101–102, 152–153, 372
ISO 9001/ISO 45001/ISO 14001 84
ISO 31000 103–104, 137, 138–140
level of abstraction 139, 147–149
LOPA 127, 150, 165–183
barriers 171–175, 179–183
conditional modifiers 176–177
enabling factors 176–177
escalation factors and secondary barriers 174–175
threats 154
Top event 170
method 134–140
PDCA 370
preventive barrier 372
primary barriers 156–163
proactive barriers 150, 371
workflows 225
recovery barrier 371, 372
risk 101–102
risk analysis 117–125
ISO 31000 117
risk assessment 369, 370
relative risk 248
risk evaluation 128–134
risk identification 30, 104–116
risk register 178–183
RM systems approach, RAGAGEP 89
threats 150, 154–156, 371, 372
workflows 228
workflows IT operations 280

Bow-Tie (BT) (*cont'd*)
workflows web-based software development 275
Top events 127, 153–154, 371, 372
barriers 157
BFA 202
domino effects 92
LOPA 170
Zone 1 127
Zone 2 127
Bow-Tie (BT) workflows 217–364
acceptability criteria 221
activities 228
AIChE-CCPS 221
ALARP
BFA 262
LOPD 248
risk matrix 222
audits 236–240
barriers 221, 223–224, 225
IT operations 281–282
web-based software development 275–279
BFA 249–265, 308, 309–317
barrier identification 255
barrier state assessment 255–258
causation analysis 258
cause-and-effect analysis 257
communication 251–252, 260–261
control barriers analysis 257
decision-making 262
event chaining 254–255
fact-finding 253–254
leadership 251–252
operational experience 344, 347–350
RCA 257, 347–348
reporting 262–263
risk treatment 261
root causes 262
SMART 262
Tripod Beta 257, 258
bottom-up approach 241–242
case study 217–242
causes 221, 222, 228
checklists
BFA 252
crowding risk assessment 282
colors 225–226
consequences 222, 223, 224, 225
LOPA 244
control 225
COVID-19 269, 271
critical barriers 231–236
fire risk assessment 358
cross-references 227
crowd risk assessment 282–288, 289
cybersecurity 220
danger 222, 223, 231
decommissioning 220
LOPA 244, 246
diagram construction 222–224
documentation attachments 226–227
drug administration 308, 318
escalation factors 225
secondary barriers 222, 223
fire in flight 269
fire risk assessment
companies managing multiple assets 348–358, 359
companies managing multiple PV plants 358–364
food contamination 269–270
FTA 230
graphics management 223–224
HLS 220
HRAs 229
hybrid approach 242
IE 225
IT operations 279–282
KPIs 221
LOPA 229–230
acceptable risk 248–249
accidents 244
conditional modifiers 248
cost-benefit assessment 249
ETA 246
FMEA 244
HAZOP 243, 246, 249
HSE 243
human error 244
IEF 244–245

IPLs 243, 246–247, 248
ISO 31000 246
PFDs 246, 248
PHA 249
shut-down 246
start-up 246
three D's 247
military helicopter operations risk assessment 288–292
MOC 224
organizational structure 227
other data 231
patient safety 292–293
performance 223
PFDs 229–230
risk matrix 222
tasks 228
primary barriers 222, 223
proactive barriers 225
process safety 293–308
reactive barriers 225
resisting capacity and ageing 266–269
risk 224
LOPA 248
web-based software development 279
risk analysis 228
risk assessment 225
risk evaluation, LOPA 248–249
risk matrix 221–222
intolerable risk 222
residual risk 222
SAP 218
SILs
BFA 262
LOPA 244, 247
stakeholders 227
tasks 228–229
taxonomies 219–221
terminology 224–225
threats 228
IT operations 280
web-based software development 275
ThyssenKrupp fire investigation 318–341, 342–343
tolerable risk
LOPA 248–249
risk matrix 222
Top events 223
top-down approach 240–241
transparency 223–224
Tripod Beta for Twente Stadium roof collapse 341, 344, 345
water treatment 344, 346
web-based software development 271–279
WI 285, 286
worked examples 265–365
BPCS. *See* basic control process system
brainstorming
barrier identification 355
BT 369–370
risk identification 104–106, 139
HAZID 112
HAZOP 110
RCA 210
risk identification 30
RM 85
BRF. *See* basic-risk factors
British Standards Institute (BSI), ISO 55000 91
BSCAT 212
BSI. *See* British Standards Institute
BT. *See* Bow-Tie
buffer overflows, BT workflows, web-based software development 277–278
bureaucratic condition, RM culture 71–73
business continuity management (BCM) 369
ISO 22301 92–94
ISO IE 27001 95
business impact assessment (BIA), ISO 22301 93

C

CAC. *See* critical administration control
calibrated risk graph, BT, risk evaluation 131–133
Casiani, O. 266
Castaldo, D. 140
causation analysis
BFA 196
BT workflows, BFA 258

causation path, BFA 195, 207–208
cause-and-effect analysis
  BFA 202–206
  BT workflows, BFA 257
  timeline 255
cause-effect diagrams (CCDs), 135, 135n1
causes
  accident investigation 189–194
  apparent cause analysis 210, 211
  BFA 202
  BT
    DC 144, 145
    LOPA 170, 171
  BT workflows 221, 222, 228
  immediate
    accident investigation 191
    BFA 207
  latent, accident investigation 192
  primary, BFA 196, 199, 208, 374
  RCA
    BFA 208–213
    BT workflows, BFA 257, 347–348
    risk identification 30
  root
    accident investigation 190, 191, 192
    accidents 53
    BFA 196, 207
    BT workflows, BFA 262
    ISO 31000 14
  secondary, BFA 196, 199, 208, 374
  tertiary, BFA 196, 199, 208, 374
  underlying 52
    BFA 202, 207
    HAZID 113
    RCA 208
CCDs. *See* cause-effect diagrams
chaining
  BT 183–184
  event
    BFA 195, 196
    BT workflows, BFA 254–255
change control, BT workflows, web-based software development 275, 277
checklists
  BT
    DC 145
    risk identification 106
  BT workflows
    BFA 252
    crowding risk assessment 282
  risk identification 33
code reviews, BT workflows, web-based software development 276
combination, BT 183–184
commission errors, THERP 387
Committee of Sponsoring Organizations (COSO), Treadway Commission 59, 60–61
Common Vulnerability Scoring System 275
communication
  BT workflows
    BFA 251–252, 260–261
    web-based software development 278
  ISO 31000, ERM 59
  RM systems approach 80
  stakeholders 368–369
complexity
  issues 251
  management, RM systems approach 78–79
  SPAR-H 389, 392
compliance
  ISO 9001 84
  ISO 19600 98–99
  RM systems approach 80–81
compliance risk
  ERM 62
  ISO 19600 98–99
conditional modifiers
  BT, LOPA 176–177, 179
  BT workflows, LOPA 248
consequence of failure score (CS), BT, critical barriers 233–234
consequences
  BT 150, 156, 371, 372
    BFA 202
    DC 145
    LOPA 170–171, 244
  BT workflows 222, 223, 224, 225
  defined 6
  HAZID 113

human factors, HAZOP 111
ISO 9001/ISO 45001/ISO 14001 82
ISO 31000
analysis 36
human factors 55
QRA 35
continuous hardware category, SPAR-H 389
continuous testing, BT workflows, web-based software development 275–276
Contos, A. 266
contractors, ISO 9001/ISO 45001/ISO 14001 86
contractor/vendor selection and management, RM systems approach 81
control
access 278
barriers 368
BPCS, BT
calibrated risk graph 132
LOPA 165
BT 370
critical barriers 234
BT workflows 225
CAC 167
change 275, 277
damage 160
defined 6
HAZID 113
ISO 9001/ISO 45001/ISO 14001 85
ISO 39001 98
ISO IE 27001 95
ISO IEC 27001 94
work 81
control barriers analysis
BT workflows, BFA 257
ISO 31000 35–36
COSO. *See* Committee of Sponsoring Organizations
cost of ownership, ISO 55000 91
cost-benefit assessment (analysis)
BT, risk evaluation 113–134
BT workflows, LOPA 249
ISO 31000
ALARP 44
risk identification 30
COVID-19, BT workflows 269, 271
critical administration control (CAC), BT, LOPA 167
critical barriers, BT workflows 231–236
fire risk assessment 358
critical systems, RM systems approach 81
cross-references, BT workflows 227
cross-site request forgery (CSRF), BT workflows, web-based software development 279
cross-site scripting, BT workflows, web-based software development 278
crowd risk assessment, BT workflows 282–288, 289
crowding risk assessment 282
cryptographic flaws, BT workflows, web-based software development 278
CS. *See* consequence of failure score
CSRF. *See* cross-site request forgery
cultural factors, ISO 31000 18, 368
culture
best practices 90
human error 379
ISO 31000, ERM 59
issues 212
RM
bureaucratic condition 71–73
between fulfilment and opportunity 62–67
generative condition 75–76
pathological condition 69–70
proactive condition 73–74
quality 67–68
reactive condition 70–71
customer loyalty, ISO 55000 92
cybersecurity
BT workflows 220
IEC 62443 90
RM systems approach, RAGAGEP 89, 90

d

damage control, BT, barriers 160
danger
BT 152–153

danger (*cont'd*)
brainstorming 104
bureaucratic condition 72
DC 144
LOPA 169–170
proactive condition 73
BT workflows 222, 223, 231
data center (DC), BT 140–145
decision-making
BT workflows, BFA 262
ISO 9001 82
ISO 31000 37
risk treatment 45
RM 15, 370
decommissioning
BT, DC 140, 142
BT workflows 220
LOPA 244, 246
deductive reasoning
BFA 206
BT, LOPA workflows 257
FTA 117
Deepwater Horizon 368
defeating factors, BT workflows 225
defense in depth
barriers 367
LOPA 243
RM systems approach, industrial safety 88
degradation factors
BT workflows 225
frequency analysis 37
human error 155
THERP 387
Deming, E., 366. *See also* Plan-Do-Check-Act
detect, decide and deflect (three D's), BT workflows, LOPA 247
Detect-Decide-Act model 56–57
Directive on Major Accidents Prevention, EU 87
domino effects 88
accidents 187
BT, Top events 92
consequences analysis 36
crowding 284
escalation factors 23
ISO 55000 92
preliminary analysis 37
risk identification 33
ThyssenKrupp fire investigation 325
drug administration, BT workflows 308, 318

e

emergency management (plan)
BT, barriers 160
BT workflows, crowd risk assessment 286, 287
ISO 19011 96–97
RM systems approach 81
EN 12973 367
enabling factors, BT, LOPA 176–177, 246
energy model 367
BT, barriers 162, 164
EN/IEC 61511-3, BT, LOPA 166–167
enterprise risk management (ERM) 4
ISO 31000 58–62
EPC. *See* error production conditions
EPGs. *See* equipment performance gaps
equipment failure
BT 101
mode criticality index 116
human error 370
RCA 212
equipment performance gaps (EPGs), RCA 211, 212
ergonomics
accident investigations latent cause 192
ISO 31000 53
lapses and slips 383
SPAR-H 390, 392
ERM. *See* enterprise risk management
error handling, web-based software development 278
error production conditions (EPC) 388
escalation factors
BT 371, 372
barriers 150, 163–165
LOPA, secondary barriers 174–175
BT workflows 225
secondary barriers 222, 223
domino effects 23

ESD, BT, critical barriers 234
ETA. *See* event tree analysis
EU
Directive on Major Accidents Prevention 87
Offshore Directive 388
event chaining
BFA 195, 196
BT workflows, BFA 254–255
event tree analysis (ETA)
BT 101, 135, 150
LOPA 167–168
risk analysis 121–125
threats 154
Top events 153
BT workflows, LOPA 246
ISO 31000 37
risk identification 30
events, BFA 199, 202, 374, 377
evidence-based methods, risk identification 33
exceptional violations 386
exceptions, BFA 196
external context
ISO 9001/ISO 45001/ISO 14001 83
ISO 31000
human factors 55
system performance review 49
RM systems approach, complexity management 78

f

fact-finding
BFA 196
BT workflows, BFA 253–254
failure modes, effects, and, criticality analysis (FMECA) 114–117
failure modes, effects, and diagnostic analysis (FMEDA) 114–117
failure modes and effects analysis (FMEA)
BT, risk identification 114–117
BT workflows, LOPA 244
HAZOP 112
falsification, accident investigation 192
Fanelli, F. 145
FARSI. *See* Functionality, Availability, Reliability, Survivability and Interactions
fault tree analysis (FTA)
BT 101, 135, 136, 150
risk analysis 117–121
threats 154
BT workflows 230
ISO 31000 37
risk identification 30
financial risk 369
ERM 61
fire in flight, BT workflows 269
fire prevention 3, 65
fire risk assessment 358
fire risk assessment, BT workflows
companies managing multiple assets 348–358, 359
for companies managing multiple PV plants 358–364
fire safety management system (FSMS), BT workflows 351, 358
Fishbone (Ishikawa) diagram 369
fitness for duty, SPAR-H 390, 392
fixation, BT workflows, BFA 257
Fleming 162
FLPPGs. *See* front-line personnel performance gaps
FMEA. *See* failure modes and effects analysis
FMECA. *See* failure modes, effects, and, criticality analysis
FMEDA. *See* failure modes, effects, and diagnostic analysis
F-N curves, BT, RN risk evaluation 128–129
food contamination, BT workflows 269–270
four Enoughs 247
frequency analysis
BT workflows, fire risk assessment 358
ISO 31000 36–37
front-line personnel performance gaps (FLPPGs), RCA 211, 212
FS. *See* function score
FSMS. *See* fire safety management system
FTA. *See* fault tree analysis
function score (FS) 233–234

functional safety
FMEDA 115, 116
IEC 61508 261
LOPA 168, 229
RM systems approach 89–91
Functionality, Availability, Reliability, Survivability and Interactions (FARSI), BT
barriers 163
critical barriers 232

g

garbage in, garbage out (GIGO) 263
GEMS. *See* Generic Eric Modeling System
generative condition, RM culture 75–76
Generic Eric Modeling System (GEMS) 381–407
Gertman 177
GIGO. *See* garbage in, garbage out
governance
BFA 370
BT 370
ISO 26000 366
ISO 31000, ERM 59
ISO 37000 366
graphics management, BT workflows 223–224
Groot, A. 266
guarantee positions 4
guidewords
HAZID 112, 113, 114
HAZOP 109, 110, 111
Guldenmund, F. 162
Guttmann 177

h

Handbook of Human Reliability Analysis with Emphasis on Nuclear Power Plant Applications (Swain and Guttmann) 177
Harbottle, E. 265
Hatch, D. 266
HAZAN. *See* hazard analysis
hazard
barriers 368
BT 101–102, 152–153, 372
defined 6
risk identification 33–34
hazard analysis (HAZAN)
BT, 135n1
ISO 9001/ISO 45001/ISO 14001 85
hazard and effects management process (HEMP), Tripod Beta 194
hazard and operability analysis (HAZOP) 111
BT
LOPA 167, 168
risk identification 108–112
BT workflows, LOPA 243, 246, 249
guidewords 109, 110, 111
ISO 9001/ISO 45001/ISO 14001 85
risk identification 30, 33
hazard identification (HAZID)
BT
hazard 153
risk identification 112–114
BT workflows
fire risk assessment 358
LOPA 243, 246, 249
guidewords 112, 113, 114
ISO 9001/ISO 45001/ISO 14001 85
risk identification 30
HAZOP. *See* hazard and operability analysis
HCR/ORE. *See* Human Cognitive Reliability
HEART. *See* human error assessment and reduction technique; Human Error Assessment and Reduction Technique
Heimplaetzer, P. 266
HEMP. *See* hazard and effects management process
HEP. *See* human error probability/potential
HFACS. *See* human factors analysis and classification scheme
HFE. *See* human failure error
high risk vulnerabilities, BT workflows, web-based software development 278
high-level systems (HLS) 82, 83
BT workflows 220
ISO 22301 366
HIPPS, BT, critical barriers 234
historical statistical data, risk identification 33
HLS. *See* high-level systems

HMI. *See* human-machine interface
Hollnagel, E. 162, 163
Hoover, J. 324
HRA. *See* human error and reliability assessment
HRAs. *See* human factor assessments
HSE, BT workflows, LOPA 243
Human Cognitive Reliability (HCR/ORE) 388
human error
  accidents 53
  BT workflows, LOPA 244
  classification 379–380
  degradation factors 155
  equipment failure 370
  GEMS 381–407
  IE 177
  ISO 31000 368
  knowledge-based performance 381–382
  mistakes 383–385
  organizational structure 53
  performance 381–382
  RCA 212
  reducing risk 386–387
  risk evaluation 387–407
  skill-based performance 381–382
  slips and lapses 382–383
  violations 385–386
human error and reliability assessment (HRA) 379–407
  mitigation barriers 152
  SLIM 388, 391–407
Human Error Assessment and Reduction Technique (HEART) 388
  ISO 31000, human factors 54
human error probability/potential (HEP) 387, 388
  AICHe-CCPS 177
  PSF 390
  SLIM 396, 397
human factor assessments (HRAs)
  BT workflows 229
  risk identification 30
human factors
  BT
    barriers 160, 164
    esclation factors 164
    threats 155
  consequences, HAZOP 111
  ISO 19011 96
  ISO 19600 99
  ISO 31000 18
    uncertainty 53–58
  RCA 210
human factors analysis and classification scheme (HFACS) 402
human failure error (HFE), SPAR-H 389–391
Human Reliability and Safety Analysis Data Handbook (German and Blackman) 177
human-machine interface (HMI). *See also* ergonomics
  ISO 31000, PSFs 58
  SPAR-H 390, 392
hybrid approach, BT workflows 242

i

IE. *See* initiating event
IEC 31010 11
IEC 61508 261
IEC 61511 89–91
IEC 62443 90
IEC 62682 90
IEF. *See* initiating event frequency
IIOT. *See* industrial internet of things
immediate cause
  accident investigation 191
  BFA 207
INAIL. *See* Italian National Institute for Workplace Incidents Insurance
incident analysis
  BFA 347–348, 370
  BT 140
  ISO 31000, human factors 55
incident barriers
  barrier life cycle 258
  BFA 375, 376
incident management
  ISO 31000 14
  RM 369
incident reporting, RM systems approach 81

independent protection layer response time (IRT), AICHe-CCPS 177
independent protection layers (IPLs)
  AICHe-CCPS 177
  BT 150
    calibrated risk graph 131, 132
    LOPA 166–167, 168, 176
  BT workflows, LOPA 243, 246–247, 248
  RM systems approach
    industrial safety 88
    RAGAGEP 89
inductive reasoning
  BFA 206
  BT, LOPA workflows 257
  HAZID 112
  risk identification 33
industrial internet of things (IIOT) 89
industrial safety, RM systems approach 87–89
industry-wide solution, BFA 261
information security. *See* ISO IEC 27001
initiating event (IE)
  AICHe-CCPS 177
  BT workflows 225
initiating event frequency (IEF)
  AICHe-CCPS 177
  BT workflows, LOPA 244–245
injection flaws 277
internal context
  ISO 9001/ISO 45001/ISO 14001 83
  ISO 31000 24–26
    human factors 55
    system performance review 49
  organizational structure 25
  RM systems approach, complexity management 78
International Organization for Standardization (ISO), 368. *See also specific ISO standards*
International Trade Centre, United Nations Industrial Development Organization 368
intolerable risk 39
  BT workflows, risk matrix 222
  LOPA PFDs 230
inverted two-pointed model 56
IPLs. *See* independent protection layers
IRT. *See* independent protection layer response time
Ishikawa (Fishbone) diagram 369
ISMS, ISO IE 27001 95
ISO. *See* International Organization for Standardization
ISO 9001 82–86, 367
  human factors 54
ISO 14001 82–86
  human factors 54
ISO 17776
  BT, hazard 152–153
  HAZID 112–113
ISO 19011 95–97
  ISO 31000, audits 48
ISO 19600 98–99
ISO 22301 92–94
  HLS 366
ISO 26000 366
ISO 31000 4, 365, 367
  adoption 14
  best available information 18
  BT 103–104, 137, 138–140, 150
    BFA 195
    LOPA 175
    risk analysis 117
  BT workflows, LOPA 246
  cultural factors 368
  customization 16–17
  dynamism 17
  ergonomics 53
  ERM 58–62
  human and cultural factors 18
  human error 368
  IEC 31010 11
  inclusiveness 17
  integral part of organizational activities 16
  ISO 9001/ISO 45001/ISO 14001 84
  ISO 55000 91
  ISO IEC 27001 94
  ISO/TR 31004 10–11, 16
  PDCA 79

PSFs 57–58
risk everywhere presence 6–10
risk life cycle 86
RM systems approach
industrial safety 89
RAGAGEP 91
structured and comprehensive 16
systems approach 78
uncertainty 368
ISO 31000 workflows
ALARP 37, 41–44
audits 48
consequences analysis 36
continuous improvement 18–23
control barriers analysis 35–36
framework improvement 29, 30
frequency analysis 36–37
human factors and uncertainty 53–58
implementation 26–27
internal context 24–26
leadership and commitment 24
monitoring and review 47–48
over time 44
preliminary analysis 37
principles 14–23
principles, framework, and processes 28–29
probability estimation 36–37
risk acceptability and tolerability 38–39
risk analysis 31, 34–38
uncertainty 38
risk assessment 29–33
risk evaluation 31, 38
risk identification 30, 31, 33–34
risk matrix 39–41
risk treatment 31, 45–47
RM processes 27–28
sensitivity analysis 38
system performance review 49–53
technical standard 10–23
understanding organization and context 24–26
workflows 23–53
ISO 37000 366
ISO 39001 (road traffic safety) 97–98
ISO 45001 82–86
human factors 54
ISO 55000 91–92
ISO 55001 91
ISO 56000 367
ISO IEC 27001 94–95
ISO/IEC 31010
BT 103
ISO 9001/ISO 45001/ISO 14001 83, 84
ISO 31000
probability estimation 37
risk assessment 30–31
ISO/TR 31004
ISO 31000 10–11, 16
RM systems approach 79
issues
bar coding 275
complexity 251
culture 212
defined 6
IPLs 246
risk analysis 220
RM tailoring 17
Swiss Cheese Model 368
IT operations, BT workflows 279–282
Italian National Institute for Workplace Incidents Insurance (INAIL) 97

j

Janseen, E. 266
Johansen, I. 160
journey planning, ISO 39001 98

k

Kakosimos, K. 108, 124
key performance indicators (KPIs)
BFA 196
BT workflows 221
fire risk assessment 358
ISO 9001/ISO 45001/ISO 14001 86
ISO 31000 14
RM systems approach 76–77
Khan, F. 160
Kletz, T. 207
knowledge management, RM systems approach 81

knowledge-based performance, human error 381–382
KPIs. *See* key performance indicators

l

lapses 382–383
latent cause, accident investigation 192
Latino, K. 213
Latino, M. 213
Latino, R. 213
lay of protection analysis (LOPA) 5–6
  BT 127, 150, 165–183
    barriers 171–175, 179–183
    conditional modifiers 176–177
    enabling factors 176–177
    escalation factors and secondary barriers 174–175
    threats 154
    Top event 170
  BT workflows 229–230
    acceptable risk 248–249
    accidents 244
    conditional modifiers 248
    cost-benefit assessment 249
    ETA 246
    fire risk assessment 358
    FMEA 244
    HAZOP 243, 246, 249
    HSE 243
    human error 244
    IEF 244–245
    IPLs 243, 246–247, 248
    ISO 31000 246
    PFDs 246, 248
    PHA 249
    shut-down 246
    start-up 246
    three D's 247
  ISO 9001/ISO 45001/ISO 14001 84, 85
  ISO 22301 94
  ISO 55000 92
  risk identification 30
  RM systems approach 80
    industrial safety 88
    RAGAGEP 89
leadership
  accident investigation 189
  BT workflows, BFA 251–252
  HAZOP 109
  ISO 31000 24
  RM systems approach 81
learning from experience (LFE) 187
level of abstraction, BT 139, 147–149
level of service 285
LFE. *See* learning from experience
Li, Y. 162
Liu, Y. 159
long-term solution, BFA 202
  BT workflows 261
LOPA. *See* lay of protection analysis
loss prevention 369
Luca, T. 145

m

major accident prevention policy (MAPP), RM systems approach industrial safety 87
major accidents 87, 220
management of change (MOC)
  BT workflows 224
    fire risk assessment 358
  ISO 9001/ISO 45001/ISO 14001 85, 86
  ISO 55000 91
  RM systems approach 81
    industrial safety 88
  validation 277
Management Oversight and Risk Tree (MORT) 367
management review, ISO 31000 49
MAPP. *See* major accident prevention policy
Marmo, L. 266
masks, BT workflows 223
maturity model, RM systems approach RAGAGEP 90
medium-term solution, BFA 202
  BT workflows 261
military helicopter operations risk assessment, BT workflows 288–292
mission statement
  governance 366

RCA 210
ThyssenKrupp fire investigation 319
mistakes 383–385
mitigation barriers
BT
consequences 152
DC 145
Zone 3 127
bureaucratic condition 72
HRA 152
PFDs 86, 179
reactive condition 71
MOC. *See* management of change
mode criticality index, FMECA 116
monitoring and review
ISO 31000 47–48
RM systems approach 79, 80, 81
MORT. *See* Management Oversight and Risk Tree
multidisciplinary group, HAZOP 111

n

near-misses 185–186
BFA 250, 369
BT 140
ISO 9001/ISO 45001/ISO 14001 83
nominal HEP (NHEP) 388, 390
non-conformities 185–186
BFA 250
BT 140
non-physical barriers, BT 161

o

Offshore Directive, EU 388
omission errors
BFA 203
THERP 387
Operational Excellence Management System 187
operational experience 186–189
BT workflows, BFA 344, 347–350
operational readiness, RM systems approach 81
operational risk 85
ERM 62
organization understanding 24
risk assessment of 66
operators, HAZOP 110
optimization violations 386
organizational activities 16
organizational structure
BT, audits 237
BT workflows 227
continuous improvement 20
human and cultural factors 18
human error 53
internal context 25
pathological condition 69
OSH risk, ISO 45001 83, 84
OSHA 87

p

PAS. *See* Publically Available Specification
passive hardware category, SPAR-H 389
pathological condition, RM culture 69–70, 72
patient safety, BT workflows 292–293
PCI DSS 271
PDCA. *See* Plan-Do-Check-Act
performance
BT, barriers 160–161
BT workflows 223
fire risk assessment 358
human error 381–382
ISO 9001/ISO 45001/ISO 14001 86
ISO 31000, ERM 59
ISO 55000 91
performance-influencing factors (PIFs), SLIM 388, 396, 397–398
performance-shaping factors (PSFs)
ISO 31000 57–58
SPAR-H 389–390, 392–393
periodic review. *See* monitoring and review
personal safety equipment, ISO 39001 98
personnel and workload, ISO 31000 58
personnel selection and training
ISO 31000, PSFs 58
RM systems approach 81
PFDs. *See* probabilities of failure on demand
PHA. *See* preliminary hazard analysis

physical barriers, BT 161
PIFs. *See* performance-influencing factors
Pitblado, R. 160
Plan-Do-Check-Act (PDCA) 18
  BFA 370
  BT 370
  ISO 9001/ISO 45001/ISO 14001 86
  ISO 31000 79
  ISO 55000 91
  ISO IEC 27001 94
  RM systems approach 76–77, 81
    industrial safety 87
    RAGAGEP 90
  Vision 2000 366
PM. *See* project management
post-crash response, ISO 39001 98
potential events
  BFA 374, 377
  consequences 156, 214
  ERM 59
  risk 11
Prashanth, I. 161
preliminary analysis, ISO 31000 37
preliminary hazard analysis (PHA)
  BT, LOPA 167
  BT workflows, LOPA 249
  RCA 210
preventive barrier, BT 135, 150, 151, 213
  ABS 171
  barrier criticality assessment 235
  causes 154
  Zone 1 127
preventive barriers, BT 372
  DC 144, 145
primary barriers
  BT 156–163
  BT workflows 222, 223
primary cause, BFA 196, 199, 208, 374
PROACT RCA 213
proactive analysis, RCA 210
proactive barriers
  BT 150, 371
  BT workflows 225
proactive condition, RM culture 73–74
proactive management, RM 369
probabilities of failure on demand (PFDs)
  BT 150
    barriers 164
    cost-benefit assessment 134
    LOPA 168, 176, 178–183, 229–230
  BT workflows 229–230
    fire risk assessment 358
    LOPA 246, 248
    risk matrix 222
    tasks 228
  Deepwater Horizon 368
  ISO 9001/ISO 45001/ISO 14001 86
  ISO 19011 96
  ISO 55000 92
  RM systems approach 80
    industrial safety 88, 89
  SPAR-H 389
probability estimation, ISO 31000 36–37
procedures and operational instructions, ISO 31000, PSFs 58
process safety management (PSM)
  BFA 308
  BT 159
  BT workflows 293–308
  functional safety 90
  LOPA 243
    PFDs 247
  major accidents 87, 220
project management (PM)
  BT preventive barriers 145
  BT workflow BFA 251
  RM 15
    culture 64
    systems approach 81
PSFs. *See* performance-shaping factors
PSM. *See* process safety management
PSV, BT
  calibrated risk graph 132
  critical barriers 234
Publically Available Specification (PAS), ISO 55000 91

q

QIQO. *See* quality in, quality out
QRA. *See* quantitative risk analysis

qualitative assessments
  BT 102
    calibrated risk graph 132
  ISO31000 RM risk analysis 35
quality assurance, ISO 55000 91
quality in, quality out (QIQO) 262
quantitative risk analysis (QRA) 35
  BT, LOPA 167–168, 176

r

RAGAGEP. *See* recognized and generally accepted good engineering practices
Rathnayaka, S. 160
Rausand, M. 160
RBD. *See* reliability block diagram
RCA. *See* root cause analysis
reactive barriers
  BT 150
  BT workflows 225
reactive condition, RM culture 70–71
Reason, J. 4, 88, 137–138
recognized and generally accepted good engineering practices (RAGAGEP), RM systems approach 89–91
recommendations, BFA 196, 199, 202, 374
  BT workflows 258–263
recovery barriers, BT 151, 371, 372
redundancy score (RS), BT, critical barriers 233–234
relative risk, BT
  LOPA 181
  risk assessment 248
reliability block diagram (RBD), FEMA 115
reporting
  BT workflows, BFA 262–263
  incident, RM systems approach 81
  ISO 31000, ERM 59
  stakeholders 368–369
residual risk
  BT workflows, risk matrix 222
  defined 6
  ISO 31000 47
  ISO IE 27001 95
  risk treatment 45
resilience
  ISO 31000 14
  RM 369
resisting capacity and ageing, BT workflows 266–269
review and revision. *See also* management of change
  ISO 31000, ERM 59
risk
  absolute, BT LOPA 181
  acceptable
    ALARP 41, 89
    BT 102
    BT workflows LOPA 248–249
    ISO 31000 38–39
    risk indices 130
    risk matrix 222
  BT 101–102
  BT workflows 224
    LOPA 248
    web-based software development 279
  compliance
    ERM 62
    ISO 19600 98–99
  defined 6
  everywhere presence 6–10
  financial 369
    ERM 61
  intolerable 39
    BT workflows risk matrix 222
    LOPA PFDs 230
  ISO 9001/ISO 45001/ISO 14001 82
  operational 85
    ERM 62
    organization understanding 24
    risk assessment of 66
  OSH, ISO 45001 83, 84
  point of view 7, 10, 11
  reduction, human error 386–387
  relative
    BT LOPA 181
    BT risk assessment 248
  residual
    BT workflows risk matrix 222
    defined 6
    ISO 31000 47

risk (*cont'd*)
ISO IE 27001 95
risk treatment 45
RM systems approach 80
strategic
ERM 61
ISO 9001/ISO 45001/ISO 14001 85
tolerable
ALARP 222, 248
BT 102
BT workflows, LOPA 248–249
BT workflows, risk matrix 222
ISO 31000 38–39
LOPA IPLs 247
risk indices 130
risk matrix 41, 222
RRFs 133
risk analysis
BT 117–125
ISO 31000 117
BT workflows 228
defined 6
ISO 22301 93
ISO 31000 12, 31, 34–38
uncertainty 38
ISO IEC 27001 94
RM systems approach, complexity management 78
risk assessment
BT 369, 370
relative risk 248
BT workflows 225
defined 6
ISO 9001/ISO 45001/ISO 14001 83
ISO 19011 96
ISO 22301 93
ISO 31000 29–33
ISO 55000 92
of operational risk 66
RM systems approach 79, 81
risk evaluation
BT 128–134
LOPA 177
BT workflows, LOPA 248–249
defined 6
human error 387–407
ISO 31000 12, 31, 38
ISO IEC 27001 94
RM systems approach 79
complexity management 78
risk identification
BT 104–116
checklists 33
defined 6
ISO 22301 93
ISO 31000 12, 30, 31, 33–34
ISO IEC 27001 94
RM systems approach 79, 81
risk indices, BT, FN risk evaluation 129–130
risk life cycle, ISO 9001/ISO 45001/ISO 14001 86
risk management (RM). *See also specific topics*
conclusion 365–377
culture 62–76
bureaucratic condition 71–73
between fulfilment and opportunity 62–67
generative condition 75–76
pathological condition 69–70
proactive condition 73–74
quality 67–68
reactive condition 70–71
decision-making 370
introduction 1–99
ISO 31000
best available information 18
continuous improvement 18–23
customization 16–17
dynamism 17
human and cultural factors 18
inclusiveness 17
integral part of organizational activities 16
principles 14–23
processes 27–28
risk everywhere presence 6–10
structured and comprehensive 16
technical standard 10–23
workflows 23–53
proactive management 369

processes 2, 4
sustainability 4
systems approach 76–99
complexity management 78
functional safety 89–91
industrial safety 87–89
ISO 9001/ISO 45001/ISO 14001 82–86
ISO 19011 95–97
ISO 19600 98–99
ISO 22301 92–94
ISO 39001 97–98
ISO 55000 91–92
ISO IEC 27001 94–95
RAGAGEP 89–91
tailoring 16–17
risk matrix
BT, risk evaluation 130
BT workflows 221–222
intolerable risk 222
residual risk 222
ISO 31000 39–41
risk priority number (RPN), FMECA 115, 116
risk register
BT 178–183
HAZOP 111
ISO 9001/ISO 45001/ISO 14001 86
risk source
BT 261
defined 6
ISO 31000 368
risk treatment
BT workflows, BFA 261
defined 6
HAZOP 111
ISO 22301 93
ISO 31000 31, 45–47
RM
culture 68
systems approach 80
risk zero 6
risk-reducing factors (RRFs)
BT, cost-benefit assessment 134
Deepwater Horizon 368
ISO 9001/ISO 45001/ISO 14001 85, 86
ISO 22301 94
RM systems approach 80
industrial safety 88, 89
RAGAGEP 89
tolerable risk 133
RM. *See* risk management
road traffic safety. *See* ISO 39001
root cause analysis (RCA)
BFA 208–213
BT workflows, BFA 257, 347–348
risk identification 30
root causes
accident investigation 190, 191, 192
accidents 53
BFA 196, 207
BT workflows, BFA 262
ISO 31000 14
routine violations 386
RPN. *See* risk priority number
RRFs. *See* risk-reducing factors
RS. *See* redundancy score
rule-based performance, human error 381–382

S
Saetre, T. I. 265
safety critical equipment (SCE), BT
critical barriers 233
LOPA 167
safety instrumental functions (SIFs), BT 131, 132
safety instrumented systems (SISs), BT
calibrated risk graph 131
LOPA 168
safety integrity levels (SILs) 89
BT
calibrated risk graph 131–133
LOPA 168
BT workflows
BFA 262
LOPA 244, 247
FEMA 115
safety life cycle (SLC) 89
safety management system (SMS)
BT, hazards and risks 101
FSM 358

safety management system (SMS) (*cont'd*)
  FSMS 351, 358
  incident barriers 376
  RM systems approach
    industrial safety 87–88
    RAGAGEP 90
SAP, BT workflows 218
SCE. *See* safety critical equipment
scribes, HAZOP 110
secondary barriers, BT
  BT, DC 144, 145
  escalation factors 165
    LOPA 174–175
    workflows 222, 223
secondary cause, BFA 196, 199, 208, 374
semi-quantitative methods, ISO31000 RM risk analysis 35
sensitivity analysis, ISO 31000 38
session management, BT workflows, web-based software development 279
Seveso Directive, RM systems approach industrial safety 87
SHIPP. *See* system hazard identification, prediction and prevention
short-term solution, BFA 202
  BT workflows 261
shut-down, BT workflows, LOPA 246
Sicari, R. 266
SIFs. *See* safety instrumental functions
significant incidents 220
Silady 162
SILs. *See* safety integrity levels
SISs. *See* safety instrumented systems
situational violations 386
skill-based performance, human error 381–382
Sklet, S. 119, 159, 162, 213
SLC. *See* safety life cycle
SLIM. *See* Success Likelihood Index Method
slips 382–383
SMART. *See* specific, measurable, agreed, realistic and timely
Smith, J. 265
SMS. *See* safety management system
socio-technical category, SPAR-H 388–389
SOPs. *See* standard operating procedures
SPAR-H. *See* Standardized Plant Risk Human Reliability Analysis
specific, measurable, agreed, realistic and timely (SMART)
  BT, safety culture 66
  BT workflows, BFA 262
stakeholders
  accident investigation 190
  BT workflows 227
  communication 368–369
  governance 366
  ISO 9001/ISO 45001/ISO 14001 85
  ISO 31000 14
    risk treatment 45, 46
    system performance review 49
  ISO 55000 91
  reporting 368–369
  RM 15
    systems approach 81
standard operating procedures (SOPs), RM systems approach 82
Standardized Plant Risk Human Reliability Analysis (SPAR-H) 388–391, 392–393
  ISO 31000, human factors 54
start-up
  BT workflows, LOPA 246
  RM systems approach 81
  SLIM 397
Statement of Applicability, ISO IE 27001 95
Stimulus-Response model 55
strategic risk
  ERM 61
  ISO 9001/ISO 45001/ISO 14001 85
strategy and objective-setting, ISO 31000, ERM 59
stress/stressors 245
  mistakes 384
  SPAR-H 389, 392
Strobhar, D. 54
Success Likelihood Index Method (SLIM) 388, 391–407
sustainability, RM 4
Swain 177
Swiss Cheese Model 4, 5, 367–368

BT 137–138
RM systems approach, industrial safety 88
SLIM 402, 404–407
system automation and demand, ISO 31000, PSFs 58
system hazard identification, prediction and prevention (SHIPP) 160
system performance review, ISO 31000 49–53

t

Tafaro, S. 265
tailoring
ISO 31000 368
RM 16–17
tasks, BT workflows 228–229
taxonomies
BFA 196, 370
BT workflows 219–221
Taylor, J. 54
teams, risk identification 33
technical specialists
accident investigation 190
HAZOP 110
Technique for Human Error Rate Prediction (THERP) 387
ISO 31000, human factors 54
tertiary cause, BFA 196, 199, 208, 374
THERP. *See* Technique for Human Error Rate Prediction
Thinking-Behavior-Result model 54
threats
BT 150, 154–156, 371, 372
BT workflows 228
IT operations 280
web-based software development 275
three D's (detect, decide and deflect), BT workflows 247
ThyssenKrupp fire investigation, BT workflows 318–341, 342–343
TIER, RCA 213
timeline
BFA 202–206
incident analysis 347–349
cause-and-effect analysis 255
RCA 210
Tripod Beta accident investigation 193
time-to-reliability correlation (T/RC) 388
tolerable mitigated event frequency (TMEF) 133
tolerable risk
ALARP 222, 248
BT 102
BT workflows
LOPA 248–249
risk matrix 222
ISO 31000 38–39
LOPA IPLs 247
risk indices 130
risk matrix 41, 222
RRFs 133
Top events
BT 127, 153–154, 371, 372
barriers 157
BFA 202
domino effects 92
LOPA 170
BT workflows 223
top-down approach, BT workflows 240–241
total unmitigated hazard cost (TUHC) 133
transparency
BT workflows 223–224
RM 15
Travers, I. 265
T/RC. *See* time-to-reliability correlation
Treadway Commission, COSO 59, 60–61
Tripod Beta
accident investigation 192–194, 195
BT workflows, BFA 257, 258
RCA 212
Twente Stadium roof collapse, BT workflows 345
TUHC. *See* total unmitigated hazard cost
Twente Stadium roof collapse, Tripod Beta 345

u

UEF. *See* unmitigated event frequency
uncertainty
BT 102

uncertainty (*cont'd*)
  ISO 31000 368
    human factors 53–58
  risk analysis, ISO 31000 38
  RM 15
underlying cause 52
  BFA 202, 207
  HAZID 113
  RCA 208
understanding violations 385–386
United Nations Industrial Development Organization, International Trade Centre 368
unmitigated event frequency (UEF) 133

V
validation
  BT, DC 145
  BT workflows, web-based software development 277
  MOC 277
  RCA 209
  risk assessment 31
Vanden Heuvel, L. 210, 212
vehicle safety, ISO 39001 98
verification, BT, barriers 162
violations 385–386
  defined 380
  reporting 264
Vision 2000 366

W
water treatment, BT workflows 344, 346
web-based software development, BT workflows 271–279
what-if (WI)
  BT, risk identification 106–108, 109
  BT workflows 285, 286
  mistakes 385
WI. *See* what-if
Willauer, H. 324
Williams, F. 324
Woods, D. 54
Work Bow-Tie, RM systems approach industrial safety 87
work control, RM systems approach 81
work organization, ISO 31000, PSFs 58
work processes
  inverted two-pointed model 56
  SPAR-H 390, 393
workflows. *See* Bow-Tie workflows; ISO 31000 workflows
World Economic Forum, Annual Global Risk Report by 7

Printed and bound by CPI Group (UK) Ltd, Croydon, CR0 4YY
16/04/2025
14658556-0004